Tracks of modularized equipment transporter show route taken by Apollo 14 astronauts on their traverse to Cone Crater.

NASA SP-272

APOLLO 14

Preliminary Science Report

PREPARED BY
NASA MANNED SPACECRAFT CENTER

Scientific and Technical Information Office   1971
NATIONAL AERONAUTICS AND SPACE ADMINISTRATION
Washington, D.C.

ACKNOWLEDGMENT

The material submitted for the "Apollo 14 Preliminary Science Report" was reviewed by a NASA Manned Spacecraft Center Editorial Review Board consisting of the following members: Philip K. Chapman (Chairman), Michael B. Duke, Helen N. Foley, Jackson Harris, Frank J. Herbert, Bob Mercer, Scott H. Simpkinson, Paul J. Stull, and John M. Ward.

Foreword

THE THIRD MANNED LUNAR LANDING, which increased to almost 200 the man-hours spent by astronauts on the Moon's surface, differed in character from previous missions. The dominant aspect of the first landing was, simply, that it was done. The second landing was notable for the precision that brought a manned spacecraft to rest 183 m from its target site, a robot spacecraft dispatched to the Moon two and a half years before. But the outstanding characteristic of the third landing, when *Antares* came down to the rolling foothills of Fra Mauro, was the exceptionally rich harvest in lunar science that the mission achieved.

At Fra Mauro, astronauts Shepard and Mitchell emplaced an automatic geophysical station that quickly began to work in harness with Station 12, already functioning 181 km to the west, forming a valuable network that permits simultaneous observation from physically separated instruments. They also made a traverse on foot of record extent in an area of extreme geologic interest and brought back to Earth data and core tubes and other geologic samples in unprecedented volume.

The preliminary scientific results reported in this publication are the product of work performed in the months immediately following the mission. Unquestionably these analyses and interpretations will be expanded and refined during the months and years to come.

GEORGE M. LOW
Deputy Administrator
National Aeronautics and Space Administration

JUNE 1, 1971

Contents

Introduction

APOLLO 14, THE THIRD MISSION during which men have worked on the surface of the Moon, was highly successful. With the understanding of the lunar environment achieved by Apollo 11 and the pinpoint-landing capability demonstrated by Apollo 12, the Apollo 14 landing could be planned for a much rougher area of the Moon and one of prime scientific interest. This mission to the Fra Mauro Formation provided geophysical data from a new set of instruments located at latitude 3°40' S, longitude 17°27' W. The Apollo 12 lunar-surface experiments package deployed in November 1969 is still functioning at latitude 3°11' S, longitude 23°23' W, in the Ocean of Storms approximately 180 km from the Apollo 14 landing site. Comparisons between data from these first two sites in the Apollo scientific network can now be made. As an example, a single known seismic event, such as the impact of the lunar module ascent stage on the surface of the Moon, resulted in positive indications at both sites.

The topography in the landing area was extremely interesting, and the geological and geochemical returns were great. Because of improved equipment, such as the modularized equipment transporter, and because of the extended time spent on the lunar surface, a large quantity and variety of lunar samples were returned to Earth for detailed examination. New information concerning the mechanics of the lunar soil was also obtained during this mission. In addition, five lunar-orbital experiments were conducted during the Apollo 14 mission, needing no new equipment other than a camera. The experiments were executed by the command module pilot in the command and service module while the commander and the lunar module pilot were on the surface of the Moon.

This report is preliminary in nature; however, it is meant to acquaint the reader with the actual conduct of the Apollo 14 scientific mission and to record the facts as they appear in the early stages of the scientific mission evaluation. As far as possible, data trends are reported, and preliminary results and conclusions are included.

Large numbers of samples and quantities of data must yet be examined and the results compared with the scientific information resulting from the Apollo 11 and 12 missions before any final conclusions can be drawn.

JAMES A. MCDIVITT
NASA Manned Spacecraft Center

Mission Description

F. J. Herbert[a] *and W. K. Stephenson*[a]

The Apollo 14 mission successfully landed the third exploration team on the lunar surface. An objective of the mission was to continue the detailed, systematic scientific investigation of the Moon begun on the Apollo 11 and 12 missions.

The space vehicle with the crew of Alan B. Shepard, commander (CDR); Stuart A. Roosa, command module pilot; and Edgar D. Mitchell, lunar module pilot, was launched from NASA Kennedy Space Center, Fla., at 4:03:02 p.m. e.s.t. (21:03:02 G.m.t.) on January 31, 1971. The launch was 40 min 3 sec later than planned because of adverse weather in the launch area.

The vehicle was inserted into an Earth parking orbit of 98.9 by 102 n. mi. 11 min 49 sec after liftoff. Earth-orbital checkout and subsequent translunar injection were accomplished with no difficulty. The planned translunar injection firing was retargeted because of the later liftoff time to permit performance of planned lunar activities in accordance with established timelines, targeting, and Sun angles. A problem was encountered when attempting to dock the command and service module (CSM) with the lunar module (LM) in the SIVB stage. After five unsuccessful docking attempts, hard dock was accomplished on the sixth attempt 4 hr 57 min after launch. A thorough inspection of the docking mechanism was conducted and the mission continued as planned. The docking probe was returned to Earth for further analysis.

After LM extraction, the SIVB was targeted to impact the lunar surface. The SIVB impacted the lunar surface at 07:40 G.m.t., February 4. Impact occurred 287 km southeast of the planned impact point, as a result of the delays in launch and docking and because of the SIVB venting and loss-of-command problems. The impact was detected by the Apollo 12 passive seismic experiment and the seismic signal lasted approximately 3 hr.

Two midcourse corrections were performed to place the spacecraft in the planned trajectory for lunar orbit insertion. The spacecraft clock was updated at 04:02 G.m.t., February 3, to compensate for the launch delay and, therefore, aline it with the flight plan. The spacecraft was inserted into an elliptical lunar orbit at 07:00 G.m.t., February 4. The descent-orbit-insertion firing, using the service propulsion system (SPS), placed the spacecraft in a 58.8- by 9.6-n. mi. lunar orbit. This was the first time the SPS was used for this purpose. During prelanding checkout and lunar descent, problems were encountered with the guidance abort switch and the landing radar, but coordination between ground-support personnel and the crew overcame the difficulties. Before the LM landing, the CSM orbit was circularized to approximately 60 n. mi. The touchdown occurred at 08:37:10 G.m.t., February 5, within 50 m (160 ft) of the target point in the Fra Mauro highlands at latitude 3°40'24" S, longitude 17°27'55" W.

The first of the two planned periods of extravehicular activity (EVA) began 5 hr 23 min after touchdown. This EVA began 40 min later than planned because of a problem in configuring the LM communications system to work with the communications system in the CDR's portable life-support system. A color television camera mounted on the descent stage provided live cov-

[a] NASA Manned Spacecraft Center.

erage of the descent of both astronauts to the lunar surface. The crew deployed the U.S. flag and the solar-wind composition experiment, erected the S-band antenna, and off-loaded the modularized equipment transporter (MET), the laser ranging retroreflector (LRRR), and the Apollo lunar-surface experiments package (ALSEP). The ALSEP was deployed 180 m west of the LM, after some difficulty in locating a level site. (The terrain was more undulatory than expected.) The LRRR was deployed an additional 30 m west of the ALSEP. The MET was used to carry the required equipment and geological samples collected during the 4-hr 49-min EVA, which included a 30-min extension of planned surface time.

The second EVA was a planned extended geological traverse to Cone Crater. At the crew's request, the EVA was begun 2 hr 27 min earlier than planned. All equipment required for the geological traverse, including the lunar portable magnetometer (LPM), was loaded on the MET. The traverse up the side of Cone Crater provided experience in climbing and working in hilly terrain in $1/6g$ conditions. More time was required for the uphill climb than anticipated, but the downslope return was accomplished relatively quickly. The samples collected and the area traversed during the EVA were documented with a 70-mm Hasselblad camera using black-and-white film and the lunar-surface closeup stereo camera using color film. Two LPM measurements were made. At the end of the EVA, the crew returned to the ALSEP site and realined the central-station antenna to improve signal reception. The second EVA lasted 4 hr 20 min, during which time the astronauts traveled approximately 3 km. The crew reported excellent mobility.

The crew stowed its gear and 43 kg of samples and performed a preascent checkout of the LM ascent stage. Liftoff occurred at 18:48 G.m.t., February 6, after 33 hr on the lunar surface. A single-orbit rendezvous was accomplished and docking was achieved without incident at 20:35 G.m.t. After crew transfer, the LM ascent stage was separated and remotely guided to impact on the lunar surface. Impact occurred at latitude 3°25′ S, longitude 19°40′ W, at 00:45 G.m.t., February 7, between the Apollo 12 and 14 seismometers. The resulting seismic signal lasted for 1.5 hr and was recorded by both instruments.

Orbital-science experiments and science photography were performed in lunar orbit. A downlink bistatic-radar experiment was conducted using both very-high-frequency and S-band signals. The S-band transponder was used to measure the CSM lunar orbit. Photography of scientifically interesting surface features was obtained in various lighting and Sun angles. Dim-light photography was conducted of diffused galactic light, zodiacal light, and the Earth dark side. Photography was also obtained of the gegenschein region. Photography of the Descartes area was obtained with a 500-mm Hasselblad after the Hycon camera became inoperative.

Transearth injection was accomplished using the SPS at 01:39 G.m.t., February 7. During the zero-g transearth coast, the crew conducted demonstrations of electrophoretic separation, heat flow and convection, liquid transfer, and composite casting. These demonstrations and a press conference were reported with live color television. Only one midcourse correction was required and the entry sequence was normal. The command module splashed down in the Pacific Ocean approximately 1 km from the target point at 20:24 G.m.t., February 9. The landing coordinates as determined by the onboard computer were latitude 27°2′24″ S, longitude 172°41′24″ W.

Summary of Scientific Results

Philip K. Chapman,[a] *Anthony J. Calio,*[a] *and M. Gene Simmons*[a]

The planned landing site for the Apollo 13 mission was in the Fra Mauro region, which is an area of prime scientific interest because it contains some of the most clearly exposed geological formations that are characteristic of the Fra Mauro Formation. The Fra Mauro Formation is an extensive geological unit that is distributed (in an approximately radially symmetric fashion around the Mare Imbrium) over much of the near side of the Moon. Stratigraphic data indicate that the Fra Mauro Formation is older than the Apollo 11 and 12 mare sites. The Fra Mauro Formation is thought to be part of the ejecta blanket that resulted from the excavation of the Imbrium Basin, which is the largest circular mare on the Moon. The Apollo 13 landing site thus offered an opportunity to sample material that had been shocked during one of the major cataclysmic events in the geological history of the Moon and thereby to determine the date of the event. Furthermore, because of the size of the Imbrium Basin, it was expected that some of the material had come from deep (tens of kilometers) within the original lunar crust. Thus, a landing at the Fra Mauro Formation, in principle, should offer an opportunity to sample the most extensive vertical section available of the primordial Moon.

After the Apollo 13 mission, which failed to achieve a lunar landing, the importance of the Fra Mauro landing site led to a decision to attempt a landing in the same area during the Apollo 14 mission. The final landing site was very close to that chosen for the Apollo 13 mission. The site was located in a broad, shallow valley between

[a] NASA Manned Spacecraft Center.

radial ridges of the Fra Mauro Formation and approximately 500 km from the edge of the Imbrium Basin. The major crater Copernicus lies 360 km to the north, and the bright ray material that emanates from Copernicus Crater covers much of the landing-site region. In the immediate landing-site area, an important feature is the young (Copernican age), very blocky Cone Crater, which is approximately 340 m in diameter and which penetrates the regolith on the ridge to the east of the landing site. Thus, Cone Crater would provide access to the excavated Fra Mauro Formation, despite the general overlying blanket of later deposits.

The lunar-surface experiments planned for the Apollo 14 mission differed somewhat from those of the Apollo 13 mission. The crew's traverse capability was improved by the addition of the modularized equipment transporter (MET), which is a light, hand-drawn cart that enabled the crew to transport tools and samples with greater ease. Two extravehicular activity (EVA) periods were planned, each of which was to last 4 hr 15 min. The principal objectives of the first EVA were to collect geological samples (including a contingency sample in case an early abort became necessary) and to deploy the Apollo lunar-surface experiments package (ALSEP). The second EVA was largely devoted to a geological sampling traverse toward Cone Crater, with several other experiments being conducted along the traverse.

This section summarizes the scientific results obtained to date from the Apollo 14 mission. It should be understood, however, that this is a preliminary report, and the results and, especially,

1

the interpretations may change after further data collection and analysis.

Geology

The Apollo 14 crew traversed a smooth-terrain unit, a Fra Mauro ridge unit, and the ejecta blanket of Cone Crater. Descriptions, photography, and samples of the Cone Crater ejecta blanket were the major geological objectives of the Apollo 14 lunar-surface activity. It was anticipated that the Cone Crater event had resulted in the penetration of the lunar regolith and excavation of material from the Fra Mauro Formation. Samples returned from the Cone Crater ejecta blanket and the photography of boulders on the ejecta blanket indicate that the Fra Mauro Formation is comprised of breccias that contain abundant dark lithic clasts and less-abundant light clasts. Clasts of breccias within breccias may represent pre-Imbrian cratering in the Imbrium Basin region.

The Apollo 14 landing site is dotted with abundant craters that range in diameter from several hundred meters to the limit of resolution of the Hasselblad data cameras. Craters between 400 m and 1 km in diameter are more numerous and more subdued than craters on the maria; this distribution is consistent with the inferred greater age (relative to the maria) of the Fra Mauro Formation. Craters along the traverse range from the old and highly subdued craters that form the rolling topography in the smooth-terrain unit to the sharp 30-m-diameter crater on the Cone Crater ejecta blanket near station C'. Rock fragments greater than 2 to 3 cm in diameter are sparse in the immediate vicinity of the lunar module (LM) landing site, but the surface of the Cone Crater ejecta blanket is characterized by abundant rock fragments that measure up to several meters in diameter.

The albedo of the fine-grained material in the vicinity of the landing site ranges from 8.2 to 15 percent, with the lower albedo values typical of the smooth-terrain unit and the higher values typical of the Cone Crater ejecta blanket. The highest albedo values yet measured on the lunar surface are the values determined for the Cone Crater ejecta blanket in the vicinity of station C1.

The albedo of the component parts of the fragmental rocks ranges from 9 to 36 percent.

Small-scale surface lineaments, although less well developed than at the Apollo 11 and 12 landing sites, are present at the Apollo 14 landing site. Two primary northwest and northeast trends and one secondary north trend agree well with the strongest trends observed at the Apollo 11 and 12 landing sites. The lineaments are less well developed on the Cone Crater ejecta blanket than on the smooth-terrain unit. Fillets of fine-grained material banked against rock fragments are common at the Apollo 14 landing site. The variety of fillet geometries suggests that the fillets are probably formed by several different mechanisms.

The large number of boulders (most of which were probably ejected from Cone Crater) offered for the first time the opportunity to study textures and structures of lunar rocks at scales from tens of centimeters to a meter. All the boulders appear to be fragmental, with abundant clasts up to 10 cm across or larger. The clastic appearance of the rocks is somewhat similar to the appearance of ejecta deposits of impact origin on Earth. Planar features (some of which are systematic in their development) are visible in all the boulders. In some of the boulders, distinct lithologic layers are evident. The boulders show different degrees of rounding by lunar erosion; differences in shape are probably caused in part by internal fracture systems and possibly by differences in lithologies.

Approximately 43 kg of lunar samples were returned by the Apollo 14 crew, including 33 rocks that weigh more than 50 g each. Thirteen of the rock samples were sufficiently documented to determine lunar locations and orientations. The locations of most of the samples have been at least approximately determined.

Nearly all the returned samples are fragmental rocks. Some rocks are composed mostly of dark clasts and a lighter colored matrix; others consist of a mixture of light- and medium-gray clasts in a dark-gray matrix. Samples of both rock types were collected from widely separated points along the traverse.

Numerous similarities can be seen between the characteristics of the returned samples and of the boulders shown in the photographs. Closely spaced fractures (in some cases, two or more intersecting fracture sets) are present in many of the returned rocks. Variations among the rocks suggest the possibility of considerable lateral or vertical variation within the Fra Mauro unit.

Returned Lunar Samples

Of the 43 kg of lunar material returned by the Apollo 14 crew, the Lunar Sample Preliminary Examination Team has examined portions of six soil samples and the surfaces of all rocks that weigh 1 g or more—approximately 150 rocks in all. Thin sections of 12 rocks were examined; optical spectroscopic analyses were performed on 16 samples; gamma-ray analyses for potassium, thorium, uranium, aluminum-26, sodium-22, and cobalt-56 were performed on 14 samples; and noble-gas analyses were performed on eight samples.

The ratio of fragmental rocks (including breccias and recrystallized clastic rocks) to igneous rocks (mostly basaltic) is approximately 9:1, which is much higher than the 1:1 and 1:9 ratios for the Apollo 11 and 12 rocks, respectively. The fragmental rocks returned from the Apollo 14 landing site also have larger and more abundant clasts and have a generally much more recrystallized matrix than the breccias from the previous missions. Three classifications of rocks from the Apollo 14 landing site were made for preliminary descriptive purposes:

(1) Rocks with predominantly lithic clasts in very friable matrices

(2) Rocks with predominantly light-colored and very abundant clasts in a matrix of moderate coherency

(3) Rocks with predominantly dark-colored clasts in a matrix of moderate to high coherency

The larger clasts are commonly clastic rocks themselves, but nonfragmental lithic clasts, which have been separated into six groups, are present. These groups are (1) clinopyroxene-plagioclase, (2) feldspar, (3) subophitic plagioclase-orthopyroxene, (4) olivine-glass, (5) granitic olivine, and (6) glass. Accessory minerals that are present include ilmenite, metallic iron, troilite, chromium spinel, ulvöspinel, metallic copper, armalcolite, zircon, apatite, and potassium-feldspar.

Only two homogeneous crystalline rocks that weigh more than 50 g were found. These rocks are basaltic, have typical igneous textures, and contain plagioclase, pyroxenes, olivine, and possibly opaques. Very-fine-grained granulitic-textured crystalline rocks were also found. Basaltic rock sample 14310, which weighs 3.4 kg, was the second largest rock returned. It is similar to terrestrial high-alumina basalts, but contains more highly calcic plagioclase and predominantly pigeonite pyroxenes. The occurrence of a solar flare on January 25, 1971, induced the formation of cobalt-56, which has been detected in the upper surface of the rock.

The soil samples collected at various locations vary in composition and grain size; and, within particular samples, a compositional variation exists among the grains of differing size. The finer grained portions have a higher glass content than the coarser grained portions. The soil collected from the immediate vicinity of Cone Crater has coarser grains than the soils from other locations and has a lower glass content and a higher content of fragmental-rock fragments. The core samples have thus far been studied only by X-ray radiography; however, definite layering and some grading of particle sizes have been observed.

Chemical analyses of the Apollo 14 material show it to be distinct from the material returned from the Apollo 11 and 12 landing sites in that lower concentrations of iron, titanium, manganese, chromium, and scandium and higher concentrations of silicon, aluminum, zirconium, rubidium, strontium, sodium, lithium, lanthanum, thorium, and uranium are present. The total carbon content of the soil samples falls within the carbon-content range found for material returned from the Apollo 11 and 12 sites. The rocks have carbon contents that range from 28 to 225 parts per million.

A noticeable difference in the noble-gas content of material from the Apollo 14 landing site was observed. The spallation-produced isotopes of these noble gases in four Apollo 14 rocks yielded exposure ages of 10 to 20 million yr, which is considerably lower than the 40- to 500-million-yr exposure ages of the rocks returned from previous missions. The solar-wind content of the fragmental rocks varies between the high values seen for the solar-wind content of fine material to essentially zero.

Soil Mechanics

Although the surface texture and appearance of the soil at the Apollo 14 landing site are similar to those at the Apollo 11 and 12 landing sites, a larger variation in the soil characteristics exists at depths of a few centimeters in both lateral and vertical directions than had previously been encountered. The walls of a trench that was dug by the commander collapsed at a shallower depth than had been predicted, evidently because of lessened soil cohesion. Calculations indicate that the soil-cohesion value at the Apollo 14 trench site may be as small as 10 percent of the values calculated for soils at previous landing sites. The grain-size distributions of most of the soil samples returned from the Apollo 14 site are similar to the grain-size distributions of soil samples from the Apollo 11 and 12 sites, with two significant exceptions: (1) a soil sample collected from near the rim of Cone Crater and (2) a soil sample from the bottom of the trench. At both of these locations, the soil was considerably coarse, with the median grain size as much as 10 times greater than at other Apollo 14 soil-sample locations.

The LM pilot was unable to push the Apollo simple penetrometer into the lunar surface near the ALSEP as deeply as had been expected. Similarly, the Apollo 14 crew encountered more difficulty in driving the core tubes than did the Apollo 12 crew. These results indicate that the soil at the Apollo 14 landing site is stronger with depth than had been previously supposed. The calculated bulk density of the soil in the lower half of the double core tube was significantly less than that determined for soil in the lower half of the Apollo 12 double core tube. Variations in the soil grain-size distribution or specific gravity (or both) may account for this difference in bulk density. The MET track observations confirm that the soil is less dense, more compressible, and weaker at the rims of small craters than in level intercrater regions.

Passive Seismic Experiment

Each lunar-landing mission to date has included deployment of a passive seismometer. The Apollo

11 instrument produced useful data for approximately 1 month; however, the Apollo 12 instrument is still functioning satisfactorily. Deployment of the Apollo 14 passive seismometer as a part of the ALSEP thus represented a major step forward in the location and interpretation of lunar seismic events by providing a second instrument, one separated from the Apollo 12 passive seismometer by a baseline of 181 km.

The Apollo 14 seismometer has detected natural seismic events at more than twice the frequency recorded by the Apollo 12 instrument, and all seismic events that have been recorded at the Apollo 12 site have also been detected by the Apollo 14 instrument. It is hypothesized that the greater sensitivity at the Apollo 14 site is a result of the thick layer of unconsolidated material that blankets this region (the Fra Mauro Formation and the overlying regolith). This layer of unconsolidated material may provide a more efficient coupling of seismic energy with the lunar surface. The Apollo 14 instrument appears to be sufficiently sensitive to detect the impacts at any location on the Moon of meteoroids that have masses in excess of approximately 1 kg.

The detected natural events can be divided into two classes (on the basis of spectral characteristics and other characteristics of the signals) that correspond to meteoroid impacts and moonquakes. Moonquake occurrences correlate strongly with the time of perigee passage of the Moon in its orbit, which leads to the hypothesis that the release of internal strain, the origin of which is unknown, is triggered by tidal stresses. It is believed that not less than nine different locations are involved in the moonquakes that have been detected by both seismometers, although more than 80 percent of the total seismic energy detected has come from a single focal zone that is located perhaps 600 to 700 km from both stations and possibly at a considerable depth within the Moon. If the depth of the focal zone is confirmed by future data, fundamentally important information about the present state of the lunar interior will be made available.

A cumulative mass spectrum for meteoroids impacting the Moon has been tentatively constructed from the data. The cumulative mass spectrum indicates that the total meteoroid mass flux is a factor of 20 less than values from previous estimates. Approximately one meteoroid impact per year of kinetic energy equal to that of the SIVB impact is predicted. The Apollo 14 SIVB impact was detected by the Apollo 12 passive seismometer, and the LM ascent-stage impact was detected at both sites. These seismic events produced the characteristic, remarkably slow decay signals that had been previously detected by the Apollo 12 instrument; the signals persisted for several hours. The passive seismometer data have greatly increased the understanding of the lunar structure and of seismic energy transmission. A simple wave-propagation model of the entire Moon has now been derived; the model involves intensive scattering of waves in the outer shell of the Moon but very low energy absorption.

Crew activities produced detectable seismic signals for the Apollo 14 instrument throughout the EVA traverses. Venting gases from and thermoelastic stress relief within the LM caused the expected seismic signals, which are continuing. In view of the long-range detectability of lunar seismic events, it appears that a scattered network of passive seismometers can produce very valuable data. The deployment of a third passive seismometer at the Apollo 15 landing site will greatly increase the capability for accurately locating natural events that occur on the Moon.

Active Seismic Experiment

For the first time in the Apollo lunar-landing program, an active seismic experiment (ASE), in addition to the passive seismometer, was deployed on the lunar surface. The Apollo 14 crew deployed a string of three geophones across the lunar surface and used a thumper device with small explosive initiators to generate seismic signals. The active seismic instrument can also be used to monitor high-frequency natural seismic activity.

The data that were generated by the thumping operation indicate the existence of a surficial layer approximately 8.5 m thick at the ALSEP site. This layer, which exhibits a P-wave velocity of 104 m/sec, may be interpreted as the regolith in this area. The 8.5-m thickness of the regolith is in good agreement with the thickness estimated

from geological studies of small craters. The seismic propagation velocity observed during the thumping operation is in remarkable agreement with the propagation velocity derived at the Apollo 12 landing site (108 m/sec), where the elapsed time between LM ascent-engine ignition and arrival of the generated seismic signal at the Apollo 12 passive seismometer was recorded.

Below the regolith at the Apollo 14 landing site is another layer, which exhibits a P-wave velocity of 299 m/sec. The thickness of this layer is estimated to be approximately 50 m. It is premature to speculate on the composition of this layer, but it is interesting to note that the estimated thickness of the layer is not in substantial disagreement with estimates of the thickness of the Fra Mauro Formation in this area. The relatively low compressional-wave velocities that have been measured in this experiment are evidence against the existence of substantial permafrost near the surface in the landing region.

The ASE also includes a rocket-grenade launcher that is capable of launching four grenades to impact at known times and at known distances (up to approximately 1500 m) from the seismometer. The rocket grenades will not be activated until data collection from the other ALSEP experiments is virtually complete.

Suprathermal Ion Detector Experiment (Lunar Ionosphere Detector)

The suprathermal ion detector experiment (SIDE), which was deployed as part of the Apollo 14 ALSEP, is essentially identical to the instrument deployed by the Apollo 12 crew. By correlation of data returned by the two SIDE instruments now operating on the Moon, discrimination is possible between moving ion clouds and temporal fluctuations of the overall ion distribution. For example, this discrimination capability enabled the interpretation of an ion event that was detected at both SIDE sites on March 19. This ion event was the passage of a large (approximately 130 km in diameter) ion cloud that moved westward at approximately 0.7 km/sec. The cloud was possibly associated with a relatively large seismic event that was recorded by the Apollo 14 passive seismometer approximately 37 min earlier.

Ions in the 250- to 1000-eV energy range have been detected streaming down the magnetosheath of the Earth as the Moon entered the magnetospheric tail. In addition, approximately 2 days after sunrise, intermittent intense fluxes of 50- to 70-eV ions with masses in the 17- to 24-amu/unit-charge range were recorded. Energy and mass spectra were obtained during the venting of the oxygen atmosphere of the the LM cabin. After ascent-stage liftoff, the Apollo 14 LM ascent-engine exhaust was detected by the Apollo 12 SIDE, beginning 1 min after the vehicle passed the Apollo 12 landing site at a minimum slant range of 27 km. The Apollo 14 SIVB and LM ascent-stage impacts both produced useful signals at the Apollo 12 SIDE.

Cold-Cathode-Gage Experiment (Lunar Atmosphere Detector)

The cold-cathode-gage experiment, which is similar to the instrument deployed as a part of the Apollo 12 ALSEP, measures the concentration of neutral atoms (i.e., the density of the lunar atmosphere) in the vicinity of the ALSEP. During the lunar night, the concentration appears to be approximately 2×10^5 cm^{-3}, although transient increases by one to two orders of magnitude are fairly frequent and last from minutes to many hours. Some of these transient increases may be caused by venting or outgassing from the LM or from other equipment at the ALSEP site. As might be expected, the neutral-atom concentration rises rapidly at sunrise (two orders of magnitude in 2 min); the concentration then decays, over a period of approximately 50 hr, to a mean daytime level of less than 10^7 cm^{-3}. Numerous gas events have also been observed during the lunar day. The mean neutral-atom levels observed are thought still to be affected by outgassing from other ALSEP equipment, but the output of neutral atoms from this source should decrease with time in an identifiable way.

Charged-Particle Lunar Environment Experiment

The charged-particle lunar environment experi-

ment (CPLEE) is designed to measure the ambient fluxes of charged particles, both electrons and ions, with energies in the range of 50 to 50 000 eV. One of the most stable features observed is the presence of low-energy electrons whenever the landing site is illuminated by the Sun. The variation in the low-energy-electron flux during the lunar eclipse of February 10 provided strong evidence that the electrons are photoelectrons liberated from the lunar surface. The solar-wind flux observed by the CPLEE has exhibited rapid time variations (periods of approximately 10 sec), both when the Moon is in interplanetary space and when it is immersed in the magnetospheric tail of the Earth. Passage of the Moon through the magnetopause and magnetospheric tail has produced some particularly interesting data, including rapidly fluctuating low-energy (50- to 200-eV) electrons, fluxes of medium-energy electrons lasting from a few minutes to tens of minutes, and electrons that have energy spectra remarkably similar to those observed above terrestrial auroras. Thus, auroral particles do appear to penetrate far into the magnetospheric tail, an observation that, if confirmed, contains important implications concerning the general topology of the magnetosphere.

After the Apollo 14 LM ascent-stage impact, two plasma clouds, which were separated in time by a few seconds, passed the CPLEE. If these plasma clouds were associated with the impact, they were traveling at approximately 1 km/sec and had diameters of 14 and 7 km.

Laser Ranging Retroreflector

Used in conjunction with the laser ranging retroreflector (LRRR) deployed during the Apollo 11 mission, the Apollo 14 LRRR array provides a capability for accurate monitoring of lunar librations in longitude. The LRRR scheduled for deployment during the Apollo 15 mission will complete the network by providing a long baseline separation of LRRR arrays in latitude as well. The deployment of a complete network is important not only because study of the lunar librations provides information about the internal structure of the Moon, but also because knowledge of lunar librations enables the errors that are introduced by them to be eliminated when range measurements are needed in other studies (such as geophysics, general relativity, etc.).

Range measurements to the Apollo 14 LRRR were successfully accomplished on the day it was deployed by the crew. Range measurements taken after the LM liftoff indicate that the ascent-stage-engine burn caused no serious degradation of the LRRR reflective properties. Continued measurements over a period of years will be required before the full scientific value of this program is realized. Results to date indicate that the LRRR-array lifetime on the Moon will be long enough for all scientific objectives to be fulfilled.

Solar-Wind Composition Experiment

The solar-wind composition experiment was similar to the investigations conducted during the Apollo 11 and 12 missions. The aluminum foil was exposed to the solar wind at the Apollo 14 site for 21 hr. Five small samples from the upper part of the foil have been analyzed to date to determine noble-gas concentrations. The helium-4 flux observed is definitely lower than the flux detected during previous missions, but the helium-4/helium-3 ratio is similar to that obtained during the Apollo 12 mission and higher than that obtained during the Apollo 11 mission. Several isotopes of neon were detected, and their concentrations were measured. For the first time, the presence of argon in the solar wind has been detected. The accuracy of the isotopic ratios for argon will be improved when larger pieces of foil have been examined.

Lunar Portable Magnetometer

The Apollo 12 ALSEP included a magnetometer that measured the unexpectedly intense, steady magnetic field of 38 gammas at the ALSEP site. The Apollo 14 ALSEP does not include a stationary magnetometer; instead, a new instrument, the lunar portable magnetometer (LPM), was used during the second EVA to measure the magnetic field at two locations separated by 1.1 km. These measurements yielded values of 103 and 43 gammas, which indicated that the magnetic anomaly at the Apollo 12 landing site (181 km

away) is by no means unique on the Moon. These results, together with the relatively high magnetic remanence found in returned lunar samples from all landing sites to date, give evidence that much of the lunar-surface material has been magnetized —perhaps even the entire crustal shell around the Moon. The LPM measurements enable a magnetic-field gradient of 54 gammas/km to be calculated, which is less than the upper limit of 133 gammas/km that was determined from measurements with the Apollo 12 magnetometer.

S-Band Transponder Experiment

The S-band transponder experiment used precision doppler tracking of the command and service module and the LM to provide detailed information about the near-side lunar gravity field. This technique was used with the Lunar Orbiter spacecraft and resulted in the discovery of the large gravity anomalies called mascons. The Apollo tracking system is capable of measuring line-of-sight velocity with a resolution of 0.65 m/sec.

Reduction of the data obtained during the Apollo 14 mission is in progress. Preliminary results show that the gravity-anomaly profile over the mascon in Mare Nectaris has a flat-topped appearance that is characteristic of a shallow plate-shaped mass anomaly. If other mascons exhibit this characteristic, it will be evidence that the mascons are near-surface features rather than deeply buried inhomogeneities. Results of the analysis of the crash orbit of the Apollo 14 LM ascent stage will be particularly interesting. Data were obtained at very low altitudes, which may enable small anomalies (a few kilometers in diameter) to be discerned.

Ancillary Experiments

In addition to the preceding experiments, several less formal investigations were conducted during the Apollo 14 mission. The details of these investigations will be published elsewhere.

The first such investigation concerned the light flashes that all crews since Apollo 11 have observed, when in the dark or when they closed their eyes, while in transit to and from the Moon and in lunar orbit. The Apollo 14 crew was briefed on this phenomenon before the mission, and an observational schedule was suggested that was intended to test the various theories of the origin of the flashes. The most significant result obtained during the flight was the discovery that it is not necessary to be dark adapted to see the flashes. This observation indicates that Cerenkov radiation from energetic cosmic rays traversing the eyeball, which had been the most widely accepted explanation for the light flashes, probably does not cause all or most of the flashes because the light from this source is quite faint. Comparison with results of recent terrestrial experiments (in which human subjects were exposed to particle beams from accelerators) suggests that some of the flashes observed in space may be caused by direct ionization interactions of cosmic rays within the retina.

During transearth coast, the crew conducted a series of simple demonstrations with the objective of gathering design data for use in future scientific and engineering applications. These demonstrations were shown on television and were filmed, and samples were returned to Earth for analyses, which are presently in progress. The results to date are as follows.

Composite Casting

Eleven samples of various immiscible compositions were heated, mixed by shaking, and allowed to solidify by cooling in zero gravity. Laboratory analysis has not yet been completed, but X-ray examination of the samples indicates that more homogeneous mixing was achieved than is possible with similar samples on Earth. Further research in this area may result in the manufacture of new combinations of materials in the space environment (such as immiscible composites and supersaturated alloys) for structural, electronic, and other applications.

Electrophoretic Separation

Results to date indicate that electrophoresis (which is a technique commonly used on Earth for analysis or separation of chemical, especially organic, mixtures) provides a sharper separation in zero gravity because of a reduction of sedimentation and thermal convective mixing. This process may allow economical purification and separation of high-value biological materials in space.

Heat Flow and Convection

It was demonstrated that surface tension can produce Bénard cells in a liquid, independently of gravity-induced convection. Zone heating of liquid samples produced an unexpected cyclic heat-flow pattern that is now under study.

Liquid Transfer

The liquid-transfer demonstration clearly showed that suitable baffles inside a tank in zero gravity permit positive expulsion of liquid contents, taking advantage of the surface-tension properties of the liquid.

1. Photographic Summary

John W. Dietrich [a]

The photographic objectives of the Apollo 14 mission were designed to support orbital and surface science, to provide data for evaluating potential sites for future lunar landings, and to document the operating characteristics of equipment in orbit and on the surface. The photographic tasks were integrated with other mission objectives to achieve a maximum return of data from the lunar voyage.

The Apollo 14 crew returned 1328 frames of 70-mm photography and 15 exposed magazines of 16-mm film. In an attempt to recover usable photographs from the two rolls of 5-in. film exposed in the lunar topographic camera (LTC) that malfunctioned, an extensive developmental analytic plan was instituted. The extended period of careful and difficult processing, coupled with concurrent research, resulted in the recovery of 193 usable photographs. All photography has been screened and indexed; but, when this report was prepared, only the photographs in and near the bootstrap site had been analyzed in detail.

The earliest planned photographic tasks for the Apollo 14 mission were scheduled to document the transposition and docking maneuvers approximately 3 hr after launch. The delay in obtaining a hard dock placed the mission behind in the time line, and photographic documentation of the repeated docking attempts increased film consumption. After withdrawal of the lunar module (LM), the SIVB stage was photographed with both 16- and 70-mm cameras as the range between the command and service module (CSM) and SIVB stage increased. The docking probe was brought into the spacecraft after docking. Although the probe appeared to be normal in every way and operated normally, it was photographed with a

[a] NASA Manned Spacecraft Center.

Hasselblad camera to document the appearance.

During the translunar-coast period, the crescent Moon and distant Earth were photographed. At a point near the Moon, the cloud of drops formed by waste water dumped from the CSM was photographed from a vantage point within the LM.

After lunar-orbit insertion, visual observations of the lunar surface and operational tasks occupied most of the first orbit. The data acquisition camera (DAC) was mounted on the sextant to record landmarks tracked during revolution 2. Near the beginning of the third revolution, the engine was fired for descent-orbit insertion.

The LTC was readied for the first of several scheduled passes over the candidate landing site near Descartes Crater. As the spacecraft approached Theophilus Crater during revolution 4, the camera was actuated to expose approximately 1 frame/sec as the spacecraft descended to less than 10 n. mi. above the terrain east of the candidate landing site. The command module pilot (CMP) reported an unusual noise—intermittent at first, then continuous—during the period of camera operation. Ground-support personnel immediately began to assess possible causes of the problem to determine the effect on remaining LTC tasks. An eating and rest period provided sufficient time for ground-support personnel to assess the problem before the next scheduled period of camera operation.

Intense preparation for undocking and separation of the LM precluded the scheduling of LM or CSM photographic tasks for two revolutions after the rest period. Undocking and separation were recorded by 70- and 16-mm cameras on both spacecraft. The CMP tracked a landmark after separation for the only test of the CSM low-altitude landmark tracking during this mission. A service propulsion system burn near the end of revolution 12 placed the CSM in a near-circular orbit.

While the CSM and the LM operated separately, the CMP followed a busy schedule. The experimental tasks included photographic assignments covering a wide range of targets and requiring varied camera-lens-film combinations, as listed in table 1–I. Most of the targets were on the lunar surface but some, such as the LM during separation and docking, were at close range. Targets far out in space were photographed at times when

the Moon shielded the CSM from both sunlight and earthshine to support studies of the gegenschein and other dim-light phenomena.

The CMP photographed selected orbital-science targets on the far side and on the eastern half of the visible hemisphere during revolution 14 as the LM crew made preparations for landing. Plans for photographing the powered-descent-initiation burn and touchdown of the LM were canceled because of the earlier malfunction of the LTC.

Landmark-tracking data, recorded with the 16-mm DAC mounted on the sextant, supported

TABLE 1–I. *Photographic Equipment Used in Command Module (CM)*

Camera	Features	Film size and type	Remarks
Hasselblad EL.......	Electric; interchangeable lenses of 80-, 250-, and 500-mm focal length	70-mm; SO–368 Ektachrome MS color-reversal film, ASA 64; 3400 Panatomic-X black-and-white film, ASA 80; SO–349 high-definition aerial film, AEI 6	Used with 500-mm lens to provide convergent stereoscopic coverage over candidate landing sites near Descartes Crater; used with appropriate lens for CSM orbital sciences
Hasselblad DC.......	Electric, Reseau plate, 80-mm lens	70-mm; 3400 Panatomic-X black-and-white film, ASA 80; SO–349 high-definition aerial film, AEI 6; 2485 black-and-white film, ASA 6000	Used for stereoscopic strip photographs of potential landing site and CSM orbital-science photographic targets
Lunar topographic camera	18-in. lens, vacuum platen, image-motion compensation.	5-in.; type SO–349 high-definition aerial film, AEI 6; 3400 Panatomic-X black-and-white film, ASA 80.	Used to obtain high-resolution photography of lunar surface near candidate Descartes Crater landing site; operating difficulties prevented scheduled convergent stereoscopic photography of the approach to the Descartes Crater site, of the landed LM near Fra Mauro Crater, and of the impact points of the Apollo 14 SIVB stage and LM
DAS...............	Interchangeable lenses of 10-, 18-, and 75-mm focal length; variable frame rates of 1, 6, 12, and 24 frames/sec	16-mm; SO–368 Ektachrome MS color-reversal film, ASA 64; SO–168 Ektachrome EF color-reversal film exposed and developed at ASA 1000; 2485 black-and-white film, ASA 6000	Bracket mounted in CSM rendezvous window to record view during transposition and docking, LM ejection, LM undocking, rendezvous and docking, and LM jettison; hand held or bracket mounted for general photography and experiment records within CSM

both operational and scientific objectives. The CMP was occupied with landmark tracking during a large part of the sunlit segments of revolutions 13, 15, 18, and 29. Photography supporting studies of the gegenschein and dim-light phenomena occupied appropriate parts of the night segments of revolutions 15, 16, 18, 19, and 26.

Long strips of overlapping zero-phase photographs were obtained during revolutions 16 and 30. The camera was pointed backward toward the zero-phase point after the CSM crossed the far-side terminator. On approaching the near-side terminator during revolution 16, the spacecraft was rotated to point the camera forward toward the zero-phase point.

During revolution 25, several orbital-science targets were photographed with a Hasselblad EL (electric) camera equipped with a 250-mm lens, while the CSM operated under constraints imposed by the bistatic-radar experiment. During revolution 26, the Hasselblad DC (data) camera was used to take a continuous strip of overlapping vertical photographs from a point near the far-side terminator across the bootstrap site near Descartes Crater, to a point southwest of Lansberg Crater.

An area north of Descartes Crater, a candidate bootstrap site for the Apollo 16 landing, was a photographic target of highest priority during this mission. The LTC coverage was designed to support the detailed evaluation of lunar-surface characteristics near possible landing points within the area. Noisy operation of the LTC on the first of three scheduled photographic passes indicated a camera malfunction. A contingency procedure for obtaining site-evaluation photography using a Hasselblad EL camera equipped with a 500-mm lens was activated. With a spacecraft maneuver providing the required image-motion compensation, convergent stereoscopic coverage of more than 100 km² in the target area was obtained during each pass. The contingency procedure was substituted for strips of vertical LTC photographs scheduled for revolution 27 and oblique LTC photographs scheduled for revolution 28. A third opportunity was scheduled during revolution 30

to increase the probability of success. During revolutions 27 and 28, the spacecraft was maneuvered about the prime target point. The ground-support personnel observing telemetry data and the CMP reported that the operations appeared smooth and satisfactory. The third maneuver, during revolution 30, was therefore targeted to the east of the prime point to increase the area covered by convergent photography. Additional orbital-science photography was taken with the Hasselblad EL and DC cameras during revolutions 28, 29, and 30 when the spacecraft orientation was not constrained by the bootstrap photography.

Shortly after the LM–CSM separation during revolution 12, the LM crew photographed the CSM above a lunar-surface scene illuminated by a high Sun. After photographing the crescent Earth rising above the lunar horizon during revolution 13, the LM crew continued preparations for the lunar landing. The bracket-mounted 16-mm DAC recorded the spectacular scene visible through the right-hand window of the LM from the time of pitchover to the landing at the Fra Mauro site 4 min later. The cameras used in the LM and on the lunar surface are listed in table 1–II.

The 33.5-hr stay on the lunar surface accommodated two extravehicular activity (EVA) periods for a total lunar-surface stay time in excess of 17 man-hr. Photography supporting the lunar-surface activities was designed to document the emplacement of experiments comprising the Apollo lunar-surface experiments package (ALSEP), to augment crew observations, to expand geologic descriptions during the extended traverse toward Cone Crater, and to record the effects of the dynamic interaction between Apollo equipment and the lunar surface.

The commander (CDR) was photographed by the LM pilot (LMP) from the LM window before he stepped to the lunar surface during the first EVA. The CDR photographed the LMP's descent to the surface with the DAC. During the LM and site inspection, the interaction between the LM footpads and the lunar surface, as well as the

TABLE 1–II. *Photographic Equipment Used in LM and on Lunar Surface*

Camera	Features	Film size and type	Remarks
Hasselblad DC, 2.....	Reseau plate, 60-mm lens	70-mm; SO–168 Ektachrome EF color-reversal film exposed and developed at ASA 160; SO–267 Plus-XX black-and-white film, ASA 278	Hand held within LM; mounted on chest bracket for EVA operations
DAC	10-mm lens; variable frame rates of 1, 6, 12, and 24 frames/sec	16-mm; SO–368 Ektachrome MS color-reversal film, ASA 64	Mounted in right LM window to record lunar-surface view during descent and ascent; hand held or mounted on modularized equipment transporter (MET) for recording EVA activities
Apollo lunar-surface closeup camera	Stereoscopic; pair of 46-mm lenses; fixed aperture and speed; built-in flash, recyclable in 10 sec	35-mm; SO–368 Ektachrome MS color-reversal film, ASA 64	Stereoscopic coverage of a 72-by 82.8-mm area with convergent angle of 9° and a base-to-height ratio of 0.16

effects of the exhaust plume, were carefully recorded photographically. Hasselblad panoramas from three points near the LM recorded the precise landing point and the relationship of surrounding topographic features. After emplacing the experiments in the ALSEP, the crew photographed the deployment site before returning to the LM. Between EVA periods, the effects of man's first traverse across the Fra Mauro site were documented by photographs from within the LM. The MET, parked in a partial shadow cast by the S-band antenna, was surrounded by tracks made by man and his equipment.

The EVA–2 photography was geologically oriented. Panoramas at key points along the extended traverse to Cone Crater and documentation photographs of specific features to support the objectives of the geology experiment are discussed in detail in section 3. The crew photographed the lunar-surface area that was visible through the LM windows after the end of EVA–2. Striking differences in the shadows recorded by the two series of pictures demonstrate the change in Sun angle between the EVA periods.

The 16-mm DAC, mounted in the LMP's window, recorded the best ascent photographs obtained to date. The 16-mm camera recorded the initial sharp interaction between the ascent-engine plume and the nearby natural or manmade objects. The motion picture records a spectacular view of the receding lunar surface as the LM first lifted and then moved westward away from the Fra Mauro site. Ascent occurred during the near-side pass of revolution 31. The docking maneuver, during revolution 32, was photographed from both spacecraft. The CSM cameras photographed the LM jettison during revolution 33. The Hasselblad EL with the 500-mm lens was substituted for the malfunctioning LTC to photograph the predicted impact point of the Apollo 14 SIVB stage during revolution 35.

After the successful transearth-injection burn near the start of revolution 35, the CSM cameras were unstowed. Several sequences of photographs were acquired with a variety of camera-lens-film combinations as the spacecraft left the Moon. Later, during the transearth coast, the DAC recorded activities within the CM. Two in-flight demonstrations (heat flow and liquid transfer) were photographed during this period.

Representative mission photographs taken between the time of docking during translunar coast and the transearth coast are shown in figures 1–1 to 1–39.

FIGURE 1–1.—The LM stowed in the SIVB stage during the transposition and docking maneuver. Gas venting from the SIVB is clearly visible (AS14–72–9920).

FIGURE 1–2.—After the delayed, but successful, docking maneuver and extraction of the LM, the SIVB stage was targeted for a crash landing on the Moon. The compartment in the SIVB formerly occupied by the LM is visible beyond the LM thrusters (AS14–72–9925).

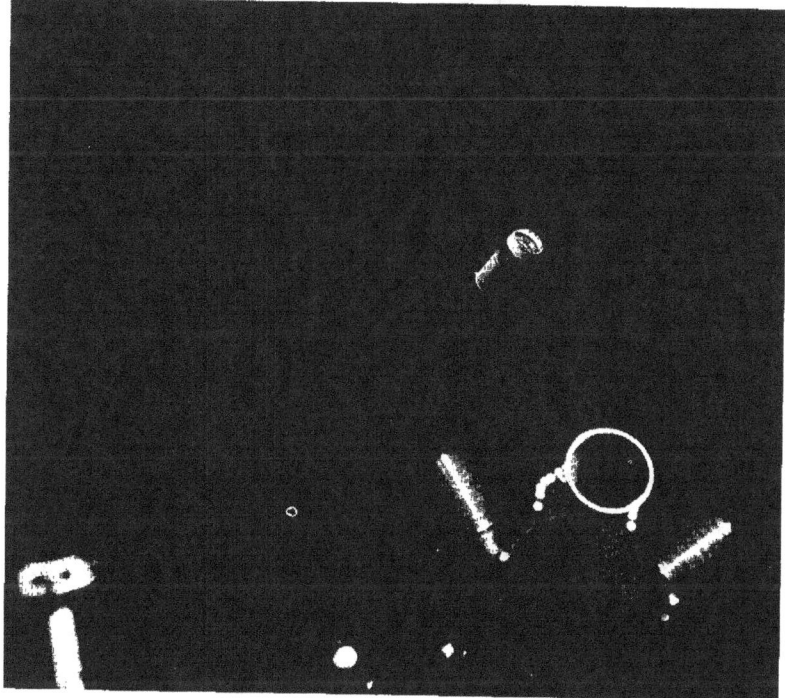

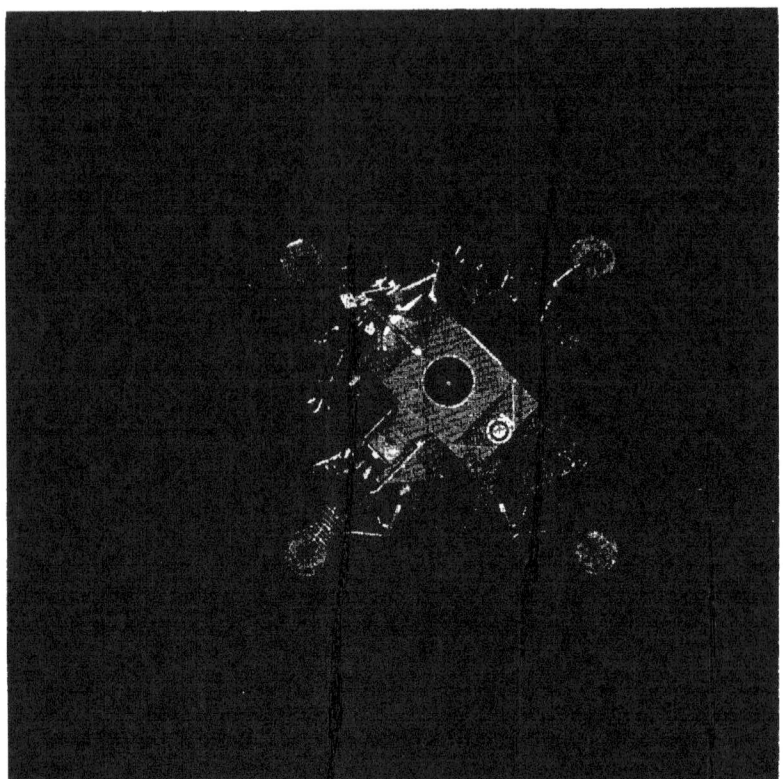

FIGURE 1-3.—The LM passing inspection after separation. The top hatch and docking target are clearly visible. The four Mylar-wrapped footpads were unfolded and locked after LM extraction (AS14–74–10206).

FIGURE 1-4.—Earthrise photographed as the LM reacquired communications with Earth one revolution before landing at Fra Mauro. Pasteur Crater, in the foreground, is centered near latitude 12° S, longitude 105° E in rugged terrain just beyond the eastern limb of the Moon as viewed from Earth (AS14–66–9228).

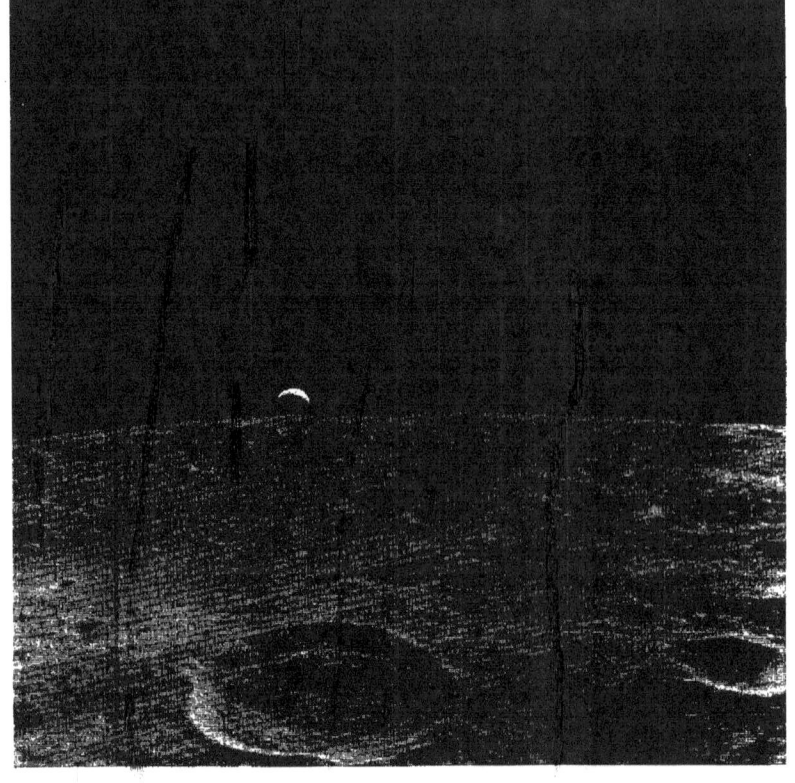

FIGURE 1-5.—The CDR steadies the flagpole near the beginning of the first EVA. The partial shadow to the right is cast by the erectable S-band antenna; the shadow of the LM extends along the left (AS14-66-9232).

FIGURE 1-6.—The LMP adjusts the color-televison camera near the beginning of the first EVA. The sheet near the lower left corner of the photograph contained samples of thermal-control coatings. The samples were photographed with the closeup stereocamera to record the degradation of optical properties after the samples were covered with dust (AS14-66-9240).

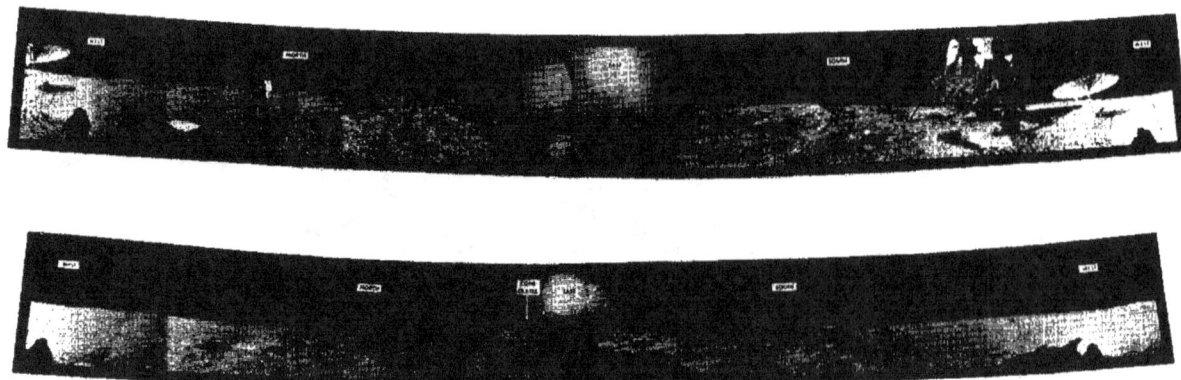

FIGURE 1-7.—Panoramas from points near the LM document the surface features at the Fra Mauro landing site before the traverse on the first EVA (S-71-25424). (a) This panorama was taken from a point northeast of the LM. (b) This panorama was taken from a point west of the LM.

FIGURE 1-8.—Composite photograph showing the texture of disturbed and undisturbed surface materials at the Fra Mauro site. The different depths of penetration by the visible footpads are the result of variations in the underlying lunar-surface materials (S-71-25423).

FIGURE 1-9.—Near-vertical view of the footpad that landed on the crater-rim deposit. The less-resistant material yielded to the load imposed by the LM and deformed the crater wall. The small Mylar-covered rod to the left of the footpad is a probe that actuates the contact light on the instrument panel when the probe touches the surface (AS14-66-9234).

FIGURE 1-10.—The LM-inspection procedure included detailed examination and documentation of the surface materials eroded by the plume of the descent engine. The grooving that indicates radial transport of material from the point below the engine is evident. The erosion did not destroy small craters under the engine bell. The prominent rock in the foreground was later collected as a sample (AS14–66–9261).

FIGURE 1-11.—After the emplacement of ALSEP instruments, numerous photographs were taken to document the deployment. The radioisotopic thermoelectric generator in the foreground produces the electrical energy for operating the central station that transmits data to Earth. The ribbons are electrical conductors that connect the control station to instruments deployed to the left and to the right. The object far beyond and to the right of the central station is the laser ranging retroreflector (LRRR) (AS14–67–9366).

FIGURE 1–12.—Closeup view of the LRRR after deployment. Fused silica retroreflectors mounted in each of the 100 holes reflect a laser beam back toward the source at an Earth-based telescope. The wire angle and the bubble level to the right of the retroreflector block are preset for the selenodetic position of the landing site to guide the proper orientation during deployment (AS14–67–9385).

FIGURE 1–13.—The aluminum-colored plastic shroud covers and protects the passive seismic experiment (PSE), which is deployed north of the central station. The red flag near the upper right is at the end of the active seismic experiment (ASE) geophone line (AS14–67–9363).

FIGURE 1–14.—After installing the line of geophones of the ASE, the LMP operates a hand-held thumper as he walks back toward the ALSEP central station. Geophones along the line detect signals generated by small explosive charges in the thumper. Footprints can be seen adjacent to the foreground crater. As the LMP moved outward, deploying the geophones, his left boot sank deeply into the soft deposits of the crater rim, but his right boot barely penetrated the undisturbed surface layer (AS14–67–9374).

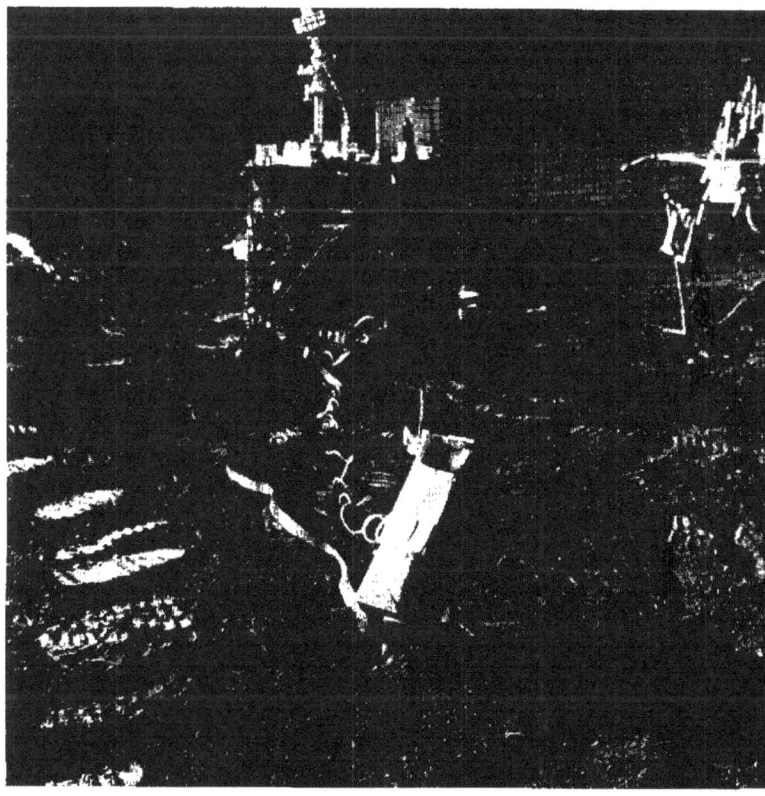

FIGURE 1–15.—The ASE will generate seismic waves with controlled explosions and monitor the response to provide data about the layering of the materials below the lunar surface. At some future date, explosive grenades from the mortar box in the foreground, launched on radio command from Earth, will provide the controlled signals (AS14–67–9361).

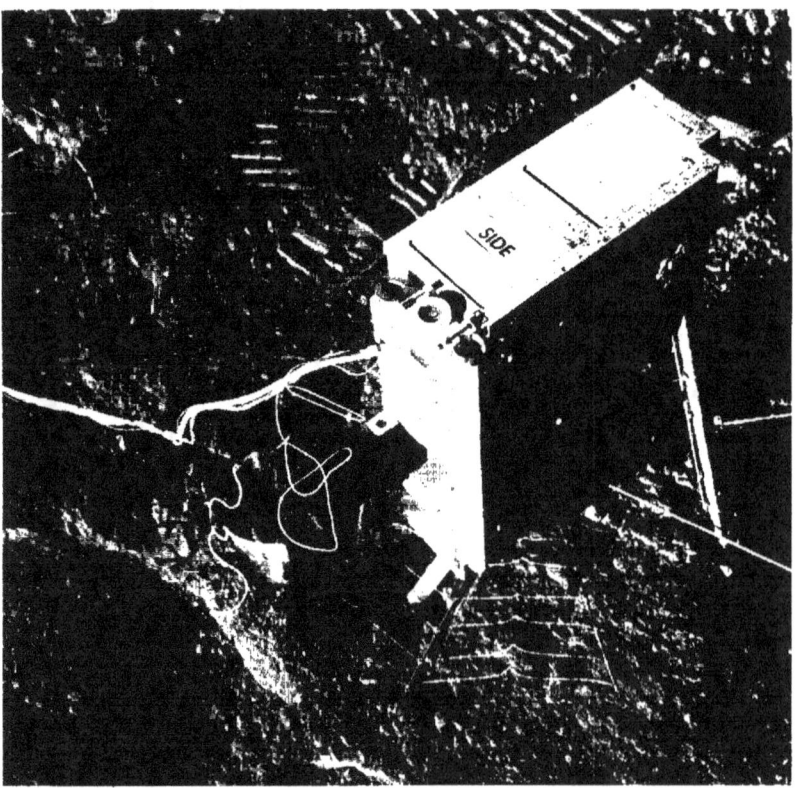

FIGURE 1–16.—Suprathermal ion detector experiment (lunar ionosphere detector) after deployment. The surrounding lunar-surface materials are highly disturbed by the astronauts' boots. The lunar soil can be seen adhering to the vertical surface of the instrument package (AS14–67–9371).

FIGURE 1–17.—The charged-particle lunar environment experiment was deployed and leveled at a position northeast of the ALSEP central station. The gold-colored broad ribbon that extends upward in the photograph is a multiconductor cable that connects the experiment with the central station. The high cohesion of the lunar-surface materials is demonstrated by the undeformed fragments of bars that retain shape after considerable shifting in the disturbed bootprint near the lower right (AS14–67–9364).

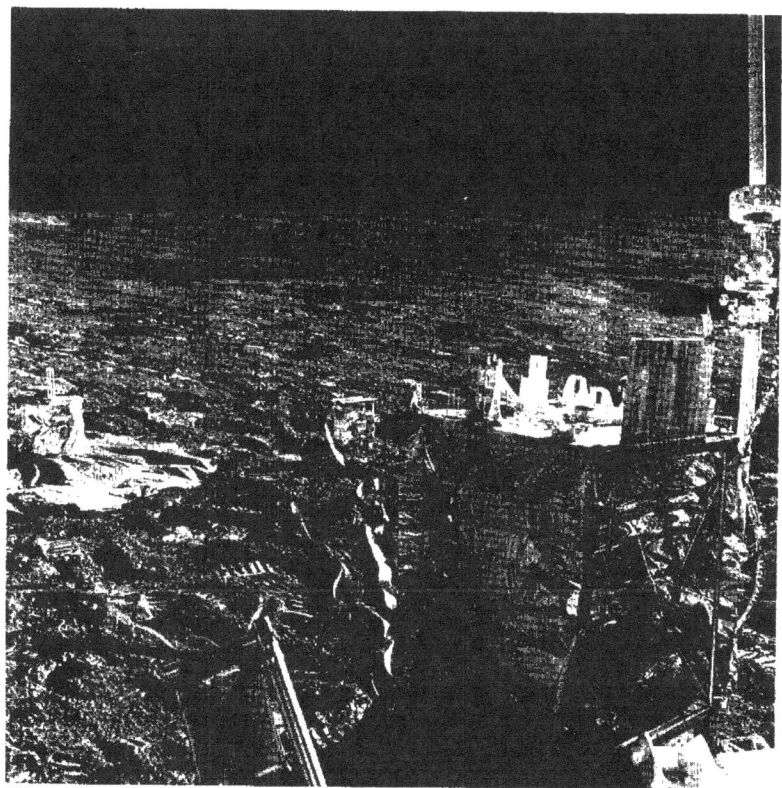

FIGURE 1–18.—The ALSEP central station transmits data collected from the instrument packages to receivers on Earth. The highly directional antenna visible at the upper right must be precisely oriented toward Earth during deployment. The ridge extending northward from Cone Crater forms the distant horizon beyond the antenna. The PSE, covered by an aluminum-colored sheet for thermal protection, is in the left center of the photograph (AS14–67–9384).

FIGURE 1–19.—A view of the LM from the ALSEP site. Lunar-surface materials exhibit a comparatively high reflectivity where they have been compacted by the MET wheels. Material sprayed out over the surface after being scuffed up by the astronauts' boots generally reduces the reflectivity of lunar-surface materials (AS14–67–9367).

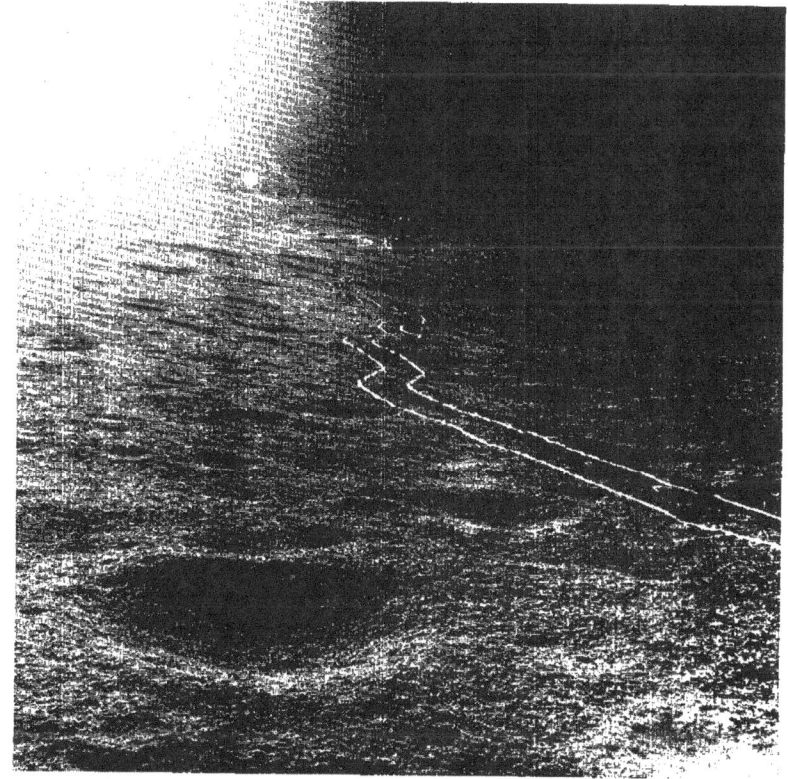

FIGURE 1–20.—Down-Sun view of the deployed ALSEP. The LMP walks away from the central station between the inbound and outbound tracks of the MET (AS14–67–9389).

FIGURE 1–21.—The CDR works with the core tube and extension handle. The MET was loaded with equipment used during the geology traverse of the second EVA (AS14–68–9405).

FIGURE 1–22.—Cross-Sun view of the triple core tube driven into the surface approximately 150 m from the LM near the beginning of the traverse of the second EVA. Soil scuffed up by the boots and tossed forward during the driving of the core tube produced the darker area around and to the right of the core tube (AS14–64–9047).

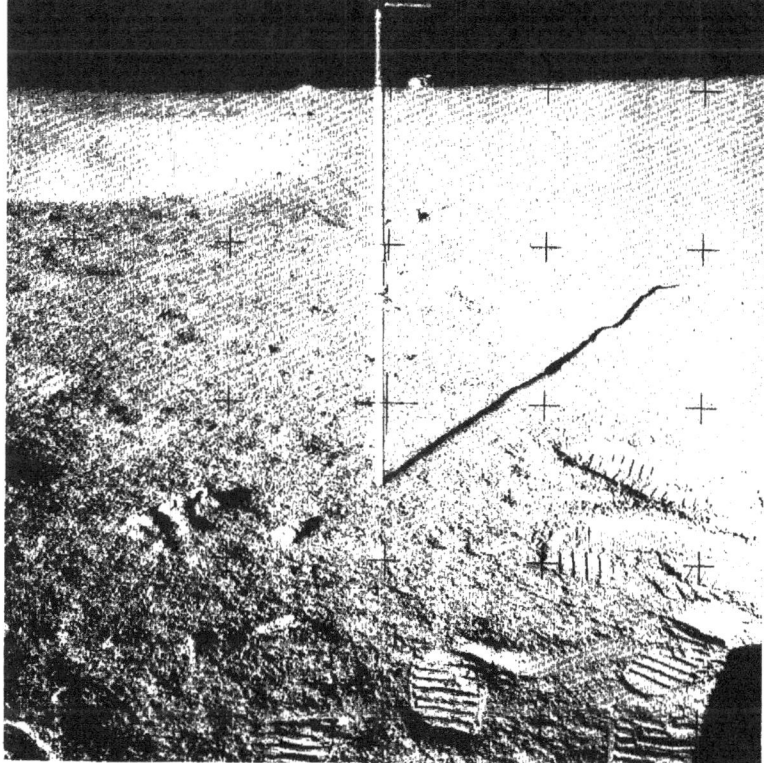

FIGURE 1–23.—View across the triple-core-tube site toward the LM. Such photographs permit the precise location of the sample areas. The area disturbed by soil thrown up by the boots does not appear dark when viewed down-Sun (AS14–64–9048).

FIGURE 1-24.—The LMP, with map in hand, walks across smooth, nearly level terrain between the LM and Cone Crater ridge during the second EVA (AS14-64-9089).

FIGURE 1-25.—Rock fragments were comparatively rare on the surface between the LM and the foot of the ridge capped by Cone Crater. Complex surface textures are present in most of the photographs of sites in this smooth terrain. The gnomon in the upper left provides orientation and scale data. A square target attached to the near leg of the gnomon provides calibration data that enhance the photometric interpretation of lunar photography (AS14-64-9073).

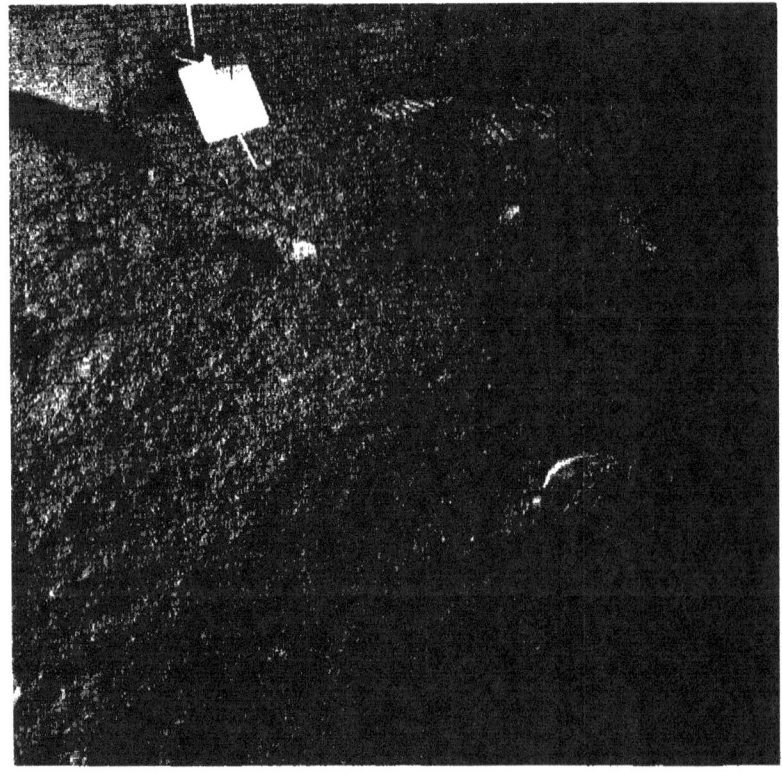

FIGURE 1–26.—Composite photograph of large rounded blocks in the boulder field high on the south flank of Cone Crater (S–71–25426).

FIGURE 1–27.—The MET in block field high on the ridge near Cone Crater. The LM, slightly more than 1450 m away and approximately 80 m lower than the MET, can be seen above and to the right of the tips of the bright rods extending from the right side of the MET (AS14–64–9121).

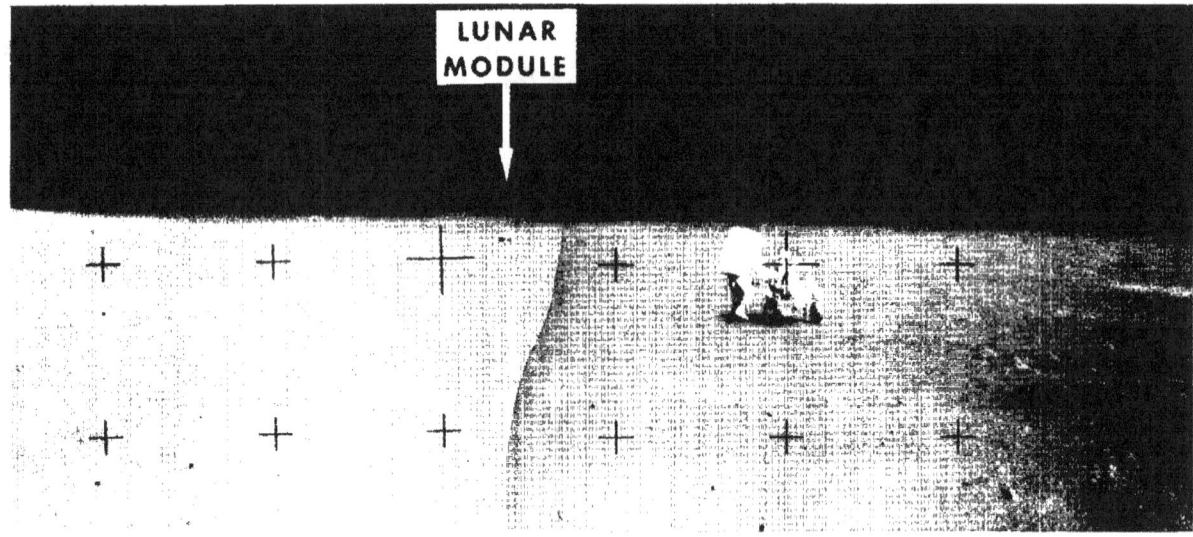

FIGURE 1–28.—Composite photograph of the LMP working by the MET at a location near Weird Crater during the second EVA (S–71–25428).

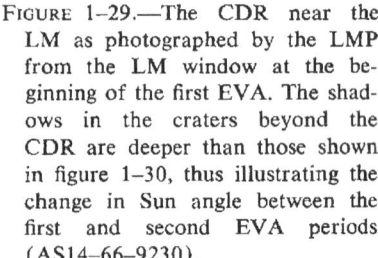

FIGURE 1–29.—The CDR near the LM as photographed by the LMP from the LM window at the beginning of the first EVA. The shadows in the craters beyond the CDR are deeper than those shown in figure 1–30, thus illustrating the change in Sun angle between the first and second EVA periods (AS14–66–9230).

FIGURE 1–30.—View through the right-hand LM window after completion of the second EVA. The instruments that comprise the ALSEP are near the upper left (AS14–66–9338).

FIGURE 1–31.—Crescent Earth hangs over the LM as the crew closes out the second EVA. (AS14–64–9194).

FIGURE 1–32.—Composite photograph of King Crater located at latitude 5.5° N, longitude 120.5° E. The CMP photographed this orbital-science target during revolution 14 with a hand-held Hasselblad EL camera equipped with a 500-mm lens. The crater center was approximately 350 km north of the spacecraft when the photographs were taken (S–71–26213).

FIGURE 1–33.—Near-vertical view of King Crater taken after the spacecraft headed toward Earth. The lengthening shadows indicate that the sunset terminator is approaching. The shadow along the western wall conceals slump-block topography that was illuminated at the time the oblique orbital photographs (fig. 1–32) were taken (AS14–71–9851).

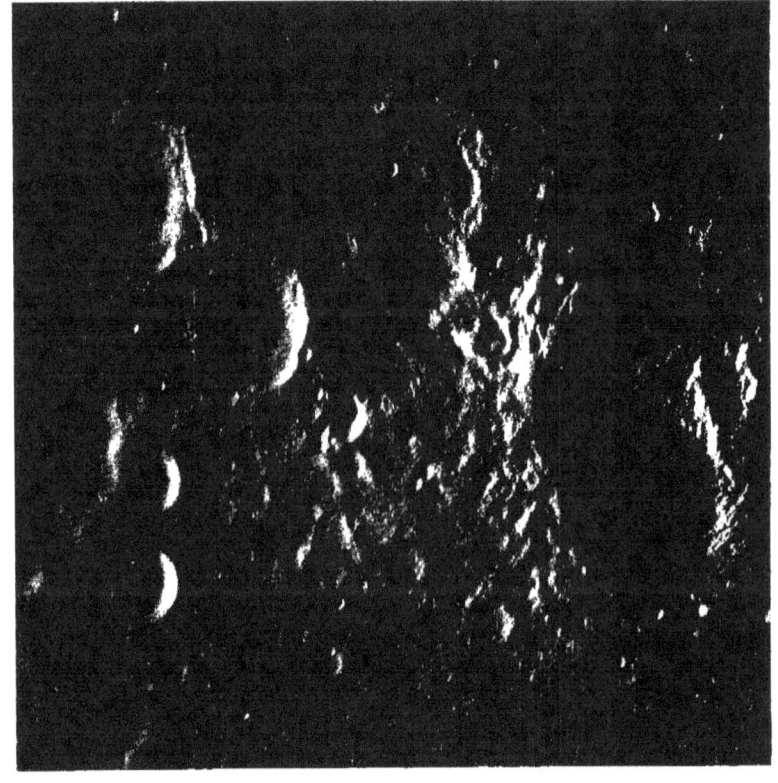

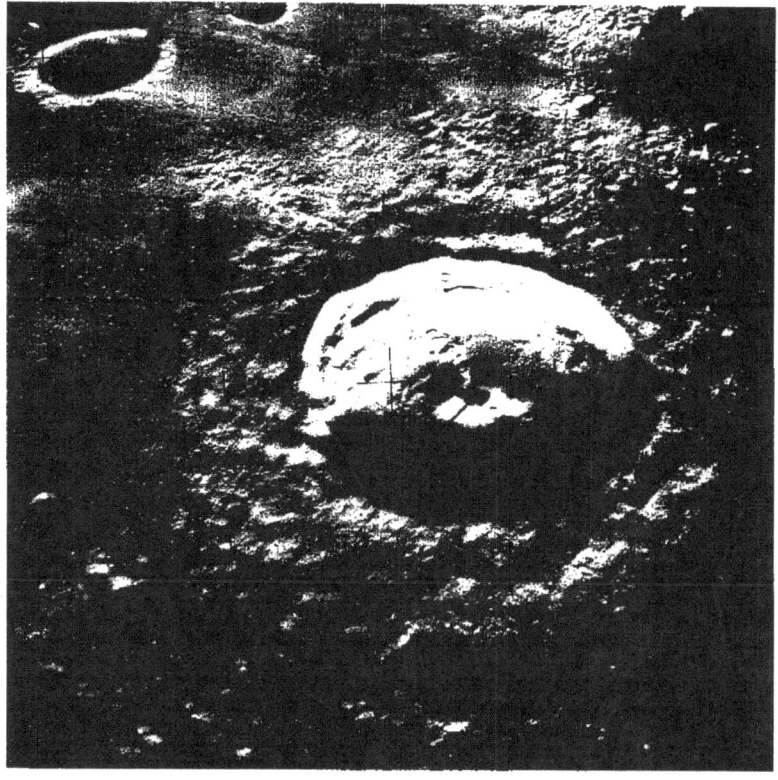

FIGURE 1–34.—Oblique view westward across Lansberg Crater taken during revolution 26 with the Hasselblad DC camera equipped with an 80-mm lens. The low-angle illumination near the terminator emphasizes gentle relief features. A sinuous channel across the smooth mare deposits at the lower left is evident (AS14–70–9825).

FIGURE 1–35.—Oblique view westward across Lansberg N Crater and the sinuous channel. Lansberg Crater is just outside the upper right corner of the frame. Thie view was photographed during revolution 26 with the Hasselblad EL camera equipped with a 250-mm lens. A sinuous channel across mare deposits in the right central part of the photograph has elevated margins that resemble the natural levees of terrestrial channels. A west-facing scarp, low side away from spacecraft, produces the sharp, slightly irregular line across mare deposits at the left center (AS14–73–10120).

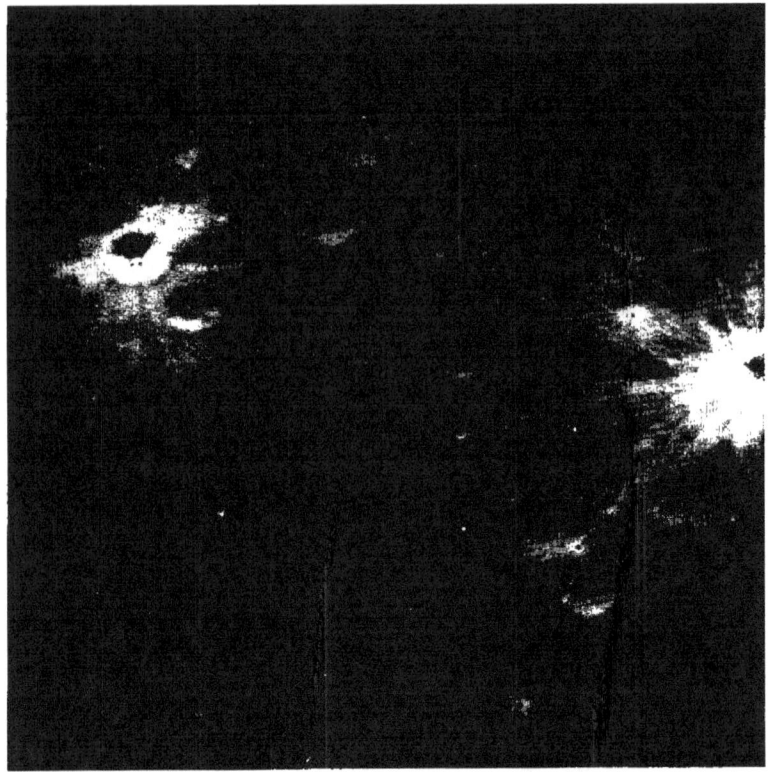

FIGURE 1-36.—Near-vertical photograph looking eastward across the candidate landing site in the highlands north of Descartes Crater. This photograph is part of the site-evaluation imagery taken with the Hasselblad EL camera equipped with a 500-mm lens (AS14-69-9527).

FIGURE 1-37.—Oblique view southwestward across Alphonsus Crater. The bright mound near the lower left is the central peak of the crater. The dark-halo craters in the western half of Alphonsus Crater have been considered potential sites for an Apollo mission. Alpetragius Crater (southwest of Alphonsus Crater) is near the center left margin. Smooth deposits in Mare Nubium extend to the horizon (AS14-73-10096).

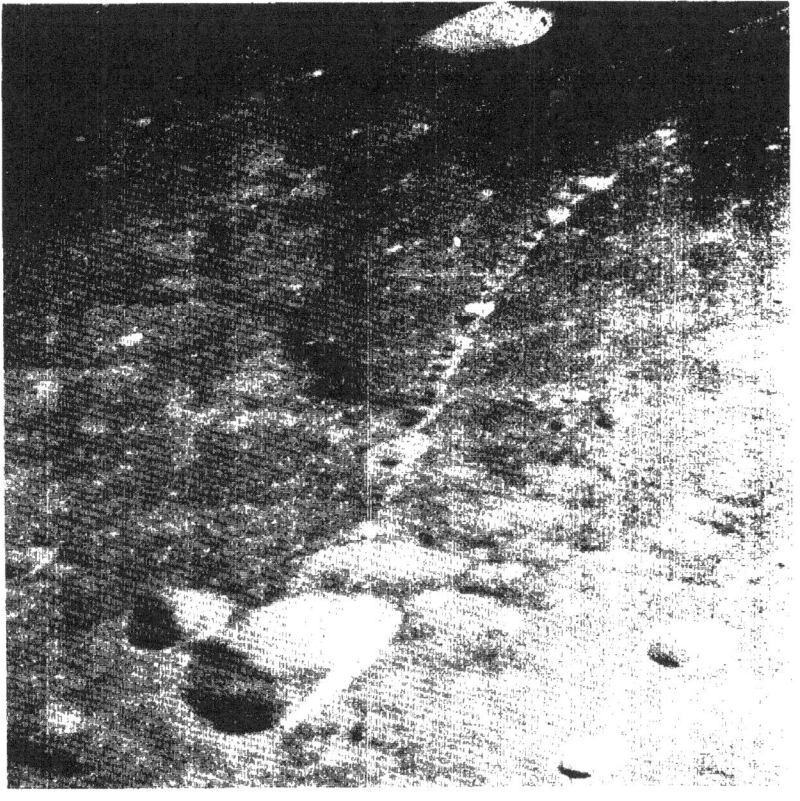

FIGURE 1–38.—Oblique view south-westward across the crater chain named Rima Davy I. The crater chain extends from the central-highlands crater in the foreground out across highland basin deposits. Smooth mare surface is evident near the upper left (AS14–73–10103).

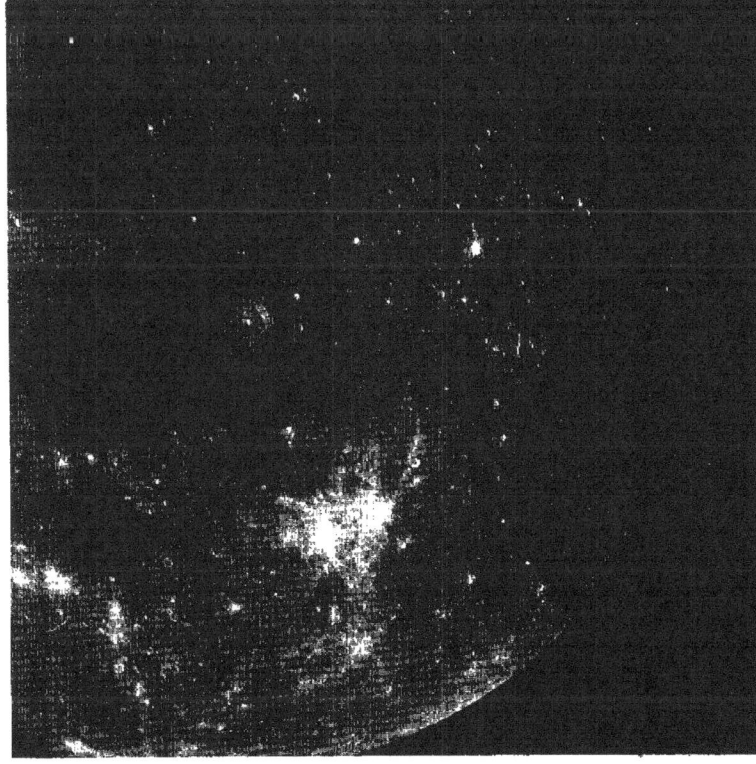

FIGURE 1–39.—High-altitude photograph of the Moon after transearth injection. Selenographic north is approximately along the diagonal toward the upper left. Langrenus Crater, photographed by the Apollo 8 crew, is the larger crater with slumped walls on the margin of Mare Fecunditatis, the larger of several lunar maria visible in this photograph (AS14–71–9906).

2. Crew Observations

Alan B. Shepard,[a] Edgar D. Mitchell,[a] and Stuart A. Roosa [a]

The Apollo 14 mission expanded the techniques and relaxed the operational limitations of previous lunar-landing missions. The specific differences between this mission and the previous missions included onboard cislunar navigation, the use of the service propulsion system for descent orbit insertion, landing in the lunar highlands, extended lunar-surface exploration time for scientific investigation, and a single-orbit rendezvous of the spacecraft. The operational use of these extended capabilities has been an important step in the systematic scientific exploration of the Moon.

This section is a summary of the scientific observations of the Apollo 14 crew. Included are comments on the appearance of the Earth and Moon during cislunar flight and lunar orbit, observations from the surface of lunar terrain features, descriptions of crew mobility in the lunar environment, and geological and soil-mechanics observations made during the two extravehicular activity (EVA) periods.

Observations During Translunar Coast

At approximately 04:00 G.m.t., February 2, 1971, we began the dim-light photography of the dark side of the Earth. We had some doubts as to the quality of the photographs obtained because of the amount of light scatter visible in the sextant.

At approximately 09:00 G.m.t., February 2, the Moon appeared to be about the size of an orange held at arm's length. It also began to show texture and to take on grayish-brown colors, as opposed to the bright appearance seen from Earth. At that time, the view of Earth had Australia at the bottom of the lighted area with the view up

[a] NASA Manned Spacecraft Center.

across the Philippines showing extensive cloud cover.

Lunar-Surface Features as Seen From Lunar Orbit

Our first good look at the lunar surface was when we went into lunar orbit. All the grays, browns, and whites that have been described by previous crews were visible. The best description that could be given was that the Moon looked like a plaster mold that had been dusted with grays and browns. All landmarks were clearly visible and very easy to distinguish. One surprising feature was the fresh appearance of many of the craters; they looked much fresher than photographs had led us to believe.

At the approach to the terminator, some relatively high crater walls and high country cast long shadows that made the surface look very rugged. We could just see the eastern rim of Gambart Crater, and the western rim was just barely lighted. The streaks that have been noted on previous missions were quite pronounced and seemed to lead back toward the Imbrian and Copernican areas.

When we were directly over the terminator and were looking down into the craters with features right on the terminator, it was difficult to determine exactly what we were seeing. Even when we knew what we were supposed to see, the features were difficult to distinguish. As we crossed the terminator, the lack of dark adaptation made it appear as though one could walk along that surface into the darkness and fall into nothing. At that point, there appeared to be absolutely nothing beyond the terminator.

After the descent orbit insertion, it appeared

that we were very low. As a matter of fact, we had the illusion that the flightpath was actually lower than some of the peaks on the horizon. The scale was so deceptive that the terrain made it appear as though the altitude could have been approximately 500 ft. The brown mountains that had appeared to have a soft blanket at the higher altitude looked very harsh at the lower altitude. However, the regolith appeared smoother and more hummocky than it had at the higher altitude. The surface appeared much smoother at altitudes at which closer detail could be seen; particularly at higher Sun angles, it appeared to be a softer surface. It was certainly an unusual experience flying at so low an altitude.

We passed over Mösting Crater and noted that it was a bench crater with an almost-vertical drop on the east side. Both the east and west sides appeared to be solid rock. The regolith nearby appeared soft and not at all rubbly. After crossing to the dark side of the terminator and without being dark adapted, we could see a very definite horizon through the optics. The craters that have bright rims were visible in the earthshine, although generally the surface was a soft gray without many distinct features.

Landing and Observations From the Lunar Module

Shortly after pitchover, Cone Crater and the landing site came into view out the window, confirming the accuracy of the flightpath. The Sun angle was good, and we were able to recognize the landing site even more rapidly than we could on the lunar-module (LM) simulator display at Kennedy Space Center. Manual descent was initiated at 350-ft altitude and approximately 2200 ft short of the desired target. We held a zero descent rate at 170 ft while performing a translation maneuver forward and to the right so contact would occur at the original targeted point. Blowing dust was first noted at 110 ft, but it was not a deterrent factor. The blowing dust appeared to be less than 6 in. deep, and rocks were readily visible through it. The LM landed on a slope with the front hatch on an azimuth of approximately 275° with respect to lunar north. The slope at the landing site was approximately 8°, but the landing site was the flattest place around.

The landing site was very close to the preplanned point but possibly a bit toward Triplet Crater. The LM appeared to be in a bowllike depression. We were looking directly toward Doublet Crater, which appeared to be 25 to 30 ft above us. A pronounced ridge beyond Doublet Crater formed the skyline and was the highest feature in front of the LM. To the north was a large old depression that formed another bowl similar to the one in which the LM appeared to have landed. We could see several ridges and rolling hills approximately 35 to 40 ft high—obviously very, very old craters that were almost indistinct—between the LM and the horizon to the north. A few boulders were scattered around between the LM and Doublet Crater. The largest boulder in view was approximately 3 ft across and located at an azimuth of approximately 350° about 150 ft away. There were two or three other craters about half as big approximately 30 to 45 ft farther away. No pattern in the arrangement of the boulders was apparent. The color of the surface was a kind of mouse brown or mouse gray and obviously changed with the Sun angle.

Numerous craters were present in the field of view—some old, very subdued; some overlapped by newer craters; and some that seemed to be relatively recent. A small pattern of four craters was located at approximately the 290°-to-350° position. The crater at the 290° position and approximately 50 ft from the LM was approximately 15 ft across and had a secondary crater in the middle of it and two or three craters grouped on the north edge. Immediately in front of the LM, approximately 15 ft away, was a 6- or 7-ft-wide crater that was pocked with a few secondary craters. The crater at the 305° position was an old crater approximately 12 ft in diameter with a fairly small, relatively fresh 1-ft-diameter crater on the southwest side. The fourth crater was at the 350° position; this crater was approximately 40 ft away and had a diameter of approximately 25 ft.

Some linear features were noted on the surface. We did not discuss the nearby lineations because they could be confused with the descent-engine pattern. However, farther to the north we noted lineations that appeared to be oriented approximately north of west to south of east. The lineations were very fine grained and were almost imperceptible except that a little shadow effect was visible, almost sand duning but not quite. Although they were plainly visible from the LM, we were not sure that the lineations would not disappear once we egressed from the LM.

First EVA Period

Lunar-Surface Visibility

The visibility on the lunar surface was generally the same as has been reported on previous lunar missions. It was noted that, when we were going down-Sun, there seemed to be a refraction around our bodies. This refraction caused a halo effect in our shadows, and we could not see surface details directly in front of us. Judgment of distance seemed to be somewhat distorted on the lunar surface, and we had a definite tendency to underestimate distances to terrain features, perhaps because of the lack of atmosphere. In addition, boulders were much smaller than they had appeared to be in photographs.

Mobility

Mobility and stability on the lunar surface were generally the same as reported on the Apollo 11 and 12 missions. With a few minutes' adjustment, each individual found a method of travel suited to him. The step-and-hop gait appeared to require a minimum of effort. There was very little tendency to overcontrol or use too much force when using tools or walking on the lunar surface. We had very little difficulty pulling the modularized equipment transporter (MET)—it proved to be quite stable. The MET bounced when it hit a small rise, but it had little tendency to overturn.

Soil Mechanics

On the surface, it was noted that the landing area appeared to be in a swale or wide valley between Triplet and Doublet Craters. The landing point was on the downhill side; the slope leveled off at a lower elevation, approximately 15 ft lower than the LM, and started back up to the rim of Doublet Crater. The area was pockmarked, as are most sections of the Moon, by an enormous number of craters.

The soil was extremely soft and fine grained, almost like a brown talcum powder. There appeared to be very few rocks of any appreciable size in the immediate area of the LM; most of the rocks were less than 2 in. in diameter. The surface on which the forward footpad landed was so soft the soil came all the way to the top of the footpad and even folded over the sides to some degree.

Very little erosion was seen directly under the engine bell. Most of the erosion seemed to be approximately 3 ft to the southeast, which was apparently where the thrust was directed when the engine was cut off; the LM drifted northwest from that point just before touchdown. The erosion pattern appeared to be no more than 4 in. deep.

Experiment Deployment

Several minor problems developed during the deployment of the experiments. We had difficulty releasing one of the Boyd bolts on the Apollo lunar-surface experiments package (ALSEP) subpallet when the guide cup became full of dirt. There seems to be no way of avoiding getting the experiments dirty during transport and deployment. On the traverse to the deployment site, the ALSEP pallets on either end of the mast oscillated vertically and the mast flexed, making the assembly rather difficult to carry. However, it is believed the present arrangement is suitable for traverses of as much as 150 yd.

During deployment of the suprathermal ion detector experiment (SIDE) and the cold cathode gage experiment (CCGE), considerable difficulty was experienced with the stiffness of the interconnecting cable. Whenever an attempt was made to move the CCGE, the cable caused the SIDE to tip over. However, after several minutes of readjusting the experiments, we managed to deploy them successfully.

The three active-seismic-experiment geophones were deployed with very little difficulty. The geophones went into the soft surface material readily; however, moving the cable pulled the second geophone out of the soil and it had to be replaced.

On the return to the LM after ALSEP deployment, we gathered samples from a selected area. Included in these samples were two Little League football-size rocks. One of the football-size rocks appeared to contain a fairly large crystal.

One thing that was noted upon return to the LM was that our footprints and the MET tracks, both to the ALSEP site and to the camera, were a darker color in the disturbed area. Both cross-Sun and down-Sun, the dirt that had been kicked up and turned over was noticeably darker than the mousy brown of the undisturbed regolith.

Second EVA Period

Observations on the Geological Traverse

At station A, the surface, although basically the same fine-grained dusty regolith, was textured and contained more small pebbles than the soil in the vicinity of the LM. The texture appeared to have been splattered with raindrops. There seemed to be a definite relationship between the texture and the small pebbles.

We deployed the lunar portable magnetometer (LPM) and relayed the readings to Mission Control Center. We had trouble when attempting to reel in the LPM cable. The set in the cable was such that, if the handle was released, the cable would unwind three or four turns. We wound it in enough to keep it off the ground and proceeded with the traverse.

The closer we progressed toward Cone Crater, the more boulders we encountered. The basic surface material was still the same fine-grained grayish brown with the raindrop effect. The MET tracks, approximately 0.25 in. deep, made a very smooth pattern reminiscent of that made by driving a tractor through a plowed field.

Station B was an area with considerably more boulders. Many of the boulders were buried or half buried, with only a few of the smaller ones lying on the surface. Most of these boulders were rounded; only a few were angular. We could see edges that had been chipped off, indicating the beginning of a smoothing process. Some of the boulders were well beyond the beginning stage of this smoothing process—most of the rough edges had fractured and fallen off the parent boulder.

Cone Crater Area

Closer to Cone Crater, the Sun angle and the slope made it difficult to determine an exact location. We reached a large boulder field that covered perhaps as much as a square mile. Most of the boulders were the same grayish brown that had been previously noted. However, we noticed one boulder that was almost white. We documented

this boulder and chipped a sample from it. We saw another boulder that had been broken open, and the broken surface appeared white while the outside was the normal brownish color. Grab samples, including another football-size sample, were taken in that area.

Just before inspecting the white boulder, we deployed the LPM for the second and last reading. After some difficulty in leveling the instrument, we relayed the reading on the voice link. The LPM was discarded at the completion of this reading. The soil in the area of the second LPM measurement appeared to have a lighter brown layer below the darker brown regolith. We attempted to obtain a core-tube sample of the color layers, but we could only get the core tube about three-quarters of the way into the surface; when the tube was withdrawn, the soil was so granular that most of it came out of the tube.

At a site approximately one crater diameter from North Triplet Crater, we collected trench samples and attempted to obtain a triple-core-tube sample. However, in two attempts, we could not get samples beyond the first core tube. A problem was also experienced with the trench sample in that the soft regolith kept falling into the trench. As the trench was dug, at least three different layers were noted: the surface layer, which was brown; a second layer that appeared to have quite a bit of black; and a third layer of very light material. The second layer was very thin (no more than 0.25 in.) and consisted of small glassylike pebbles. Before leaving the Triplet site, we gathered documented samples and several grab samples, including one of the whitish rocks.

On the return to the LM, the commander made a side trip to the ALSEP to verify the alinement of the antenna on the ALSEP central station. After making a small alinement correction, we collected a few samples and returned to the LM, retrieved the solar-wind composition foil, stowed the samples and equipment, and prepared for liftoff.

Orbital Operations

The orbital photographic and scientific experiments conducted by the command module pilot simultaneously with the surface exploration included the gegenschein photography, the bistatic-radar experiment, the S-band transponder experiment, surface landmark tracing, and bootstrap photography using the Hycon and Hasselblad cameras.

The gegenschein photography was performed on orbit 16. The new window shade worked very well; to be doubly safe about inadvertent light strikes, the floodlight near the window was taped.

The bistatic-radar and S-band transponder experiments were performed on schedule. The spacecraft was configured as planned and the experiments appeared to have been satisfactory.

The landmark tracking was accomplished with no significant difficulties. On the first pass over the Apollo 14 site, I saw a white spot that was obviously foreign to the typical lunar surface. I suspected the spot was the LM and my suspicion was verified when I saw the LM shadow. The first day I tracked the LM, the Sun angle was still fairly low and the shadow of the LM was plainly visible. The shape of the LM could not be seen as such, but there was no doubt it was there. The next day, while doing landmark tracking, I had one landmark just before the Fra Mauro region and one after it, so I looked for the LM again. This time I saw the LM without any difficulty. The LM shadow had diminished considerably; but, by that time, the ALSEP had been deployed and the glint coming off it was easy to see.

The bootstrap photography was accomplished using the 70-mm Hasselblad camera with a 500-mm lens after the failure of the Hycon lunar topographic camera. By using the crewman-optical-alinement-sight maneuver to hold the camera on target, I apparently obtained some good stereophotographs of the Descartes landing area. Three passes were made over the Descartes area to obtain stereostrips covering the region.

Ascent From the Lunar Surface

When the hot-fire test of the jets was performed before liftoff, the erectable S-band antenna blew over. The liftoff was smooth and the ascent engine performed quite satisfactorily. When we pitched over and looked at the landing site, we could see the gold-and-silver Kapton insulation flying out from the descent stage in a radial pattern and parallel to the lunar surface, just as had been reported by the Apollo 11 and 12 crews.

The last of the scientific objectives was accomplished when we made another series of dim-light photographs of the dark side of the Earth. On the transearth coast, the same problem with scattered light was experienced as during trans-lunar coast. Upon completing the photography, we began preparing for the final phase of our return to Earth.

3. Preliminary Geologic Investigations of the Apollo 14 Landing Site

G. A. Swann,[a][†] N. G. Bailey,[a] R. M. Batson,[a] R. E. Eggleton,[a]
M. H. Hait,[a] H. E. Holt,[a] K. B. Larson,[a] M. C. McEwen,[b]
E. D. Mitchell,[b] G. G. Schaber,[a] J. P. Schafer,[a] A. B. Shepard,[b]
R. L. Sutton,[a] N. J. Trask,[a] G. E. Ulrich,[a] H. G. Wilshire,[a]
and E. W. Wolfe [a]

The Apollo 14 lunar module (LM) landed at latitude 3°40′24″ S, longitude 17°27′55″ W, in the Fra Mauro region. The landing site is 1230 km south of the center of the Imbrium Basin and 550 km south of the southern rim crest of the basin. This site was selected to study a lunar stratigraphic unit called the Fra Mauro Formation, which covers a substantial part of the earthward lunar surface. The formation is a broad belt surrounding Mare Imbrium and is believed to be material excavated by a large impact that formed the Imbrium Basin.

The LM landed about 1100 m west of Cone Crater, which is located on a ridge of Fra Mauro Formation. Cone Crater is a sharp-rimmed, relatively young crater approximately 340 m in diameter that ejected blocks of material up to 15 m across, which were derived from beneath the regolith. Sampling and photography of these blocks were primary objectives of the mission. Rays of blocky ejecta from Cone Crater extend westward beyond the landing site. The landing took place on a smooth terrain unit recognized in premission Lunar Orbiter and Apollo orbital photography. Sampling and description of this unit were other main objectives of the mission.

During the first period of extravehicular activity (EVA), the crew traversed westward over the smooth terrain for a round-trip distance of ap-proximately 550 m (fig. 3–1) and deployed the Apollo lunar-surface experiments package (ALSEP). Sixty-nine rock samples for which locations have been determined were collected by the crew: seven in the contingency sample, 29 in the comprehensive sample, 31 in the bulk sample, and two small football-size rocks.

The crew covered a round-trip distance of approximately 2900 m (eastward from the LM during the second EVA (fig. 3–1)). During the traverse, they crossed the smooth terrain, the Fra Mauro ridge unit, and a section through the continuous ejecta blanket of Cone Crater to within 20 m of the crater rim crest. Forty-eight rock samples, the locations of which have been determined, were collected at points along the traverse. The modularized equipment transporter (MET) was used to transport the samples and collection tools.

Detailed analysis of surface photographs of boulders ejected from Cone Crater and comparison of these photographs with returned samples indicate that the Fra Mauro Formation is mainly composed of moderately coherent breccias in which dark lithic clasts up to 50 cm or more across and less abundant light clasts are set in a light matrix. Subordinate rock types that may be part of the Fra Mauro Formation include coherent breccias with about equal amounts of light and dark clasts and breccias with irregular bands of very light clastic rock.

Boulders ejected from Cone Crater record a

[a] U.S. Geological Survey.
[b] NASA Manned Spacecraft Center.
[†] Principal investigator.

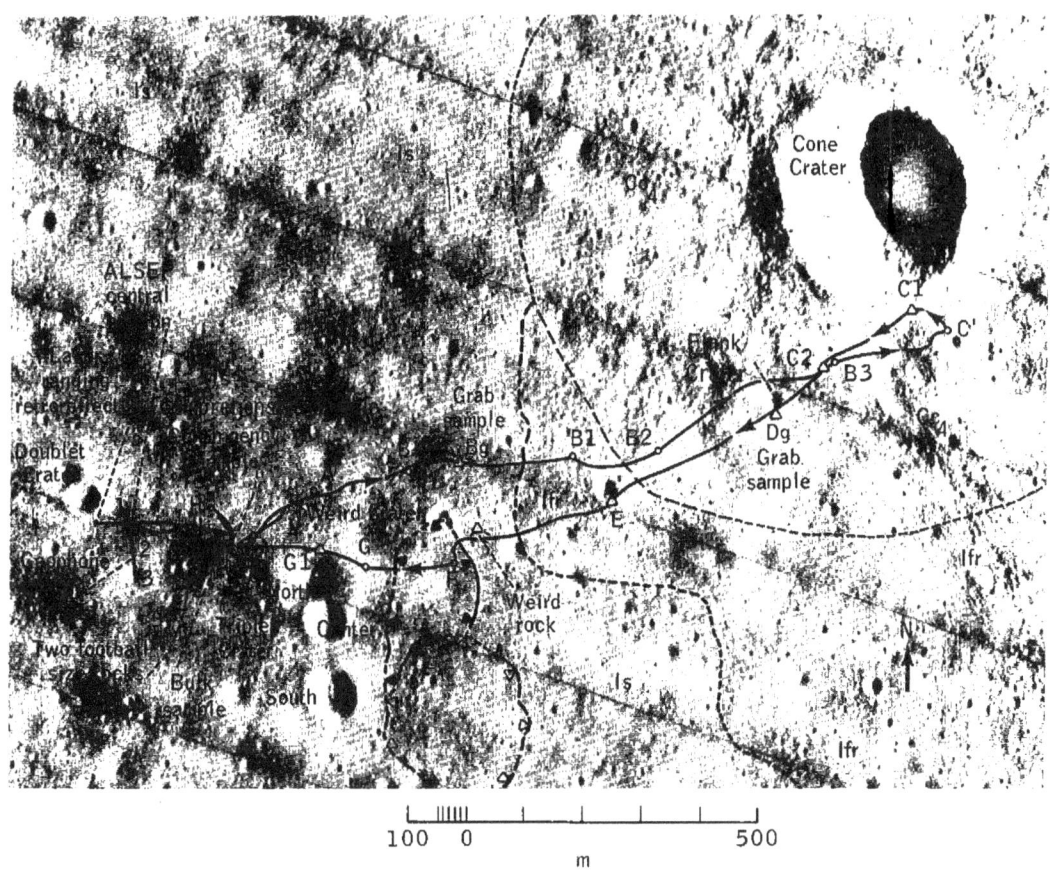

100   0   500
m

Explanation

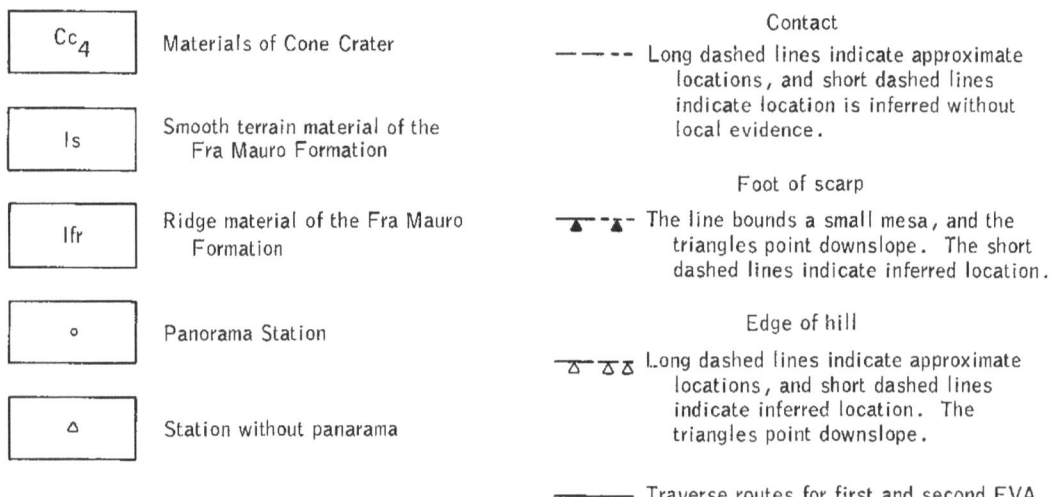

Cc4	Materials of Cone Crater	
Is	Smooth terrain material of the Fra Mauro Formation	
Ifr	Ridge material of the Fra Mauro Formation	
o	Panorama Station	
Δ	Station without panarama	

Contact
— — -- Long dashed lines indicate approximate locations, and short dashed lines indicate location is inferred without local evidence.

Foot of scarp
The line bounds a small mesa, and the triangles point downslope. The short dashed lines indicate inferred location.

Edge of hill
Long dashed lines indicate approximate locations, and short dashed lines indicate inferred location. The triangles point downslope.

———— Traverse routes for first and second EVA.

FIGURE 3–1.—Map of major geologic features in the Apollo 14 traverse area.

FIGURE 3–2.—Panorama 1, taken from the LM before the first EVA.

complex history in which the youngest structures (several sets of intersecting fractures and planar, glass-lined sheeting structures that cross clasts and matrices alike) may have resulted from the cratering event. Earlier events, presumably relating to the origin of the Fra Mauro Formation or older ejecta blankets, include lithologic layering, deformation, and induration of the breccias. Clasts of breccia within the breccias may represent pre-Imbrian cratering in the Imbrium Basin region.

Photographic surveys taken during the Apollo 14 lunar stay were designed to accomplish the following tasks:

(1) Locate and illustrate topographic features at each major geologic station

(2) Record the surface characteristics of each sample area and determine the orientation and location on the lunar surface of the samples at the time of collection

(3) Document geologic targets of opportunity

Other photographic surveys were taken to document the deployment of the ALSEP and the soil mechanics experiment.

Four hundred and seventeen photographs were taken on the lunar surface with the Hasselblad Electric data camera during the Apollo 14 mission. Fifteen panoramas, consisting of 275 photographs, were taken for major station location and general geologic documentation, of which 10 are included in this report (figs. 3–2 to 3–11). Forty-nine pictures were taken for sample documentation, and 27 pictures were taken to document ALSEP deployment. The remaining pictures were of miscellaneous targets of opportunity.

The major geologic stations were located on a rectified copy of Lunar Orbiter 3 frame H–133 by feature correlation between the Hasselblad and Lunar Orbiter photographs and by resection.

Geology of the Fra Mauro Landing Site

The surface of the Moon is grossly divided into relatively dark, low-lying plains or maria, and the brighter, generally more rugged areas of the terra. The maria, sites of the Apollo 11 and 12 landings, are densely covered with craters from several centimeters to a few hundred meters in diameter, with a scattering of larger craters up to a few tens of kilometers in diameter. Much of the terra, site of the Apollo 14 landing, is densely covered with craters several tens of kilometers in diameter; before the Apollo 14 mission, Lunar Orbiter and Apollo photographs showed that small craters are also present on the terra in large numbers down to sizes of a few meters.

Regional Geologic Setting

The Fra Mauro Formation (refs. 3–1 and 3–2) is an extensive blanketlike deposit lying in a broad band around the Imbrium Basin (fig. 3–12; refs. 3–3 and 3–4) and is interpreted as ejecta from the impact that formed the basin. Stratigraphic relationships around the margin of the Imbrium Basin show that a significant number of geologic events occurred between the formation of the basin and its later filling by mare material (refs. 3–2 and 3–5). These events included formation of large craters such as Archimedes,

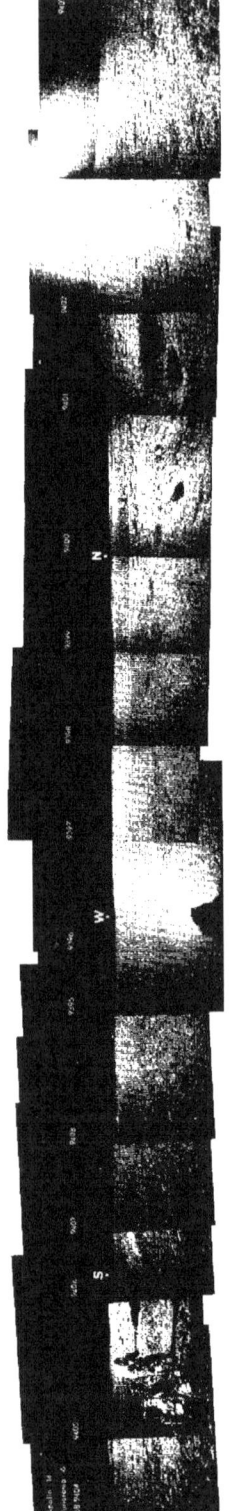

FIGURE 3–3.—Panorama 6, taken at station A.

FIGURE 3–4.—Panorama 7, taken at station B.

FIGURE 3–5.—Panorama 8, taken at station B1.

FIGURE 3-6.—Panorama 9, taken at station B2.

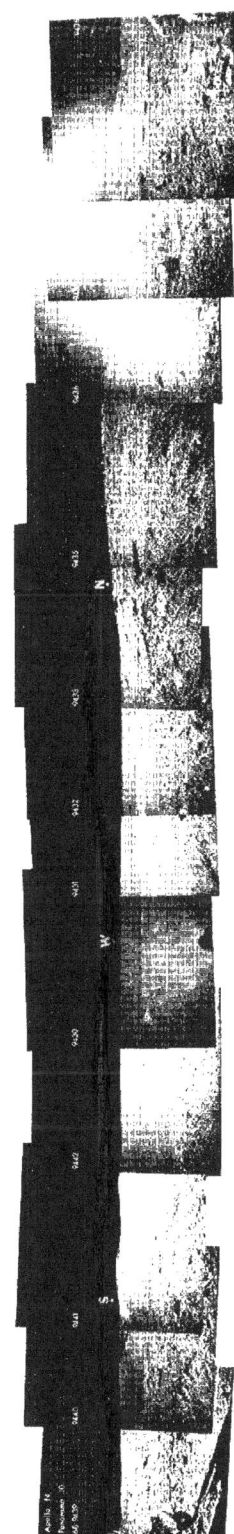

FIGURE 3-7.—Panorama 10, taken at station B3.

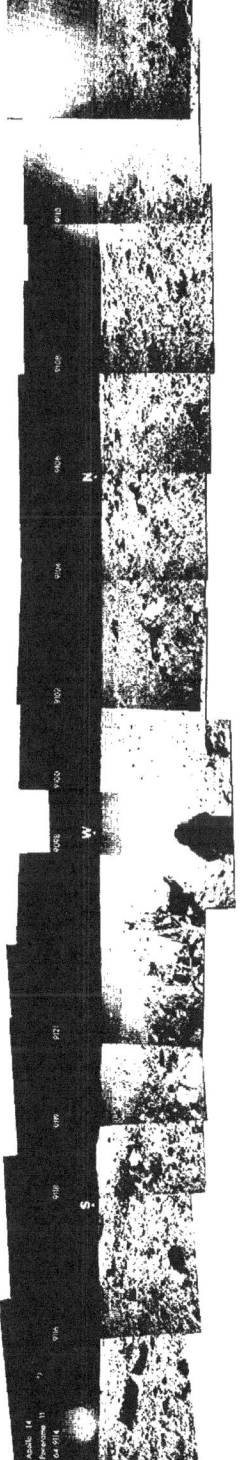

FIGURE 3-8. Panorama 11, taken at station C'.

FIGURE 3-9.—Panorama 12, taken at station F.

FIGURE 3-10.—Panorama 13, taken at station G.

FIGURE 3-11.—Panorama 14, taken at station H.

emplacement of the relatively light terra plains materials, and, in the western part of the Moon, formation of the Orientale Basin. The Fra Mauro Formation is, therefore, older than the mare materials sampled by the Apollo 11 and 12 crews.

The Apollo 14 LM landed near the outer edge of the Fra Mauro blanket. Part of the periphery of the Fra Mauro Formation, including the Apollo

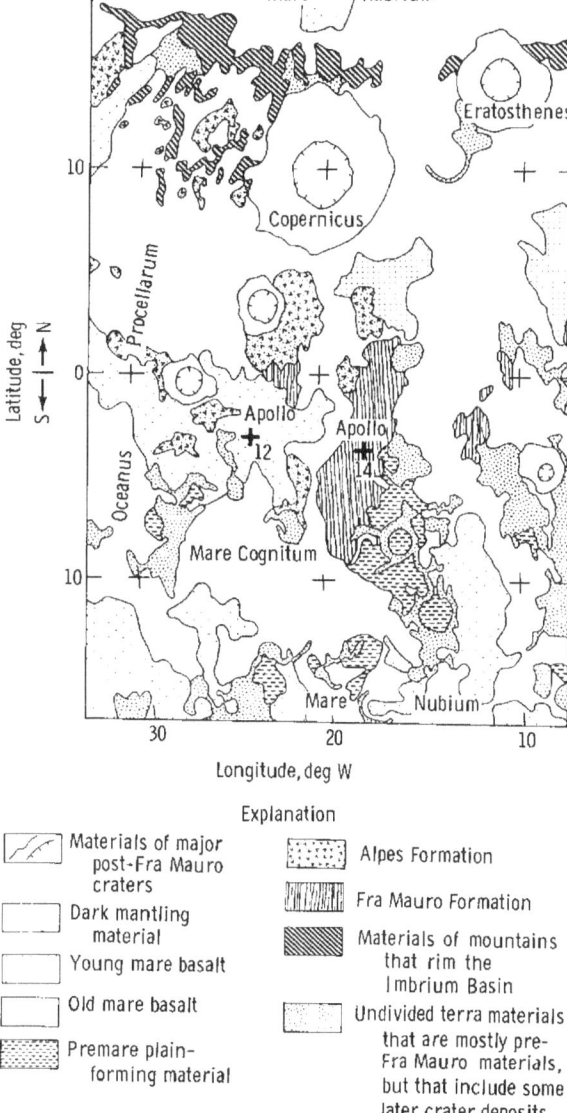

FIGURE 3–12.—Regional geologic map of the area surrounding the Apollo 14 landing site.

14 landing site, was photographed by Lunar Orbiter 3 and has been mapped (ref. 3–6) at scales of 1:250 000 and 1:25 000. In this area, the formation generally grades southward from a ridged deposit to a complexly cratered one. The ridges are the most characteristic feature of the Fra Mauro Formation. Locally, somewhat flatter tracts, typically measuring a few kilometers across, have slightly lower albedos than those of the ridges and occur in shallow surface depressions of the Fra Mauro Formation. Some of these are mapped as a smoother Fra Mauro component and others as possible overlying accumulations of volcanic, probably pyroclastic, material. Elsewhere, distinct plains-forming units of the terra that are obviously younger than the Fra Mauro Formation have been mapped, and these are, in turn, overlapped by still younger dark plains of the maria (fig. 3–12). Some hills in the area appear to be volcanoes superposed on the Fra Mauro Formation, and other similar but more heavily cratered hills may be volcanoes that were formed before the Fra Mauro Formation was deposited.

Ridges of the Fra Mauro Formation in the vicinity of the landing site are mostly 1 to 4 km wide, a few to several tens of meters high, and from five to 10 times as long as they are wide. The ridges are slightly sinuous and roughly radial to the Imbrium Basin. Comparison of the Imbrium Basin with the younger and better preserved Orientale Basin suggests that the ridges were formed largely by flowage of material (probably fragmental rock debris) radially along the ground during excavation of the basin. Fracturing of the pre-Fra Mauro Formation rocks in a pattern radial to the Imbrium Basin may also have contributed, at least locally, to the relief of the ridges.

The major geologic objectives of the mission were to describe, photograph, and sample the ejecta blanket of the 340-m-diameter Cone Crater (fig. 3–1). This crater is situated on one of the ridges that may be Fra Mauro Formation material that flowed radially outward from the Imbrium Basin. Lunar Orbiter photographs indicate that the crater penetrates below the fine-grained lunar regolith into a blocky or bedrock substrate. It was anticipated that this substrate would be Imbrium Basin ejecta more or less in the original form.

Local Geologic Setting

Three principal photogeologic map units were traversed during the two Apollo 14 periods of EVA (fig. 3–1): A smooth terrain unit on which the LM landed, slopes of a cratered ridge of the Fra Mauro Formation, and the blocky rim deposit of Cone Crater. The smooth terrain unit is grossly level over distances of one to several kilometers, but is densely populated with subdued crater forms several tens of meters to several hundred meters across and generally several meters to several tens of meters deep, which cause the surface to be undulating. The Fra Mauro ridge, which extends several kilometers north of Cone Crater, has slopes of 10° to 15° covered with patterned ground in the vicinity of the sec-

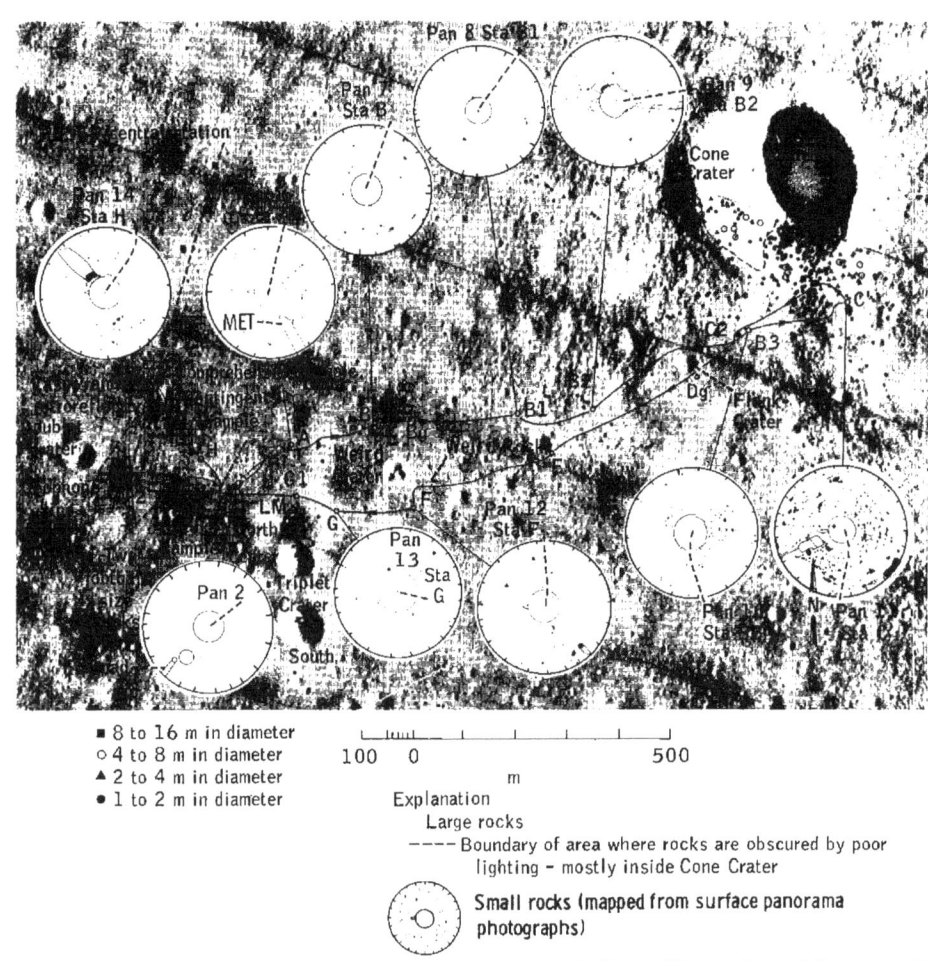

- ■ 8 to 16 m in diameter
- ○ 4 to 8 m in diameter
- ▲ 2 to 4 m in diameter
- ● 1 to 2 m in diameter

100   0   500

m

Explanation

Large rocks

---- Boundary of area where rocks are obscured by poor lighting – mostly inside Cone Crater

Small rocks (mapped from surface panorama photographs)

Circular insets on the map represent areas of 10-m radius surrounding photographic panoramic stations. Hachures in circular insets show directions of individual photographs that constitute the panoramas.

Distribution of rocks larger than 10 cm in diameter within 10-m radius of panorama station is shown by solid pattern, drawn to scale. Large rocks marked by black dots and triangles shown in insets of

panoramas 8, 12, and 13 are not part of the panoramic views. The locations were mapped from Lunar Orbiter photography and apply to the map beneath the circular insets (not to the insets themselves).

Fine lines within circular insets show areas occulted from the camera by large rocks or the MET; or areas in shadow. Inner circles shown by fine lines indicate the near limiting field of view of the panoramic photographs.

Traverse and station symbols are the same as in figure 3-1.

FIGURE 3–13.—Rock-distribution map.

ond EVA. At least four, old, moderately subdued, 200- to 1000-m-diameter craters, which are older than Cone Crater, are cut into the Fra Mauro ridge north, east, and south of Cone Crater within several hundred meters of the rim crest of Cone Crater. Rim deposits of these older craters are essentially unrecognizable photogeologically, but some unmodified remnants are to be expected at depth under the regolith formed since deposition of these rim materials. The interiors of these craters have slopes of 10° to 15° as do the slopes of the Fra Mauro ridge. Cone Crater is approximately 340 m in diameter. The rim of Cone Crater is moderately to densely strewn with 2- to 15-m blocks as seen in Lunar Orbiter 3 high-resolution photography (fig. 3–13). Spacings between blocks of a few to several meters are common in several dense patches extending as far out as 125 m from the rim crest. In the remainder of the mapped rim deposit, spacings as much as several tens of meters between blocks 2 m across and larger are common.

In addition to these major units, the Apollo 14 landing site is dotted with abundant craters ranging in diameter from several hundred meters down to the limit of resolution of the hand-held cameras and in morphology from relatively fresh to almost completely obliterated. Craters in the size range from 400 m to 1 km are both more numerous and more subdued than craters in the same size range in the lunar maria; this distribution is consistent with the inferred greater age of the Fra Mauro Formation. The slope of the cumulative crater-size frequency-distribution curve on the Fra Mauro Formation between diameters of 1 km and 400 m is approximately −2 and lies close to the theoretical steady-state curve suggested in references 3–7 and 3–8 and in figure 3–14. In sizes below the 400-m category, fewer craters are located on the Fra Mauro Formation than on the mare material, an anomaly probably caused by a combination of thicker regolith and higher slopes at the Fra Mauro site. Small craters in the lunar regolith probably are being destroyed at a faster rate by downslope movement of loose debris on the rolling hills of the Fra Mauro area than on the more level surfaces. The fact that the walls of the trench dug by the Apollo 14 commander (CDR) caved in quickly and his comment that many small craters in the area appeared to be

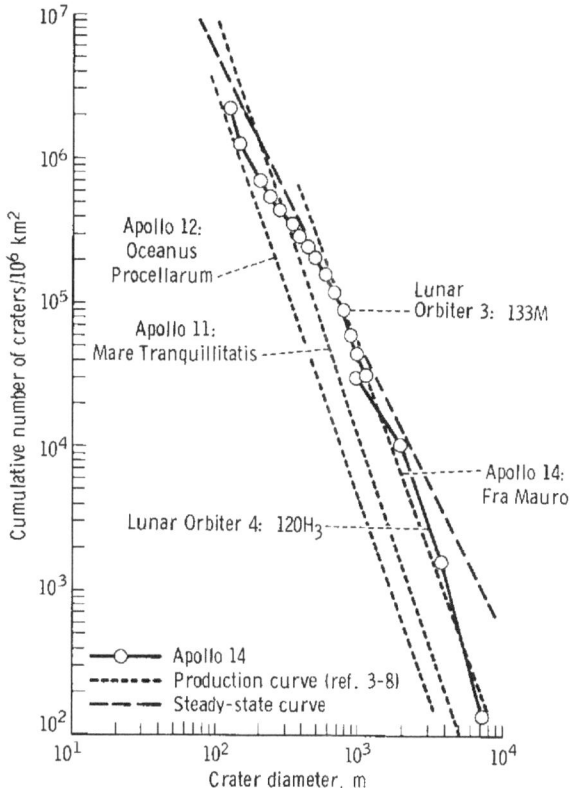

FIGURE 3–14.—Cumulative size-frequency distribution of craters on the lunar surface at the Apollo 14 site compared to crater distributions at the Apollo 11 and 12 sites.

slumped are significant in this regard. From a consideration of the diameters of craters having blocky ejecta blankets, the regolith in the Fra Mauro region is estimated to range from 10 to 20 m in thickness.

Variations in morphology of craters in the site indicate a homologous series of craters of different ages. The age sequence of craters along the traverses from oldest to youngest (ref. 3–6 and fig. 3–13) is interpreted as follows:

(1) Highly subdued craters expressed as very gentle depressions at the landing site of the LM, west of the LM in the area of ALSEP deployment, and north of station A

(2) The crater designated "North Triplet," the moderately subdued 50-m crater east of station F, and the moderately subdued 10-m crater at station A

(3) Cone Crater and the sharp 45-m crater at station E

(4) The sharp 300-m crater at station C′ and the small 10-m crater next to which a football-size rock was collected on the first EVA

Preliminary Interpretations

Twelve 360° panoramas were taken during the first and second EVA periods. Three of these were taken within a triangle around the LM during the first EVA, and nine were taken along the traverse route during the second EVA. Lunar-surface characteristics as seen in the panoramas are markedly different in the vicinity of Cone Crater from those at panorama stations more than a crater diameter away from Cone Crater. The distribution of boulders in the traverse area and the distribution of rocks greater than 20 cm in diameter at the panorama stations are shown in figure 3-13. The LM landing point and station A are in areas where rock fragments larger than 2 or 3 cm in diameter are sparse. Stations B, F, and G are in areas where rock fragments up to 10 to 20 cm in diameter are common, and stations B2, B3, and C′ are in areas of abundant rock fragments greater than 20 cm in diameter. Station H has a moderate number of rock fragments greater than 20 cm in diameter. Rock fragments up to large boulder size are common east of station B1 and become increasingly abundant from just east of station B1 to B2 to C′. The continuous ejecta blanket from Cone Crater extends from the rim crest west to between stations B2 and B1 and is probably only patchy in the vicinity of station B1. Farther west across the landing site, Cone Crater ejecta probably occur only as isolated patches or along rays.

Surface material is noticeably finer grained at the LM and at stations A, B1, B2, F, G, and H than at stations B3 and C′. The topography where the surface material is finer is broadly undulating at wavelengths ranging from tens to hundreds of meters and amplitudes up to approximately 10 m. This topography characterizes old eroded craters, mostly of Eratosthenian and early Copernican ages. The topography at stations B3 and C′ is undulating at wavelengths ranging from several centimeters up to several meters and amplitudes

up to a meter or two. The undulatory surface topography in the vicinity of stations B3 and C′ probably reflects the original hummocky ejecta deposits from Cone Crater.

The Apollo 14 crew briefly described a raindrop pattern, similar to that previously described by the Apollo 12 crew, in the vicinity of station A. This pattern is readily seen in all the panoramas, except B3 and C′, and in some of the sample documentation photographs. The patterns are best illustrated by NASA photographs AS14–67–9390 to AS14–67–9393, which were taken at low Sun angle during the first EVA. The raindrop patterns on the fine-grained surface material appear to consist of small craterlets up to 4 cm in diameter. Shadows from very small fragments on the surface tend to enhance falsely the raindrop appearance in some of the photographs. The raindrop patterns are probably formed by impact of small meteorites and by secondary particles from these impacts. A raindrop craterlet 1 cm deep would be destroyed by subsequent impacts in about 3 million yr (ref. 3–8), so that a fresh surface should become more or less saturated by 4-cm craterlets in this time span. The rounding of boulders ejected from Cone Crater suggests that it is much older than 3 million yr and also that small impacts are numerous; yet the raindrop pattern is much less evident in the ejecta blanket than in the western part of the area. Downslope movement of material, which would tend to destroy the craterlets, appears to have occurred on the steeper slopes on which the ejecta blanket was laid. Furthermore, the formation of craterlets would be impeded by the coarseness of the debris on the ejecta blanket. Fine-grained hummocks a few centimeters in wavelength and in amplitude tend to obscure the craterlets. Also, the higher Sun angle during the time the crew spent on the ejecta blanket as compared to the lower Sun angle during the first EVA makes the craterlets less conspicuous than those photographed during the first EVA.

Small lineaments similar to those seen in the Apollo 11 and 12 photographs[1] were described

[1] G. G. Schaber and G. A. Swann: Surface Lineaments at the Apollo 11 and 12 Landing Sites. Proc. Apollo 12 Lunar Sci. Conf. (Houston), Jan. 11–14, 1971. To be published in Geochim. Cosmochim. Acta.

by the crew and are visible in some of the photographs. These lineaments are described in more detail in another portion of this section.

Optical Properties of Surface Materials

Optical properties of selected lunar materials were measured from black-and-white photographs taken along the geologic traverse during the second EVA. Comparisons of optical properties of the returned lunar samples with the in situ optical properties provide a basis for recognition of similar materials in lunar-surface photographs.

Photometric data were obtained from a first-generation film positive by microdensitometry with a 100-μm-diameter circular aperture. Film luminances were calculated from the sensitometric step wedge on the processed film and the reported camera settings used during lunar photography. Photographs that include the gnomon and photometric chart have their own internal calibration because the sensitometric luminances of the gnomon and chart steps are compared with the luminances computed from the preflight goniophotometric calibration of the gray steps. Scene luminances measured from the film were adjusted for frame shading and then compared to the expected luminance derived from the lunar photometric angles, a lunar photometric function, and an assumed solar irradiance of 13 000 lm.

Scene luminances from the fine-grained surface material were measured from documented sample photographs, selected panoramas, and selected photography of prominent rocks. Measurements were made at the lowest phase angles possible for best accuracy because lunar reflectance decreases rapidly with increasing phase angle and the uncertainties in the local lunar photometric function at larger phase angles increase the errors. Comparisons of reflectance from lunar materials are made at the same phase angle wherever possible. All general comparisons between areas of measurements are made by projecting the measured luminance to the zero-phase-angle luminance (albedo) by means of the lunar photometric function.

The albedo variation of undisturbed fine-grained material along the geologic traverse ranges from 8.2 to 15 percent. The albedo variation of the lower and smoother areas (stations A, B, G, and G1) ranges from 8.2 to 9.1 percent. Additional measurements of the albedo of the lower smooth-terrain unit were made from panoramic photographs taken toward the west from station C'. These photographs indicate that the smooth-terrain unit albedos vary from 8.2 to 8.4 percent. The small albedo variation (8.2 to 9.1 percent) over the undulating smooth-terrain unit, excluding bright ray or crater-wall materials, is similar to the albedos of typical mare landing areas (i.e., the Apollo 12 and Surveyor 3, 5, and 6 sites).

At all previous lunar-landing sites, disturbed regolith reflected less light than adjacent undisturbed material; the decrease in reflectivities ranged from 5 to 26 percent. Disturbed materials around the Apollo 14 LM and over the smooth unit show decreased reflectivities of 5 to 11 percent. The disturbed materials on the flank of Cone Crater commonly show little, if any, decreased reflectivity.

The measured and extrapolated normal albedo of selected rocks varies from 9.4 to 36.4 percent. The range of rock albedo measured from Surveyor photographs is 9 to 22 percent (ref. 3–9), whereas Apollo 12 measurements range from 12 to 18 percent. The highest reflectance ever measured or observed on the lunar surface is around station C1.

The rocks in photographs AS14–68–9448 to AS14–68–9453 (shown later in figs. 3–20 and 3–29(a)) exhibit large variations in reflectance from their component parts, ranging from approximately 9 to 36 percent luminance variation. The large rock shown in photograph AS14–68–9449 contains dark clasts and a light band in a gray matrix and exhibits extrapolated albedos of 16 percent from dark clasts, 20 percent from gray matrix, and approximately 36 percent from the light-gray band.

Lineaments

Preliminary evaluation of small-scale surface lineaments at the Apollo 14 site revealed two primary (northwest and northeast) and one secondary (north) azimuthal trends (figs. 3–15 and 3–16). The lineament systems agree well with the strongest trends observed at the Apollo 11 and

12 landing sites. Small-scale lineaments are noticeably less well developed at the Apollo 14 site, which may be due to the presence of a relatively thick, 15- to 20-m fragmental debris layer. Long individual lineaments ranging to 100 m or more are present.

Three general types of patterned ground are apparent at the Fra Mauro landing site as observed from the surface photography:

(1) A widespread northwest-trending texture discussed by the crew during their pre-EVA description. (This type of patterning is impossible to map as individual lineaments but is a distinct impression on the regolith surface.)

(2) Isolated, intermediate-scale (a few meters maximum) lineaments that are mappable and consist of either shallow grooves or small alined craterlets, or both

(3) Unusually long (up to 100 m or more), straight, well-defined linear scarps and chains of small craterlets indicating, respectively, (a) relatively recent, minute vertical displacement and (b) possible sifting of fine-grained material down active fractures into a more coherent, jointed, substrate at depth

All types of lineaments appear to vary in abundance along the Apollo 14 traverse route but, in general, are more pronounced in photographic panoramas taken around the LM during the first

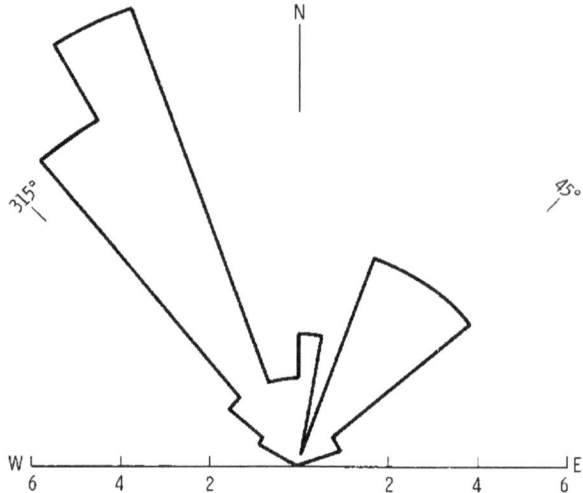

Figure 3–16.—Azimuth-frequency diagram of 49 lineaments in the Apollo 14 traverse area.

EVA. The Sun angle was low (about 13°) during the first EVA and may have considerably enhanced lineament observability. Within the Cone Crater ejecta blanket, lineaments with amplitudes and wavelengths of a centimeter or less are poorly developed and difficult to see, probably because of an abundance of fragments in this same size range.

The regolith may not be sufficiently thin within the Apollo 14 traverse area to permit well-developed small-scale lineaments. These were found at the Apollo 12 site, however, in the thin, firm regolith on the rim crests of old subdued craters such as Surveyor and Middle Crescent craters. Firm, compact soil was mentioned by the Apollo 14 crew as being only between stations B1 and B2 (fig. 3–1). In pre-LM liftoff debriefing, it was mentioned that this firm ground was a small isolated patch in a generally powdery surface.

A preliminary evaluation of regional scale lineaments around the Apollo 14 site has been reported.[2] A reasonable correlation exists between the local and regional linear trends (fig. 3–17).

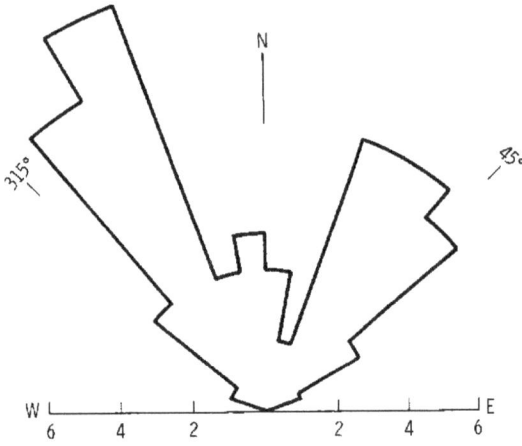

Figure 3–15.—Azimuth-frequency diagram of 70 lineaments in the Apollo 14 traverse area.

[2] G. G. Schaber and G. A. Swann: Surface Lineaments at the Apollo 11 and 12 Landing Sites. Proc. Apollo 12 Lunar Sci. Conf. (Houston), Jan. 11–14, 1971. To be published in Geochim. Cosmochim. Acta.

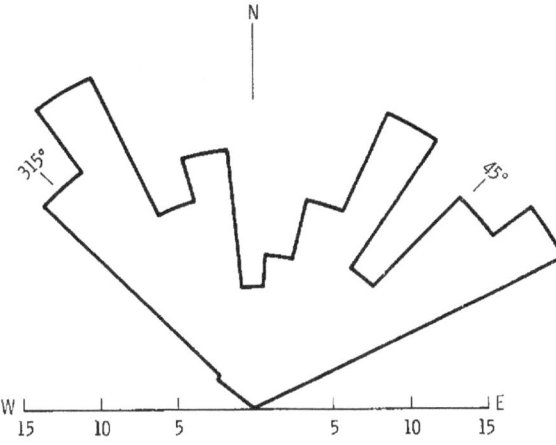

FIGURE 3–17.—Azimuth-frequency diagram of 183 lineaments in a 243-km² region surrounding the Apollo 14
landing site.

Fillets

Fillets are defined as embankments of fine-
grained material partially or entirely surrounding
larger fragments (ref. 3–10), and as accumulations of fragments on uphill faces of rocks (ref.
3–11). Apollo 14 photographs contain a variety
of fillets that can be studied with respect to the
sizes, shapes, textures, orientations, and locations
of rocks. It is evident that these factors influence
the development of fillets and that, in turn, fillets
may provide a decipherable record of lunar-
surface erosion and deposition.

During the second EVA, the crew described
fillets at stations A, B, and C2. They photographed examples of these features with the
closeup stereocamera at station A and with
Hasselblad cameras at stations A, B2, C1, C2, at
Weird Rock near station F, and at the boulder
field north of the LM at station H. All the boulders
at these stations appear to be breccias or coarsely
clastic rocks. The one nonfilleted rock at White
Rocks has conspicuous light and dark layers.

The types of contacts between rock and soil
are classified into three general categories based
on the attitude of the rock surface: low angle or
shallow, steep, and overhanging (fig. 3–18).
Fillet characteristics for each of the major rocks
radially outward from Cone Crater follow, beginning with the rocks in the White Rocks area.

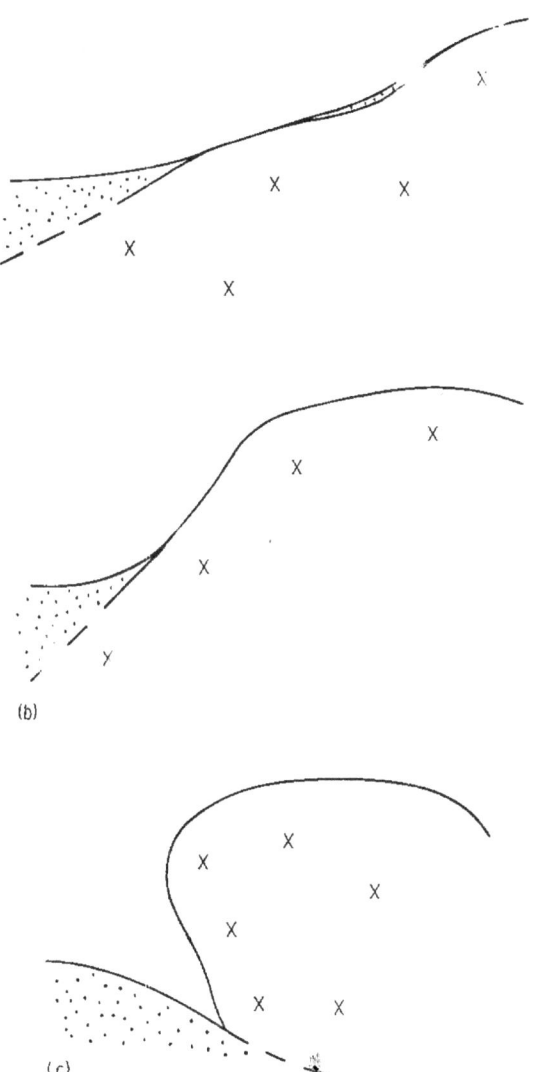

(b)

(c)

FIGURE 3–18.—Types of fillet contacts between rock and
soil. (a) Shallow, with thin pockets of dust high on
rock surface. (b) Steep, with concave fillet against
base of rock. (c) Overhanging, with little or no resulting fillet.

White Rock. The White Rock (fig. 3–19) from
which sample 14082 was taken is approximate
1.5 m wide, has steep to overhanging sides, and a
poorly developed fillet that is disturbed by soil
from the LM pilot's (LMP) feet.

Saddle Rock. The biggest of the White Rocks,
Saddle Rock (fig. 3–20), is the largest block that
was observed closely. The southeast side reveals

extensive burial or filleting at the base with a thin coating of loose, fine, granular material extending well up onto the more gently sloping surfaces. The fine and granular materials are darker than most of the fresh rock surfaces but are slightly lighter than the dark clasts. Lineations on the dust-covered, finely fractured rock surface are not re-flected in the lower fillet surface. Subangular fragments several centimeters across are abundant on the fillet surface. Some fragments are partially buried. A large, loose fragment is lying on the south slope of the block. The north side of the block, toward Cone Crater, is not exposed, but the east edge appears to drop off steeply as though the north side were steep to overhanging, and less burial or filleting is evident.

Contact Rock. Contact Rock (fig. 3–21) at station C1 is unusual in that it has no evident

FIGURE 3–20.—Closeup view of fillet on Saddle Rock.

fillet but rather is in a depression with a raised rim. The visible sides of the rock are steep to overhanging, and the height-to-width ratio is 0.7, which is the highest of any of the large boulders studied, further indicating that very little of it is buried. The lower white portion is angular, and the upper darker portion is rounded. No obvious fine material occurs on the upper surfaces. Larger rocks surrounding the base are distinctly filleted

FIGURE 3–19.—Closeup stereophotograph of fillet on White Rock near station C1.

FIGURE 3–21.—Closeup view of fillet on Contact Rock.

and the rounded ones several inches high are largely blanketed by fine soil. Smaller fragments several centimeters across have few fillets.

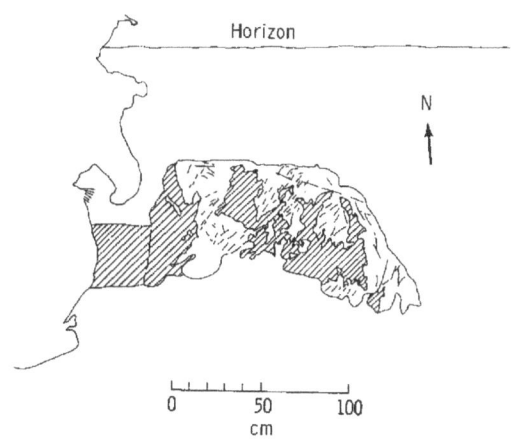

FIGURE 3–23.—Fillet on Big Rock at station B2.

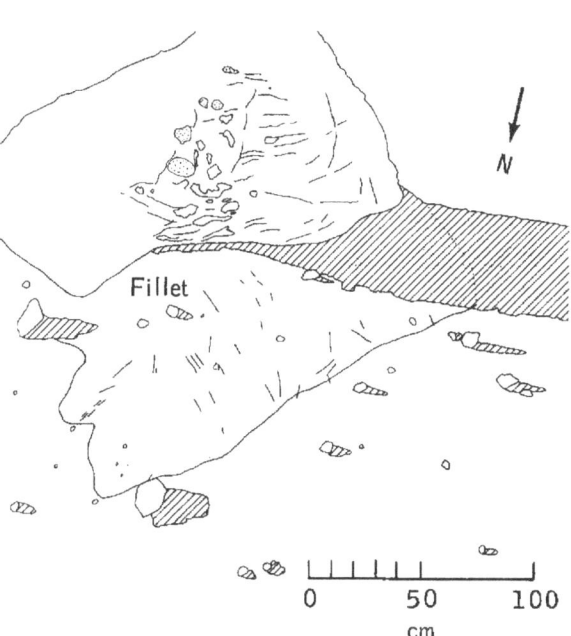

FIGURE 3–22.—Fillet on Filleted Rock at station C2.

Filleted Rock. Filleted Rock (fig. 3–22) at station C2 is 110 m south of the rim of Cone Crater. The rock is approximately 2 m long and 1 m high and has a broad fillet extending toward the camera and toward Cone Crater for approximately 3 m. The slope is gentle at this location, and a few lineaments or grooves trend roughly northwest-southeast. The rock is a very rounded and friable-looking coarse breccia from which a crystalline rock (sample 14053) was collected.

Big Rock. Big Rock (fig. 3–23) at station B2 is an unusual flat-topped rock 400 m southwest of Cone Crater. The lower portion is partially covered by a dark coating of fines. It has a very clean upper surface and steep sides with a 1-m-wide fillet against the southeast base. Three or four rounded fragments several centimeters wide are partially exposed on the fillet surface. Big Rock has a height-to-width ratio of 0.4, the lowest of the large rocks studied.

Weird Rock. Weird Rock (fig. 3–24) is 700 m southeast of Cone Crater rim. In places, the edges slope steeply inward. Very thin filleting occurs on the south and west sides, but the rock has a pronounced ridge of darker fines and fragments trending about north-south against the eastern end that extends for several meters to the south and east. This thick fillet, as shown on more distant photographs, contains numerous small clasts and appears to be the margin of an ejecta blanket that banked against the eastern side of Weird Rock. A few patches of dark fine material are evident on the southward-sloping surface about midway up the rock. The height-to-width ratio of

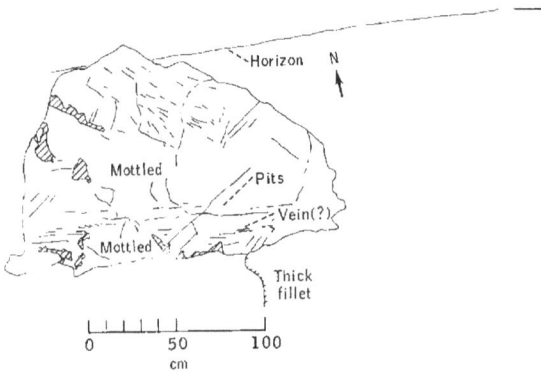

FIGURE 3–24.—Fillet on Weird Rock near station F.

0.65 is high, indicating that the rock may not be very deeply buried.

Turtle Rock. Turtle Rock (fig. 3–25) near station H is well rounded and has a well-developed fillet extending more than a meter from the base. The rock appears to be a coarse breccia. Two fragments were collected from the upper surface of the rock, and two were collected from the fillet. Rock fragments encircle Turtle Rock beyond the intervening 10- or 20-cm band of the outward sloping fillet. Removal of samples from the top of the rock showed the underlying surface to be darker than the surrounding rock surface. This dark surface may be caused by trapped fines, or it may be a rough fracture surface that has not been exposed to lunar weathering.

Sloping Rock. Near station A, two closeup stereophotographs were taken of the contact between a fillet and a gently sloping rock surface. The Sloping Rock and fillet are shown from approximately 15-m distance in Hasselblad photograph AS14–68–9409; the dark fillet trends about east-west on the light rock surface. A closeup photograph of the contact of the fillet with the dusty boulder surface at the top is shown in figure 3–26. Small, closely spaced linear depressions are observed to cut across the contact in the clumpy fines. The fines are medium gray, very fine-grained with clots up to 0.5 cm, and contain very sparse white particles, the color of the underlying rock.

Summary. The size and shape of fillets around the larger rocks in this area may be controlled by the shape and friability of the host rocks. The more rounded (and thus perhaps more friable) rocks with lower height-to-width ratios have bigger and more completely developed fillets. At the Apollo 12 site (ref. 3–12), this type fillet is represented by the mounds, which are very friable and rounded and exhibit maximum fillet development. Rocks with gently sloping sides commonly have more extensive fillets than rocks with steep sides. Most rocks with overhanging sides have few or no fillets developed on those exposures. The nonfilleted Contact Rock at station C1 is the best example of this relationship; however, complete lack of a fillet on Contact Rock suggests that the rock was recently deposited in that position relative to the adjacent heavily filleted rocks. Conversely, Contact Rock may be harder than the other rocks in the area and therefore is not extensively eroded because of its high resistance.

Boulders

Boulders larger than 1-m diameter are abundant in the Cone Crater ejecta blanket and diminish in number farther away from the crater rim (fig. 3–13). Boulders at the Fra Mauro site provide the first opportunity to study the geology of large lunar bedrock segments, although the boulders have almost certainly been rotated from their original positions in the lunar bedrock.

The fragmental nature of all the boulders is evident in the photographs. Many clasts range in size from 2 to 10 cm. The clastic appearance of the rocks is somewhat similar to that of terrestrial deposits of ejecta derived from impact events (fig. 3–27). Planar features are visible in all the boulders. In some, distinct lithologic layers are evident, whereas, in others, evidence for lithologic differences between layers is subtle or absent. The rocks also show one or more sets of well-developed, systematic planar fractures and a variety of less systematic, mostly curved fractures.

The irregular shapes of the boulders are con-

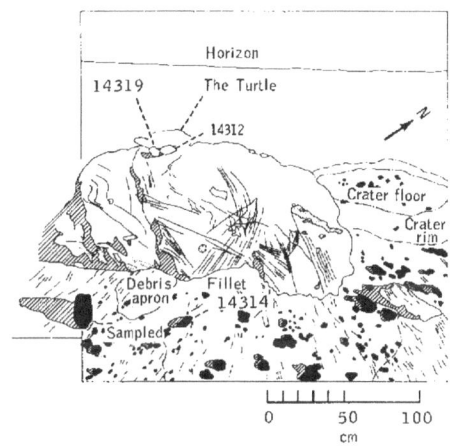

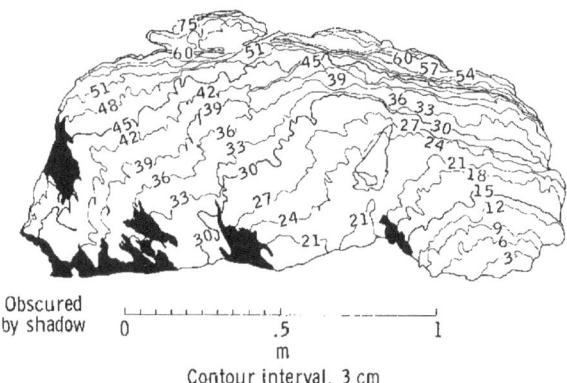

FIGURE 3–25.—Fillet on Turtle Rock at station H.

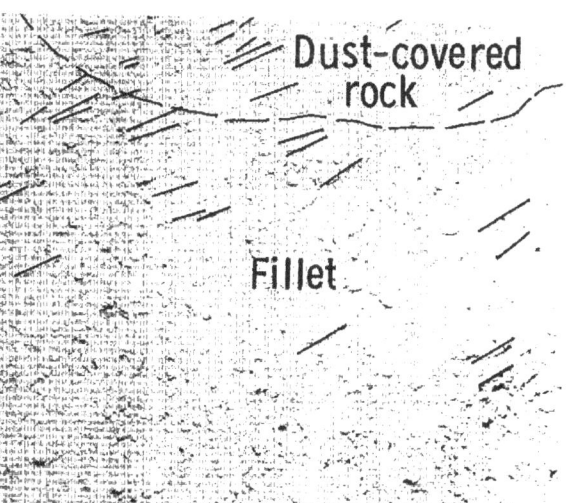

FIGURE 3–26.—Closeup photograph of fillet on Sloping Rock.

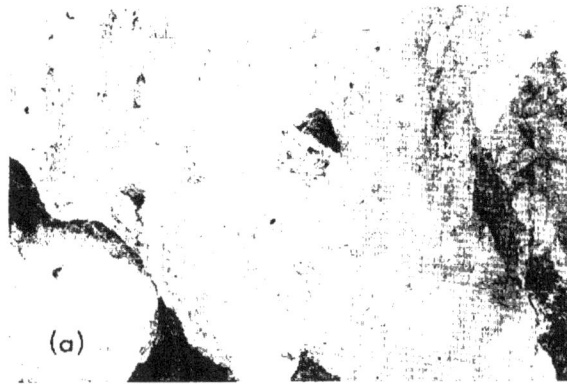

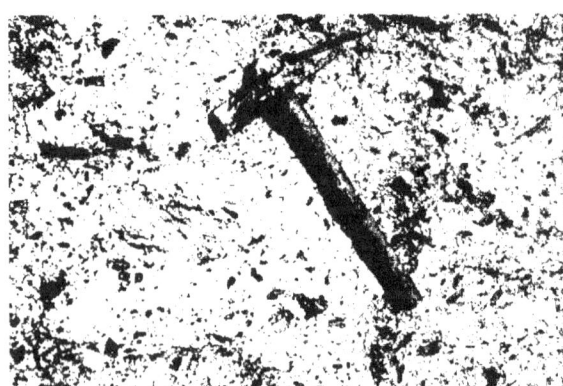

FIGURE 3–27.—Comparison of clastic appearance of terrestrial and lunar rocks. (a) Quarry wall in fragmental debris ejected from Meteor Crater, Arizona. (b) Turtle Rock at Station H. (c) Quarry wall in Suevite deposit, Ries Crater, Otting, Germany.

trolled in part by internal structures. Erosion has rounded the corners and resistant areas are etched into relief. At least some surfaces are intricately patterned by pits (probably of impact origin—the

so-called zap pits) that are generally 1 cm or less in diameter. Distinct raised rims are visible in which the pits are suitably illuminated.

White Rocks. The White Rocks (figs. 3–28 and 3–29) at station C1 are the largest boulders ex-

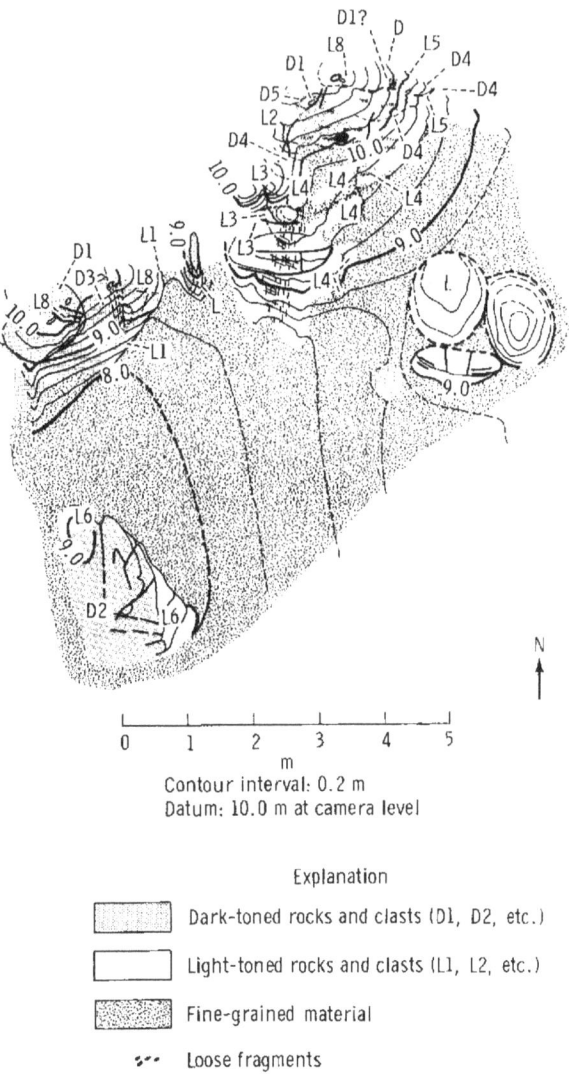

Contour interval: 0.2 m
Datum: 10.0 m at camera level

Explanation

Dark-toned rocks and clasts (D1, D2, etc.)
Light-toned rocks and clasts (L1, L2, etc.)
Fine-grained material
Loose fragments
Contact between rock types
Internal layering
Joints, dotted where buried
Closely spaced fractures

Note: Table 3-I contains rock descriptions.

FIGURE 3–28.—Preliminary geologic map of White Rocks area.

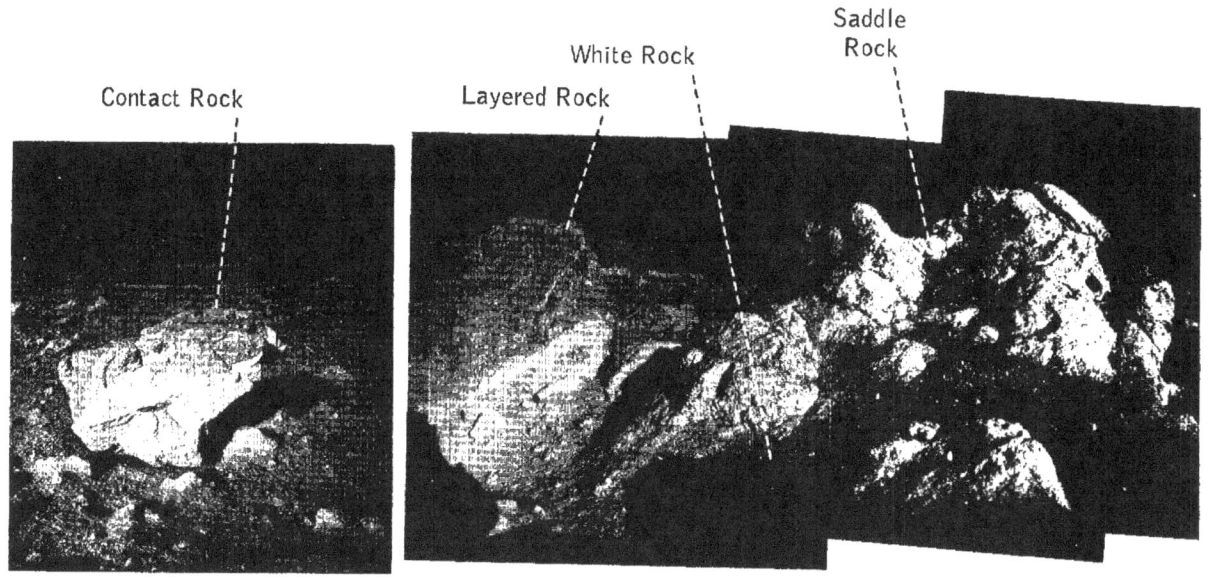

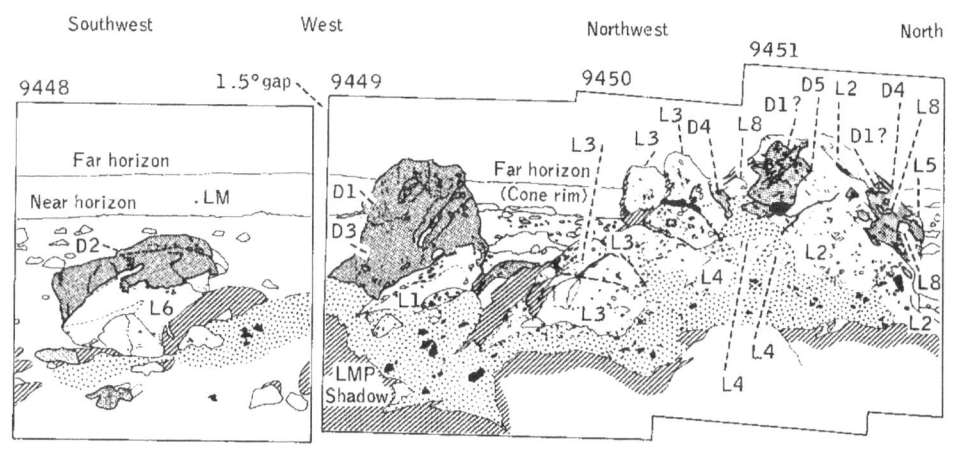

Explanation

Rock materials   Structural surfaces

Dark-toned rocks and clasts (D1, D2, etc.)   Layering: N45W strike, northeast dip; spaced tens of centimeters apart

Light-toned rocks and clasts (L1, L2, etc.)   Close-spaced fractures: N10E, subvertical; locally
                                                                          emphasizes layering; spaced a few centimeters apart

Fine-grained material   Close-spaced fractures: N70W, subvertical; pervasive;
                                                                          spaced 1 cm or less apart

Loose fragments   Joints: N30E, subvertical; irregular spacing

                                                                          Joints: N70E, dip northwest; spaced 0.25 m apart

Note:
    Table 3-I contains rock descriptions.

FIGURE 3–29.—White Rocks.

amined by the LM crew. The White Rocks exhibit a wide variety of stratigraphic and structural features. For convenience, four rocks were named: White Rock, Saddle Rock, Layered Rock, and Contact Rock.

Photogeologic rock types were defined on the basis of photographic tone and surface textures such as roughness and apparent erosional resistance. The overall twofold preliminary classification is based on the light-dark types shown in figure 3–30. Thus, Saddle Rock is dominantly light with dark patches, Layered Rock is the darkest, Contact Rock is roughly half and half, and White Rock is dominantly light. The major rock types and their characteristics are summarized in table 3–I.

The rock types have been crosscut by a variety of planar surfaces including layering, well-defined joints, and closely spaced fractures. General patterns of these surfaces are shown in figures 3–28 and 3–29.

White Rock. White Rock (fig. 3–19) at the east side of White Rocks boulder field is 1.25 m long oriented east-west. White Rock has a rather

FIGURE 3–30.—Light and dark differentiation of White Rocks.

blocky shape, and consists of one apparent rock type with a prominent dark clast at the east end. Sample 14082, chipped from the top surface of White Rock by the LMP, was the only sample chipped from a boulder.

Saddle Rock. Saddle Rock (fig. 3–20), at the north end of the White Rocks boulder field, is 4.5 m long from the north to the south end. The crestline forms nearly a right angle pointing northwest which may reflect structural control. The irregular surface of the rock has dark hackly patches, resistant pinnacles, and a variety of light-to-dark clasts. Saddle Rock shows evidence of at least five sets of planar surfaces. The most prominent surface, which is interpreted as layering, is inclined to the right and is expressed as a series of parallel indentations and discontinuous resistant ribs. This set of surface controls the shape of the east face of the pinnacle, immediately south of the saddle. The second most prominent set consists of subvertical fractures spaced a few centimeters apart, trending about north-northeast. These are expressed as closely spaced, near-vertical shadow lines on the short resistant ribs previously mentioned.

Layered Rock. Layered Rock (fig. 3–29), at the west end of the White Rocks boulder field, is approximately 3 m long and 2 m high. It is characterized by a prominent west-dipping, light-toned, clast-rich layer at the base overlain by a series of dark, nearly vertical layers. The top appears to be a single clast approximately 1.7 m long, which in turn consists of dark clastic fragments up to 40 cm long.

Contact Rock. Contact Rock at the south end of the field is approximately 3 m long. The most striking feature is the irregular contact between dark rock above and light rock below (fig. 3–29). The upper dark portion is rounded and knobby, and the lower portion is angular. The light layer contains what appears to be fine fractures, or thin layers, which are subparallel to the contact in the rock. Irregularities of the contact between layers are similar in appearance and scale to layers within the ejecta blanket of Meteor Crater, Arizona (fig. 3–31).

TABLE 3–I. *Summary of Photogeological Rock-Type Characteristics*

Code	Location	Surface texture	Erosional resistance	Clasts (color and size)	Occurrence
			Light Rocks		
L1	Layered rock	Smoothly undulating	High	Light; from <1 cm to several cm	Layer
L2	Saddle rock	Smooth	High	Light and dark; from <1 cm to several cm	Well to poorly layered
L3	Saddle rock	Knobby, lumpy	High	Light; a few cm	Layer
L4	Saddle rock	Moderately smooth	Moderate	Light; ≃1 cm	Underlies a surface that slopes south
L5	Saddle rock	Moderately rough	Moderate	Dark; ≃1 cm	Irregular clasts
L6	Contact rock	Finely rough	Moderate	Light; from <1 cm to a few cm	Irregular layer
L7	White rock	Granular	Moderate	Light; ≃1 cm	Block
L8	All rocks	Unknown	Moderate to high	Unknown	Clasts
			Dark Rocks		
D1	Layered rock	Smooth	High	Unknown	Clasts
D2	Contact rock	Finely rough	Moderate	Light and dark; ≃1 cm	Layer
D3	Layered rock	Bumpy	Moderate	Light and dark (several cm)	Layer
D4	Saddle rock	Coarsely hackly	Low	Light; from ≃1 cm to tens of cm	Irregular area
D5	Saddle rock	Finely hackly	Low to moderate	Light and dark; vesicular or pitted areas	Poorly defined irregular area

Filleted Rock. Filleted Rock (fig. 3–22) at station C2 is approximately 2 m wide at the base and 1 m high. The rock is roughly equant and in gross aspect has extensive, smooth, facetlike surfaces meeting in rounded corners. The surface is rough in detail, with abundant irregular angular protuberances up to a few centimeters in size. The shadowed surface is distinctly mottled with irregularly shaped light-and-dark patches that appear to be clasts up to approximately 10 cm in diameter surrounded by a generally dark material.

The light clasts, which appear to be granular, are at a scale of 1 cm or less. Sample 14053, a 7- by 5- by 3-cm basaltic rock collected from the surface of Filleted Rock at station C2, is probably a clast from the boulder. The lenticular shape of the light clasts imposes a weak, horizontally layered fabric on the shadowed surface. A few deep, crudely linear shadows that dip about 50° to the left (parallel to the surface of the sunlit face) may represent the traces on the shadowed face of a fracture system with 10- to 20-cm spacing.

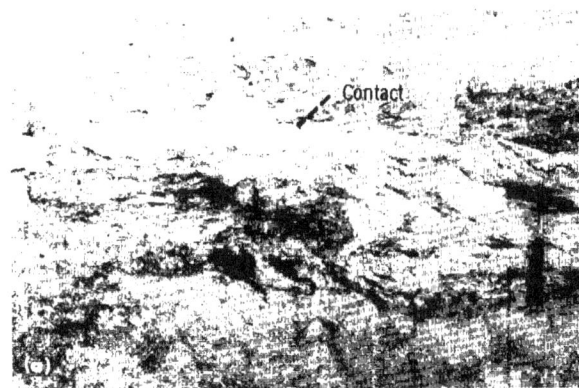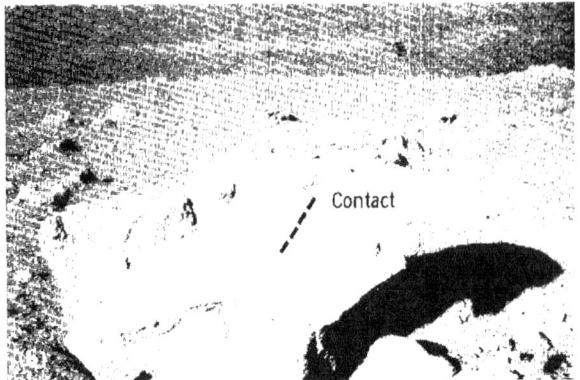

FIGURE 3-31.—Comparison of terrestrial and lunar contacts. (a) Quarry wall in fragmental debris ejected from Meteor Crater, Arizona. (b) Contact Rock at station C1.

Big Rock. Big Rock (fig. 3-23) at station B2 is approximately 1.5 m wide and 0.6 m high. Big Rock is an irregular boulder with rounded protuberances and depressions with wavelengths and amplitudes of approximately 10 to 30 cm. The relatively planar top may be a fracture surface. Fine light- and medium-gray mottling at 1-cm or finer scale gives the surface a granular appearance that is due in part to intricate pitting of the rock surface and may be partially due to clasts of light-colored rock in a medium-gray matrix. A distinctive structural grain on the rock surface is caused by a subparallel set of abundant, discontinuous, alternating light-and-dark bands that dip approximately 50° to the left with approximately 1-cm spacing. In places, the light-colored bands form ridges; elsewhere, the relief is not evident and banding may represent a true color variation related to compositional layering.

Weird Rock. Weird Rock (fig. 3-24) near station F is approximately 2 m wide at the base and 1.25 m high. It is roughly equant block with an irregular surface characterized by angular to rounded protuberances and depressions with amplitudes up to approximately 10 cm. Groups of subparallel clefts and ridges on the rock surface reflect discontinuities or zones of weakness. The surface is granular or mottled because of roughly equant to irregular light-gray clasts up to approxi-

mately 10 cm in diameter. Much of the surface is pitted. The pits are generally less than 1 cm in diameter, and some have raised rims that are especially evident in low-angle lighting. Some of the apparent mottled texture may be a result of shadow distribution on pitted rock surfaces rather than of lithologic differences within the rock.

Three major sets of planar surfaces and at least two minor irregular sets are present. The most prominent of these is outlined by sharp, apparently horizontal clefts with 1- to 10-cm spacing in the lower part of Weird Rock. An irregular white layer, approximately 1 cm thick, resembles a vein along the extension of a cleft in the right-hand portion of the rock. The horizontal surfaces do not transect but are terminated or are deflected by a semicircular mottled zone (perhaps a large clast) in the lower left-hand part of the rock. A second prominent set of surfaces similar to the first dips to the left intersects the first set at an apparent angle of approximately 40°, and is well displayed at the right-hand side of the block. The third prominent set of surfaces occurs in the upper part of the rock and dips to the right at an apparent angle of 35°. Surfaces in this set are less prominent and include many small ridges and fewer sharp depressions. Subtle tonal banding parallel to this set occurs in the upper part of the rock. The set may represent textural or compositional layering etched into low relief. Less prominent groups of surfaces, probably fractures, dip steeply to the right or form long, gently curving, nearly vertical clefts.

Turtle Rock. Turtle Rock (fig. 3–25) is the largest boulder in a field of rocks at station H. The boulder derives its name from the turtlelike feature on the upper surface. The LMP collected two loose-lying rocks (samples 14312 and 14319) from the upper surface of Turtle Rock and two chips (one identified as sample 14314) from the fillet adjacent to the boulder. The boulder is approximately 1.5 m wide and protrudes above the fillet approximately 0.75 m. The depth or north-south dimension visible in lunar-surface photographs is 0.75 m.

Turtle Rock contains abundant centimeter-size clasts, with a few up to 10 cm, in a nonresolvable matrix. The clasts are dark gray to white. The lighter clasts commonly have dark rims. Most are approximately equant, but tubular, ellipsoidal, and contorted forms are common. Many of the white clasts are in depressions. The white wedge-shaped area may not be a clast but a concentration of white material that occurs along fracture planes. Hundreds of zap pits are visible on the surface of Turtle Rock. They range in diameter from a centimeter down to the limit of photographic resolution. Several circular depressions up to 4-cm-diameter sizes are probable impact pits. One of these, with a large spall zone, is indicated in the lower center of the rock (fig. 3–25). The shape of the rock is controlled largely by fracturing. The most prominent structural feature is the apparent layering that dips toward the lower right in the photograph. Several platy, angular fragments lying on the surface of the rock have broken away from these southeast-dipping layers. The layering may be due to a combination of mineral banding, as seen in the central and left side of the rock, and fracturing, best seen below and to the right of the wedge-shaped white area. In the central area of the rock, the bands are deformed into lenticular and highly contorted shapes outlined sharply by a dark layer up to several millimeters thick. Many of the layers converge and are similar in appearance to crossbedding. Closely

spaced, parallel planar features trend northwest and dip steeply northeastward. The planes are spaced a few millimeters to a centimeter apart and are marked by barely discernible white lines in the photographs. Some of the white lines are bordered by equally thin dark lines on both sides. These fractures cut across clasts and other fractures. A third set of fractures occurs in the lower left-hand corner of the rock. These fractures dip northward, perhaps 40° to 50°, and are spaced 5 mm to 2 cm apart. These fractures are visible only where etched into relief.

Split Rock. At station C', a large boulder (fig. 3–32) has broken into two large pieces. This well-rounded boulder, along with others in the strewn field, was probably ejected from Cone Crater. The edges of the fracture that separates the boulder pieces are sharp. The rounding of the

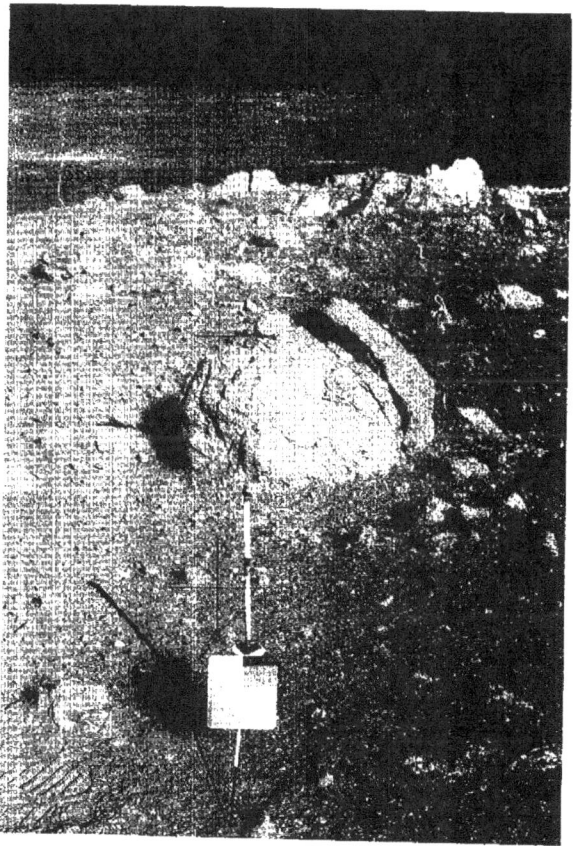

FIGURE 3–32.—Split Rock on Cone Crater ejecta blanket near station C'.

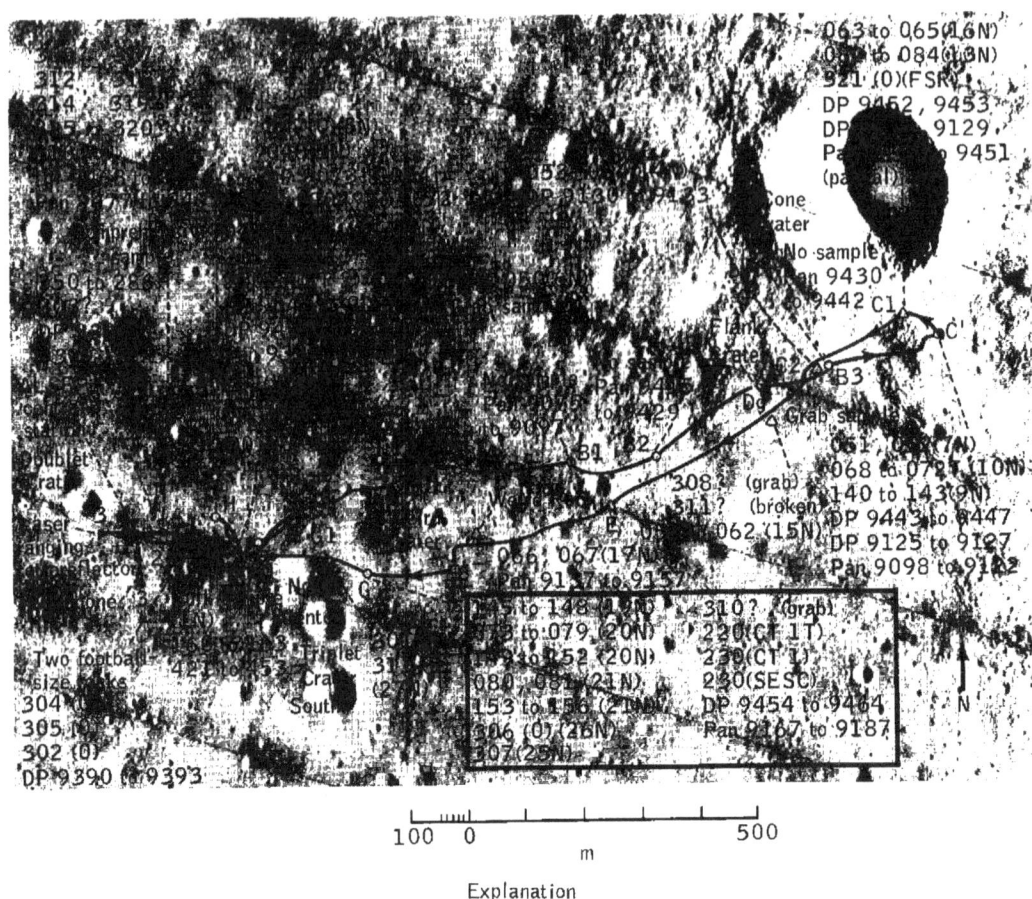

```
     L  uuul    l      l      l      l
    100    0                        500
                    m
```

Explanation

Bag 6N	Prenumbered sample bag
14307	Sample for which location is "known" by reference to sample bags used at the time of collection
14306 (0)	Sample for which location and lunar orientation known
14318 (T)	Sample for which location is tentative, based on identification of sample in lunar-surface photographs
14310?	Sample for which location is tentative based on description by the Apollo 14 crew, the process of elimination of known samples, or the possibility of sample mixing during transfer between or within weigh bags
(Grab)	Sample that was not photographed before sampling or put into prenumbered bags
CT 1T	Core tube number 1 with a tab
DP 9409-	Sample documentation photograph numbers
Pan 9049-	Panorama photograph numbers
SESC	Special environmental sample container

Traverse and station symbols are same as on figure 3-1.

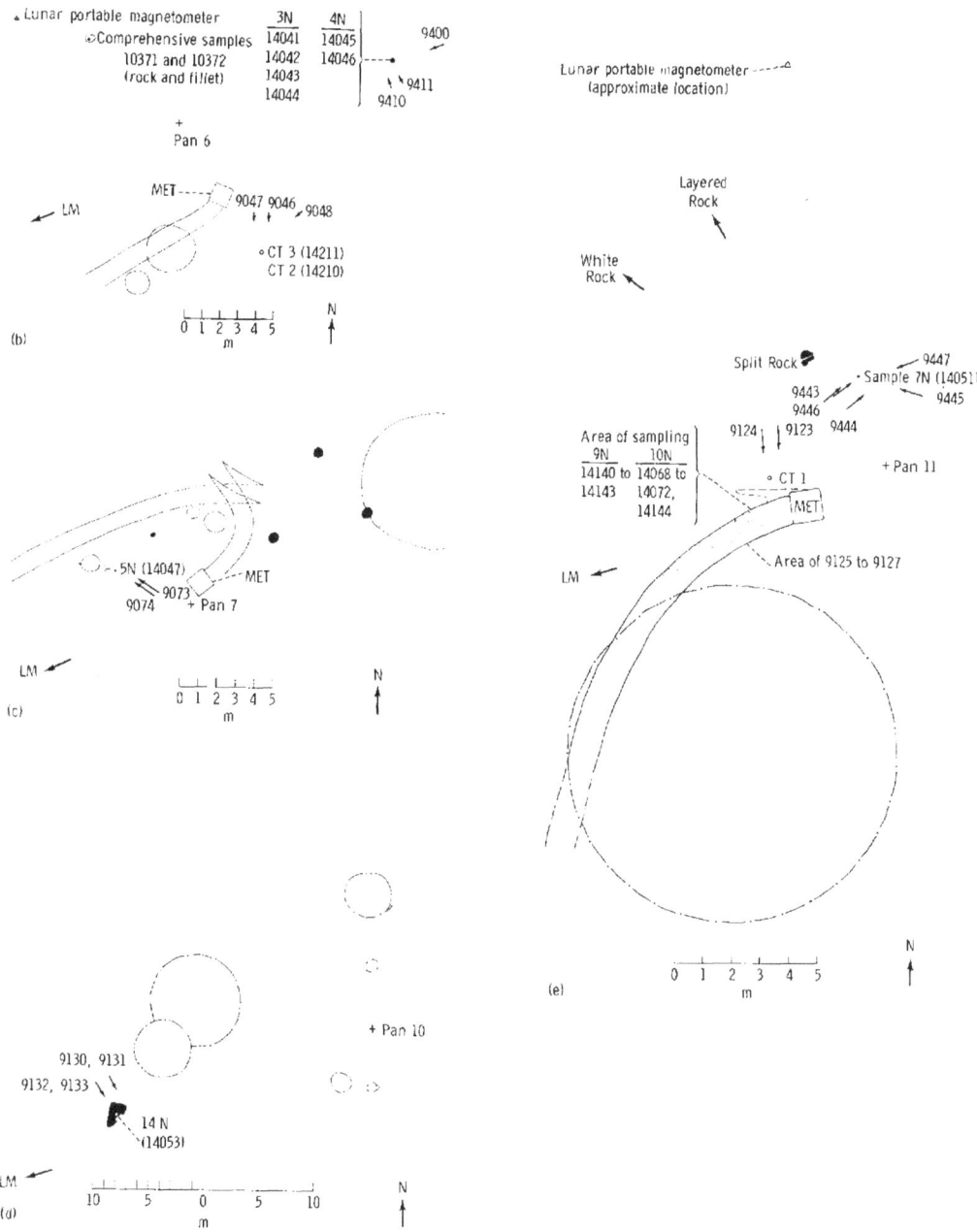

FIGURE 3–33.—Traverse map showing sample and photograph locations. (a) Overall view. (b) Planimetric map of station A. (c) Planimetric map of station B. (d) Planimetric map of station B3 and C2. (e) Planimetric map of station C'.

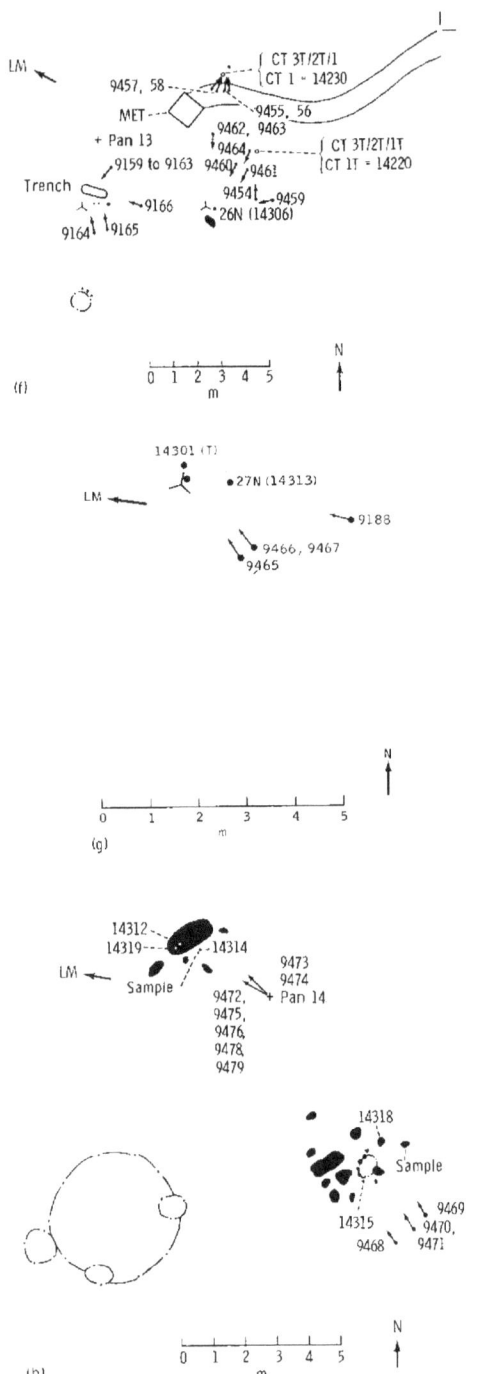

FIGURE 3–33. (Concluded).—Traverse map showing sample and photograph locations. (f) Planimetric map of station G. (g) Planimetric map of station G1. (h) Planimetric map of station H.

boulder is similar to that of most of the other boulders on the Cone Crater ejecta blanket. This rounding probably occurred after the boulder was ejected from the crater. The sharpness of the fracture, however, suggests that the fracture is relatively recent and occurred long after the boulder assumed the present rounded form. This type fracture is probably caused by meteorite impact and should contribute significantly to the erosion of rocks by exposing larger surface areas to micrometeorite bombardment.

Samples

The Apollo 14 crew returned approximately 43 kg of lunar samples. The samples include 35 rocks that weigh more than 50 g each and approximately 30 smaller rocks that weigh from 10 to 50 g each, as well as numerous rock chips and fines collected with the scoop. Four core tubes of fines were collected, two of which were driven together as a double core. Sample locations in tables 3–II and 3–III are keyed to the traverse maps (fig. 3–33). The samples are listed by traverse stations where collected (table 3–II). Sample locations are keyed to a consecutive listing of Lunar Receiving Laboratory (LRL) sample numbers in table 3–III.

The location of samples is established by relating returned samples to the documentary sample bags, lunar-surface photographs, and Apollo 14 crew descriptions that pertain to the given samples. Lunar orientations of rock samples are also determined from lunar-surface photographs. Location and orientation are established with various degrees of confidence, from the case in which both are accurately known to the case in which neither is known.

Sample locations and documentation are cross-referenced in table 3–IV between LRL sample numbers, locations on the traverse map, and lunar-surface documentary photographs; the table also contains a brief description of samples from a study of the rocks and photographs of the rocks in the LRL. The table is divided into four groups according to the degree of confidence in locating

TABLE 3–II. *Sample Locations by Station*

Station designation	Sample container	LRL sample no.	Remarks
EVA–1			
Contingency.........	None	14001 to 14012	Locations known by reference to the sample bags used at the times of collection.
Comprehensive.......	Weigh bags	14250 to 14259	
	Weigh bags	14260 to 14289	
Bulk...............	Bag 1N	14160 to 14163	
	Bag 1N	14421 to 14453	
Football-size rock....	None	14304	Locations and lunar orientations are known.
	None	14305	
	None	24302 (broken from end of 14305)	
EVA–2			
A.................	Bag 3N	14041 to 14044	Locations known by reference to the sample bags or core tubes used at the times of collection.
	Bag 4N	14045 and 14046	
	Double core (tube 3)	14211	
	Double core (tube 2)	14210	
B.................	Bag 5N	14047	Location and lunar orientation known.
	Bag 5N	14048	Location known by reference to the sample bag used at the time of collection.
B₆.................	Bag 6N	14049 and 14050	Locations known by reference to the sample bag used at the times of collection.
B1.................	No sample	....	
B2.................	No sample	....	
B3.................	No sample	....	
C'.................	Bag 7N	14051	Location and lunar orientation known
	Bag 7N	14052	
	Bag 9N	14140 to 14143	Locations known by reference to the sample bags used at the times of collection.
	Bag 10N	14068 to 14072	
	Bag 10N	14144	
C1.................	Bag 13N	14082 to 14084	
	Bag 16N	14063 to 14065	
	Football-size rock	14321	Location and lunar orientation known.
C2.................	Bag 14N	14053 and 14054	Locations known by reference to the sample bag used at the times of collection.
Dg................	Grab sample	14308(?)	Location tentative, based on astronaut descriptions, the process of elimination of known samples, or the possibility of sample mixing during transfer between or within weigh bags; samples broken after collection.
	Grab sample	14311(?)	

TABLE 3–II. *Sample Locations by Station*—Continued

Station designation	Sample container	LRL sample no.	Remarks
E..................	Bag 15N	14055 to 14062	⎱Locations known by reference to the sample bags used
F..................	Bag 17N	14066 and 14067	⎰ at the times of collection.
G................	Bag 19N	14145 to 14148	Location known by reference to the sample bag used at the time of collection; trench samples; material taken from the top of the trench.
	Bag 21N	14080 and 14081	⎱Location known by reference to the sample bag used
	Bag 21N	14153 to 14156	⎰ at the time of collection; trench samples; material taken from the middle of the trench.
	Bag 20N	14073 to 14079	⎱Location known by reference to the sample bag used
	Bag 20N	14149 to 14152	⎰ at the time of collection; trench samples; material taken from the bottom of the trench.
	SESC	14240	Location known by reference to the sample container used at the time of collection.
	Core tube 1	14220	Core sample; location tentative; based on tentative identification of the sample from lunar-surface photographs.
	Core tube 1	14230	Core sample; location known by reference to the core tube used at the time of collection.
	Bag 25N	14307	Location known by reference to the sample bag used at the time of collection; rock sample.
	Bag 26N	14306	Location and lunar orientation known; rock sample.
	Grab sample	14310(?)	Location tentative, based on astronaut descriptions, the process of elimination of known samples, or the possibility of sample mixing during transfer between or within weigh bags; rock sample.
G1................	Bag 27N	14313	Location and lunar orientation known; rock sample.
	Grab sample	14301	Location tentative; based on tentative identification of the sample from lunar-surface photographs; rock sample.
H................	Grab sample	14312	⎫Locations and lunar orientations known; rock
	Grab sample	14314	⎬ samples.
	Grab sample	14315	⎭
	Grab sample	14316(?)	⎫Locations tentative, based on astronaut descriptions, the process of elimination of known samples, or
	Grab sample	14317(?)	⎬ the possibility of sample mixing during transfer between or within weigh bags; rock samples.
	Grab sample	14318	⎱Locations and lunar orientations known; rock
	Grab sample	14319	⎰ samples.
	Grab sample	14320(?)	Location tentative, based on astronaut descriptions, the process of elimination of known samples, or the possibility of sample mixing during transfer between or within weigh bags; rock sample.

Samples From Unknown Locations

	Weigh bag 1027	14303, 14165 to 14189	Probably comprehensive samples.
	Weigh bag 1031	14190 to 14202	⎱Probably fines and chips fallen from or collected with
	Weigh bag 1031	14309	⎰ samples during EVA-2.
	Weigh bag 1038	14290 to 14297	Probably fines and chips fallen from or collected with loose bagged samples 14312 to 14321.

TABLE 3–III. *Sample Locations by Sequential LRL Sample Number*

LRL sample no.	Traverse station	Sample container no.	LRL sample no.	Traverse station	Sample container no.
14001 to 14012 [a]....	Contingency sample	None	14250 to 14259 [a]....	Comprehensive sample	
14041 to 14046 [a]....	A [b]	3N and 4N [c]	14260 to 14289 [a]....	Comprehensive sample	
14047 [d] and 14048 [a]..	B	5N	14290 to 14297 [a]....	Not known, EVA-2	Weigh bag 1038
14049 and 14050 [h]...	Bg [e]	6N			
14051 [d] and 14052 [a]..	C'	7N	14301 [g]	G1	
14053 and 14054 [a]...	C2	14N	14302 [d]	Plus 14305, EVA-1 FSR [a]	
14055 to 14062 [a]....	E	15N			
14063 to 14065 [a]....	C1	16N	14303(?) [i]	Not known, comprehensive sample	Weigh bag 1027
14066 and 14067 [a]...	F	17N			
14068 to 14072 [a]....	C'	10N			
14073 to 14079 [a]....	G	20N	14304 [d]	EVA-1 FSR [h]	Weigh bag 1027
14080 and 14081 [a]...	G	21N	14305 [d]...........	Plus 14302, EVA-1 FSR [b]	Weigh bag 1027
14082 to 14084 [e]....	C1	13N			
14140 to 14143 [a]....	C'	9N	14306 [d]	G	26N
14144 [a]	C'	10N	14307 [a]	G	25N
14145 to 14148 [a]....	G	19N	14308(?) [i]	Dg, plus 14311	
14149 to 14152 [a]....	G	20N	14309(?) [i]	Not known, probably broken from EVA-2 grab sample	Weigh bag 1031
14153 to 14156 [a]....	G	21N			
14160 to 14163 [a]....	Bulk [f]	1N			
14165 to 14189 [a]....	Not known, comprehensive sample(?) collected on EVA-1	Weigh bag 1027	14310(?) [i]	G	
			14311(?) [i]	Dg, plus 14308 [e]	
14190 to 14202 [a]....	Not known, collected on EVA-2	Weigh bag 1031	14312 [d]	H	
			14313 [d]	G1	
			14314 [d]	H	
14210 [a]	A	Core tube 2	14315 [d]	H	
14211 [a]	A	Core tube 3	14316(?) [i]	H	
14220 [a]	G	Core tube 1 (tab)	14317(?) [i]	H	
			14318 [d]	H	
14230 [a]	G	Core tube 1	14319 [d]	H	
14240 [a]	G	SESC	14320(?) [i]	H	
			14321 [d]	C1	
			14411 [a]	A core bit	
			14414 [a]	G core bit	
			14421 to 14453 [a]....	Bulk	1N

[a] Location known by reference to sample bags used at the time of collection.

[b] Photographs or documented sampling stations.

[c] Prenumbered sampling bags used for collecting samples.

[d] Location and lunar orientation known.

[e] Grab sample location.

[f] Scooped fines taken late in EVA-1.

[g] Location tentative, based on tentative identification of sample in lunar-surface photographs.

[h] Football-size rock.

[i] Location tentative, based on description by the Apollo 14 crew, the process of elimination of known samples, or because of the possibility of sample mixing during transfer between or within the weigh bags.

and orienting samples. Group I, that of highest confidence, includes samples for which locations and lunar orientations are well known. Group II includes samples for which locations are either known or tentative, but for which orientations are known only tentatively or not at all. Group III includes samples for which locations are known or tentative, but for which orientations are not known. Group IV includes samples that have not been located.

TABLE 3–IV. *Sample Locations and Documentation*

Station no.	Sample no.	Lunar-surface photograph		Sample description
		No.	Type	
Group 1: Samples photographed **before** collection (location and orientation known)				
EVA-1: Football-size rocks	14304	AS14-67-9390 AS14-67-9391	}Before sample	The sample is a blocky, subangular rock cut by a few poorly developed, irregular fractures. Zap pits are not prominent, and all surfaces appear immature. The rock is a coherent breccia with a moderate percentage of angular to subrounded, blocky to slabby, melanocratic clasts in a very light-gray matrix. A very small percentage of leucocratic clasts is present.
	14305 14302	AS14-67-9392 AS14-67-9393	}Before sample	The sample is a blocky, subangular rock with a poorly developed set of planar fractures. Two nearly planar faces of the rock appear to be controlled by splitting along planar fractures. Zap pits are inconspicuous, and all surfaces appear immature. The rock is a coherent breccia with a moderate proportion of subrounded melanocratic clasts and subordinate leucocratic clasts in a very light-gray matrix.
EVA-2: B.	14047	AS14-64-9073 AS14-64-9074	}Before sample	The sample is a blocky, subangular rock with about 10 percent of the surface coated by vesicular glass. Irregular, slightly rounded surfaces are lightly covered by glass-lined zap pits. One nearly planar bounding face of the rock has well-developed slickensides. Multiple sets of irregular fractures occur at one end of the specimen. The sample is a friable, fine-grained clastic rock and has a small percentage of subangular leucocratic clasts in a medium-gray matrix.
C'.	14051	AS14-68-9443 AS14-68-9444 AS14-68-9445 AS14-68-9446	}Before sample	The sample is a blocky, subrounded rock with all surfaces lightly covered by zap pits with or without glass linings. Spall-like fractures occur locally. Irregular to rounded cavities 1 to 3 mm across may be clast molds. The sample is a friable, fine-grained clastic rock and has a small percentage of subrounded leucocratic and subordinate melanocratic clasts in a medium-gray matrix.
		AS14-68-9447	}After sample	
C1 (football-size rock)	14321	AS14-64-9128 AS14-64-9129	}Before sample	The sample is a blocky, subrounded rock with a moderately dense covering of glass-lined zap pits on all surfaces. Multiple irregular fractures are well developed along one edge of the sample. The rock is a coherent breccia with about 40 percent of blocky, angular to well-rounded clasts of which the great majority are melanocratic. The matrix is very light gray.
G.	14306	AS14-68-9459 AS14-68-9460 AS14-68-9461	}Before sample	The sample is a blocky, subangular rock with 1 face lightly covered by glass-lined zap pits and the remaining rounded faces more densely covered by zap pits. A prominent planar fracture, lined by vesicular glass, makes an angle of approximately 20° with the long axis of the sample. The rock splits along this fracture exposing part of the fracture surface and the glass coating. A poorly developed set of planar fractures lies at an angle of 65° to the prominent fracture. The rock
		AS14-68-9462 AS14-68-9463 AS14-68-9464	}After sample	

Group	Sample	Photographs		Description
G1	14313	AS14-64-9188, AS14-68-9465, AS14-68-9466, AS14-68-9467	Before sample	is a coherent breccia having 25 or more percent of irregular, blocky to slabby, angular to subrounded leucocratic clasts in a medium-gray matrix. The glass-lined fracture appears to cut matrix and clasts alike. The sample is a blocky, subangular rock with a prominent notch produced by spalling along 2 sets of fractures intersecting at an angle of 105°. All surfaces have a light to moderate density of glass-lined zap pits. The rock is a coherent breccia having about 10 percent of well-rounded to subangular clasts in a medium-gray matrix. Both leucocratic and melanocratic clasts occur, but fewer melanocratic clasts are present.
H	14312	AS14-68-9472, AS14-68-9473, AS14-68-9474, AS14-68-9475, AS14-68-9478, AS14-68-9479	Before sample / After sample	Eastern rock from top of Turtle Rock: This sample is a blocky, angular rock cut by irregular fractures parallel to the long axis of the rock. Glass-lined zap pits are moderately dense on all surfaces. The rock is a coherent breccia with a moderate percentage of angular melanocratic clasts that tend to blend with the light-gray matrix. A very few leucocratic clasts are present.
H	14314	AS14-68-9472, AS14-68-9473, AS14-68-9474, AS14-68-9475, AS14-68-9476	Before sample / After sample	Rock taken from fillet below Turtle Rock: This sample is a slabby, angular rock with no apparent zap pits. All surfaces appear immature. Several irregular fractures are parallel to the flat surface of the slab. The slabby shape of the rock appears to be controlled by fractures. The rock is a coherent breccia with a medium-gray matrix and a moderate percentage of light and dark clasts. Light clasts appear to predominate.
H	14315	AS14-68-9468, AS14-68-9469, AS14-68-9470, AS14-68-9471	Before sample / After sample	The sample is a domical, blocky rock with one nearly flat nonpitted side and the rest rounded and heavily pitted. A set of closely spaced fractures makes an angle of from 10° to 15° with the flat surface of the rock. The rock is a coherent breccia in which leucocratic clasts are dominant. The estimated percentage of clasts is 40 percent. The matrix is medium gray.
H	14318	Same as 14315		The sample is a blocky, angular rock, heavily pitted on all sides. A series of well-developed parallel fractures is parallel to one surface of the rock and parallel to the long axis of the sample. The rock is broken along one of these fractures and no pits are present on the broken surface; 30 percent of the exposed fracture surface is coated with vesicular glass. The glass-lined fractures appear to cut clasts and matrix alike (fig. 3–50). The rock is a tightly coherent breccia with an estimated 50 percent clasts. Of these, 60 percent are judged to be leucocratic, and 40 percent mesocratic and melanocratic. One leucocratic clast has a dark clast within it, and several melanocratic clasts contain light clasts.
H	14319	AS14-68-9472, AS14-68-9473, AS14-68-9474, AS14-68-9475, AS14-68-9476, AS14-68-9478, AS14-68-9479	Before sample / After sample	Western rock from top of Turtle Rock: The sample is a blocky, angular rock with a highly irregular surface. There is a low density of glass-lined zap pits on 3 rock faces that are somewhat rounded. The rest of the surface has no pits. One face is extremely fresh. Several irregular fractures cut the rock at a variety of angles. The rock is a coherent breccia that is broken apart along fractures. Clasts make up 30 percent of the rock and melanocratic clasts are by far the dominant type. Some of these have white clasts within them.

TABLE 3–IV. *Sample Locations and Documentation*—Continued

Group 2: Samples photographed before collection (location known or tentative (T), orientation tentative (T) or not known (N))

Station no.	Sample no.	Lunar-surface photograph No.	Type	Sample description
EVA 1: Comprehensive sample	14250 to 14255 14256 to 14259 14260 to 14263 14264 to 14286	AS14-67-9388 AS14-67-9389	Location photographs	No samples documented individually. May be partly from contingency sample.
EVA-2: A	14041 to 14043 (T) 14044	AS14-68-9409 AS14-68-9410 AS14-68-9411 AS14-68-9412 AS14-68-9413	Before sample / After sample	The sample is a blocky, angular rock with a small percentage of the surface coated by vesicular glass. The rock is cut by wide-spaced, irregular fractures intersecting at high angles; the rock breaks readily along these fractures. Some surfaces are lightly covered by glass-lined zap pits. The rock is mostly a friable, fine-grained clastic rock with less than 5 percent of subrounded leucocratic clasts in a medium-gray matrix. A piece of this rock (14043) has a considerably higher proportion of clasts but is otherwise similar.
A	14045 (T) 14046	Same as 14041 to 14043		The sample is a blocky, subangular rock with a rough, hackly surface. Glass-lined zap pits occur on all but one surface. The sample has very poorly developed, irregular internal fractures, but one face of the sample has broad, parallel steps suggestive of fracture control. The sample is a friable, fine-grained clastic rock with very sparse subangular leucocratic clasts in a medium-gray matrix.
C'	14140 to 14143 14068 to 14072 (T) 14144	AS14-64-9125 AS14-64-9126 AS14-64-9127	Before sample / After sample	All samples are blocky, angular to subrounded rocks with very rough surfaces. Sample 14068 appears to be shattered on one side; but, otherwise, all the rocks in this set are unfractured. All these rocks lack zap pits, but irregular vugs are moderately developed. The samples are fine-grained crystalline rocks with sparsely scattered, large (to 1 mm), white grains.
C1	14082 (N) 14083 (N)	AS14-68-9452 AS14-68-9453 (After chipping but before bagging, rock may be seen on fillet in AS14-68-9452)		The sample is a blocky to slightly slabby, angular rock with a very rough surface. Glass-lined zap pits are very sparsely distributed over one surface. There are no fractures. The sample is a very friable, fine-grained clastic rock with a very small percentage of subrounded melanocratic clasts in a very light-gray matrix. Clasts are locally concentrated in a thin layer at one end of the specimen (14083).
G	14145 to 14148			Samples taken from top of trench.
G	14080 to 14081 14153 to 14156			Samples taken from middle of trench. The samples are extremely irregular, angular rocks with very sparse zap pits. The rocks are not fractured, but some surfaces have slickensides. The samples are composed of fragments of fine-grained clastic rocks with sparse leucocratic clasts, loosely bonded by high vesicular glass.
G	14073 to 14079 14149 to 14152			Samples taken from bottom of trench. Samples 14073, 14074, 14078, and 14079 are blocky, subangular to subrounded rocks lacking fractures and pits. The samples are light-gray, equigranular, fine-grained, crystalline rocks.

	Sample	Photographs		Description
G	14240	AS14-64-9158 to AS14-64-9166		Sample 14076 is a blocky, subangular, smooth-surfaced rock lacking zap pits. One set of fractures cuts the rock parallel to the long axis. The sample is a coherent heterogeneous clastic rock. One end consists of breccia with a moderate abundance of melanocratic clasts in a light-gray matrix, and the other end is a fine-grained clastic rock with sparse melanocratic clasts in a light-gray matrix. The contact between the two lithologies is sharp but irregular.
G1 (T)........	14301 (T)	AS14-64-9188 AS14-68-9466 AS14-68-9467	Before sample After sample	Sample 14077 is a blocky, subrounded rock with a moderately rough surface. No fractures or zap pits occur. Irregular vugs are sparsely distributed over the surface. The sample is a light-gray, fine-grained, inequigranular crystalline rock with sparse, large, white grains. Bottom of trench; SESC sample. Trench documentation.
H (T).........	14316	....		The sample is a blocky, subrounded rock with a moderately dense cover of glass-lined zap pits. Several irregular fractures cut the rock, and spalling along two intersecting fractures left a V-shaped protuberance on one side of the rock. The sample is a coherent, medium-gray clastic rock with sparse subangular leucocratic clasts and less abundant melanocratic clasts in a fine-grained matrix.
				The sample is a subslabby, subangular rock with one flat surface free of pits and the rest rounded and irregular with numerous glass-lined pits. Planar to sub-planar, glass-lined fractures are parallel to the flat surface of the rock and the rock has broken along one of these. The rock is a coherent breccia with an estimated 20 percent of blocky, subangular to rounded clasts in a medium-gray matrix. The clasts are dominantly leucocratic. One medium-gray clast contains white clasts, probably clastic feldspar. One light clast contains lighter clasts.
H (T).........	14317	....		The sample is a slabby, angular rock with no apparent zap pits. All surfaces appear immature. A few irregular fractures are parallel to the flat surfaces of the slab. The rock is a coherent breccia with a small percentage of leucocratic clasts up to 3 mm across. The matrix is fine grained and gray.
H (T).........	14320	....		The sample is a slabby, angular rock. One side appears fresher than the rest but all sides have about the same high density of glass-lined pits. Several irregular fractures occur at odd angles to the long axis of the rock. The rock is a coherent breccia with a moderate percentage of clasts. Most of the clasts are melanocratic with a moderate percentage of clasts are melanocratic; only a small percentage is leucocratic; the matrix is light gray.
A	14211 (Core tube 3)	Core tube samples AS14-64-9046		
	14210 (Core tube 2)	AS14-64-9047 AS14-64-9048 AS14-68-9454		
G	14220 (Core tube 1 (tab))			
G	14230 (Core tube 1	AS14-68-9455 AS14-68-9456 AS14-68-9457 AS14-68-9458		

TABLE 3–IV. *Locations and Documentation—Continued*

Station no.	Sample no.	Lunar-surface photograph		Sample description
		No.	Type	
				Group 3: No surface photographs (location known or tentative (T))
EVA-1:				
Contingency sample	14001 to 14005 14006 to 14012 14260 to 14263 14160 to 14163 14421 and 14422 14425 to 14453			May be partly from comprehensive sample.
Bulk sample 1N				
EVA-2:				
Bg	14049 and 14050			Grab sample: The sample is a blocky, subrounded rock lacking zap pits and having only a few very poorly developed, irregular fractures. The sample is a very friable, fine-grained clastic rock with less than 1 percent subrounded leucocratic clasts in a medium-gray matrix.
C2	14053			The sample is a blocky, subrounded rock with glass-lined zap pits on only one side. Vugs, lined by a light-colored mineral, are present. The sample is an equigranular, fine-grained crystalline rock.
E	14055 to 14062			Samples 14055 and 14058 are blocky, subangular to subrounded rocks lightly covered by glass-lined zap pits. A few poorly developed irregular fractures occur. Sample 14055 has approximately 15 to 20 percent of the surface coated by vesicular glass. The samples are friable, fine-grained clastic rocks with 5 to 15 percent of subrounded leucocratic clasts in a medium-gray matrix. Samples 14056, 14057, 14059, 14060, and 14061 are blocky, subrounded rocks, mostly lacking zap pits and fractures. The samples are very friable, fine-grained clastic rocks with less than 5 percent subrounded leucocratic clasts in a medium-gray to brownish-gray matrix.
C1	14063 and 14064			The samples are blocky, subangular to subrounded rocks with a light to moderate density of glass-lined zap pits. Irregular fractures are poorly developed and numerous subrounded clast molds occur in both rocks. The samples are friable breccias with approximately 40 percent of subangular to subrounded clasts in a very light-gray, fine-grained matrix. Melanocratic clasts are subordinate to leucocratic clasts.
F	14066			The sample is a blocky, subrounded rock that has rounded faces that are heavily covered by glass-lined zap pits. The rock is very hackly at one end and has a few other irregular fractures. The sample is a moderately friable breccia with 15 to 20 percent of subangular melanocratic clasts and a few leucocratic clasts in a fine-grained, light-gray matrix.

G.	14307	The sample is a blocky, slightly slabby, agular rock cut by multiple irregular fractures. The sample is a moderately coherent breccia with about 20 percent subangular to subrounded leucocratic clasts in a fine-grained, medium-gray matrix. Seriate-size distribution of leucocratic clasts is apparent. There appears to be a weak foliation of clasts approximately parallel to the flat side of the rock.
Dg (T) (football-size rock)	14308 14311 (T)	Grab sample: The sample is a blocky, subrounded rock broken into 4 pieces along irregular fractures. Irregular vugs are sparsely distributed through the rock. The sample is a coherent clastic rock with a small percentage of subangular clasts, mostly leucocratic, in a fine-grained, crystalline groundmass.
G (T) (football-size rock)	14310 (T)	The sample is a blocky rock with 2 rounded surfaces heavily covered by zap pits and the remaining faces free of zap pits and joining along sharply angular edges. Irregular vugs are sparsely distributed through the rock. The sample is a fine-grained, medium-gray equigranular crystalline rock.

Group 4: Samples for which location has not been determined

	14165 to 14189 14190 to 14202 14293 to 14297 14303	The sample is a blocky, subrounded rock, lacking zap pits and having only a few, very poorly developed, irregular fractures. The sample is a very friable, fine-grained clastic rock with less than 1 percent of subrounded leucocratic clasts in a medium-gray matrix.
	14309	The sample is a slabby, subrounded rock but has a few irregular fractures. Only a few zap pits are present. One face is irregular and may be a freshly broken surface. The rock is a moderately coherent breccia with a moderate percentage of subrounded melanocratic clasts in a light-gray matrix. A few feldspar clasts (up to 3 mm long) are present. (Location unknown; returned in weigh bag 1031 with other grab samples from EVA–2.)

Locations are known with some degree of confidence for all but two of the 35 largest rock samples. Group I contains eight rocks for which locations and orientations are known with greatest confidence (samples 14047, 14051, 14302, 14304, 14305, 14306, 14313, and 14321). Samples 14302 and 14305 broke apart after sampling. Group II includes nine rocks that have been identified or tentatively identified with orientations tentatively established (samples 14041, 14042, 14045, 14301, 14312, 14314, 14315, 14318, and 14319). Three of these samples (14041, 14042, and 14045) are the largest fragments from a fractured clod that broke apart when collected by the LMP. Sixteen additional rocks weighing more than 50 g are listed in groups II and III. Locations are known or suggested for these rocks, but the rocks have not yet been precisely oriented. (These 16 rocks are samples 14049, 14053, 14055, 14063, 14064, 14066, 14082, 14264, 14265, 14267, 14271, 14307, 14308, 14310, 14311, and 14320; orientations of these rocks are not known because of the lack of lunar-surface photographs or because of the small-sized images in the photographs. Exposure and burial surfaces are known, however, for samples 14053, 14082, and 14310, based on weathering characteristics.) Samples 14264, 14265, 14267, and 14271 were collected in the comprehensive sample. Samples 14308 and 14311 broke apart after sampling. Samples 14169 and 14303 are the only rocks weighing more than 50 g for which locations within the traverse area are considered to be very tentative.

Sample 14303. Sample 14303 is the largest rock for which a location has not been reasonably well established from photography or crew comments. It was returned in the same weigh bag (1027) with the first EVA football-size rocks (samples 14304 and 14305) collected by the CDR who was probably carrying weigh bag 1027 at the time. This weigh bag also included samples 14165 to 14189, of uncertain location. By association, it appears likely that all these samples were collected during the first EVA, although some repacking of bags was done at the end of the second EVA to alleviate sample-storage problems. After gathering the comprehensive sample, the CDR reported picking up two small football-size rocks, both photographed by stereopairs before collection and both recognized from the photographs as samples 14304 and 14305 (including sample 14302 before breaking). The comprehensive sample is the most reasonable place in which to include samples 14303 and 14165 to 14189. The LMP may have just picked up sample 14303 when he said, "Hey, here—don't close it, here is one in here for that . . . one in here I picked up," and put it into the CDR's weigh bag.

Sample 14047. Sample 14047 was collected at station B from the west rim of a sharp 5-m-diameter crater during the second EVA. It is a poorly indurated clod that was located on a gently sloping and moderately undulating area. Sample 14047 is approximately 10 cm long and is one of the few rocks this size seen in the stereopair (AS14–64–9073 and AS14–64–9074), although the panorama taken at station B shows many rocks and boulders larger than 10 cm in the general

FIGURE 3–34.—Sample 14047 before collection.

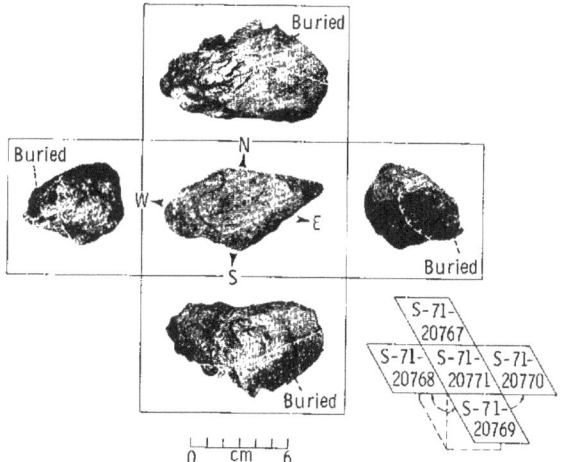

FIGURE 3-35.—Orthogonal views of sample 14047 in approximate lunar orientation.

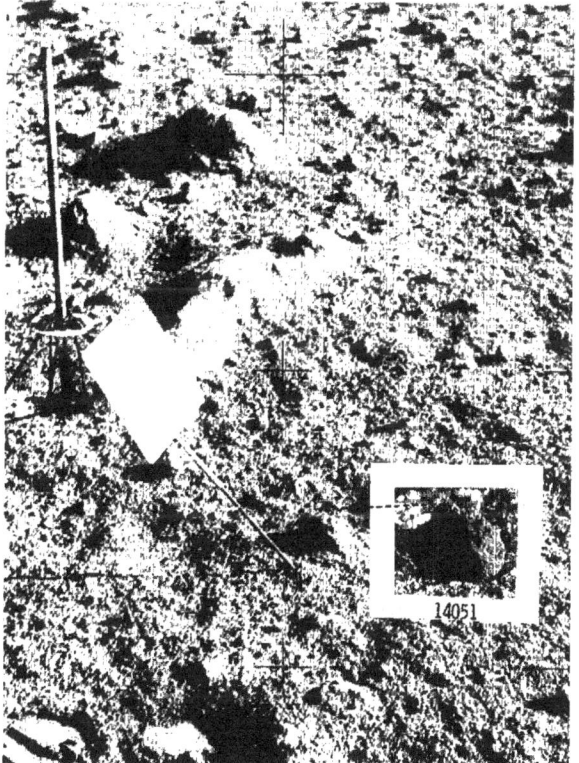

FIGURE 3-36.—Sample 14051, showing approximate lunar orientation reconstructed in the LRL.

area. Sample 14047 has poorly developed fillets on the south and east sides. The degree of filleting appears the same for the few other rocks in the

immediate vicinity. The lunar-surface orientation of the sample is shown in figures 3-34 and 3-35. It should be noted that the glass-covered surface was buried at the time the sample was collected.

Sample 14051. Sample 14051, a rock approximately 7 cm long, was collected at station C′ on the southeast slope of Cone Crater. The sample was located on a fairly flat but gently undulating area heavily strewn with rocks and boulders. Rocks in the area range in size from a centimeter to boulders several meters long. Most appear to be partially buried and most are subangular to rounded. Sample 14051 was only slightly buried. The lunar-surface orientation is shown in figures 3-36 and 3-37.

Sample 14053. Sample 14053 was collected from the side of a large filleted boulder at station C2 (photographs AS14-64-9130 to AS14-64-9133). The sample, which has not been recognized in the photographs, is a crystalline rock that the CDR picked from the eastern, weathered surface of the boulder. The sample is heavily pitted on several faces but appears to be freshly broken on what must have been the part embedded or lying on the rock. Numerous, fairly coarse, light and dark clasts are visible in the shadowed portion of the filleted boulder. Sample 14053 is probably a clast weathered from the boulder. A

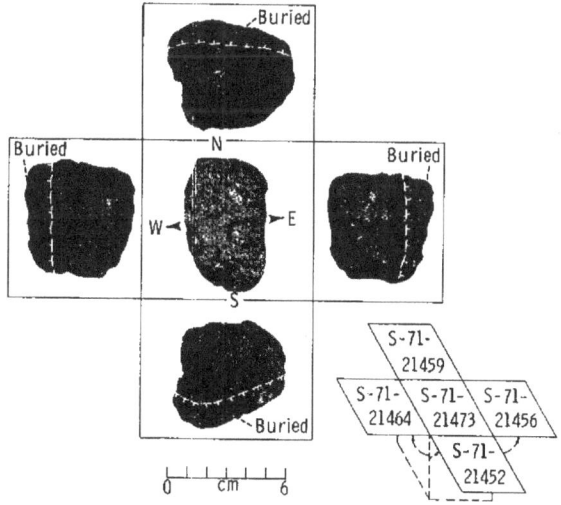

FIGURE 3-37.—Orthogonal views of sample 14051 in approximate lunar orientation.

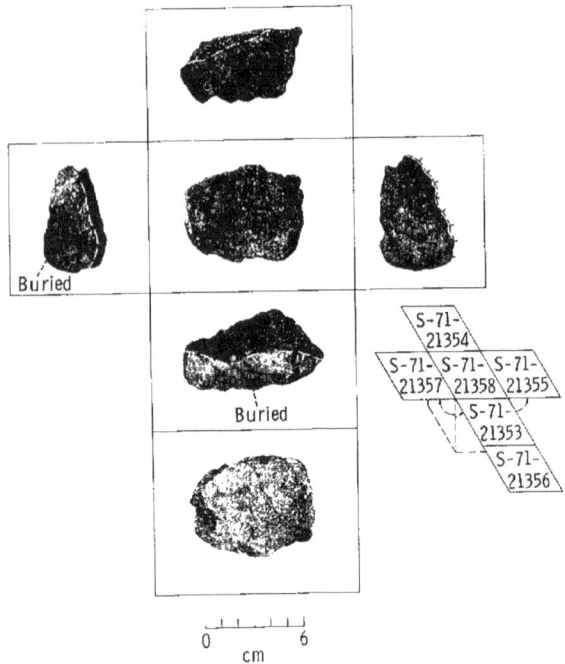

FIGURE 3-38.—Orthogonal views of sample 14053.

complete orthogonal view of sample 14053 is shown in figure 3-38.

Samples 14068 to 14072. Samples 14068 to 14072 are documented small samples, all less than 50 g in weight, that were collected at station C'. In the presampling photograph (AS14-64-9126), the fragments appear to lie on the surface, between the MET tracks, at the northeast side of an 8- or 10-cm-diameter crater. This may be what the CDR referred to as a "secondary impact that disrupted the surface" and had penetrated the lighter gray material below. The small fragments resemble one another and may be pieces of a larger fragment broken by a low-energy impact. Probable correlations of the samples on the lunar surface (AS14-64-9126) with LRL photographs are shown in figure 3-39.

Sample 14082. Sample 14082 is an example of the so-called White Rocks. The sample was apparently chipped from a 1-m boulder at station C1. Other boulders in the immediate vicinity are as much as 10 m across. The boulder from which the sample was chipped is a fragmental rock with 1-cm dark-and-light clasts set in a very light-gray matrix. One indistinct subrounded dark clast is

6 cm across. Numerous probable micrometeorite impact pits are visible. A photograph of the boulder (fig. 3-19) shows what is believed to be the piece chipped off and the indentation left by the sampling.

Sample 14305 (and 14302). Sample 14305 is the second of the two small football-size rocks photographed and collected during the first EVA. A small part of the rock that broke off sometime between stowage on the lunar surface and unpacking in the LRL was numbered 14302 in the LRL. Before sampling, the combined rock was on a very gently undulating surface similar to that where sample 14304 was situated. An area approximately 1 m around the rock was mapped (fig. 3-40) from a stereopair of photographs taken with the Hasselblad camera (AS14-67-9392 and AS14-67-9393). Sample 14305 is the largest rock in the immediate area. This sample is approximately 15 cm across, whereas the next

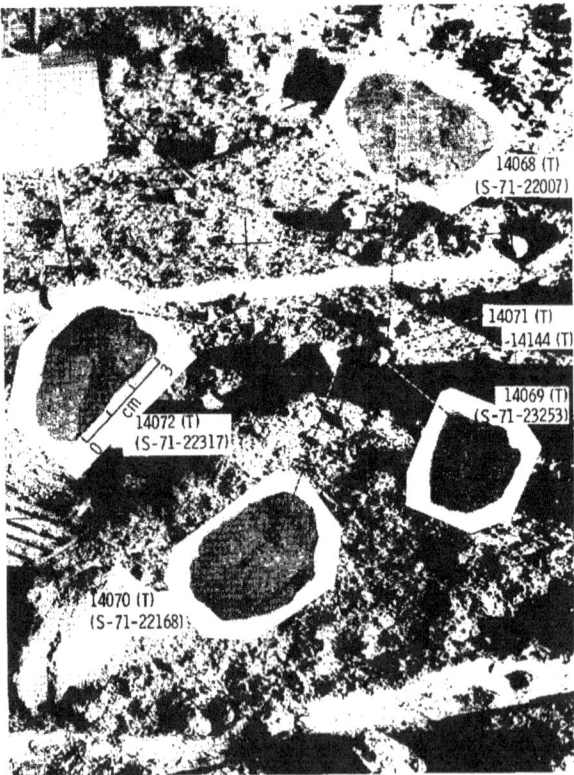

FIGURE 3-39.—Samples 14068 to 14072 and sample 14144 before collection.

FIGURE 3–40.—Sample 14305 (includes 14302 before breakage) before collection; also shown in approximate lunar orientation reconstructed in the LRL using oblique lighting.

by the rock rather than banked against the rock. The freshness of the crater formed by the rock and the angularity of the crater edges suggest that the rock has been in this position on the lunar surface for a relatively short period of time. It is highly probable that this rock landed in the present position as a result of a recent impact that formed a nearby crater, not as a result of the formation of any of the large older craters in the area such as Doublet, Triplet, or Cone. The lunar-surface orientation of rock 14305 is shown in figure 3–41 and 3–42.

Sample 14306. Sample 14306 was collected at station G, 70 m east of the moderately subdued

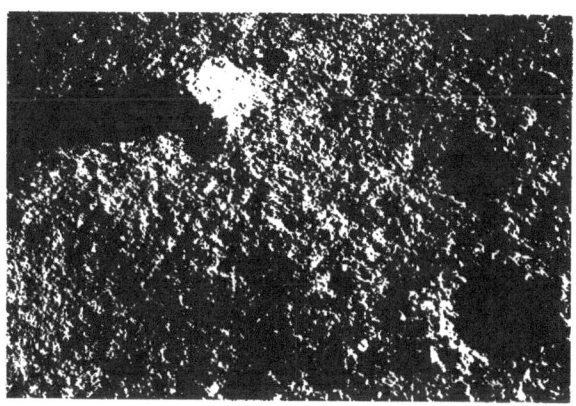

largest rock seen in the photographs is approximately 4 cm in diameter.

Sample 14305 lies on the northern edge of a small, sharp, irregular depression. This depression appears to have been formed by impact of rock 14305, the lower southwest edge of the rock having formed the major part of the depression. The vertical southwest edge of the rock appears to have formed the northeast-trending grooved part of the depression by the sliding of the rock to the northeast. A small fillet approximately 1 cm deep is present against the southeast edge of the rock as though the rock dug into the fine-grained surface material as it slid into the present position. The fillet partially fills the northwest side of two 3-cm raindrop-type craterlets, which indicates that the fillet material moved in a direction away from the rock rather than toward the rock; such would be the case if the material were pushed out

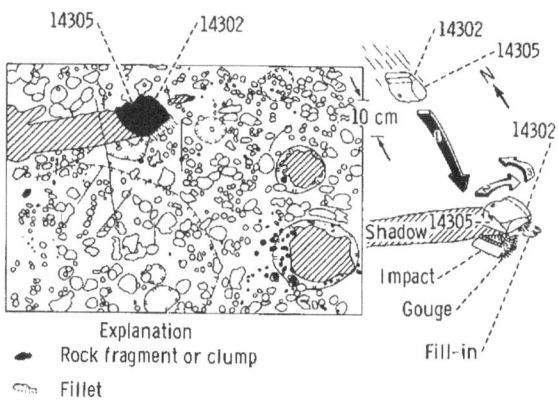

Explanation

- Rock fragment or clump
- Fillet
- Sharp crater with raised rim
- Subdued crater
- Raindrop depression
- Shallow trough or lineament
- Shadow cast by rock fragment

FIGURE 3–41.—Photograph, sketch map, and diagrammatic sketch of impact trajectory of sample 14305.

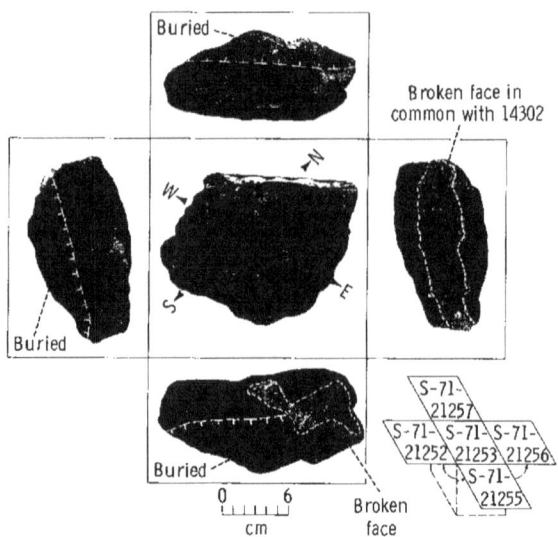

FIGURE 3-42.—Orthogonal views of sample 14305 in approximate lunar orientation.

FIGURE 3-43.—Sample 14306, showing approximate lunar orientation reconstructed in the LRL using oblique lighting.

North Triplet Crater. The surface in the immediate vicinity of sample 14306 is very gently undulating with subdued craters ranging from a few centimeters to 2 m in diameter. Other rocks or clods ranging from the limit of resolution of the photographs to 10 cm in the longest dimension are common but not abundant. One rock approximately 60 cm in the longest dimension lies about 50 cm south of sample 14306. The rock has a poorly developed fillet approximately 0.5 cm high against the two visible sides. The larger rock just south of rock 14306 has a well-developed fillet 2 to 3 cm high and appears to have a small amount of fine-grained material contained in a fracture on the east side; all visible surfaces of rock 14306 appear to be free of fine-grained loose material. The orientation of rock 14306 on the lunar surface is shown in figures 3-43 and 3-44. The rock is notable because it has a glass-filled 2-mm-wide fracture that cuts cleanly across one side (fig. 3-45).

Sample 14313. Sample 14313 was collected at station G1 on the north rim of North Triplet Crater. The rock has an unusual L-shape and lay on the lunar surface in the orientation shown in figures 3-46 and 3-47. Other rocks or clods from 6 to 10 cm across are moderately abundant on the surrounding regolith.

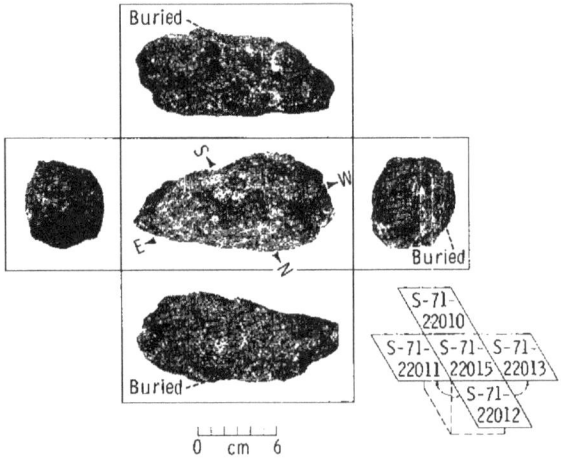

FIGURE 3-44.—Orthogonal views of sample 14306, shown in approximate lunar orientation.

FIGURE 3–45.—Sample 14306 showing prominent glass-filled fracture.

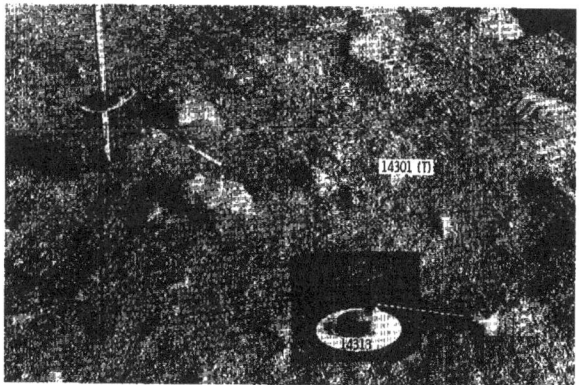

FIGURE 3–46.—Samples 14301 and 14313; approximate lunar orientation of latter sample reconstructed in the LRL using oblique lighting.

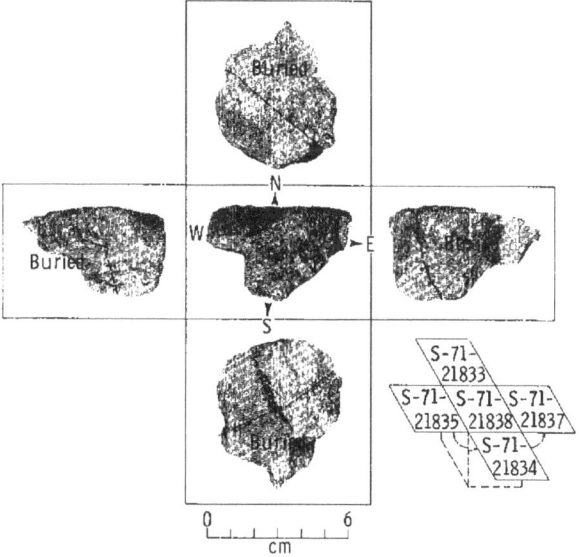

FIGURE 3–47.—Orthogonal views of sample 14313 in approximate lunar orientation.

Sample 14318. Sample 14318 is identified as coming from the south end of the north boulder field, station H, and is the largest of the four labeled rocks shown in the foreground of photographs AS14–68–9468 to AS14–68–9471 (fig. 3–48). Four more samples were taken from Turtle Rock (fig. 3–49). Sample 14318 is a different type of breccia from that represented by sample 14321 (shown later in fig. 3–52). Very light-gray and light-to-medium gray clasts are set in a darker fine-grained matrix (fig. 3–50). One fracture surface that extends into the rock is partially coated with vesicular glass. Sample 14318 is shown in figure 3–51 in appropriate lunar orientation by orthogonal views using LRL photographs.

Sample 14321. Sample 14321 is the large football-size rock collected during the second EVA near station C1. An area approximately 1 by 1.5 m around the rock was mapped using stereophotographs (AS14–64–9128 and AS14–64–9129) taken with the Hasselblad Electric data camera just before the rock was collected. The rock was situated on the hummocky ejecta blanket of Cone Crater. Other rocks ranging in size from the limit of resolution of the Hasselblad photog-

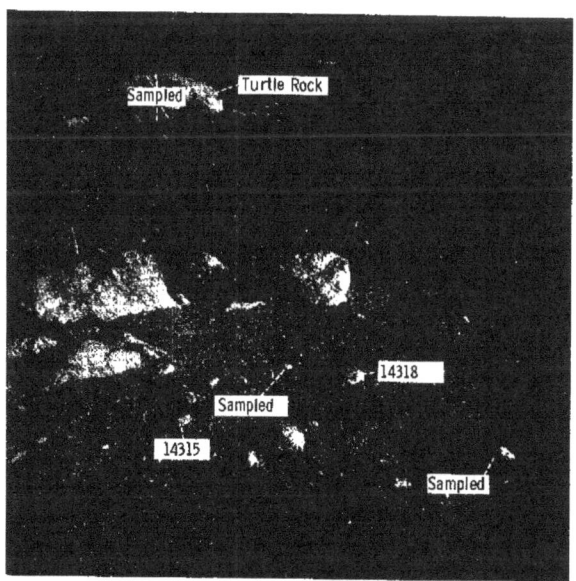

FIGURE 3–48.—View of portion of north boulder field at station H, showing locations of samples 14312, 14315, and 14318.

FIGURE 3–49.—Turtle Rock; samples from rock surface and fillet.

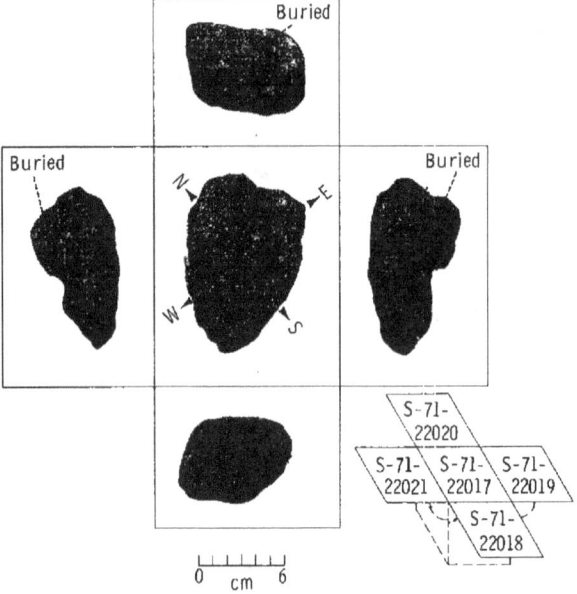

FIGURE 3–51.—Orthogonal views of sample number 14318 in approximate lunar orientation.

FIGURE 3–50.—Sample 14318, showing light- and medium-gray clasts in a dark-gray matrix.

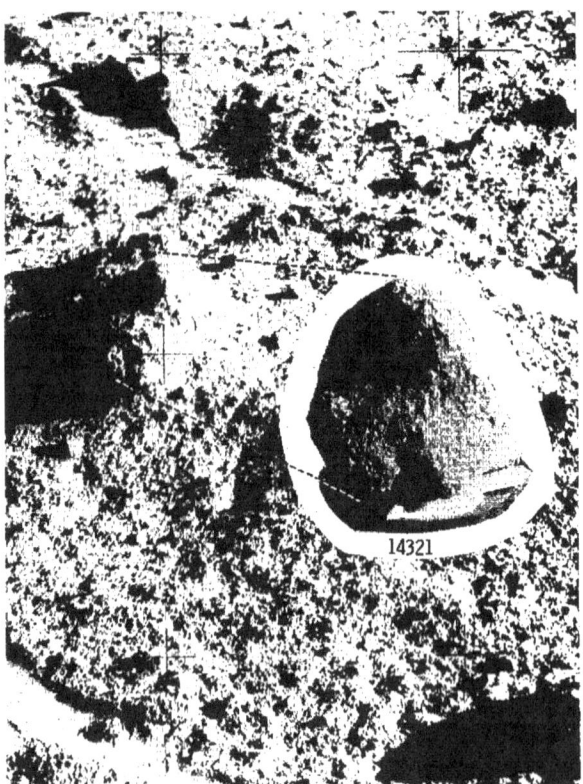

FIGURE 3–52.—Sample 14321 before collection; also shown in approximate lunar orientation reconstructed in the LRL using oblique lighting.

raphy to large boulder size are common in the area (figs. 3–52 and 3–53); rock fragments ranging from the resolution of the photography to 5 cm in diameter are common in the immediate vicinity of the sample. This rock was the largest within the area mapped; the next largest rock is approximately 20 cm in the longest dimension. A well-developed fillet approximately 2 cm deep is present against the visible sides of rock 14321.

Sample 14321 is similar in physical appearance to numerous other rocks of approximately the same size on the ejecta blanket of Cone Crater. The rock was well rounded and partially buried

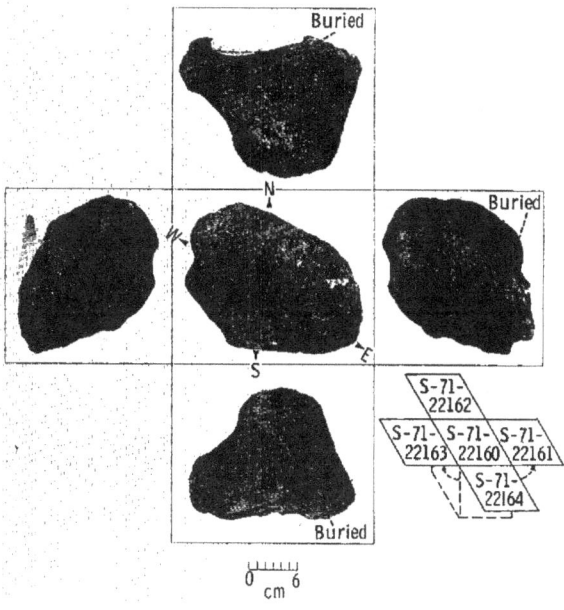

FIGURE 3–54.—Orthogonal views of sample 14321 in approximate lunar orientation.

with a well-developed fillet and appears to be as old as other rocks tentatively associated with the Cone Crater event.

The approximate lunar-surface orientation of sample 14321 is shown in figures 3–52 to 3–54. An LRL photograph (fig. 3–55) shows abundant dark, subrounded clasts as much as 4 cm across set in a lighter matrix.

Summary. By far, most of the returned samples are fragmental rocks (table 3–IV). Several different types are present and these are described in detail in section 5 of this report. Some of the fragmental rocks are friable and others are more coherent. Among the more coherent fragmental rocks, some can be seen in LRL photographs and by direct examination in the LRL to be composed mostly of dark, dense clasts in a lighter colored matrix; others consist of a mixture of light- and medium-gray clasts in a dark-gray matrix with light clasts predominating in most examples. Both types of coherent breccias were collected from widely separated points in the area traversed by the Apollo 14 crew.

Comparison of the samples with the boulders shown in many of the surface photographs indi-

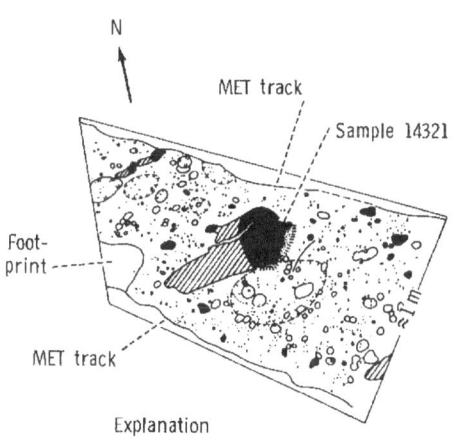

Explanation

- Rock fragment or clump
- Fillet
- Sharp crater with raised rim
- Subdued crater
- Raindrop depression
- Shallow trough or lineament
- Shadow cast by rock fragment

FIGURE 3–53.—Sample 14321 and immediate vicinity.

FIGURE 3–55.—Sample 14321 showing dark clasts and lighter matrix.

cates numerous similarities. Closely spaced fractures are conspicuous in many of the coherent breccias returned to Earth, and the shape of many of these rocks has been controlled by the fracture sets. Some returned rocks contain two or more sets of intersecting fractures. Similar fracture sets are prominent in all the large boulders appearing in the surface photographs. Both light and dark clasts, some as long as 50 cm, are recognizable in all the boulders. Dark, dense clasts are especially prominent in boulders at the White Rocks area at station C1. Because of lack of resolution, it is not possible to identify the boulders as either of the two types of coherent breccia just described. However, many of the boulders have a rough hackly appearance that is more typical of the dark-clast-rich breccias than it is of those breccias in which light clasts are predominant. In addition, one sample of dark-clast-rich breccia (14321) was described by the LM crew as typical of most boulders in the block field on the rim of Cone Crater. Overall, more samples of breccia with dark clasts in a light matrix were returned than were samples in which light clasts predominate. It is, therefore, inferred that dark-clast-rich breccia is the dominant rock type in the block field surrounding Cone Crater, including the raylike extension west as far as the north boulder field.

Breccias with a dark-gray matrix in which relatively light clasts are more abundant than darker

clasts were less abundant in the returned samples than dark-clast-rich breccias. It is not yet known whether both types of coherent breccias are representative of the Fra Mauro Formation. If they are, they indicate considerable lateral or vertical variation within that unit. An alternative interpretation is that the coherent breccias with abundant light clasts represent a unit underlying the Fra Mauro Formation that was excavated from depth by Cone Crater and earlier large impacts.

Conclusions

The Apollo 14 LM landed on an area of lunar terra in what premission studies indicated was part of the ejecta blanket deposited during the formation of the Imbrium Basin. This material has been named the Fra Mauro Formation. The following conclusions derived from premission studies set the geologic context in which the results of the mission can be placed.

(1) The Fra Mauro Formation is older than any mare material on the Moon, including that at the Apollo 11 and 12 landing sites.

(2) Craters in the 400-m to 1-km size range are two to three times more abundant in the Fra Mauro region than in the mare material of the Apollo 11 and 12 landing sites.

(3) Consistent with this more intense cratering history, the regolith in the Fra Mauro region is significantly thicker than that on the mare material.

(4) The eastern part of the site, including the area of a relatively young, 340-m-diameter crater named Cone Crater, contains one of a family of north-south trending ridges that are characteristic of the Fra Mauro Formation. The western part is much more level except for superposed craters. The possibility exists that the materials forming the bedrock of the ridge and level areas may differ.

(5) Much of the eastern part of the landing site is covered with the rubbly ejecta blanket of Cone Crater. The rubble in this ejecta blanket was excavated from beneath the thick regolith by the impact that formed Cone Crater.

Geologic analysis of the mission and LRL photography, the voice transcript of the mission, the crew debriefings, and the observation of returned samples by members of the Geology Experiment Team led to the following conclusions concerning the geology of the landing site.

The site is densely covered with craters in all stages of destruction. Some craters as much as 400 m across have undergone nearly complete destruction, and the overlapping of relatively large, very gentle depressions gives the topography at the site a strongly undulating aspect. In contrast, the largest craters that have undergone nearly complete destruction at the Apollo 11 and 12 landing sites are on the order of 50 to 100 m in diameter.

The lunar regolith at Fra Mauro is obviously thicker than at the mare sites, although estimates of the thickness are still preliminary. The surface material is finer grained in the western portion of the site away from the Cone Crater ejecta blanket than in the continuous ejecta blanket itself. Rock fragments larger than a few centimeters across are rare in the western part of the site and become progressively more abundant toward Cone Crater. The regolith appears to be looser and less cohesive than that developed on the mare material; downslope movement of this loose debris has caused the eradication of small craters on slopes and extensive slumping of crater walls.

Boulders up to 15 m across are present on the rim of Cone Crater; photographs of these boulders provide the first dramatic glimpse of relatively large segments derived from lunar bedrock and of detailed rock structures. Smaller boulders occur throughout the Cone Crater ejecta blanket and as isolated occurrences on raylike extensions of the ejecta blanket.

All the boulders for which stereophotographs are available appear to be coherent breccias, some with discrete clasts up to 150 cm across, larger than any of the returned samples. Both light and dark clasts are recognizable. Resistance of the breccias to the weathering effects of the lunar environment varies considerably; some breccias have weathered to smooth, resistant surfaces and others to hackly, rough surfaces that may be rubbly.

Significant and striking features within the boulders are sets of parallel fractures spaced at several millimeters to approximately 1 cm. Several intersecting sets of differently spaced fractures are present in some boulders.

Portions of some boulders close to the rim of Cone Crater are crudely layered with very light material that forms irregular bands from 25 to 40 cm thick. The light bands contain both lighter and darker clasts up to 10 cm across, and the host rock of the bands contains light clasts up to 10 cm across. Irregular parts of other boulders are also very light, but a layered relationship is not evident. Boulders containing light layers occur only near the rim of Cone Crater and hence may come from deeper levels in the crater.

Most of the large blocks have fillets of lunar fines and fragments embanked against the basal edges. The size of a fillet is commonly proportional to the size, degree of rounding, and apparent friability of the host rock. Fillets are preferentially developed against outward-sloping rock surfaces and contain coarse fragments spalled off the host rock. Burial of rocks is a combined product of (1) ejecta blanketing by adjacent impact events of all sizes, particularly on well-rounded rocks the tops of which are close to the ground, and (2) self-burial by micrometeorite and thermal erosion of the exposed rock surfaces.

Two well-developed sets of surface lineaments have the northwest and northeast trends observed at the Apollo 11 and 12 sites. A secondary set trends north. The large number of very long, straight lineaments is unique to the Apollo 14 site. These may be the result of very small, recent, vertical displacements along fractures, or sifting of fine-grained material down into fractures that were propagated to the surface from a more coherent jointed substrate.

The returned lunar samples consist dominantly of fragmental rocks that vary significantly in the degree of toughness and in the proportions and character of included fragments. Any attempts to reconstruct the bedrock geology of the site from an analysis of the distribution of the various rock types must take into account the intense cratering history that the area has undergone and must recognize that cratering events both comminute existing bedrock and form new rocklike material from previously comminuted debris. Several tentative conclusions regarding the relationship of the samples to the local geology may be stated as follows.

(1) Moderately well-indurated breccia, in which the dominant clasts are dark and are set in a lighter matrix, is present throughout the

traverse area and may be the dominant rock type in the Cone Crater ejecta blanket. Material of this type occurs in raylike extensions of the ejecta field as far west as the north boulder field (station H). Some fragments of this material have been ejected to the surface more recently than the Cone Crater event. Because of its relative abundance, it seems probable that material of this type is representative of the Fra Mauro Formation.

(2) Another type of coherent breccia different from that discussed in the preceding paragraph was returned in lesser amounts from the lunar surface. This type breccia consists of a mixture of light and dark clasts in nearly equal amounts in a tough, coherent, gray matrix. This breccia type was collected at points throughout the traverse including the north boulder field. It is either another part of the Fra Mauro Formation, indicating considerable vertical or horizontal differences in that unit, or it may represent rocks that underlie the Fra Mauro Formation and were excavated by Cone Crater and older impacts.

(3) The boulders reveal a complex sequence of events, not all of which can be related to the Cone Crater event. Lithologically different layers within the boulders appear to change strike and dip radically over short distances within single boulders, indicating deformation of the layers. Although the rocks are heterogeneous breccias, they have become sufficiently coherent to sustain planar fractures that cross matrices and clasts alike. Some such fractures, as shown by returned samples, have vesicular glass linings. The youngest structures, the glass-lined fractures, are the best candidates to represent the Cone Crater event, but they were preceded by the formation of lithologic layering, deformation, and induration. These earlier events may be related to the emplacement of the Fra Mauro Formation as well as to pre-Fra Mauro events in the early history of the Moon.

(4) Some clasts within breccias are themselves fragmental rocks, and these may be the product of impact events that predate emplacement of the Fra Mauro Formation.

References

3-1. EGGLETON, R. E.: Preliminary Geology of the Riphaeus Quadrangle of the Moon and Defini-tion of the Fra Mauro Formation. Astrogeol. Studies, Ann. Prog. Rept., Aug. 1962–July 1963. Pt. A, U.S. Geol. Survey open-file Rept. 1964, pp. 46–63.

3-2. WILHELMS, DON E.: Summary of Lunar Stratigraphy—Telescopic Observations. Contributions to Astrogeology, U.S. Geol. Survey Prof. Paper 599–F, 1970, pp. F1–F47.

3-3. EGGLETON, R. E.; AND MARSHALL, C. H.: Notes on the Apenninian Series and Pre-Imbrian Stratigraphy in the Vicinity of Mare Imbrium and Mare Nubium. Astrogeol. Studies Semiann. Prog. Rept., Feb. 25, 1961, to Aug. 24, 1961. U.S. Geol. Survey open-file Rept. 1964, pp. 132–137.

3-4. WILHELMS, D. E.; AND McCAULEY, J. F.: Geologic Map of the Near Side of the Moon: U.S. Geol. Survey Misc. Geol. Inv. Map I–703, 1971.

3-5. BALDWIN, RALPH B.: The Measure of the Moon. Univ. of Chicago Press, 1963.

3-6. EGGLETON, R. E.; AND OFFIELD, I. W.: Geologic Maps of the Fra Mauro Region of the Moon. U.S. Geol. Survey Map I–708, 1970.

3-7. TRASK, NEWELL J.: Size and Spatial Distribution of Craters Estimated From Ranger Photographs: Rangers VIII and IX. Pt. II of Experimenters' Analyses and Interpretations. Preliminary Geologic Map of a Small Area in Mare Tranquillitatis, Tech. Rept. 32–800, Mar. 1966, pp. 252–263.

3-8. SHOEMAKER, E. M.; HAIT, M. H.; SWANN, G. A.; SCHLEICHER, D. I.; ET AL.: Origin of the Lunar Regolith at Tranquility Base. Physical Properties. Vol. III of Proc. Apollo 11 Lunar Sci. Conf., A. A. Levinson, ed., Geochim. Cosmochim. Acta Suppl. 1, Pergamon Press, 1970, pp. 2399–2412.

3-9. HOLT, H. E.; AND RENNILSON, J. J.: Photometry of the Lunar Regolith, as Observed by Surveyor Cameras. Television Observations From Surveyor. Science Results. Surveyor Project Final Rept., sec. III, par. H, Pt. II of JPL Tech. Rept. 32–1265, 1968, pp. 109–113.

3-10 MORRIS, E. C.; AND SHOEMAKER, E. M.: Fragmental Debris. Geology. Television Observations From Surveyor. Science Results. Pt. II of Surveyor Project Final Rept., sec. III, par. G3, JPL Tech. Rept. 32–1265, 1968, pp. 69–86.

3-11. GAULT, D.; COLLINS, R.; GOLD, T.; GREEN, J.; ET AL.: Lunar Theory and Processes. Scientific Results. Pt. II of Surveyor III Mission Rept., sec. VIII, JPL Tech. Rept. 32–1177, 1967, pp. 195–213.

3-12. SHOEMAKER, E. M.; BATSON, R. M.; BEAN, A. L.; CONRAD, C., JR.; ET AL.: Geology of the Apollo 12 Landing Site. Sec. 10 of Apollo 12 Preliminary Science Report, Pt. A, NASA SP–235, 1970.

ACKNOWLEDGMENTS

Appreciation is extended for the discussions and manuscript reviews by W. R. Muehlberger, L. T. Silver, and V. L. Freeman. Also, gratitude is expressed to the Technical Support Unit personnel of the Center of Astrogeology, U.S. Geological Survey, in preparing the illustrations, and to B. S. Reed, R. L. Tyner, L. B. Sowers, and J. S. Loman for their support. This report was prepared under NASA contract T–6523G.

4. Soil Mechanics Experiment

J. K. Mitchell,[a][†] *L. G. Bromwell,*[b] *W. D. Carrier III,*[c] *N. C. Costes,*[d]
and R. F. Scott[e]

The objectives of the Apollo 14 soil mechanics experiment are (1) to obtain data on the compositional, textural, and mechanical properties of lunar soils and the variations of these properties with depth and lateral displacement at and among the three Apollo landing sites; and (2) to use these data to formulate, verify, or modify theories of lunar history and lunar processes; develop information that may aid in the interpretation of data obtained from other surface activities or experiments (e.g., lunar field geology, passive and active seismic experiments, and modularized equipment transporter (MET) performance); and develop lunar-surface models to aid in the solution of engineering problems associated with future lunar exploration. The in situ characteristics of the unconsolidated lunar-surface materials can provide an invaluable record of the past influences of time, stress, and environment on the Moon. Of particular importance are such characteristics as particle size, particle shape, and particle-size distribution, density, strength, and compressibility.

The soil mechanics experiment relies heavily on observational data such as are provided by photography, astronaut commentary, and examination of returned lunar samples. Quantitative data sources are limited; however, semiquantitative analyses are possible, as shown in reference 4–1 for Apollo 11 and in reference 4–2 for Apollo 12.

Such analyses are strengthened through terrestrial simulation studies[1] (ref. 4–3).

The results of the Apollo 11 and 12 missions have generally confirmed the lunar-surface soil model developed by Scott and Roberson (ref. 4–4); that is, the lunar soil is a predominantly silty fine sand, is generally gray-brown in color, and exhibits a slight cohesion. Evidence of both compressible and incompressible deformation has been observed. The lunar soil erodes under the action of the lunar module (LM) descent-engine exhaust during lunar landing, kicks up easily under foot, and tends to adhere to most objects with which it comes into contact. The value (or range in values) of the in situ bulk density of the lunar soil remains uncertain, although Apollo 11 and 12 core tube data and core tube simulations[2] give a best estimate of 1.7 to 1.9 g/cm^3. Limited direct evidence before the Apollo 14 mission suggested that some increase in soil strength and density occurs with depth beneath the lunar surface.

Observations at five Surveyor landing sites and the two previous Apollo landing sites indicate relatively little variation in surface soil conditions with location. Core tube samples from the Apollo 12 mission exhibited a greater variation in grain-

[a] University of California at Berkeley.
[b] Massachusetts Institute of Technology.
[c] NASA Manned Spacecraft Center.
[d] NASA Marshall Space Flight Center.
[e] California Institute of Technology.
[†] Principal investigator.

[1] N. C. Costes, G. T. Cohron, and D. C. Moss: Mechanical Behavior of Simulated Lunar Soils Under Varying Gravity Conditions. Proc. Apollo 12 Lunar Sci. Conf. (Houston), Jan. 11–14, 1971. To be published in Geochim. Cosmochim. Acta.
[2] W. David Carrier III, Stewart W. Johnson, Richard A. Werner, and Ralf Schmidt: Disturbance in Samples Recovered With the Apollo Core Tubes. Proc. Apollo 12 Lunar Sci. Conf. (Houston), Jan. 11–14, 1971. To be published in Geochim. Cosmochim. Acta.

size distribution with depth than had been found for Apollo 11 core tube samples.

The Apollo 14 mission has provided a greater amount of soil mechanics data than either of the previous missions for two reasons. The crew covered a much greater distance during the extra-vehicular activity (EVA) than in previous missions, and the Fra Mauro landing site represented a topographically and geologically different region of the Moon. In addition, three features of particular interest to the soil mechanics experiment were new to the Apollo 14 mission—the Apollo simple penetrometer (ASP), the soil mechanics trench, and the MET. Each of these has been used to shed new light on lunar soil characteristics, and each is discussed in detail in this section.

Although the analyses and results presented in this report are still preliminary in nature, certain conclusions are already apparent.

(1) At the Apollo 14 landing site, a greater variation in soil characteristics exists laterally and within the upper few tens of centimeters than had been previously encountered.

(2) Although the lunar-surface material (uniform gray, fine silty sand) appears and behaves about the same at all locations, much coarser material (medium- to coarse-sand size) may be encountered at depths of only a few centimeters.

(3) Core tube penetrations, measurements with the ASP, and analyses of the interaction between the MET and the lunar surface have been useful for estimating soil properties and, in conjunction with the observations at the soil mechanics trench, for establishing that the lunar-surface soil strength increases with depth.

(4) In this report, as in previous reports, the lunar soil properties have been derived from estimates of penetration and force. The variation of soil properties indicated by the Apollo 14 mission reinforces the need for more quantitative and definitive measurements.

Descent and Landing

The Apollo 14 LM descended more steeply in the final stages of lunar approach than did the Apollo 11 and 12 spacecraft, although the descent profile was similar to that of Apollo 12 (ref. 4–2).

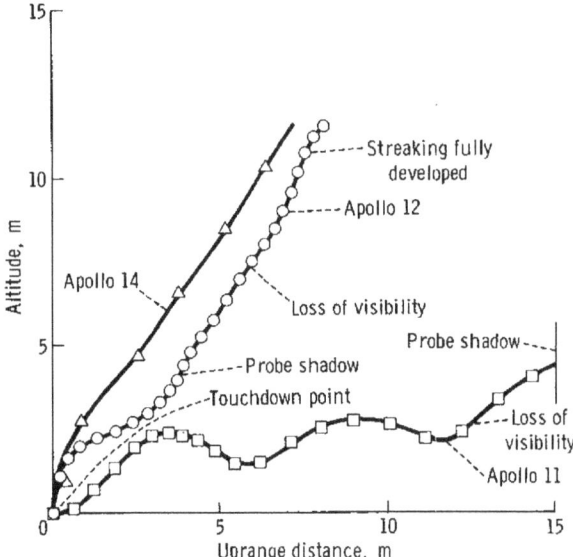

FIGURE 4–1.—A comparison of the final-approach profiles of the Apollo 11, 12, and 14 LM's.

A comparison of the final-approach profiles of all three spacecraft is presented in figure 4–1.

The final stage of the Apollo 14 descent proceeded from a pause of approximately 20 sec at an altitude of approximately 55 m above the lunar surface, while the spacecraft moved westward approximately 120 m until it was almost above North Triplet Crater. The altitude then decreased to approximately 30 m in the next 30 sec, as the LM continued another 120 m westward. The final approach of 35 sec took place at almost a 45° angle to the lunar surface. When the footpad probes made contact, the spacecraft was moving slightly west of north, according to the marks made on the surface by the probes (fig. 4–2). The landing was soft, and the astronauts reported little or no stroking of the shock absorbers. Approximately 2 sec after footpad touchdown, the descent engine was shut down. When the spacecraft came to rest, the $+Z$ leg (on which the ladder is mounted) was oriented approximately west 16° north, and the spacecraft was tipped forward in this direction (pitch) approximately 2°. At right angles to this direction, the LM was tilted down to the north, or in the $+Y$ direction (roll), approximately 7°.

FIGURE 4-2.—Position of $+Y$ footpad embedded in the rim of a 2-m-diameter crater. The track of the contact probe on the surface appears somewhat east of south (AS14–66–9258).

The astronauts commented that blowing dust was first observed at an altitude of approximately 33 m and that the quantity of dust from that altitude down to the surface seemed less than had been encountered during the Apollo 11 and 12 landings. The dust apparently caused no visibility difficulties for the Apollo 14 crew. The Sun angle at landing was higher for the Apollo 14 mission than it had been for the Apollo 12 landing. A comparison of the descent motion pictures confirms the astronaut observations, in that the appearance of the blowing lunar-surface material during the Apollo 14 descent seems qualitatively similar to that observed during the Apollo 11 landing. Dust was first observed at altitudes of 24, 33, and 33 m for the Apollo 11, 12, and 14 landings, respectively. These observations occurred 65, 52, and 44 sec before touchdown, respectively. Because of the effect of Sun angle and spacecraft orientation, however, the appearance of the dust in the motion pictures may not be a reliable indication of the quantity of material removed.

After the landing, the astronauts reported that the lunar surface gave evidence of the greatest erosion in an area approximately 1 m southeast of the region below the engine nozzle. They noted that as much as 10 cm of surface material may have been removed during the landing. A distinct erosional pattern is visible in figure 4–3, which shows the area below the descent-engine nozzle. Except for a disturbed area in the left middle distance, the surface gives the appearance of having been swept by engine gases in the same way as had occurred on previous missions. The disturbed area may have developed as a consequence of a grazing contact of the $+Y$ footpad contact probe during the landing. It was noted in a previous report (ref. 4–1) that such a disturbance to the lunar surface breaks up the surface material and renders it more susceptible to engine-exhaust erosion.

In the Apollo 14 descent motion pictures, it is evident that the lunar surface remains indistinct for a number of seconds after descent-engine shutdown. This event was probably caused by venting from the soil of the exhaust gas stored in the voids of the lunar material during the final stages of descent. The outflowing gas carries with it fine soil particles that obscure the surface.

FIGURE 4–3.—Area below the descent-engine nozzle showing erosional features caused by the exhaust gas. The $-Y$ footpad can be seen in the distance (AS14–66–9262).

Some of the tilting of the spacecraft can be explained by an examination of the footpads. The −Y (fig. 4–3) and −Z footpads have penetrated the surface only to a depth of 2 to 4 cm, whereas the +Y (fig. 4–2) and +Z footpads have penetrated to a depth of 15 to 20 cm. The +Y footpad penetration mechanism is clearly visible in figure 4–4; the footpad contacted and plowed into the rim of a 2-m-diameter crater. The motion of the footpad through the soil caused a buildup of a mound of soil on the north side of the pad. The height of the mound is somewhat higher than the actual penetration depth of the footpad. The astronauts reported that the +Z footpad, which is in the shadow of the LM, also landed on the rim of a small crater and exhibited appearance and penetration characteristics similar to the appearance and penetration characteristics of the +Y footpad. The mechanical properties of the soil, which are inferred from the response of the soil to the landing (which occurred with little or no shock-absorber stroking) and from the appearance of the soil in the footpad photographs, appear to be similar to the mechanical properties of the lunar material on which the Apollo 11 and 12 LM's landed.

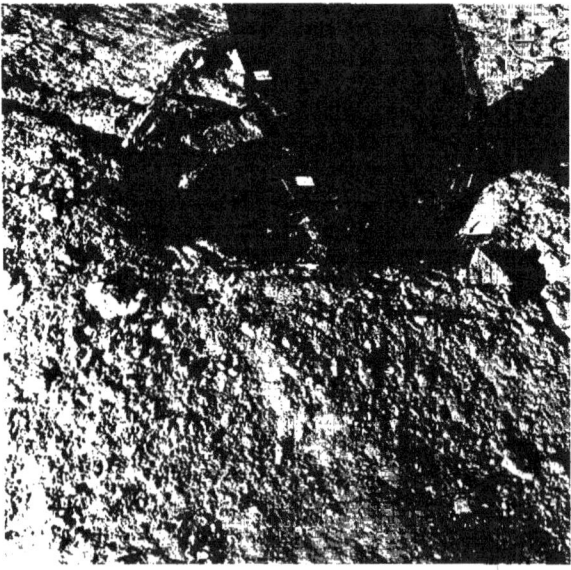

FIGURE 4–4.—The +Y footpad embedded in the lunar soil. The gold foil surrounding the landing leg is probably the protective wrapping on the MET (AS14–66–9234).

The penetration of the +Z and +Y footpads caused 1° to 1.5° of LM tilting in the westerly and northerly directions. Consequently, at the landing site, the strike of the lunar-surface slope is approximately west 16° north, and the dip is approximately 5.5° in the direction north 16° east.

Extravehicular Activities

General Observations

The behavior of the surface soil, as observed during the two EVA periods, was in many respects similar to that observed during earlier missions. The soil could be kicked up easily during walking but would also compress underfoot. Footprints ranged in depth from 1 to 2 cm on level ground to 10 cm on the rims of fresh, small craters. The MET tracks averaged 2 cm and ranged up to 8 cm in depth. Soil conditions evidently had less influence on the mobility of the astronauts and the MET than did the topography.

Patterned ground ("raindrop" pattern) was fairly general, except near the top of Cone Crater. The factors responsible for the development of this surface texture are not yet known, although the texture is probably related to the impact of small particles on the lunar surface.

As has been observed during previous missions, disturbed areas appear darker than undisturbed areas as may be seen in the background of figure 4–5. Smoothed and compressed areas (e.g., MET tracks and astronauts' footprints) are brighter at some Sun angles, as shown in figure 4–6. In some instances, it was difficult for the astronauts to distinguish between small, dust-covered rocks and clumps or clods of soil. The tops of many of the large rocks were free of dust, although fillets of soil were common around the bottom. The astronauts commented that the major part of most large rocks appeared to be buried beneath the surrounding lunar surface. They also noted no obvious evidence of natural soil-slope failure on crater walls.

Adhesive and Cohesive Behavior

As on previous missions, dust was kicked up by walking. Objects brought into contact with the

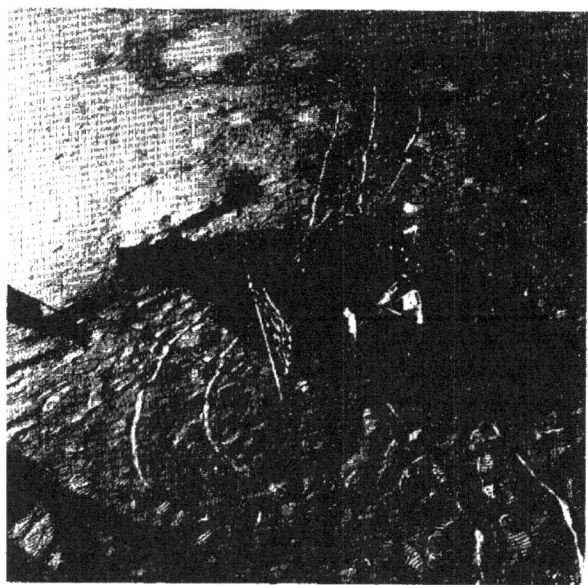

FIGURE 4-5.—The MET tracks in the immediate vicinity of the LM. The track depth in the soft spots in the foreground and along the rim of the small crater in the background should be noted (AS14-66-9325).

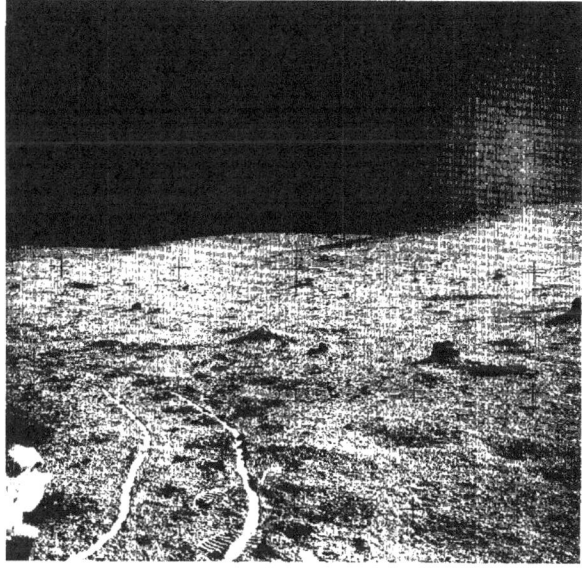

FIGURE 4-6.—The MET tracks on relatively level, firm soil in the vicinity of station B during EVA-2 (AS14-64-9058).

lunar surface or with flying dust tended to become coated, although the layer generally remained thin. The astronauts reported that the MET

sprayed dust around; however, thick layers did not form on any of the component parts, and no "rooster tail" dust plume was noted behind the MET wheels.

Although dust adhered readily to the astronauts' suits, it could be brushed off easily, except for that part that had been "rubbed in" the fabric. Similarly, dust sprinkled onto the thermal degradation sample (TDS) array was easily brushed off, although that portion of the dust that filled in the number depressions of the TDS tended to cohere, as may be seen in figure 4-7. In this case, the dust has been formed into the pattern of the numbers shown; then, when the TDS was tapped, the dust remained intact and bounced out of the depression.

Adhering dust caused some difficulty during deployment of the suprathermal ion detector experiment and the lunar portable magnetometer. Overall, however, dust appears to have been much less of a problem during the Apollo 14 mission than during previous missions. Although it is possible that the dust at the Fra Mauro site is, in fact, less adhesive, it should be noted that greater precautions were taken to minimize dust prob-

FIGURE 4-7.—The TDS array after lunar soil had been sprinkled on it; blocks 9 to 12 are visible. The array was moved slightly just before this photograph was taken, and the soil filling some of the inset numbers has jiggled out, while still retaining the shape of the numbers (AS14-77-10367).

lems on this mission. The crewmen brushed each other off before ingressing the LM, thereby lessening housekeeping difficulties on the return trip.

That the surface soil at the Fra Mauro site possesses a small cohesion is demonstrated in several ways:

(1) Clumping of soil in the vicinity of the LM footpads and elsewhere

(2) The numbers formed by dust on the TDS array

(3) The behavior of soil in astronaut footprints and MET tracks—the deformed shapes would not be retained by a soil without cohesiveness.

Evidence exists, however, that at least in some areas the cohesion is less than would be anticipated from the results of previous missions. In particular, a sample fell out of the single core tube (tube 2043) near Cone Crater, and the soil mechanics trench caved in at a considerably shallower depth than had been anticipated. As noted subsequently, the soil at relatively shallow depth (a few centimeters) is coarser than at the surface in some areas.

The source of lunar soil cohesion has not yet been determined; however, several sources are possibilities (e.g., primary valence bonding between freshly cleaved surfaces, electrostatic attraction, and bonding through adsorbed layers). In any case, it would be reasonable to expect that different minerals probably have different cohesion, that different-sized particles of the same mineral have different cohesion, and that soils of differing sizes contain different minerals. As one illustration, study of the Surveyor 3 soil mechanics surface sampler that was returned by the Apollo 12 crew has shown that small glass spheres found in the lunar soil adhered better to painted surfaces than angular fragments.

If a single mineral type and a fixed particle shape are considered, then increased cohesion with decreasing particle size can be accounted for in terms of increased numbers of interparticle contacts per unit volume. Such an interpretation, however, would require the assumption that the attractive force per contact is essentially independent of particle size or normal force at the contact. More detailed study of lunar soil cohesion is planned.

Evaluation of Lunar Soil Mechanical Properties From MET-Track Data

The MET (fig. 4–8) is a two-wheeled, ricksha-type vehicle equipped with pneumatic tires. The MET was used to carry instruments, geological tools, and photographic equipment during both periods of EVA. The maximum weight of the vehicle and the payload in lunar gravity was 128 N (28.9 lb) including a maximum payload weight of rock and soil samples of 38.7 N (8.7 lb) (lunar gravity). During the second EVA, the MET traversed a distance of approximately 2.8 km, as shown in the geologic traverse map (fig. 3–1, sec. 3). During this traverse, the total weight of the MET and the payload fluctuated between 89 and 120 N depending on the amount of sample collected.

The crew reported that the MET performed very satisfactorily. It was more stable than had been expected and could traverse the surface over a range of speeds without loss of control. No appreciable soil adhesion was noticed on the MET tires or other structural components, and only a little sideways spraying or ejection of fine-grained material was observed while the MET was pulled along the surface. The only difficulty encountered

FIGURE 4–8.—The MET during the second EVA (AS14–68–9404).

in pulling the MET developed while attempting to climb relatively steep grades. This difficulty was evidently not one of inadequate soil traction. Near Cone Crater, it was easier for both astronauts to carry the MET than for one of them to pull it uphill alone. As it rolled on level surface or downhill at relatively high speeds, the MET bounced; however, bouncing on the Moon was less than that observed on Earth in $1/6g$ simulation tests.

The MET tracks at various locations along the geologic traverse can be seen in figures 4–5, 4–6, and 4–9. The astronauts estimated the track depths to vary between 0.5 and 2 cm and noted that the largest sinkages occurred around soft crater rims. A closeup photograph of a MET track in the immediate vicinity of station A is shown in figure 4–10. The estimated track depth at this location is 1.0 cm.

The facts that (1) the MET was equipped with pneumatic tires of known geometry and load-deflection characteristics, (2) the lunar soil at the Apollo 14 site is predominantly granular, with a large percentage of grains in the sand-size range, and (3) the soil possesses only little cohesion render MET-track data amenable to the dimen-

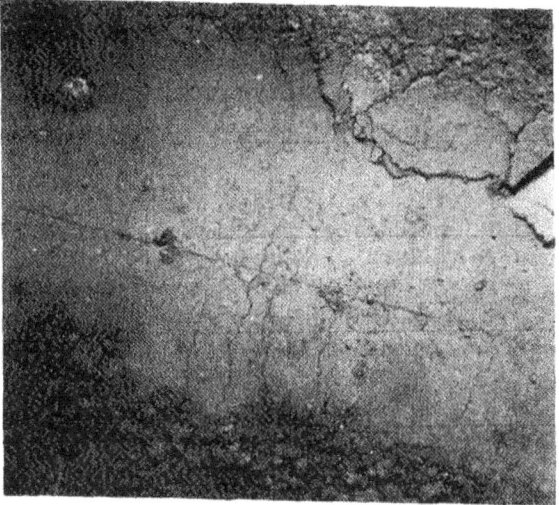

FIGURE 4–10.—The MET has evidently rolled across slightly firmer soil, because the track is narrower than before, which indicates less sinkage. In addition, there has been considerably less tendency for the soil to adhere to the tire, although fine cracks cross the track transverse to the direction of motion. A small white rock has been pushed into the soil by the MET tire in the upper left-hand corner of the photograph. The indention at the upper right-hand corner of the photograph was probably produced by the edge of the camera while it was being positioned (AS14–77–10358).

FIGURE 4–9.—The MET tracks at station C'. The track through the small hummock in the background and the segment along which the MET apparently bounced are noteworthy (AS14–64–9129).

sional analysis developed by Freitag (ref. 4–5) that relates to the performance of pneumatic tires on soft granular soils. Accordingly, if the assumption is made that the functional relationships describing the tire-soil interaction under lunar environmental conditions and low velocities are essentially the same as the relationships describing the corresponding tire-soil interaction on Earth, the normalized sinkage z/d, in which z is the penetration depth of the tire and d is the tire diameter, is related to the dimensionless quantity

$$N_{SM} = \frac{G(bd)^{3/2}}{W} \frac{\delta}{h} \qquad (4-1)$$

where

G = average change of cone penetration resistance with depth

b = tire-section width

W = tire load

δ = tire deflection

h = tire-section height

by the same functional relationship as the corresponding quantities that have been found by Freitag to relate to sand under terrestrial environmental conditions.

The dimensionless quantity N_{SM} has been termed the sand-mobility number (ref. 4–5), which relates the soil strength and density characteristics, the tire geometry and deformation characteristics, and the tire load.

Based on the previous assumptions, the average penetration-resistance gradients of the lunar soil at various locations along the geological traverse were determined on the basis of MET-track-depth data (estimated from Hasselblad camera stereopairs) and known tire characteristics by using the z/d versus N_{SM} diagram in reference 4–5. The results of this analysis are presented in table 4–I. If,

at the scale of the MET wheels, it is assumed that cohesion of the lunar soil is negligible, then the G values for the lunar soil would vary in direct proportion to the gravity level. With this assumption, G values for the lunar soil under terrestrial-gravity conditions were computed (designated as G_E in table 4–I).

The gradation of the lunar soil at the Apollo 14 landing site appears to be similar to that of an LSS used at the U.S. Army Engineer Waterways Experiment Station for lunar roving vehicle (LRV) mobility-performance simulation studies (ref. 4–6). Thus, the G_E values were used to estimate in-place soil-void ratios at corresponding locations from available G versus bulk density diagrams for the LSS (WES mix), based on the assumption that the behavior of the LSS and the

TABLE 4–I. *Lunar-Soil Mechanical Properties Evaluated*

Location	Local topography	Sinkage, z, cm	z/d[a]	N_{Sm}[b]	G_L, MN/m³ [c]
Immediate vicinity of LM..	Level, firm, soft spots.........	1.3 to 2.0	0.032 to 0.049	22 to 13	1.7 to 1.0
		3.8 to 5.1	.094 to .126	8 to 5	.63 to .40
ALSEP site.............	Level, firm...................	1.3 to 2.5	.032 to .062	21 to 11	1.7 to .87
Between LM and station A; immediate vicinity of TV camera.	Level, firm................	1.3 to 1.5	.032 to .037	21 to 17	1.7 to 1.3
A......................	Level, firm; crater rim........	.5 to 1.3	.012 to .032	60 to 21	4.7 to 1.7
		8.0	0.197	2	0.16
B......................	Level, firm...................	1.3 to 1.5	.032 to .037	21 to 17	1.7 to 1.3
B1.....................	Severe maneuver............	5.1	0.126	5	0.40
B2.....................	Level, firm...................	1.3 to 2.5	.032 to .062	21 to 11	1.7 to .87
B2.....................	Level, soft...................	2.5 to 5.1	.062 to .126	11 to 5	.87 to .40
B3.....................	Level, firm...................	1.3 to 3.8	.032 to .094	21 to 8	1.7 to .63
C'.....................	Level, firm; small hummock...	1.3 to 2.0	.032 to .049	21 to 13	1.7 to 1.0
		2.5 to 7.6	.069 to .187	11 to 2	.87 to .16
C'.....................	Level, firm; small hummock...	1.3 to 2.5	.032 to .062	21 to 11	1.7 to .87
		3.8	0.094	8	0.63
G......................	Level, firm...................	1.3 to 2.5	.032 to .062	21 to 11	1.7 to .87
Average................	Level, firm terrain...........	1.96	0.048	13	1.03
Average................	Crater rims, soft spots........	5.13	0.126	5	0.40

[a] Sinkage number (ref. 4–5).

[b] Sand-mobility number, $\dfrac{G(bd)^{3/2}}{W}\dfrac{\delta}{h}$ (ref. 4–5)

where G = penetration-resistance gradient (average rate of the cone index with depth)
  b = tire-section width, 9.65 cm
  d = tire diameter, 39.88 cm
  W = tire load, 44.48 to 62.28 N in lunar gravity
  δ = tire deflection, 0.635 to 0.864 cm for the tire load range indicated

h = tire-section height, 8.64 cm

[c] G_L = penetration-resistance gradient under lunar-gravity conditions.

[d] G_E = penetration-resistance gradient under terrestrial-gravity conditions, if the assumption is made that $G_E = 6G_L$.

[e] e_E = void ratio of lunar soil simulant (LSS) (Waterways Experiment Station (WES) mix) (specific gravity

actual lunar soil will be the same if both are at the same void ratio. These values are listed in table 4–I as e_L. The specific gravity of lunar soil at the Apollo 11 and 12 landing sites has been found to be 3.1. If the specific gravity of the soil particles throughout the Apollo 14 landing area is uniform, then the bulk density values of the lunar soil in situ would be as indicated under ρ_L in table 4–I. It may be seen that the average in situ density near the surface for level firm terrain would be 1.87 g/cm³ and on crater-rim soft spots, 1.79 g/cm³.

From the analytical procedures developed by

[3] N. C. Costes, G. T. Cohron, and D. C. Moss: Mechanical Behavior of Simulated Lunar Soils Under Varying Gravity Conditions. Proc. Apollo 12 Lunar Sci. Conf. (Houston), Jan. 11–14, 1971. To be published in Geochim. Cosmochim. Acta.

Costes,[3] the in-place angle of internal friction was estimated by using the relation

$$\frac{G}{\rho g} = N_q \qquad (4\text{--}2)$$

where

g = acceleration of gravity

N_q = bearing-capacity factor (ref. 4–7) that depends on the angle of internal friction ϕ of the soil when G and ρ are as defined previously

Values of the friction angle ϕ have been determined from knowledge of N_q obtained by using equation (4–2) and are listed in table 4–I as ϕ_D and ϕ_S. Those values designated as ϕ_S correspond to N_q values for rough-surfaced cones at shallow depths; whereas those denoted by ϕ_D correspond

From MET-Track Data Obtained During Apollo 14 EVA

G_E, MN/m³ [d]	$e_E = e_L$ [f]	ρ_L, g/cm³ [g]	$\dfrac{G_L}{\rho_L g_L}$ [h]	ϕ_D, deg [i]	ϕ_S, deg [i]	Stereopairs
10 to 6.2	0.63 to 0.66	1.90 to 1.87	547 to 327	38.8 to 36.8	45.9 to 43.8	66–9324, 25
3.8 to 2.4	.69 to .73	1.83 to 1.79	210 to 137	34.9 to 33.0	41.8 to 39.8	66–9324, 25
10 to 5.2	.63 to .67	1.90 to 1.86	547 to 286	38.8 to 36.2	45.9 to 43.2	67–9375, 76
10 to 8.1	.63 to .64	1.90 to 1.89	547 to 421	38.8 to 37.8	45.9 to 44.9	Closeup, 77–10358
28 to 10	.57 to .63	1.97 to 1.90	1459 to 547	42.7 to 38.8	50.6 to 45.9	68–9406, 07
0.95	0.80	1.73	56.6	28.8	35.6	68–9406, 07
10 to 8.1	.57 to .64	1.90 to 1.89	547 to 421	38.8 to 37.8	45.9 to 44.9	64–9057, 58
2.4	0.73	1.79	137	33.0	39.8	64–9075, 76
10 to 5.2	.63 to .67	1.90 to 1.86	547 to 286	38.8 to 36.2	45.9 to 43.2	68–9423, 24
5.2 to 2.4	.67 to .73	1.86 to 1.79	286 to 137	36.2 to 33.0	43.2 to 39.8	68–9428, 29
10 to 3.8	.63 to .70	1.90 to 1.82	547 to 212	38.8 to 35.0	45.9 to 41.8	68–9442
10 to 6.2	.63 to .66	1.90 to 1.87	547 to 327	38.8 to 36.8	45.9 to 43.8	64–9128, 29
5.2 to .95	.67 to .80	1.86 to 1.73	286 to 56.6	36.2 to 28.8	43.2 to 35.6	64–9128, 29
10 to 5.2	.63 to .67	1.90 to 1.86	547 to 286	38.8 to 36.2	45.9 to 43.2	64–9125, 26
3.8	0.73	1.82	212	35.0	41.8	64–9125, 26
10 to 5.2	.63 to .67	1.90 to 1.86	547 to 286	38.8 to 26.2	45.9 to 43.2	64–9175, 76
6.16	0.66	1.87	337	36.9	43.9	
2.39	0.73	1.79	137	33.0	39.8	

of solids = 2.89); obtained by entering G_E values in the G versus e diagram for LSS (WES mix) (ref. 4–6).

[f] e_L = void ratio of the lunar soil; assumed to be the same as that of LSS (WES mix) at comparable g levels.

[k] ρ_L = estimated lunar soil bulk density, if the assumption is made that the specific gravity of soil particles at the Apollo 14 site is 3.1 (i.e., the same as that determined from lunar soil samples returned from Apollo 11 and 12 missions).

[h] G_L = acceleration of gravity on the lunar surface, 1.62 m/sec².

[i] ϕ_D, ϕ_S = in-place angle of internal friction of the lunar soil; estimated by assuming that G_L is the rate of change in bearing capacity of lunar soil with depth and by using bearing-capacity theory as applied, respectively, to rough-surfaced cones at large depths and rough-surfaced cones at shallow depths (both cones with apex angle $2\alpha = 30°$ (ref. 4–7)).

to N_q values for rough-surfaced cones at great depths. The former solution would be considered the more applicable to the present case. The values of friction angle using the shallow-cone theory are significantly higher than those estimated previously (ref. 4–4) for soil at the lunar surface. It may be noted that the relationship developed by Freitag (ref. 4–5), which relates sinkage number to sand-mobility number, was derived for cohesionless soils. It would be expected for a soil of given friction angle and some cohesion that sinkage would be less than for the same soil without cohesion. As a consequence, estimated values of sand-mobility number would be too high, with the result that estimates of density and friction angle would also be too high.

Values of friction angle (by using triaxial compression tests) were found (ref. 4–8) for an LSS to be on the order of 37° for a void ratio corresponding to the density values estimated from the MET tracks. With a modified LSS, which was developed on the basis of grain-size distributions from Apollo 11 and 12 samples, Mitchell et al. (ref. 4–3) report friction angles of up to 48° from the results of plane-strain tests.

Apollo Simple Penetrometer

During deployment of the Apollo lunar-surface experiments package (ALSEP), the geophone/thumper anchor was used as a penetrometer to obtain three two-stage penetrations into the lunar surface. After completion of these tests, the device was used to anchor the geophone cable when the cable was placed in position for the active seismic experiment. The location of the three penetration tests was at a site approximately 8 m south of the ALSEP (fig. 3–28, sec. 3). After connecting the ASP to the extension handle, the LM pilot (LMP) first pushed the ASP into the lunar soil as far as possible with one hand, called out the penetration depth, and then pushed it deeper with both hands. This procedure was repeated twice more at points located approximately 4 m apart.

A photograph of the ASP is shown in figure 4–11. The ASP may be seen in the left middle ground of figure 4–12 in position as anchor for the geophone cable. The penetrometer consists of a 68-cm-long aluminum shaft, which is 0.95 cm in diameter and has a 30° cone tip (apex angle) on one end and a connection for the extension handle on the other end. The penetration depths are noted in figure 4–11 and in table 4–II. When the LMP pushed with both hands, no difficulty was experienced in penetrating deeper into the lunar surface, which indicates that the penetrometer had not struck a rock during the one-hand portion of each test. In the first test, the LMP achieved a depth of 62 cm by using both hands

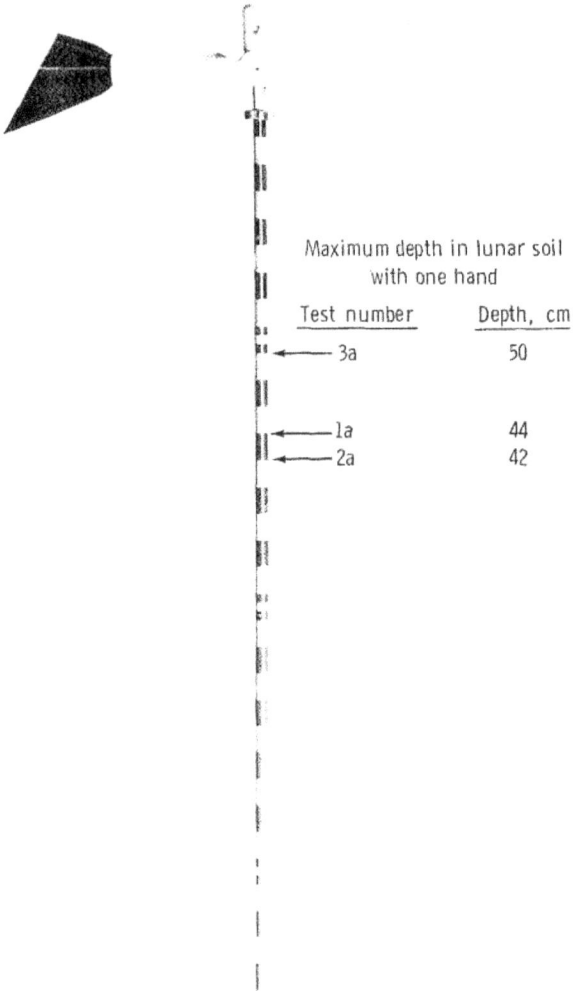

Maximum depth in lunar soil with one hand

Test number	Depth, cm
3a	50
1a	44
2a	42

FIGURE 4–11.—Apollo simple penetrometer. The LMP pushed the ASP into the lunar soil three times. The arrows indicate the maximum depths he achieved while pushing with one hand. The full black-and-white stripes are each 2 cm long and provide a depth scale (S–70–34922).

FIGURE 4–12.—The ASP, in position as anchor for the geophone cable, is shown in the left middle ground. Approximately 15 cm of the penetrometer shaft are visible above the lunar surface. The apparently greater softness of the soil in the vicinity of the small crater in the foreground, as indicated by the deep footprints, should be noted (AS14–67–9376).

able range of forces exerted on the ASP by the LMP.

The resistance of the lunar soil to penetration by the ASP consists of two components: point resistance and skin friction along the shaft

$$F = F_p + F_s \qquad (4\text{–}3)$$

where

F = lunar soil resistance
F_p = point resistance
F_s = skin friction

Both components are functions of soil strength, density, and compressibility and of frictional characteristics at the penetrometer-soil interface. Of the two types of resistance, point resistance is by far the more sensitive to variations in soil properties. A general expression for skin friction is

$$F_s = f_s \pi D z \qquad (4\text{–}4)$$

where

f_s = unit skin friction
D = shaft diameter
z = penetration depth

The unit skin friction is commonly calculated by using the equation

$$f_s = \gamma z K \tan \alpha \qquad (4\text{–}5)$$

where

γ = unit weight of the soil
K = coefficient of lateral Earth pressure
$\tan \alpha$ = coefficient of friction between the shaft and the soil

and achieved full penetration (68 cm) in the last two tests. The forces associated with the one-handed and two-handed modes shown in table 4–II were determined from simulations performed by a suited subject flying in a $1/6g$ trajectory in a KC–135 airplane. The data indicate the prob-

TABLE 4–II. *ASP Penetrations*

Test	Depth, z, cm	Mode of force application	Force, F, N	Unit penetration resistance, F/A, N/cm² [a]
1:				
a..............	44	1-handed	71 to 134	100 to 188
b..............	62	2-handed	134 to 223	188 to 314
2:				
a..............	42	1-handed	71 to 134	100 to 188
b..............	68	2-handed	<134 to 223	<188 to 314
3:				
a..............	50	1-handed	71 to 134	100 to 188
b..............	68	2-handed	<134 to 223	<188 to 314

[a] Cross-section area of the penetrometer, 0.71 cm²; all force is assumed to be carried in point bearing ($F_s = 0$).

Analysis of data in reference 4–9 for cohesion-less soils shows that actual unit skin-friction values can be as much as five times greater than values calculated by using equation (4–5) and that a ratio of unit skin friction to point bearing of 0.3 percent is the least that can reasonably be assumed. If this were the case for the ASP measurements, then a significant portion of the applied force would have been carried by shaft friction. However, the value of F_s has also been investigated by laboratory simulations and has been found to be negligible compared with the uncertainty associated with the forces exerted by the astronaut. Thus, in the following analysis, values have been determined two ways; that is, (1) $F_s = 0$ and (2) $f_s/(F_p/A) = 0.3$ percent, where A is penetrometer cross-sectional area.

The strength properties of lunar soil have so far been determined quantitatively only by means of the soil mechanics surface sampler experiment on Surveyors 3 and 7. On the basis of bearing-capacity theory for a shallow flat plate, Scott and Roberson (ref. 4–4) calculated the following parameter values:

Parameter	Value
Friction angle, ϕ, deg	37 to 35
Cohesion, c, kN/m²	0.35 to 0.70
Bulk density, ρ, g/cm³	1.5

The bearing-capacity theory for a penetrometer such as the ASP is not nearly as well developed as that for a shallow flat plate. Still, the penetrometer theory can be used to gain some insight into the shear-strength parameters of the in situ soil at the Apollo 14 landing site.

The bearing capacity of the ASP is given by

$$\frac{F_p}{A} = cN_c + \rho g z N_q \qquad (4-6)$$

where N_c and N_q are functions of ϕ, and g is the acceleration of gravity.

A unique set of values of ϕ and c cannot be obtained from this equation, because there are two unknowns. However, for each value of ϕ, there is only one value of c, and a curve of ϕ as a function of c can be calculated. This has been done (fig. 4–13) using a value of $\rho = 1.8$ g/cm³ (which has little effect on the calculations) and

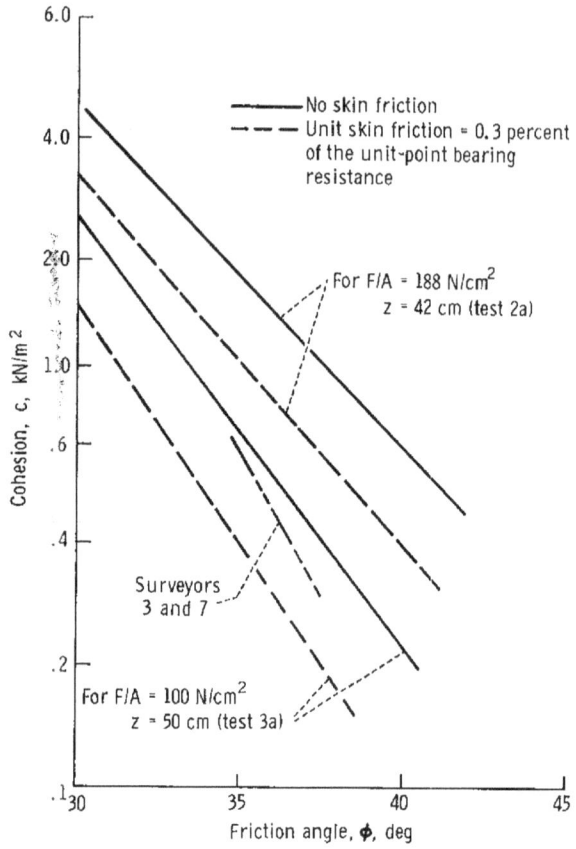

FIGURE 4–13.—Cohesion as a function of friction angle. The values of ϕ and c at the Apollo 14 ALSEP site probably lie within the indicated bands. The parameters were calculated from the ASP data by means of the bearing-capacity theory.

values for N_c and N_q (incompressible deformation assumed) from Meyerhof (refs. 4–7 and 4–10). Two bands of values are shown. For one band of values, the assumption has been made that all the applied force is carried in point bearing, and the other band is based on the assumption that the unit friction-to-bearing ratio is 0.3 percent.

The bands of values for ϕ as a function of c are a result of the range of depths for the three one-handed tests and of the uncertainty associated with the force applied by the astronauts (table 4–II). If the bearing-capacity theory for the penetrometer is assumed to be correct, then the set of values for ϕ and c for the in situ soil at the Apollo 14 ALSEP site must lie somewhere within these bands. Also plotted (fig. 4–13) is c as a function of ϕ for data from the Surveyor 3 and 7 sites. The band for the ASP parameters, if no skin friction is assumed, falls above and to the right of the Surveyor parameters, which implies that the values of the soil-strength parameters at the Apollo 14 landing site may be greater than those at the two Surveyor sites. However, if friction is a significant factor, then the parameter values at the different landing sites may be approximately the same. The former assumption is not unreasonable, however, because the Surveyor data are from surface tests, and the ASP data are from subsurface tests. The lunar soil strength probably increases generally with depth.

The variation of unit penetration resistance with depth in the lunar soil is probably very complex, as a result of layers and pockets in the soil with varying bulk density, grain-size distribution, cohesion, etc. The self-recording penetrometer, planned for use on the Apollo 15 and 16 missions, will return continuous force versus depth readings. It is anticipated that measured variations in penetration resistance with depth will indicate different soil depositional units. An infinite number of possible distributions of unit penetration resistance as a function of depth would fit the ASP data obtained during the Apollo 14 mission, because only endpoint values are known, although the LMP did indicate that penetration resistance increases with depth. Three simple examples labeled A (linear) and B (nonlinear) and C (bilinear) are shown in figure 4–14. These three examples demonstrate further that the data are insufficiently sensitive to reveal lateral inhomogeneities among the three penetration sites. Detailed investigation of lateral inhomogeneity will be performed on the data that are expected to be returned from the Apollo 15 and 16 missions.

Soil Mechanics Trench

At station G (fig. 3–1, sec. 3) on the return portion of the second EVA traverse, the commander (CDR) attempted as planned to excavate a 60-cm-deep trench with one vertical sidewall into the lunar surface. The purposes of this excavation were (1) to expose the in situ stratigraphy, (2) to provide a means for estimation of soil-strength parameters by using stability analyses, (3) to provide additional data on soil texture and consistency, (4) to enable sampling at depth, and (5) to determine the ease with which surface material can be excavated.

The site chosen for trenching was the western rim of a small crater approximately 6 m in diameter and 0.75 m deep. The lunar trenching tool (fig. 4–15) consisted of a 21.6-cm-long (8.5 in.) by 15.2-cm-wide (6 in.) blade oriented 90° to a handle. The tool was used like a backhoe by the CDR.

The trench site is shown before excavation in figure 4–16, where the edge of the small crater may be seen in the left of the photograph. The CDR stood in this crater and faced west while digging. It appears from the gnomon shown in figure 4–16 and in other photographs that the lunar surface slopes uphill from the crater toward the west.

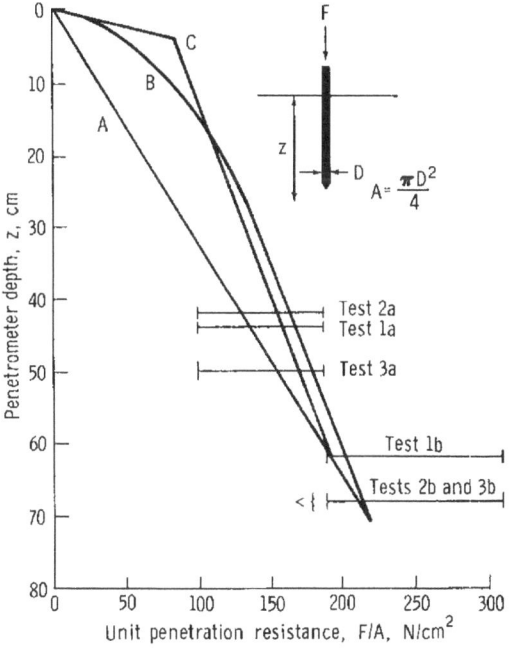

FIGURE 4–14.—Penetration resistance as a function of depth for the ASP. The horizontal bars indicate the data shown in table 4–II and indicate the uncertainty associated with the force applied by the astronaut. The curves labeled A, B, and C are examples of distributions of penetration as a function of depth that fit the ASP data.

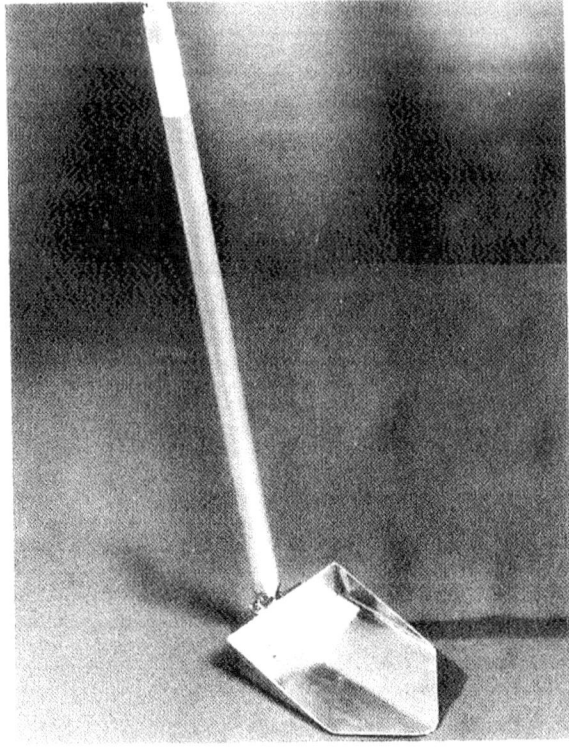

FIGURE 4–15.—Lunar trenching tool (S–70–34925).

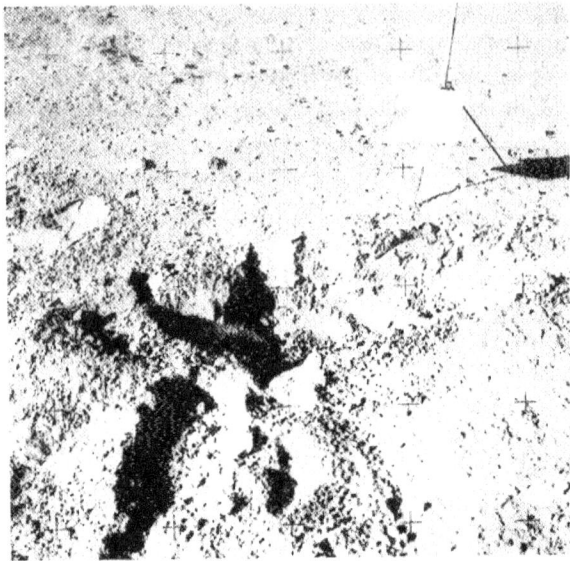

FIGURE 4–17.—View of the trench from the northeast. The steepest sidewalls are estimated at 65° to 80°. A pile of excavated material is at the end of the trench at the left in the photograph (AS14–64–9161).

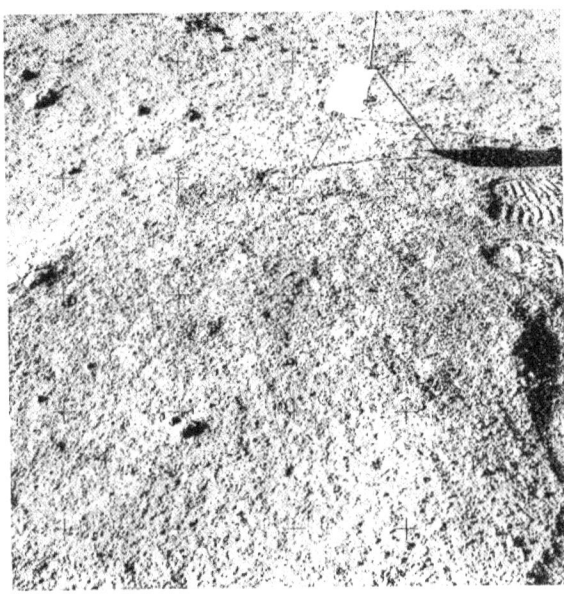

FIGURE 4–16.—Trench site before excavation. A portion of the 6-m-diameter crater in which the CDR stood while digging is visible near the left edge of the photograph (AS14–64–9158).

FIGURE 4–18.—Down-Sun view of the trench. The steep-walled side of the trench is not visible because of the bright Sun (AS14–64–9166).

A view of the completed trench from the northeast is shown in figure 4–17, from the east (down-Sun) in figure 4–18, and from the southwest in figure 4–19. A plan view of the trenching site is shown in figure 4–20. The steepest trench wall is visible in figure 4–17. The CDR reported that digging was easy and estimated his first cut to be a depth of approximately 15 cm with sidewalls at an angle of 70° to 80°. With the next cut, the walls were steepened to 80° to 85°, but, at that point, they started caving in. It is not clear from the various transcripts and debriefings just how steep the walls remained overall; but it appears that, for the remainder of the excavation, slopes steeper than approximately 60° could not be maintained.

The excavation passed through three distinct layers. The upper 3 to 5 cm were dark brown and fine grained. Next, a very thin layer (0.5 cm thick or less) of black, glassy particles was encountered. Beneath this layer was a much lighter colored and coarser grained material. This stratification is not readily visible in figures 4–17 to 4–19; however, the difference in grain size is marked (as seen by the distribution curves for samples from the upper and lower layers in fig. 4–21).

Based on previous estimates of lunar soil friction angle and cohesion at other sites, it was anticipated that trenching to a depth of 60 cm with near-vertical sidewalls should be possible without difficulty. If one assumes conservatively a homogeneous soil with a density of 1.9 g/cm³, a cohe-

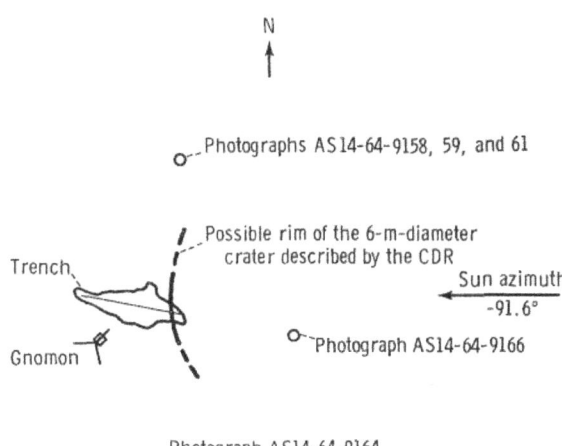

Scale: 1 cm = 0.5 m

FIGURE 4–20.—Plan view of the trench area.

sion of 0.35 kN/cm², and a friction angle of 35°, then a stability analysis indicates that a vertical cut should be possible to a depth of 85 cm before sidewall failure develops. Although the CDR estimated a maximum trench depth of 45 cm, analysis of the Apollo 14 photographs shows a maximum trench depth in the range of 25 to 36 cm, which depends on the sidewall slope angle (fig. 4–22). These estimates are based on the known dimensions of the gnomon and camera photographs. (It should be noted in this regard that the CDR stopped digging not because he could not go deeper, but because the constant crumbling of the trench walls made the retrieval of an uncontaminated sample from the trench bottom impossible.)

Combinations of cohesion c, density γ, slope height H, slope angle β, and friction angle ϕ that correspond to incipient failure are shown in figure 4–23. These curves are based on the assumption that the curved (circular) failure surfaces pass through the toe of the slope and that the soil is homogeneous. A band of values is shown in figure 4–23 that probably encompasses the conditions in the Apollo 14 trench.

For the slope angles and corresponding trench depths shown in figure 4–22, an assumed unit weight of 3 kN/m³ in the lunar gravity field, and

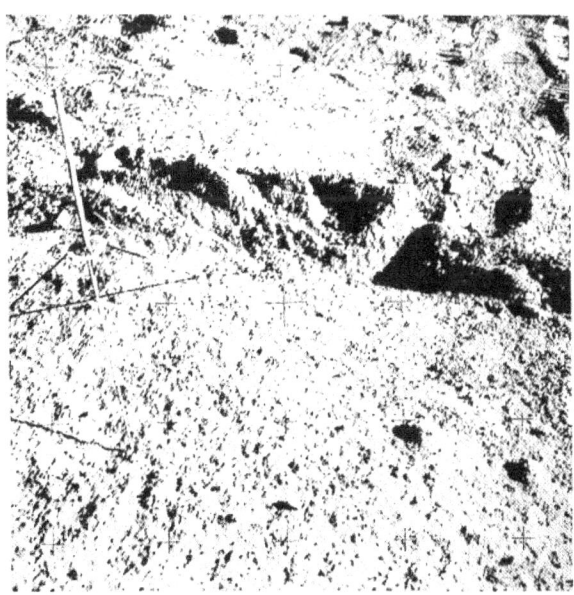

FIGURE 4–19.—View of the trench from the southwest (AS14–64–9164).

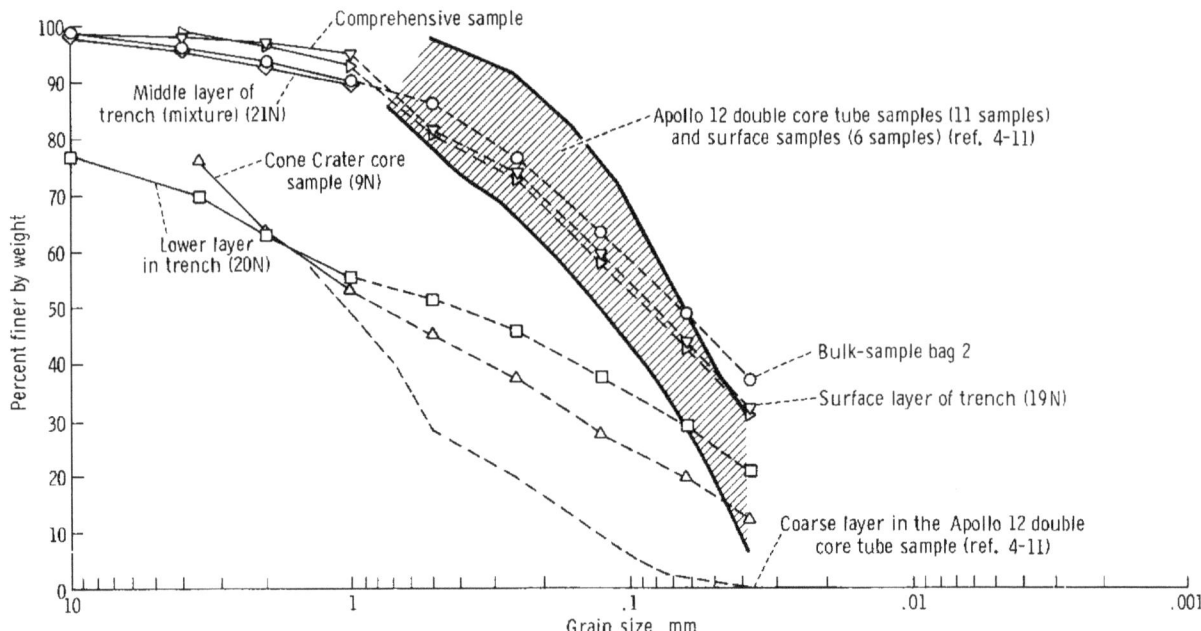

FIGURE 4–21.—Lunar soil grain-size distributions for several Apollo 12 and 14 samples. Apollo 14 sieve analyses were performed by the Lunar Sample Preliminary Examination Team. The solid portions of the Apollo 14 curves are for large samples (50 to 8000 g), and the dashed portions are for small (<1 mm) subsamples (0.036 to 0.113 g) of the large samples.

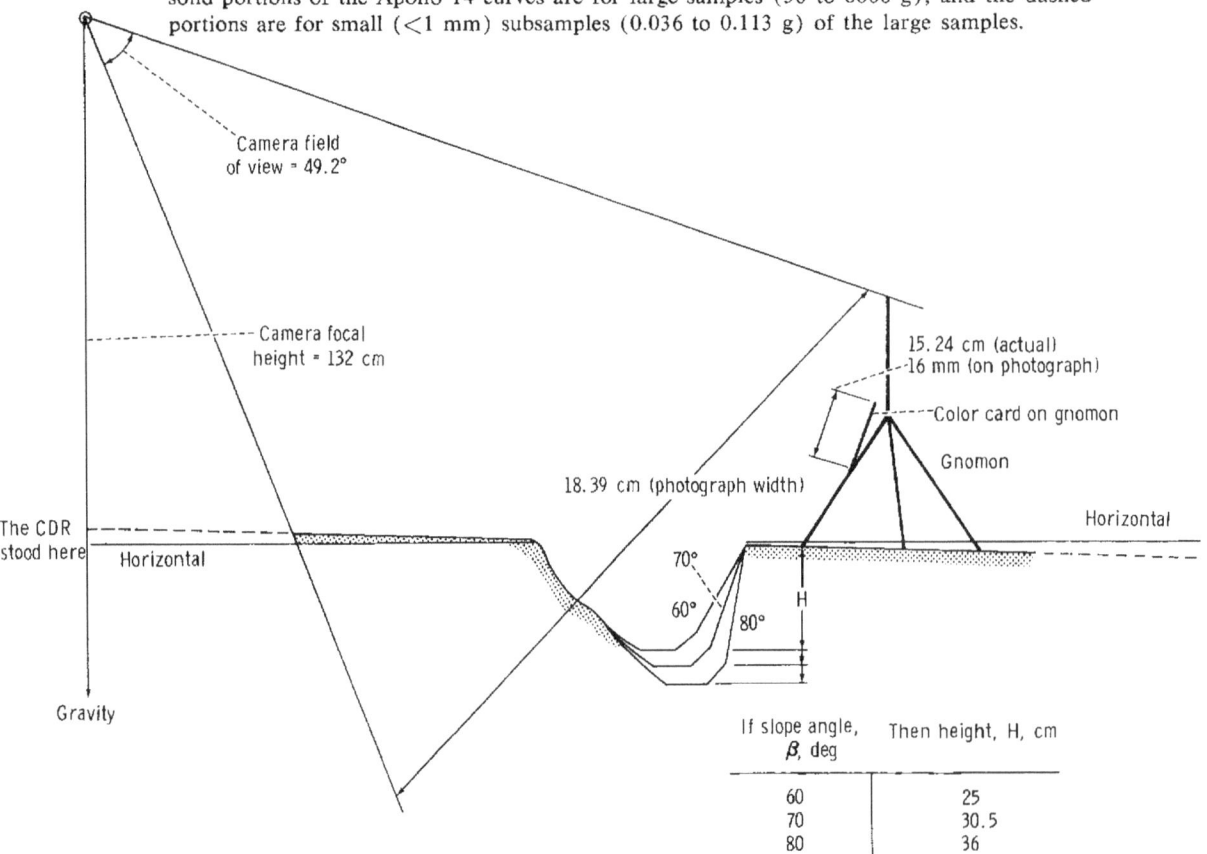

FIGURE 4–22.—Trench cross section at the point of maximum depth.

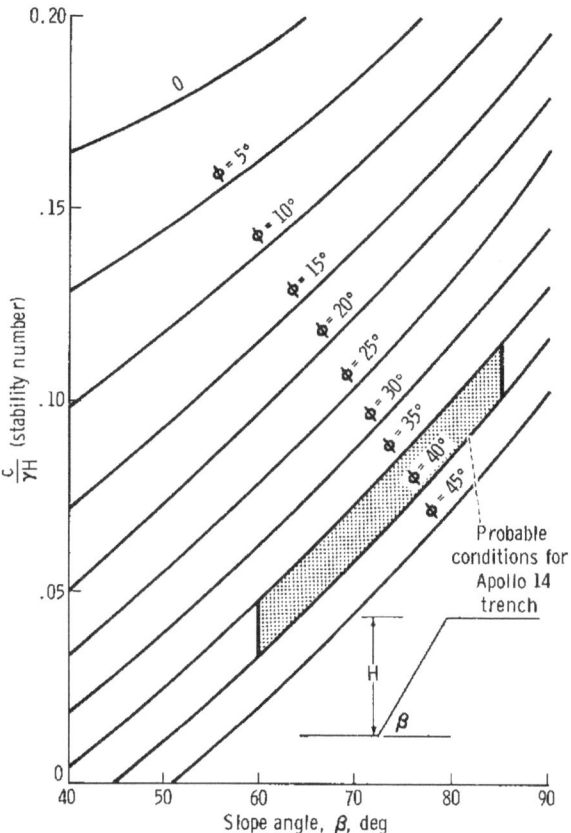

FIGURE 4-23.—Stability numbers for homogeneous slopes.

an assumed friction angle of 35°, the required cohesion values for stability fall in the range of approximately 0.03 to 0.10 kN/m². These values are only on the order of 10 percent of the values estimated using previous lunar-surface data. If the higher values of friction angle (40° to 45°) deduced from the MET-track data are representative of the soil in the trench, then the values of cohesion required are even less. It should be noted, however, that the computed values of cohesion are lower bound estimates; that is, they are the minimum values required to maintain stability for the conditions shown. Nonetheless, it does appear that the soil beneath the surface at the location of the Apollo 14 trench is less cohesive than would have been anticipated on the basis of previous data. This lower value for cohesion is compatible with the cohesion of the coarser grain sizes encountered.

In Situ Particle Size Versus Frequency Distribution

A preliminary analysis has been made to compare behavior during surface penetration and trenching with that predicted based on the Apollo 11 and 12 particle-size versus frequency distribution curves deduced in references 4-12 and 4-13. For the analysis, the assumptions have been made that particle-size distribution does not vary with depth and that the volumetric ratio of variously sized particles equals the area ratio observed at the surface or,

$$\frac{A_i}{A_T} = \frac{V_i}{V_T} \qquad (4\text{-}7)$$

where

A_i = area of particles of size i in the surface sample
A_T = total surface area of the sample
V_i = volume of particles of size i in the sample
V_T = total volume of the sample

Sufficient data are not available at present to determine the probable variations in size versus frequency distribution with depth. The soil mechanics trench provides a possible means of determining vertical variations in particle size at selected sites. The heat flow experiment on the Apollo 15 mission should also provide such information, provided that excessive disturbance does not occur during the drilling operation.

A computer program has been written to calculate the total area of influence of all particles large enough to interfere with a penetrating probe. The influence area is given by the following equation

$$A_I = \sum_{i=i_c}^{i=n} \frac{\pi}{4} \left(D_i + D_p \right)^2 N_i \qquad (4\text{-}8)$$

where

A_I = total influence area of all particles larger than i_c
i_c = smallest particle size that will stop penetration
D_i = diameter of particle i
D_p = diameter of the probe
N_i = number of particles in size range i (850 size ranges used to describe the Apollo

12 size versus frequency distribution curve)

The ratio of this influence area to the total sample area represents the probability of striking a critical particle in a unit volume. The analysis is conducted to any desired depth and the cumulative probability of not encountering a critical obstructing particle is calculated.

The probability analysis was used to determine the likelihood that the three ASP penetrations could be made to the full depth of the instrument (68 cm) without encountering a rock equal to or larger than the diameter of the instrument (0.95 cm). The probability for the three events was calculated to be 51 percent. (The probability of a single ASP penetration not encountering a sizable obstruction is approximately 80 percent.)

The four core tube events were analyzed in a similar manner. The combined probability of the four tubes being driven to their respective depths without encountering a particle equal to or greater than the core tube diameter (1.95 cm) was 67 percent. In actual fact, one of the core tubes (tube 2022) hit an obstruction. The probability that one or more core tubes would hit an obstruction was only 33 percent.

This type of analysis using the size versus frequency distribution to predict and to interpret the success of subsurface probings offers promise for future missions, which will include instrumented penetrometer experiments and the heat flow experiment with the associated lunar-surface drill. To make the analysis more effective, an evaluation of the size versus frequency distribution should be made at each landing site. In addition, detailed study of trenches and deep core samples may give a more accurate model of the size versus frequency distribution with depth.

The Apollo 11 and 12 size versus frequency distribution curves were also used to predict the number of particles of a given size that should be found in the soil mechanics trench. The results indicate that only a few particles larger than 1 cm would have been expected in the excavated soil. However, the CDR reported many rocks in the 1- to 2-cm range, and numerous particles of this size can be observed in the photographs of the trench. Furthermore, the grain-size distributions deter-

mined in the Lunar Receiving Laboratory (LRL) for material from the bottom of the trench (fig. 4–23) showed more than 20 percent of the sample to be coarser than 1 cm. This is strong evidence that the particle size versus frequency distribution counts made at the surface of the Apollo 11 and 12 sites are not good indicators of the size distributions likely to occur with depth at other sites.

LRL Observations

More than 13 kg of lunar soil were returned in two bags (inside the first sample return container) representing the bulk and comprehensive samples. In addition, two single and one double core tube samples were obtained, as listed in table 4–III, and bag samples were returned from the soil mechanics trench and the material brought up but not retained in a core tube driven in the vicinity of Cone Crater. Although laboratory study of these samples is still in a preliminary stage, some observations relative to the physical and mechanical characteristics of the lunar soil are possible.

General Characteristics

The soil from the Fra Mauro site is generally lighter in color than that at the Apollo 11 and 12 sites (except for the very light sample from the Apollo 12 site). The color is more brown than black, probably because there is more plagioclase in the Apollo 14 soil than in earlier samples. More variation in color exists from sample to sample than was observed in the samples from the Apollo 11 and 12 sites. Thus far, a significant proportion of glass beads has not been found.

The Apollo 14 bulk sample appears to be less cohesive than the soil from previous Apollo sites, even after sieving through a 1-mm sieve to remove the coarse fraction. Whether this lesser cohesion reflects a coarser grain-size distribution, different composition, or the effects of prolonged exposure to atmosphere is not yet known.

Grain-Size Distribution

Grain-size distributions thus far obtained are shown in figure 4–21. Ranges in grain-size distribution for Apollo 12 samples are shown for com-

TABLE 4–III. *Preliminary Data on Apollo 14 Core Tube Samples*

Core tube no.	LRL sample no.	Returned sample weight, g	Returned sample length, cm	Bulk density, g/cm³	Core tube depth, cm
2045 [a]	14211	39.5	....	....	13;[b] 64 [c]
2044 [a]	14210	169.7	31.9	1.75	15; [b] <36 [c]
2022	14220	80.7	....	....	23
2043 [d]	14230	76.0	....	....	45

[a] Double core tube.
[b] Depth before final driving.
[c] Crew estimates (no photograph taken).

[d] This core tube was driven twice: First, on Cone Crater, where some or all of the sample fell out; and, second, at North Triplet Crater during the second attempt at a triple core, where some of the sample fell out.

parative purposes. It can be seen that the bulk sample, which is composed of surface and near-surface material, and the surface layer from the trench have particle sizes comparable to that of the Apollo 12 samples. However, the material that fell out of the core tubes near Cone Crater and the material from the lower layer of the trench are considerably coarser, although still somewhat finer than the coarse layer found in the Apollo 12 double core tube (ref. 4–11). The granularity of the sample from near Cone Crater could account for the fact that it was not retained in the core tube.

Core Tube Samples

In terms of sample recovery and quality, the Apollo 14 core tubes are poorer than those returned from the Apollo 12 mission. During the first part of the second EVA, the CDR obtained a double core tube (tubes 2045 and 2044, table 4–III) at station A (fig. 4–24). Near the rim of Cone Crater at station C′, his attempt to take a single core tube (tube 2043) in a rock-strewn area failed when the soil sample (some or all) fell out of the core tube. Later at Triplet Crater (station G), two attempts to recover a triple core sample were thwarted, and the crew had to settle for two single core tube samples. In the first attempt, the LMP reported that he had struck rock and had been unable to penetrate deeper than one core tube. Examination of the bit from this core tube (tube 2022) confirmed this, because the edge of the stainless-steel bit was visibly dented and

burred. In the second attempt, the LMP moved approximately 9 m north of the previous location to a site that appeared similar. Here, he was able to drive the tube (tube 2043, the same tube used on Cone Crater) deeper than before (one and a half tubes), but he was still unable to achieve a triple core. Furthermore, when the LMP was disassembling the core tubes, part of the sample

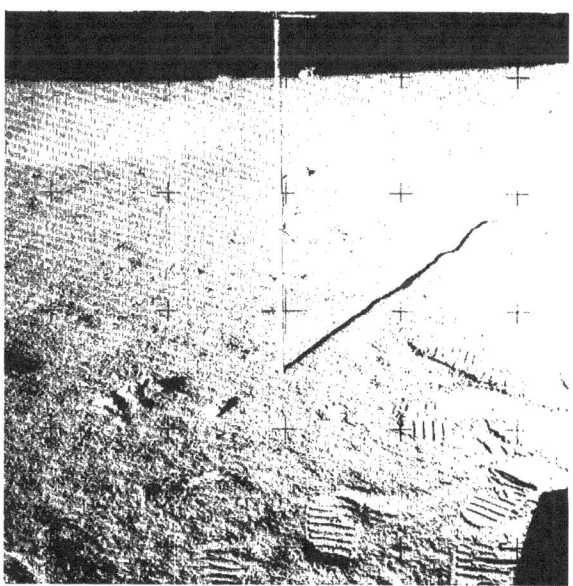

FIGURE 4–24.—Double core tube. The double core tube sample was taken at station A during the second EVA. The tube was driven to a depth of 64 cm (1⅞ core tube lengths) although when this photograph was taken, the depth was 13 cm. The LM is visible on the horizon (AS14-64-9048).

and a Teflon follower fell out from between the lower and middle core tubes.

Although the core tube samples had not been examined in detail at the time of this report, the limited data now available indicate the following:

(1) The calculated bulk density of the soil in the lower half of the double core tube (tube 2044, table 4–III) is 1.75 g/cm³. This value is significantly less than the value determined for the lower half of the Apollo 12 double core (1.96 g/cm³). This difference will be investigated further to determine if differences in the grain-size distribution or the specific gravity of the two soils (or both) can be responsible. The tare weight of the core tube used in the calculation will also be checked.

(2) The grain-size distribution for the sample from the bottom of the trench indicates a much coarser soil at depth in the vicinity of Triplet Crater, where the two triple cores were attempted. This coarser soil could account for the driving difficulties encountered in that area.

(3) The sample from the second attempt with a triple core tube (tube 2043) was severely disturbed. The angle between the tubes and the lunar surface varied significantly in the course of the driving, which reflects the LMP's efforts to drive it as deeply as possible. Not only did some of the sample fall out the back end of the tube, but some may have fallen out the front end as well; furthermore, the sample that remained was not restrained from moving in the tube, because the Teflon follower also fell out. Finally, the sample from Cone Crater may also be mixed in with the sample from Triplet Crater, because this tube was used at both locations.

(4) The double core sample offers the best chance for determining the original depth of the samples in the lunar surface by means of the method developed by Carrier et al.[4]

(5) Unfortunately, lunar-surface core tube documentation photography was incomplete for two of the four core tubes. The postsampling photograph that was intended to show the maximum depth of penetration was accidentally omitted in

the case of the double core and the first attempt at a triple core. Thus, proper interpretation of the core tube data is made difficult.

(6) Two core tubes (tubes 2023 and 2024) were left on the Moon because the crew believed, probably correctly, that they were empty. Had the tubes been returned, however, they would have been helpful in interpreting the sample in tube 2043. For example, the presence of dust on the liner or the location of the Teflon follower inside the tube might have been an indication of how much sample had entered the second core during the second attempt at the triple core. (The sample fell out when the tubes were disconnected.)

Conclusions

The following conclusions are made for the Apollo 14 mission on the basis of the observations and analyses completed thus far:

(1) Although the surface texture and appearance at the Apollo 14 landing site are similar to those at the Apollo 11 and 12 landing sites, a greater variation exists in the characteristics of the soil at shallow depths (a few centimeters) in both lateral and vertical directions than had previously been supposed. Soil with a particle-size distribution in the medium- to coarse-sand range may be encountered at shallow depths.

(2) Blowing dust apparently caused less impairment of visibility during landing than on previous missions.

(3) Evidence exists that venting from the soil voids of descent-engine exhaust gas after engine shutdown caused fine particle movement.

(4) The mechanical properties of the soil, as exhibited by response to the LM landing, appear similar to those of the lunar material on which the Apollo 11 and 12 spacecraft landed.

(5) Patterned ground (raindrop pattern) was fairly general, except near the top of Cone Crater.

(6) As had been the case in previous missions, dust was easily kicked up and tended to adhere to any surface contacted; however, overall dust was less of a problem than on previous missions.

(7) The soil at shallow depth is somewhat less cohesive than the soil observed in previous missions.

(8) Analysis of MET-track depths has indi-

[4] W. David Carrier III, Stewart W. Johnson, Richard A. Werner, and Ralf Schmidt: Disturbance in Samples Recovered With the Apollo Core Tubes. Proc. Apollo 12 Lunar Sci. Conf. (Houston), Jan. 11–14, 1971. To be published in Geochim. Cosmochim. Acta.

cated a probable in-place density of 1.8 to 1.9 g/cm³ and angles of internal friction in the range of 40° to 45° (if a cohesionless soil is assumed) for soil near the surface. The MET-track observations show that the soil is less dense, more compressible, and weaker at the rims of small craters than in level intercrater regions.

(9) Values for cone index gradient G deducted from MET-track depths average 0.4 MN/m²/m on soft crater rims and 1.0 MN/m²/m on level firm terrain. These values are compatible with those assumed for LRV studies but are significantly less than those corresponding to penetration depths of approximately 50 cm, as determined from ASP data.

(10) The results of ASP measurements suggest that the soil in the vicinity of the Apollo 14 ALSEP may be somewhat stronger than soil at the landing sites of Surveyor 3 and 7.

(11) Comparison of the results of analysis using ASP, MET-track, bootprint, and trench data shows that soil strength increases with depth.

(12) No difficulty was encountered in digging a trench into the lunar surface. Because of unexpectedly low cohesion of the soil at the trench site, the trench sidewalls caved in at somewhat smaller trench depths than had been predicted.

(13) The stratigraphy at the trench site showed a dark, fine-grained material (to a depth of 3 to 5 cm) that is underlain by a very thin glassy layer that, in turn, is underlain by a material of medium- to coarse-sand gradation.

(14) Computations of soil cohesion at depth at the trench site result in lower bound estimates considerably lower than expected (0.03 to 0.10 kN/m² as opposed to 0.35 to 0.70 kN/m²).

(15) Based on surface-particle size versus frequency distribution counts from the Apollo 11 and 12 sites, the probability of the three ASP and four core tube penetrations encountering obstructions were 51 percent and 33 percent, respectively.

(16) The surface-particle size versus frequency distribution counts from the Apollo 11 and 12 sites are not good indicators of the size distributions likely to occur with depth at other sites.

(17) Preliminary observations in the LRL have shown that (a) soil from the Fra Mauro landing site is generally lighter in color than at the Apollo 11 and 12 landing sites, (b) the bulk sample is less cohesive than the soil obtained from previous Apollo landing sites, and (c) core tube sample quality was poorer than that obtained from previous missions.

(18) The calculated bulk density of the soil in the lower half of the double core tube was 1.75 g/cm³. This density is comparable to the density estimated from MET-track data but is less than that determined for the lower half of the Apollo 12 double core tube sample.

(19) The values for different soil properties presented in this report have been deduced from estimates of force and penetration. The variation in these values, as well as the uncertainties in the analyses, reinforces the need for more quantitative and definitive measurements in subsequent missions. The self-recording penetrometer scheduled for use on the Apollo 15 and 16 missions should be an invaluable source of such quantitative soil data.

References

4–1. COSTES, N. C.; CARRIER, W. D.; MITCHELL, J. K.; AND SCOTT, R. F.: Apollo 11 Soil Mechanics Investigation. Sec. 4 of Apollo 11 Preliminary Science Report, NASA SP–214, 1969.

4–2. SCOTT, R. F.; CARRIER, W. D.; COSTES, N. C.; AND MITCHELL, J. K.: Mechanical Properties of the Lunar Regolith. Sec. 10 of Apollo 12 Preliminary Science Report, pt. C, NASA SP–235, 1970.

4–3. MITCHELL, J. K.; HOUSTON, W. W.; VINSON, T. S.; DURGUNOGLU, T.; ET AL.: Lunar Surface Engineering Properties and Stabilization of Lunar Soils. Univ. of California at Berkeley, Space Science Laboratory, NASA Contract NAS 8–21432, 1971.

4–4. SCOTT, R. F.; AND ROBERSON, F. I.: Soil Mechanics Surface Sampler. Science Results. Pt. II of Surveyor Project Final Rept., TR 32–1265, Jet Propulsion Laboratory, June 15, 1968, pp. 195–207.

4–5. FREITAG, D. R.: A Dimensional Analysis of the Performance of Pneumatic Tires on Soft Soils. Tech. Rept. 3–688, U.S. Army Engineer Waterways Experiment Station, Vicksburg, Miss., Aug. 1965.

4–6. GREEN, A. J.; AND MELZER, J. K.: Performance of Boeing-GM Wheels in a Lunar Soil Simulant (Basalt). Tech. Rept. M–70–15, U.S. Army Engineer Waterways Experiment Station, Vicksburg, Miss., 1970.

4-7. MEYERHOF, G. G.: The Ultimate Bearing Capacity of Wedge-Shaped Foundations. Proc. Int. Conf. Soil Mech. Found. Eng. 5th, vol. 2, 1961, pp. 105–109. (Also available as TM–72, Canada Research Council Associate Committee on Soil and Snow Mechanics, Jan. 1962.)

4-8. MITCHELL, J. K.; AND HOUSTON, W. W.: Lunar Surface Engineering Properties Experiment Definition. Final Rept. (June 20, 1968, to July 19, 1969), vol. 1, Mechanics and Stabilization of Lunar Soils, Univ. of California at Berkeley, Space Science Laboratory (NASA Contract NAS 8–21432), NASA CR–102963, Jan. 1970.

4-9. VESIC, A. B.: Bearing Capacity of Deep Foundations in Sand. Highway Research Board Rept. 39, National Research Council, 1963, pp. 112–153.

4-10. MEYERHOF, G. G.: Some Recent Research on the Bearing Capacity of Foundations. Can. Geotech. J., vol. 1, no. 1, Sept. 1963, pp. 16–26.

4-11. LINDSAY, J. F.: Sedimentology of Apollo 11 and 12 Lunar Soils. J. Sediment. Petrology (will appear in Sept. 1971 issue).

4-12. SHOEMAKER, E. M.; BAILEY, N. G.; BATSON, R. M.; DAHLEM, D. H.; ET AL.: Geologic Setting of the Lunar Samples Returned by the Apollo 11 Mission. Sec. 3 of Apollo 11 Preliminary Science Report, NASA SP–214, 1969.

4-13. SHOEMAKER, E. M.; BATSON, R. M.; BEAN, A. L.; CONRAD, C., JR., ET AL.: Geology of the Apollo 12 Landing Site. Sec. 10 of Apollo 12 Preliminary Science Report, pt. A, NASA SP–235, 1970.

5. Preliminary Examination of Lunar Samples

The Lunar Sample Preliminary Examination Team [a]

The surface of the Moon can be divided into the dark mare areas and the bright highland regions. The mare regions cover approximately one-third of the near side of the Moon and make up a small fraction of the far side. These mare areas are recognized as the areas of most recent widespread rock formation on the lunar surface. The first three groups of samples returned from the Moon to Earth (i.e., the samples from the Apollo 11, Apollo 12, and Luna 16 missions) all come from typical mare regions.

Detailed chemical and petrographic studies of the samples from the three widely separated mare regions show that the dark regions of the Moon are probably underlain by basaltic rocks, which are iron-rich and sodium-poor (relative to similar terrestrial rocks). Absolute ages determined for basaltic rocks from the Apollo 11 and 12 sites and for crater densities on nearby mare surfaces suggest that the final filling of most mare basins took place between 3.0 and 4.0 billion yr ago.

The stratigraphic and petrographic studies of the mare samples lead to two general inferences regarding the Moon—that the internal temperatures of at least parts of the Moon reached the melting point of basalt less than 1 billion yr after the formation of the Moon and that the evolution of most of the lunar surface took place very early in the history of the planet. Gilbert (ref. 5–1) and several other authors have suggested that one of the last major events in the evolution of the pre-mare lunar surface was the formation of the Mare Imbrium Basin and the ejection of an extensive blanket of material from this basin by the Imbrian impact. This blanket was named the "Fra Mauro Formation" after a region north of the crater

Fra Mauro where a portion of the presumed ejecta blanket appears to protrude above the surrounding mare basalt flows (ref. 5–2). On February 5 and 6, 1971, the Apollo 14 lunar-landing crew visited a site approximately in the center of this island of premare material.

FIGURE 5–1.—View of boulders near the rim of Cone Crater. The white rock (sample 14082) was chipped from the boulder in the foreground, from a point just below the end of the hammer handle (40 cm long). The large dark clasts in this boulder and other inhomogeneities in other boulders are noteworthy (AS14–68–9453).

[a] Team composition listed in "Acknowledgments" at end of this section.

The commander and the lunar module pilot landed at a site (latitude 3°40'19" S, longitude 17°27'46" W) 1230 km south of the center of the Imbrium Basin on February 5. Approximately 43 kg of rock and soil samples were returned from points on traverses that spanned a distance more than 1 km long. Further details on the regional and local geology are given in section 3 of this report. Some samples collected during the two extravehicular activity (EVA) periods were returned in two containers, which were designated Apollo lunar sample return container (ALSRC) 1007 (from the first EVA) and ALSRC 1006 (from the second EVA). Most of the larger rock and soil samples were returned in three Teflon bags, which were designated "totebags." A summary inventory of the entire collection of samples returned from the Apollo 14 mission is listed in table 5–I. The lunar locations of several of the more interesting samples are shown in figures 5–1 to 5–3. The total collection includes 33 rocks that weigh more than 50 g each (table 5–II) and approximately 30 smaller rocks that weigh from 10 to 50 g each. Four core tubes of lunar soil were collected, two of which were driven together as a double core tube.

FIGURE 5–3.—One of the large boulders in the northwest boulder field (northwest of the lunar module (LM)). Rocks that were collected by the astronauts are identified; the white clasts that are visible in the boulder are noteworthy (AS14–68–9474).

FIGURE 5–2.—Sample 14082, a fragmental rock with a light-colored matrix. The rock contains few lithic clasts and many feldspar clasts (S–71–21481).

The preliminary examination of these rocks began on February 11 and continued until March 26. This study included biological, mineralogical, chemical, and isotopic measurements that were intended to (1) provide data necessary to divide and distribute the samples for more thorough and detailed investigations and (2) determine whether the samples posed a threat to the terrestrial biosphere. Techniques used were, in general, the same as for previous missions, with the exception that no analyses for organic compounds were performed. Rocks ranging from approximately 1 g to the largest rock returned were classified by the use of low-power binocular microscopes. Limited sampling of selected rocks was made for chemical analysis and thin-section preparation, and these small samples should not necessarily be considered to be representative of the whole rock from which they originated.

All samples were processed in pure nitrogen (less than 10 parts per million (ppm) of con-

TABLE 5–I. *Apollo 14 Sample Inventory*

Material [a]	Mass, g	Container
Contingency sample: Fines	1 077.9	Teflon bag
Bulk sample:		
Fines	8 358.6	ALSRC 1007
Rocks	69.5	ALSRC 1007
Documented Samples:		
Fines	591.5	ALSRC 1006
Rocks	2 337.7	ALSRC 1006
Special environment sample container (fines?)	168	Stainless-steel can
Core tubes:		
Double core tube	209	
2 single core tubes	157	Aluminum core liner
2 core bits	11	
Documented and grab samples (from totebags):		
Fines	969.1	Teflon bag
Rocks	25 552.9	Teflon bag
Comprehensive sample:		
Fines	2 856.8	ALSRC 1007
Rocks	568.5	Teflon bag
Summary:		
Fines	13 853.9	
Rocks	28 528.6	
Special samples	545.0	
Total	42 927.5	

[a] Fines <1 cm; rocks >1 cm.

taminant gases) in stainless-steel glove cabinetry designed to provide biological containment. Contaminant levels of the gases hydrogen, oxygen, carbon monoxide (CO), carbon dioxide (CO_2), methane (CH_4), and argon (Ar) were monitored by gas chromatography and oxygen and water (H_2O) monitors. Monitors were exposed before sample processing to establish controls for particulate contaminants.

The procedures within the Lunar Receiving Laboratory (LRL), the reasons for the sample quarantine, and the tools used by the astronauts are given in reference 5–3.

Mineralogy and Petrology

Fragmental rocks with complex textural rela-tionships predominate in the Apollo 14 samples. Lesser amounts of homogeneous crystalline rocks, some with basaltic textures and some with fine-grained granulitic textures, were also returned. Most of the larger fragmental rocks are distinct from fragmental rocks returned from the Apollo 11 and 12 missions. Also, the fragmental rock to basaltic rock ratio (by number of rocks) is ap-proximately 9 to 1 in contrast to previous missions where the ratios were approximately unity (for Apollo 11) and 1 to 9 (for Apollo 12). Both the homogeneous crystalline rocks and the fine-grained granulitic rocks have their counterparts as clasts in fragmental rocks, and the possibility exists that some or all of them are clasts dislodged from fragmental rocks.

TABLE 5–II. *Apollo 14 Rock Samples Greater Than 50 g*

Sample no.	Weight, g	Location	Orientation	Comments
14041	166	Station A	Tentatively known	Surface photograph
14041	103	Station A	Tentatively known	Surface photograph
14045	65	Station A	Tentatively known	Surface photograph
14047	242	Station B	Known	Surface photograph
14049	200	Grab sample after station B	Unknown	Approximate location only
14051	191	Station C	Known	
14053	251	Station C2	Unknown	Location known
14055	111	Station E	Unknown	
14063	135	Station C1	Unknown	
14064	108	Station C1	Unknown	
14066	510	Station F	Unknown	Location known
14082	63	Station C1	Known	
14169	79	Unknown	Unknown	Location unknown
14264	118	Comprehensive-sample area	Unknown	Location known
14265	66	Comprehensive-sample area	Unknown	Location known
14267	55	Comprehensive-sample area	Unknown	Location known
14271	97	Comprehensive-sample area	Unknown	Location known
14301	1361	Station G1	Tentatively known	Partially buried rock at North Triplet Crater
14303	898	Station H (northwest boulder field)	Unknown	
14304	2499	EVA–1 football-size rock	Unknown	
14305 (14302)	2498	EVA–1 football-size rock	Unknown	
14306	585	Station G	Known	
14307	155	Station G	Unknown	Location near trench
14310	3439	Grab sample (possibly at station G)	Known	Orientation known from gamma ray
14311	3204	Grab sample	Unknown	3 additional broken pieces belonging to sample 14311
14312	299	Station H	Known	
14313	144	Station G1	Known	
14314	116	Station H	Tentatively known	
14315	115	Station H	Unknown	
14318	600	Station H	Tentatively known	
14319	212	Station H	Known	
14320	65	Station H	Unknown	
14321	8996	Station C1	Known	

Fragmental Rocks

The fragmental rocks consist of a variety of rock and mineral clasts set in matrices that range from a friable, fine-grained, clastic mass to a fine-grained, very coherent, crystalline mosaic of interlocking crystals. Relative to the breccias returned from the Apollo 11 and 12 sites, some of the rocks from the Apollo 14 site are lighter in color, more friable, and contain fewer and smaller clasts; but most are more coherent, contain less glass, and have larger and more abundant clasts. Matrices range from those with exclusively clastic textures (such a matrix termed one with a low degree of crystallinity in this report) to matrices that contain euhedral to subhedral crystals and have textures that resemble those of some metamorphic rocks (with a high degree of crystallinity). In rocks with intermediate degrees of crystallinity, some ilmenite and plagioclase grains commonly exhibit crystal faces.

An attempt was made to classify the fragmental rocks on the basis of (1) coherence, (2) percentage of clasts greater than 1 mm in diameter, (3) color of the lithic clasts, and (4) degree of matrix crystallinity. Limited thin-section study suggests that the degree of crystallinity of the matrix will prove to be a significant parameter for classification when a more extensive microscopic survey of the clastic rocks has been made. A working classification that is descriptive was developed on the basis of hand specimen study. The fragmental rocks fall into three groups, each of which is represented by approximately one-third of the rocks.

The samples in the first group are those that have friable matrices and contain approximately 5 percent clasts greater than 1 mm, among which lithic clasts predominate. In more than half of these rocks, light-colored clasts make up more than 90 percent of the total clasts larger than 1 mm; the remainder has less than 50 percent dark-colored clasts. Most of the clasts are themselves fragmental. Angular glass fragments and broken and intact glass spheres, which are mostly medium brown in color, are present. A number of the group 1 samples bear the marks of the thin aluminum foil used to wrap and protect them; they crumble very easily and resemble weakly indurated soil.

The samples in the second group are characterized by light-colored clasts (in excess of darker clasts) and moderate coherency. The abundances of clasts are high, and nearly 80 percent of these rocks have more than 5 percent of their clasts greater than 1 mm across. In approximately 60 percent of these rocks, leucocratic clasts make up more than 90 percent of the total lithic clasts. There are some clasts of basaltic rock, some of a very-fine-grained melanocratic rock, and a few monomineralic granular olivine clasts.

The samples in the third group are characterized by dark-colored lithic clasts in excess of light-colored clasts. Two subgroups of this group have moderately coherent and coherent matrices. Ninety percent of the rocks in this group have 5 percent or more of their clasts greater than 1 mm across. Most of the clasts are crystalline, aphanitic, dark-gray to black fragments with recrystallized matrices. Basalt and "dunite" clasts are present in some rocks.

Most lithic clasts are equant, but some are elongate or tabular. Rounded clasts are common, although angular clasts predominate (fig. 5–4).

FIGURE 5–4.—Sample 14321 (the largest rock returned from the Apollo 14 landing site; 9.0 kg). The rock is fragmental with a medium-gray matrix and contains abundant dark fragmental clasts. One clast is 10 cm across (S–71–22985).

Lithic clasts range in size from a few micrometers to 16 cm (sample 14321). The largest clast seen in lunar-surface photographs measures more than 1 m across. Some of the larger clasts are rimmed by bands of matrix of finer median grain size and generally lighter color than that of the bulk matrix of the rock. Lithic clasts can be classified as frag-

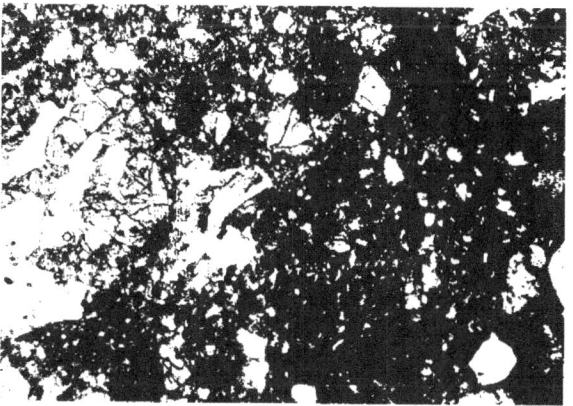

FIGURE 5–5.—Photomicrograph of a thin section of sample 14304 in which a fragmental rock that contains a large clast with basaltic texture and many mineral and fragmental clasts is shown. The scale is 3.1 mm across the field of view (S–71–25497).

mental clasts (fragments that themselves are composed of fragmental material) and homogeneous, largely crystalline clasts. Most fragmental clasts contain the same range of clast types as the host rock and are distinguishable only by differences in the matrices. The degrees of crystallinity of the clasts are equal to or greater than that of the matrix. Boundaries of clasts in a single rock may range from sharp to diffuse, both of which types of boundaries may be found in the same rock. As many as three generations of fragmental rocks within clasts have been observed.

A short description of the nonfragmental lithic clasts in order of abundance follows:

(1) Clinopyroxene-plagioclase clasts with basaltic texture are common and generally contain minor amounts of ilmenite, metallic iron, and troilite (fig. 5–5). Plagioclase grains commonly show a fine mosaic; mosaicism in pyroxene is less abundant.

(2) Feldspathic clasts, which consist principally of plagioclase with only minor amounts of other phases (pyroxene, olivine (possibly), ilmenite, metallic iron, and zircon (possibly)), are fairly abundant. Plagioclase grains are commonly enclosed by a very-fine-grained feldspathic matrix. Many plagioclase grains are deformed, and fine plagioclase mosaics are not

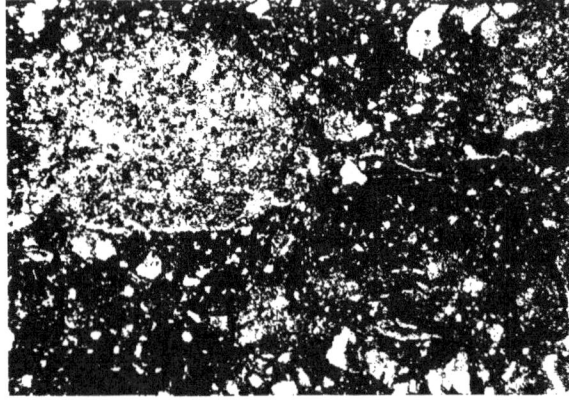

FIGURE 5–6.—Photomicrograph of a thin section of sample 14301 in which a fragmental rock that contains abundant lithic clasts is shown. The large clast is an example of the fine-grained granular feldspathic clasts. Other lithic clasts include a broken glass spherule and a dark-matrix fragmental clast. The scale is 3.1 mm across the field of view (S–71–25490).

uncommon. Some feldspar grains show no deformation and are composed of plagioclase with a granulitic texture (fig. 5–6).

(3) Plagioclase-orthopyroxene clasts with subophitic texture contain intergrown plagioclase and orthopyroxene with minor clinopyroxene, ilmenite, and iron. Euhedral plagioclase is accompanied by anhedral pyroxene, and, in some cases, the orthopyroxene is poikilitic.

(4) Olivine-glass clasts are rare and consist of skeletal olivine crystals in glass.

(5) Clasts consisting almost entirely of olivine with a granulitic texture are present in some rocks.

(6) Angular glass clasts and glass spheres are present in those rocks that do not have a crystalline matrix. The glass varies in color, amount and type of inclusions, and heterogeneity. Some glass clasts are apparently devitrified, and one type consists of relic angular plagioclase and pyroxene grains in an extremely-fine-grained groundmass that possibly contains skeletal olivine microlites. All glass clasts in the recrystallized fragmental rocks are at least partially devitrified and contain needlelike crystals, possibly of feldspar.

The types and relative abundances of mineral clasts vary widely in the several fragmental rocks that have been studied microscopically. These minerals are anorthitic plagioclase, clinopyroxene (augite and pigeonite), orthopyroxene, ilmenite, olivine, metallic iron, troilite, chromium spinel, ulvöspinel, native copper, Armalcolite, zircon, apatite, potassium (K) feldspar (possibly), and several unidentified phases. A pink, high-relief, isotropic mineral in several of the fragmental rocks may be garnet or spinel. The plagioclase-to-clinopyroxene ratio ranges very roughly from 2:1 in many fragmental rocks to approximately 5:1 in some of the lighter colored fragmental rocks. A brief description of the more abundant minerals follows.

Plagioclase occurs as single grains and commonly also as fine-grained mosaics that were probably produced from single crystals. Some maskelynite is present. A few clasts consist of an equigranular mosaic of plagioclase crystals. Clinopyroxene is abundant, with both pigeonite and augite represented. Some clinopyroxene grains are large and undeformed and contain fine, very regular exsolution lamellae of pyroxene; however, deformation and mosaic texture are common. Orthopyroxene, which may contain lamellae of clinopyroxene up to 5 μm wide, is commonly undeformed.

Olivine occurs as angular single crystals, and a few grains are subhedral. Some clasts up to 1 mm in diameter are an equigranular mosaic of olivine grains. Several olivine clasts are rimmed by an intergrowth of pyroxene and ilmenite.

Ilmenite is the most abundant opaque mineral, although it is less abundant than in previously returned lunar material. Ilmenite commonly occurs as small angular to subrounded grains, and small (10-μm) laths occur in the fragmental rocks that show some degree of crystallinity. A few large (up to 250 μm) subrounded grains occur; many of these show deformation twin lamellae.

Metallic iron is widely dispersed throughout the rocks and occurs in grains ranging from less than 1 to 250 μm in diameter. The grains range in shape from rounded to ragged and are highly irregular. The abundance of metal is commonly less than that in Apollo 12 basaltic rocks. The association of metal with troilite is less pronounced than in the Apollo 11 and 12 rocks.

Troilite is less common than in the Apollo 11 and 12 rocks and occurs as small rounded to angular grains that may contain inclusions of native iron. Chromium spinel, ulvöspinel, native copper, and Armalcolite occur as angular grains in extremely minor amounts. Most of the mineral clasts are such as might be expected from comminution of the lithic fragments. However, the sparsely occurring single-mineral fragments of clinopyroxene and orthopyroxene with exsolution lamellae and fragments of larger olivine and ilmenite grains have not been seen in lithic clasts.

Matrices of fragmental rocks vary in crystallinity (one extreme type contains abundant glass) and, in these matrices, the fragments range in size to less than 1 μm. At the other extreme, some matrices appear to be totally crystalline. In these, the matrix consists of lath-shaped plagioclase, anhedral pyroxene, platelike ilmenite, and small metal grains. The bulk of fragmental rocks appears to exhibit some degree of crystallinity, and the examples with totally crystalline matrices would be classed as crystalline rocks of fragmental origin.

Some of the fragmental rocks studied in thin sections contain dark-brown glass that occurs in two ways: as a filling material for the fractures and veins (up to 5 mm thick) that may crosscut matrices and clasts, and as a coating material for some clast surfaces. Most of the glass veins and coatings contain mineral inclusions and spherules of metallic iron, and some of the veins and coatings are vesicular. The rocks tend to break along glass-filled fractures.

Homogeneous Crystalline Rocks

The homogeneous crystalline rocks can be divided into two groups: those with basaltic textures and those that have very-fine-grained granulitic textures. The only crystalline rocks with masses greater than 50 g are samples 14053 and 14310. These basaltic rocks have typical igneous textures and are composed principally of plagioclase, brown and yellow-brown pyroxenes, olivine, and opaque minerals. These rocks differ from the basaltic rocks returned from the Apollo 11 and 12 sites in being much richer in plagioclase and poorer in ilmenite and in having much lighter colored pyroxenes.

The basaltic rocks are fine and even grained, are aphyric, and have textures that range from intersertal to intergranular to ophitic. The grain size ranges from fine to medium and is commonly coarser than that of the homogeneous nonbasaltic crystalline rocks. The basaltic rocks fall readily into two subgroups: those with approximately 40 to 45 percent plagioclase (samples 14053, 14071, and 14074) and those that have approximately 60 to 70 percent plagioclase (samples 14073, 14078, 14079, 14276, and 14310).

Sample 14310 is a fine-grained basaltic rock with scattered small cognate inclusions that are finer grained than the body of the rock (fig. 5–7). Rock 14310 has an intergranular to intersertal texture and consists of euhedral plagioclase laths and anhedral pyroxene crystals. Modal analyses indicate approximately 66 percent plagioclase, 31 percent clinopyroxene, and lesser amounts of ilmenite, troilite, metallic iron, chromium spinel, and ulvöspinel. The mesostasis, which amounts to a few volume percent, consists of complex fine-grained material, including apatite (possibly);

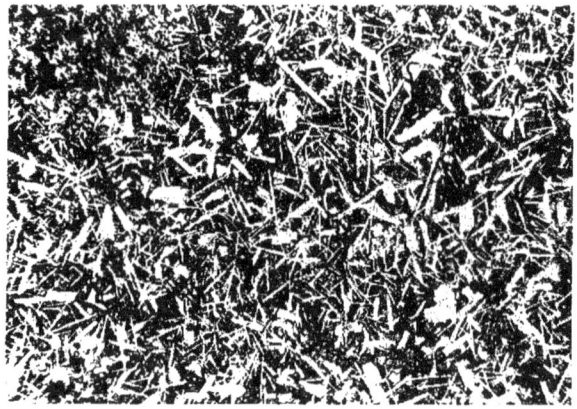

FIGURE 5–7.—Photomicrograph of a thin section of sample 14310 in which the intergranular texture of the plagioclase and pigeonite and a cognate inclusion in the lower right corner are shown. The scale is 3.1 mm across the field of view (S–71–25551).

ilmenite; troilite, which contains metallic-iron blebs; possible alkali feldspar; an unidentified orange-brown, transparent, isotropic mineral with high relief and a high index of refraction; and clear glass, which contains opaque to reddish-brown material, possibly glass spherules. This sample is of special interest because the chemistry and modal mineralogy differ from those of previously described lunar basaltic rocks. Rock 14310 resembles some terrestrial high-alumina basalts in major-element content (see tables 5–IV and 5–V later in this section), but highly calcic plagioclase (approximately An_{89}) and a predominance of pigeonite among the pyroxenes distinguish this rock from terrestrial high-alumina basalts.

Sample 14053 has an ophitic texture and is coarser grained and more mafic than sample 14310. The rock is moderately inhomogeneous and appears to be subtly layered, with modal plagioclase approximately 40 percent in one part and approximately 60 percent in the remainder. One thin section of this rock is composed of pyroxene (51 percent, mainly pigeonite), bytownitic or anorthitic plagioclase (41 percent), and a few percent olivine that is commonly present as anhedral grains in cores of pigeonite grains. Phases in minor abundance include metallic iron, ilmenite, troilite, and ulvöspinel with ilmenite lamellae along octahedral planes. Cristobalite (approximately 2 percent) occurs in vugs and is associated

with the mesostasis. The mesostasis is complex and includes spongy masses of metallic iron, dendritic ilmenite, fayalite (possibly), and vermicular mixtures of colorless and black glasses.

The fine-grained holocrystalline rocks are commonly inequigranular and contain 1 to 5 percent of irregular large mineral grains, commonly plagioclase, but less commonly olivine or pyroxene. All but two rocks in this group are light colored and consist principally of plagioclase and light-gray pyroxene. The two dark rocks (samples 14006 and 14440) are aphanitic, and the mineralogies are not yet known because no thin sections have been examined. Both the light and the dark rock types appear to be common as clasts in the fragmental rocks. Lithologic similarity to clasts and the small size support the idea that these rocks were once clasts in fragmental rocks. The paucity of dark rocks in this group is consistent with the small percentage of friable, easily disaggregated breccias in which dark clasts are dominant.

Shock Metamorphism

Shock features are best developed within individual clasts of fragmental rocks and in soil particles and are largely absent from basaltic rocks. Within fragmental rocks, the evidence for shock decreases with increasing crystallinity of the matrix. As in previous sample returns, abundant evidence of multiple shock events exists. In fact, as many as three generations of clasts within a fragmental rock may contain evidence of shock metamorphism.

Weak shock. Most lithic and mineral inclusions within the fragmental rocks are intensely fractured and shattered. Extreme mosaicism is common and is one of the most outstanding features in the fragmental rocks. No deformation structures unequivocally indicative of static deformation were observed. Thus, it is concluded that the abundant mosaicism is to a large extent, if not exclusively, caused by weak shock effects below the Hugoniot elastic limit.

Moderate shock. Shock features indicative of solid-state deformation are present in fragmental rocks and soils. These include planar features in plagioclase, pyroxene, and olivine that are associated with a decrease in refractive index, pressure twinning in plagioclase and pyroxene, and diaplectic feldspar glasses. Despite the high abundance of lithic and mineral clasts, however, these features are as rare as they were in the Apollo 11 and 12 materials.

Strong shock. Shock-fusion products are present as heterogeneous schlieren-rich glass fragments and glass spheres, which in some cases contain mineral relics. Some lithic clasts contain veins of shock-melted glass that contains numerous metallic spherules. In addition, glass spatter of various shapes and sizes was observed on the surfaces of some rocks. Thus, as in samples returned from previous missions, widespread evidence exists of shock-induced fusion.

Rock Surface Features

In general, the surface features observed on Apollo 14 rocks are similar to those reported for rock samples from previous missions (refs. 5–3 and 5–4).

Micrometeoroid impact craters are very common and are characterized by a central glass-lined depression (pit) and a concentric area of conchoidal fractures (spall zone). Some exceptionally large pit craters (up to 5-mm pit diameter) were observed. Some friable rocks have craters without glass-lined central pits and with no pronounced spall zone. The crater shapes are typical for craters in slightly cohesive targets (ref. 5–5). The most friable rocks do not display any signs of cratering; such features may have been destroyed in transit.

Melt droplets and small glass spatter of various shapes, sizes, and colors are common on rock surfaces. Large glass coatings (1 to 3 cm² in area) are rare. When present, they seem to be thinner and much more vesicular than the spatter observed on rocks returned from previous missions, and they range from approximately 0.01 to 1 mm in thickness. In addition, glass fills fractures in some rocks and, in one case, metallic spatter (a few square millimeters in area) was observed.

A few rocks display planar surfaces (10 to 50 cm² in area) that are characterized by parallel grooves a few millimeters apart and up to 1 mm deep. These surfaces are otherwise free of relief and resemble slickensides. A soil line that separated a dust-covered bottom surface from a dust-free top surface was observed. Less cohesive fragmental rocks with a high percentage of lithic inclusions have hackly surfaces that are characteristic of clast molds. In general, the surface features observed on the moderately cohesive and tough rocks are similar to the surface features reported for rocks returned from previous missions.

Soil

Seven soil samples were returned from the Apollo 14 site; data concerning six of these samples are contained in table 5–III. The grain-size distributions for six of these samples are shown in figure 5–8. Four samples (samples 14148, 14163, 14156, and 14259) have grain size distributions similar to those of the Apollo 11 and 12 soils, except that the soils from the Apollo 14 site are more poorly sorted. Two samples (14149 and 14141) are coarser than any lunar soil samples previously examined, with the exception of the coarse layer in the Apollo 12 double core (sample 12028).

TABLE 5–III. *Soil Samples Returned From the Apollo 14 Site*

Sample no.	Mass, g	Median grain size, mm	Location and name
14259	2833.44	0.050	Comprehensive sample [a]
14163	7881.57	.065	Bulk soil sample (weigh bag 2 at the end of EVA–1)
14141	53.15	.735	Cone Crater
14148	76.95	.087	Surface of the trench
14156	153.85	.068	Middle of the trench
14149	158.72	.410	Bottom of the trench

[a] The comprehensive soil sample was skimmed from the upper 1 cm of the surface.

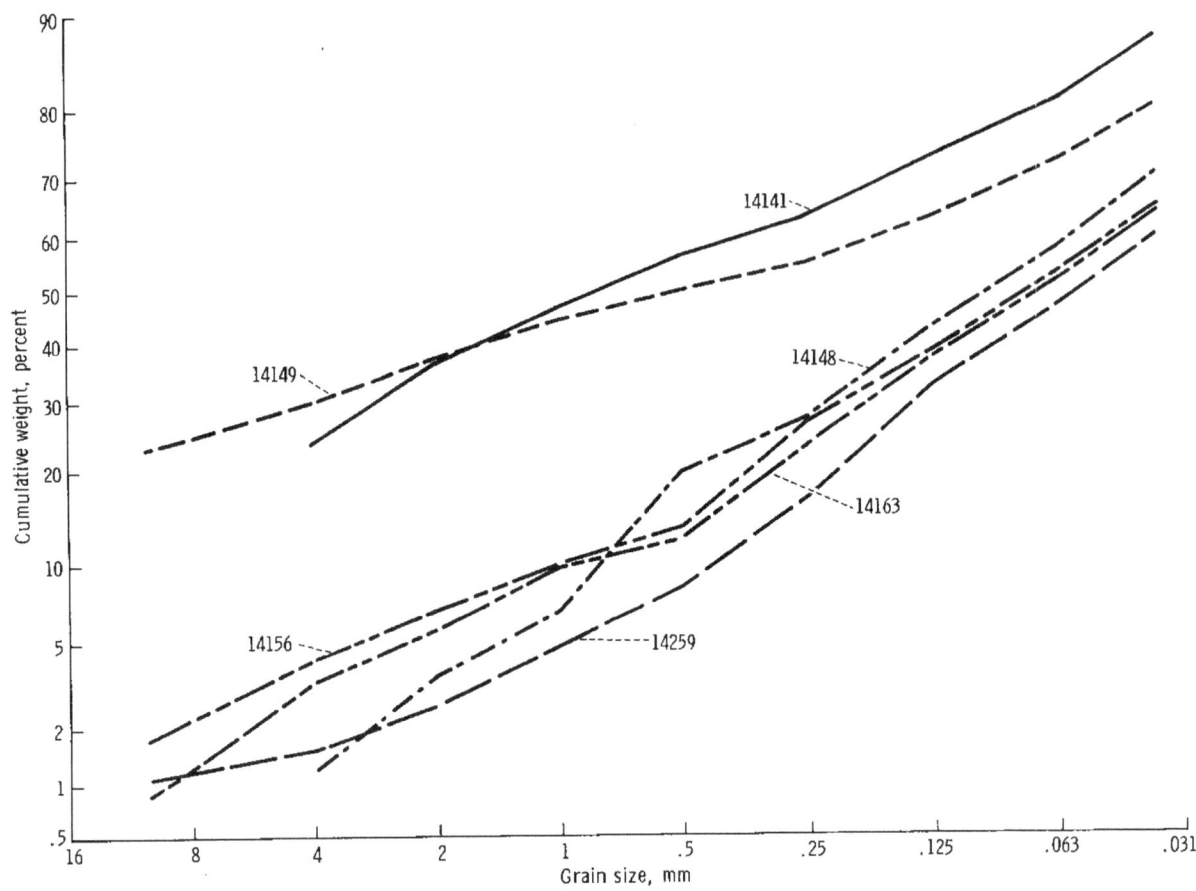

FIGURE 5-8.—Grain size distributions of six Apollo 14 soil samples.

Fractions of the finer grained soil less than 1 mm across are characterized by (1) high glass content, (2) lithic fragments and glass more abundant than mineral fragments in the coarser material of this fraction (and the reverse in the smaller sizes), (3) plagioclase and pyroxene as the most abundant minerals, and (4) a variation of plagioclase-to-pyroxene ratios.

Soil samples 14163, 14148, and 14149 contain 40 to 75 percent glass, whereas sample 14141, the Cone Crater soil, contains less than 10 percent glass. In the fraction with grain sizes less than 62.5 μm, the glass fragments are generally highly angular and colorless to pale green. With increasing grain size, dark-brown glass fragments, many of which appear to be agglutinates, are more abundant. Glass spheres are present, but are not common. Several examples of compound glass spheres (one glass sphere enclosing another) were observed in samples 14148 and 14163.

The proportion of lithic fragments in the soils increases with grain size and reaches 70 to 100 percent in the 1- to 2-mm fraction. Lithic fragments include both fragmental and homogeneous crystalline rocks. The fragmental rock particles predominate over the crystalline rock particles and, in some samples, crystalline rock fragments are absent. The lithic fragments are similar (in modal variability and range of types) to the lithic clasts that were found in the fragmental rocks described previously. Fragments of crystalline rocks compose approximately 40 percent of the 1- to 2-mm fraction sampled from the bottom of the trench. Most of these crystalline rocks are plagioclase-orthopyroxene rocks with subophitic textures. Clinopyroxene, opaque minerals (mostly ilmenite), and mesostasis are also present.

Mineral fragments are most abundant in the 37- to 62.5-μm fraction, consisting of angular grains of plagioclase, pyroxene, olivine, and opaque minerals. The two most abundant minerals are plagioclase and clinopyroxene. The plagioclase-to-clinopyroxene ratios range from 16:1 to 1:2; the ratio varies not only from sample to sample, but from one size fraction to another. Olivine and orthopyroxene are the next most abundant minerals, and opaque minerals are least abundant. In terms of compositional range and deformational features, all the minerals are similar to those described as mineral clasts in the fragmental rocks.

The Cone Crater soil sample (14141) is distinctly different from the other soils. As mentioned previously, it is coarser grained (with a median grain size of approximately 0.74 mm), and the glass content is very low (less than 10 percent in the 0.5-mm fractions). The Cone Crater soil sample contains a high content of fragmental-rock fragments (40 to 60 percent), even in the finest size fractions (62.5 μm). The remainder of the material is glass-bonded fragmental rock.

Core Tube Samples

Four core tube samples, which consist of a double and two single core tube samples, were returned from the Fra Mauro site. None of the core tubes has been opened for study, but stereo X-radiographs have been taken to determine the sample, quantity, texture, and layering, and the amount of disturbance in each core.

The double core tube sample (39.5 cm long and 209 g), which was collected at station A during the second EVA, shows textural changes that may represent at least seven and possibly nine layers, which range from 1 to 13 cm thick. Overall, a pronounced decrease in grain size occurs from the bottom to the top of the core. Near the base, 10 to 15 percent of the rock fragments are greater than 5 mm long, whereas, near the top, only a few particles are larger than 2 mm. The largest rock fragment visible in the X-radiographs is 1.2 cm long.

A single core tube sample (16.5 cm long and 81 g), which was collected at Triplet Crater, exhibits, on the basis of textural changes visible in the X-radiographs, at least three layers that range from 0.5 to 5 cm thick. Ten to 20 percent of the soil appears to consist of fragments greater than 2 mm long in a matrix of fine-sand- or coarse-silt-size particles. In this case, as in the double core tube sample, a crude grading of particles larger than 2 mm is seen from the base to the top of the core.

Approximately 15 percent of another single core tube sample (12.5 cm long and 71 g) collected at Triplet Crater consists of rock fragments greater than 2 mm long in a fine-grained matrix. The sample is loose in the tube and is highly disturbed.

Significant variations in the mechanical behavior of the lunar soil were observed at the Fra Mauro site compared with observations at previous Apollo sites. A detailed discussion of the lunar soil mechanical properties can be found in section 4 of this report.

Emission Spectrographic Analysis

Chemical analyses of some samples were made by the use of three optical-emission spectrographic methods: (1) a general method for the analysis of major and minor elements by the use of strontium (Sr) as the internal standard and the buffer; (2) a method for the detection of minor and trace elements by the use of palladium (Pd) as the internal standard; and (3) a method for the determination of K, lithium (Li), rubidium (Rb), and cesium (Cs) by the use of sodium (Na) as the internal standard and the buffer. The analytical methods used were similar to those used previously (refs. 5–3 and 5–4). Usually, the method that used a buffer system with burning conditions nearest the conditions of the element being measured was given the most consideration. The overall precision of the determination is ±5 to ±10 percent of the amount present. Accuracy was controlled by the use of international rock-standard samples (G–1, W–1, SY–1, BCR–1, AGV–1, GSP–1, PCC–1, and DTS–1) for calibration. Another calibration sample that was added was lunar sample 10084, which is a homogeneous fine-grained sample that was analyzed by many principal investigators.

A total of 16 samples was analyzed. The weights of the samples used for analysis were typically 100 to 150 mg. In most cases, to protect the integrity of the rocks, the samples for analysis were from chips in the same sample bag as the rocks or were chips that had fallen off the rock during preliminary sample processing. The analyses presented may, therefore, not be representative of the whole rocks. Most of the rocks are heterogeneous and, thus, may have considerable variations in composition.

TABLE 5–IV. *Elemental and Mineral Abundances in Lunar Samples*

(a) Elemental Abundance

Element	Sample no.											
	14053	14321,14	14049	14310	14321,9	14042	14301	14065	14066	14305	14259	Fines (average)

Elemental abundance, percent

Si.............	22.6	22.4	22.9	23.5	23.5	24.0	22.9	22.6	24.0	23.0	22.6	22.5
Al.............	6.4	7.4	9.0	10.6	9.5	8.5	9.0	11.1	8.0	8.5	9.5	9.3
Mg............	5.0	7.2	6.6	4.8	6.6	5.2	6.6	5.0	5.7	7.8	5.5	5.9
Fe.............	12.6	10.1	7.6	6.0	7.0	7.4	7.6	5.3	7.3	7.4	7.8	8.0
Ca.............	8.5	6.1	6.4	7.8	5.8	7.4	6.3	8.5	7.1	5.3	7.8	7.4
Ti.............	.9	1.4	1.0	.8	.9	1.1	1.0	.6	1.1	1.0	1.1	1.1
Na............	.28	.30	.63	.47	.42	.36	.58	.68	.42	.63	.38	.42
K.............	.12	.28	.44	.44	.46	.52	.60	.83	1.0	1.0	.42	.43
Mn............	.22	.20	.14	.11	.12	.12	.15	.09	.12	.14	.14	.15
Cr.............	.30	.29	.13	.11	.11	.12	.12	.07	.09	.12	.14	.14

Elemental abundance, ppm

Ba.............	190	380	670	630	730	820	920	820	960	930	570	638
Co.............	48	33	40	31	32	56	44	19	39	32	39	44
Cu.............	13	13	16	11	7	19	17	6	7	13	14	18
La.............	10	40	63	36	65	70	92	32	72	54	46	49
Li.............	11	18	20	19	19	19	20	20	25	23	18	21
Ni.............	14	180	260	165	240	280	230	60	210	205	320	304
Nb............	19	22	52	43	46	68	63	57	60	49	40	48
Rb............	2	7	14	15	14	14	17	33	29	31	10	13
Sc.............	90	43	25	20	16	30	31	16	24	22	21	24
Sr.............	180	140	200	250	180	210	240	250	220	200	170	206
V.............	135	85	48	35	32	74	63	46	52	52	50	51
Yb............	10	20	28	30	28	27	33	27	31	28	24	24
Y.............	90	160	220	180	220	110	260	200	250	210	170	210
Zr.............	310	670	880	930	860	1030	1000	980	970	900	720	922

(b) Mineral Abundance, Percent

SiO_2............	48	48	49	50	50	51	49	48	51	49	48	48
Al_2O_3...........	12	14	17	20	18	16	17	21	15	16	18	18
MgO...........	8.4	12	11	8.0	11	8.6	11	8.3	9.5	13	9.2	9.9
FeO...........	16	13	10	7.7	9.0	9.5	9.8	6.8	9.4	9.5	10	10
CaO...........	12	8.5	8.9	11	8.2	11	8.8	12	10	7.4	11	11
TiO_2...........	1.5	2.4	1.7	1.3	1.5	1.8	1.7	.95	1.9	1.6	1.8	1.8
Na_2O...........	.38	.40	.85	.63	.58	.48	.78	.92	.58	.85	.52	.57
K_2O...........	.14	.33	.53	.53	.56	.63	.72	1.0	1.2	1.2	.50	.52
MnO...........	.29	.26	.18	.14	.15	.16	.19	.12	.16	.18	.18	.19
Cr_2O_3...........	.44	.42	.19	.16	.16	.18	.17	.10	.13	.18	.20	.20
ZrO_2...........	.04	.05	.07	.13	.12	.14	.07	.13	.13	.12	.10	.12
Total.......	99.2	99.4	99.5	99.5	99.4	99.5	99.3	99.3	99.0	99.1	99.5	100.3

The results of chemical analyses are given in table 5–IV. Included are all data obtained from analyses made on rock samples and an average value for the analyses of five soil samples. Sample 14259 (the comprehensive soil sample) is included as representative of the fines.

Elements detected in sample 14259 in abundances greater than 0.1 percent (by weight) in decreasing order of concentration are silicon (Si), aluminum (Al), iron (Fe), calcium (Ca), magnesium (Mg), titanium (Ti), K, Na, manganese (Mn), and chromium (Cr). Minor elements (between 0.02 and 0.1 percent by weight)) are zirconium (Zr), barium (Ba), nickel (Ni), yttrium (Y), and Sr. Trace elements (less than 200 ppm) are vanadium (V), lanthanum (La), niobium (Nb), cobalt (Co), scandium (Sc), ytterbium (Yb), Li, copper (CU), and Rb.

The compositions of the samples from the Apollo 14 site, with the exception of that of

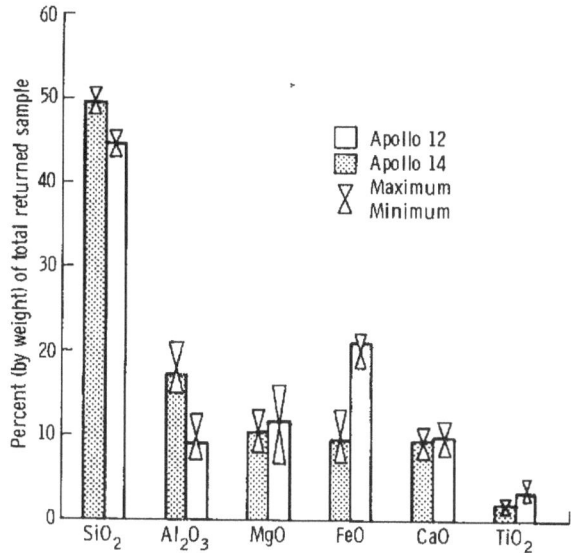

FIGURE 5–9.—Comparison of mineral contents for Apollo 12 and 14 rocks. The top of the bar represents the mean composition.

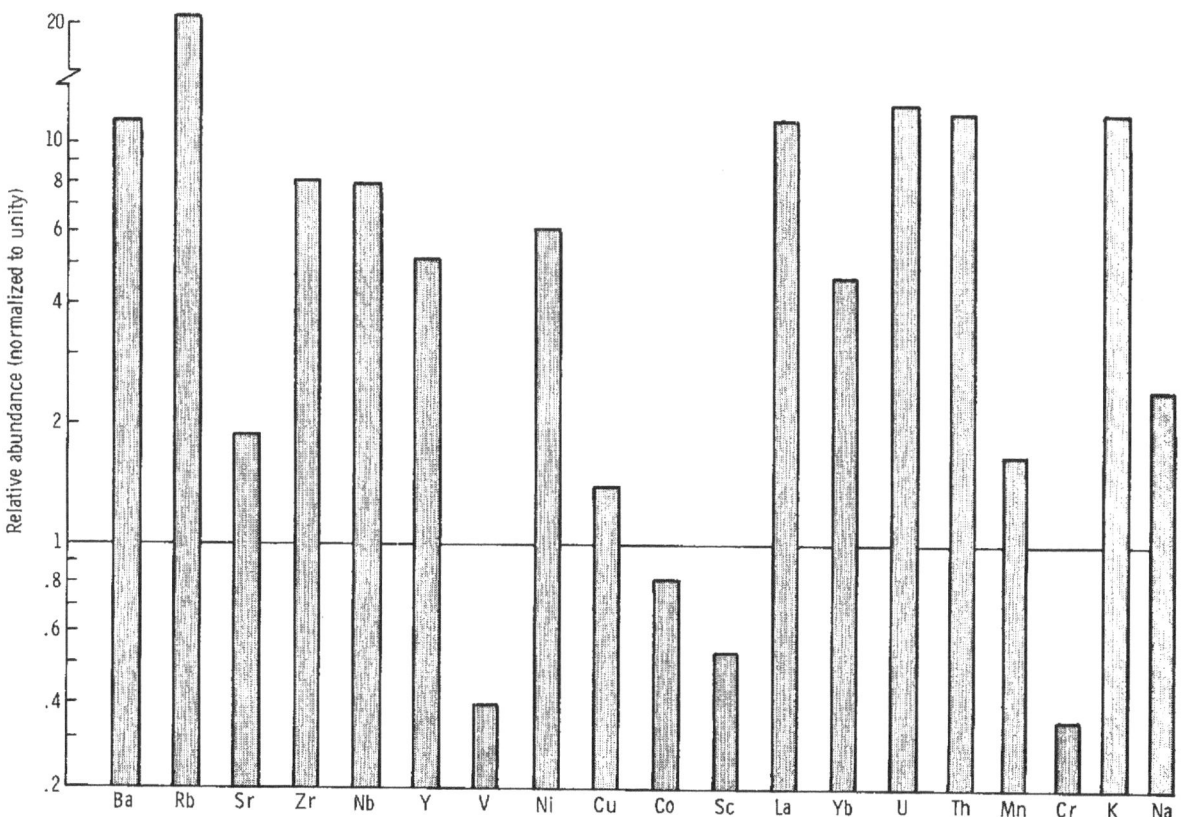

FIGURE 5–10.—Relative abundances of minor and trace elements in Apollo 14 rocks with respect to Apollo 12 rocks (normalized to unity).

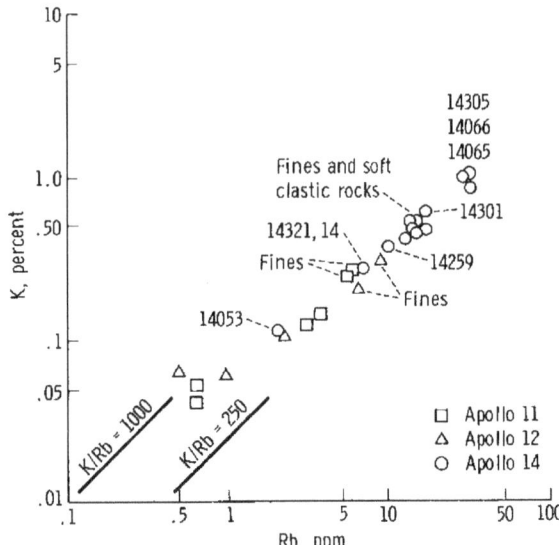

FIGURE 5-11.—The K/Rb relationship for Apollo 14 rocks and soils compared with selected Apollo 11 and 12 rocks and average soils.

sample 14053, are clearly distinct from those of the Apollo 11 and 12 and Luna 16 samples. Considering average values for the abundances of the elements reported, the concentrations of Fe, Ti, Mn, Cr, and Sc are lower than in samples returned from Mare Tranquillitatis, Oceanus Procellarum, or Mare Fecunditatis. Silicon, Al, Zr, K, Rb, Sr, Na, Li, and La are generally more abundant than in previously examined samples. Barium and Y are more abundant in the rocks from the Apollo 14 site compared with previous lunar rocks, but their concentrations in the soil are similar to the concentrations found in fines from the Apollo 12 site. A comparison of Apollo 12 and 14 compositions is shown in figures 5-9 and 5-10.

Potassium-to-rubidium values are remarkably constant and average 325±50 in the fragmental rocks and 330±25 in the fines (fig. 5-11). Unlike the samples returned from the Apollo 12 site, the samples from the Apollo 14 site have Rb/Sr ratios higher in the rocks than in the soil. Both values (0.10±0.05 for rocks and 0.06±0.001 for fines) are higher than in previously analyzed lunar mare basalts. Iron-to-nickel ratios in the soil average 260±25 and show remarkable constancy. In the fragmental rocks, more variation exists, and the Fe/Ni ratios average 525±260. The values for

the two basaltic rocks (samples 14310 and 14053) are 360 and 9000, respectively. Except for rock sample 14053, Fe/Ni ratios are the lowest yet obtained for lunar material. The Ni content of sample 14310 (165 ppm) is the highest yet obtained for a lunar basaltic rock and is higher than that of most terrestrial oceanic basalts. Nickel is more abundant in soil than in fragmental rocks, presumably because of greater meteoritic admixture. It is noteworthy that the average Ni content of fragmental rocks from the Apollo 11 site is similar to that of the soils from the Apollo 11 site, whereas the Ni content of the Apollo 14 fragmental rocks is significantly lower with respect to the Apollo 14 soil.

The soils and fragmental rocks are very similar in bulk composition. However, the average values of K, Ba, La, and Yb in the soil are somewhat lower than the average values for these elements in the rocks, which suggests that a small amount of exotic material is present in the soil. Rock sample 14310 is very similar to the fragmental rocks in composition, which raises interesting questions about its origin. The most chemically distinct sample analyzed is rock sample 14053, which mineralogically and chemically resembles more closely the basaltic rocks from the Apollo 12 site than other rocks or soil from the Fra Mauro site. Apart from the K/Rb correlation mentioned previously and a similar correlation between K and Ba, little correlation is evident in abundances of pairs of elements. Relationships may appear, however, when individual clasts within fragmental rocks are analyzed.

The Rb/Sr chemistry of the material returned from the Apollo 14 site is clearly different from those of the basaltic rocks returned from the Apollo 11 and 12 and Luna 16 sites. Differences between Apollo 14 material and basaltic rocks returned from the maria are such that rock from the Apollo 14 site must be regarded as a type of lunar material distinct from the basaltic rocks of mare origin.

The previously returned lunar materials closest in composition to the rocks from the Apollo 14 site are the "norite," which are rocks enriched in KREEP (potassium, rare-Earth elements, and phosphorus) (ref. 5-6) component (or "mottled-gray" fragments described by some investigators[1] from the coarse fines returned from the Apollo 12 site), the dark portion of rock 12013 (ref. 5–7), and Luny Rock 1 from the Apollo 11 site (ref. 5–8). The relatively constant compositions of the Apollo 14 fragmental rocks suggest either that the rocks were derived from an extremely homogeneous source or, more likely, that efficient mixing occurred.

The samples appear to be free from inorganic contamination. Niobium, which is present in the skirt of the LM descent engine, is found in similar abundance levels in samples taken near the spacecraft and at locations considerable distances away. Niobium was also found at approximately the same levels in both rock and fines samples. Indium (In), which is present in the seals of the rock boxes, was not detected.

Concentrations of Long-Lived Radioactive Elements

At the Radiation Counting Laboratory (RCL) of the LRL, one crystalline rock, seven fragmental rocks, and five samples of fines were analyzed by the use of nondestructive gamma-ray spectroscopy. Sample analysis was begun 51 hr after splashdown. Analyses were performed by using the sodium iodide (thallium activated) (NaI(Tl)) dual-param-

[1] The following papers are to be published in the Proc. Apollo 12 Lunar Sci. Conf. (Houston), Jan. 11–14, 1971, in Geochim. Cosmochim. Acta:

E. A. King, J. C. Butler, and M. F. Carmen: The Lunar Regolith as Sampled by Apollo 11 and 12: Gram Size Analyses, Modal Analyses, Origins of Particles.

D. McKay, D. Morrison, J. Lindsay, and G. Ladle: Apollo 12 Soil and Breccia.

J. A. Wood, U. Marvin, J. B. Reid, G. J. Taylor, et al.: Relative Proportions of Rock Types and Nature of the Light-Colored Lithic Fragments in Apollo 12 Soil Samples.

C. Meyer, Jr., F. K. Aitken, P. R. Brett, D. S. McKay, et al.: Rock Fragments and Glasses Rich in K, REE, P, in Apollo 12 Soils: Their Mineralogy and Petrology.

W. Quaide, V. Oberbeck, T. Bunch, G. Polkowski, et al.: Investigations of the Natural History of the Regolith at the Apollo 12 Site.

T. Anderson, Jr., R. C. Newton, and J. V. Smith: Apollo 12 Mineralogy and Petrology: Light-Colored Fragments, Minor-Element Concentrations, Petrologic Development of the Moon.

eter low-background gamma-ray spectroscopy system and stainless-steel sample containers described previously (refs. 5–3 and 5–4). Fines were contained in aluminum right-circular cylinders that were enclosed in the stainless-steel containers. The data acquisition followed the procedures developed for the studies of samples returned from the Apollo 11 and 12 sites (refs. 5–3 and 5–4).

The activities of the radionuclides were determined by the method of least squares and by a computer-assisted spectrum-stripping method. The standards used for data reduction were either right-circular cylinders, as described previously, or phantoms (replicas that contain known amounts of radioactivity) of samples returned from previous missions. The results are summarized in table 5–V. The errors that are indicated in table 5–V include the statistical errors, but most of the errors are caused by differences in thickness and shape between the samples and the standards.

Five Apollo 14 soil samples have uniform K contents of approximately 0.5 percent. These samples all have similar thorium (Th) and uranium (U) contents, which are approximately 14 and 4 ppm, respectively. The concentrations of K, Th, and U in Apollo 14 soils are higher than those in Apollo 11 and 12 soil samples. The K/U ratios average 1100 and are somewhat lower than those in Apollo 11 and 12 samples, but the Th/U ratios, which average 3.7, are comparable to the Th/U ratios of previously returned samples.

The fragmental rocks, with the exception of sample 14082 (White Rock), have K values that range from 0.3 to 0.7 percent, Th values that range from 9 to 16 ppm, and U values that range from 2.5 to 4.1 ppm. These values, which are comparable to those of sample 12034, are higher than those of most of the samples returned from previous missions. The K/U ratios for the fragmental-rock samples average 1300 and are thus lower than the Apollo 12 ratios. The Th/U ratios, which average 3.7, are close to the average ratios for samples from the Apollo 11 and 12 sites. The Th, U, and K contents of rock sample 14082 are lower than those of the other six clastic rocks by factors of 2 to 3.

Crystalline rock sample 14053 has a K content similar to that of the low-K group of Apollo 11

TABLE 5–V. *Gamma-Ray Analysis of Lunar Samples*

Sample no.	Weight, g	K, percent (by weight)	Th, ppm	U, ppm	^{26}Al, dpm/kg	^{22}Na, dpm/kg	^{56}Co, dpm/kg	Remarks
			Clastic rocks					
14045	65	0.36±0.04	13.8±1.4	3.7±0.5	130±40	83±25	....	
14066	510	.69± .07	15.3±1.5	4.1± .6	110±20	52±10	....	
14082	63	.18± .02	4.6± .5	1.4± .2	140±30	68±14	....	White clasts
14301	1361	.55± .05	12.8± .5	3.6± .5	53±11	36± 7	....	
14302	381	.55± .05	14.3±1.4	3.8± .6	85±17	52±10	....	
14315	115	.30± .03	9.1± .9	2.5± .4	160±30	60±12	....	
14318	600	.49± .05	12.8±1.3	3.3± .5	120±20	36± 7	....	
			Crystalline rocks					
14053	251	0.088±0.009	2.24±0.22	0.64±0.10	98±20	59±12	....	Ge(Li) detector used.
14310	3425	.49 ± .06	13.7 ±1.7	3.7 ± .6	80±20	55±15	25±5	
			Fines smaller than 1 mm					
14163	491	0.48 ±0.05	13.9±1.4	3.9±0.6	78±16	45± 9	....	Bulk fines
14259	495	.42± .04	13.4±1.3	3.8± .6	220±40	84±17	....	Comprehensive fines
14148	70	.41± .04	14.9±1.5	4.1± .6	190±40	70±14	....	Top of trench
14156	136	.40± .04	14.5±1.4	3.9± .6	180±40	66±13	....	Middle of trench
14149	85	.44± .40	14.8±1.5	3.9± .6	150±30	58±12	....	Bottom of trench

rocks and has Th and U contents intermediate between those of Apollo 11 and low-K and high-K crystalline rocks. Sample 14053 has the low K/U ratio that is typical of Apollo 14 samples and the usual lunar Th/U ratio.

Potassium, Th, and U contents were determined for igneous rock sample 14310 by the use of a 40-cm³ germanium (lithium-activated) (Ge(Li)) detector. This rock contains the highest K, Th, and U contents of any crystalline lunar rock thus far examined. The K/U and Th/U ratios for this rock are similar to those for the Apollo 14 fragmental rocks. Abundances of K in Apollo 14 material are somewhat less than those in most terrestrial basalts, whereas the abundances of Th and U are enriched by a factor of approximately 3 over those of previously analyzed lunar material and are enriched by approximately the same factor over the Th and U abundances in most terrestrial basalts.

Carbon Content

The total carbon (C) contents of the lunar samples were determined by oxygen combustion followed by gas-chromatographic detection of the CO_2 produced. Samples that weighed approximately 200 mg were used. The system was calibrated with the National Bureau of Standards steel standard 348. Samples of this standard, which contain from 20 to 90 ppm of C, were analyzed under the same conditions as were the lunar samples. All other techniques were as previously described (ref. 5–9). The total C results give no indication of the specific chemical species present.

The results of the analyses are given in table 5–VI. The fines samples have values for C content that range from 70 to 180 ppm. These values fall within the range of total C found in samples from the Apollo 11 and 12 sites. The trench samples do not appear to show a significant difference in C content with depth.

TABLE 5–VI. *Total Carbon Abundances*

Sample no.	C, ppm	Remarks
Soils		
14163....	145±10	Bulk soil (<1 mm)
14163....	150±10	Bulk soil (<1 mm)
14163....	70±10	Bulk soil (<1 mm)
14163....	120±10	Bulk soil (<1 mm)
14259....	160±10	Comprehensive soil (<1 mm)
14148....	160±10	Trench soil—top (<1 mm)
14156....	180±10	Trench soil—middle (<1 mm)
14149....	135±10	Trench soil—bottom (<1 mm)
14141....	80±10	Cone Crater soil (<1 mm)
Fragmental rocks		
14042....	225±10	
14047....	210±10	
14311....	200±10	
14049....	190±10	
14313....	130±10	
14066....	90±10	
14063....	80±10	
14301....	50±10	
14305....	32±8	
14321....	28±8	
Crystalline rock		
14310....	35±8	Fine-grained basaltic rock

The fine-grained fragmental rocks from the Apollo 14 site range in total C content from 28 to 225 ppm. The lowest C content was found in the largest rock (sample 14321), which has pronounced crystallinity. The fragmental rock samples that were analyzed may not be representative of the whole rocks because of the small sample sizes (200 mg), the heterogeneity of the rocks, and the surface-soil contamination problems. The only homogeneous crystalline rock (sample 14310) that was analyzed has a total C content of 35 ppm, which is a value similar to the values found for the basaltic rocks from the Apollo 11 and 12 sites.

Induced Radioactivities

Sodium-22 and ^{26}Al were analyzed in the same samples that were analyzed for K, Th, and U. The

techniques used were the same as reported previously. Results are listed in table 5–V. These results are preliminary; therefore, error limits are large. Manganese-54, ^{56}Co and ^{46}Sc were detected in some samples.

In the soil, the ^{26}Al specific activity ranges from 78 to 220 disintegrations per min per kg (dpm/kg), with the lower limit of the range representing the bulk fines and the higher limit representing the comprehensive fines that were collected from the top centimeter of the lunar surface. The ^{22}Na specific activity ranges from 45 to 84 dpm/kg for bulk fines and comprehensive fines, respectively.

In the rocks, ^{26}Al ranges from 53±11 dpm/kg for rock sample 14301, which was almost buried in the lunar regolith, to 157±31 dpm/kg for rock sample 14315. Sodium-22 ranges from 36±7 dpm/kg (for rock samples 14318 and 14301) to 83±25 dpm/kg (for rock sample 14045).

Determinations of Rock Orientations From Solar-Flare Data

A solar flare occurred on January 25, and its effects on lunar samples permit a unique opportunity to study solar-flare-induced-radioactivity products and allow the recent (post-January 24) surface orientation of certain rocks to be determined. Solar-flare protons produce ^{56}Co (with a half-life of 77 days) by proton-neutron reaction on Fe. The ^{56}Co is predominantly produced on the side of the rock that faces the Sun. The top and bottom of rock sample 14310 were determined by scanning the rock surface with a 40-cm^3 Ge(Li) detector to detect the emitted radiation of ^{56}Co.

Noble Gases

Abundances and isotopic compositions of the five stable noble gases were determined by mass spectrometry in one soil sample and in seven fragmental rocks (table 5–VII). The concentrations of noble gases in fines (sample 14259) and in rock samples 14047, 14049, and 14301 are approximately two orders of magnitude higher than in the other four rocks measured. The noble gases are predominantly of a solar-wind-implantation origin and occur in concentrations similar to those measured in soil and breccia returned by

TABLE 5–VII. *Total Noble-Gas Contents* [a]

Sample no.	Weight, mg	^{3}He	^{4}He	^{22}Ne	^{38}Ar	^{84}Kr	^{132}Xe	^{20}Ne/^{22}Ne	^{22}Ne/^{21}Ne	^{36}Ar/^{38}Ar	^{40}Ar/^{36}Ar
			$\times\,10^{-6}\,cm^3/g$ (at STP)			$\times\,10^{-9}\,cm^3/g$					
14063,5 (fragments)..	3.21	0.37	1 770	[b]0.54	2.31	6.8	3.2	15	14.6	5.21	99.2
14063,5 (whole rock).	17.08	.50	1 980	1.40	2.31	.80	[b].19	12.8	22.9	5.24	61.8
14066,2..........	12.86	.19	2 020	[b].22	.52	.87	.39	11.5	8.77	4.64	382
14305,9..........	30.18	.168	2 510	[b].062	[b].060	[b].23	[b].14	6.96	2.40	2.28	2205
14321,13..........	27.40	.204	1 313	[b].189	[b].41	[b].56	[b].28	10.7	5.80	4.15	260
14301,7..........	1.89	4.49	12 300	18.6	38.0			12.0	29.5	4.94	16.2
14301,7..........	18.65				22.3	9.7	2.9			5.18	14.9
14047,2..........	1.80	23.1	54 400	73.4	283			13.0	26.6	5.25	1.38
14047,2..........	17.0				216	144	23			5.27	1.37
14049,3..........	1.77	20.0	47 600	71.6	252			12.6	25.0	5.24	1.90
14049,3..........	15.7				188	119	21			5.28	1.81
14259,10..........	1.55	36.8	77 700	95.4	328			12.7	27.7	5.28	1.37
14259,10..........	17.12				202	348	92			5.31	1.25

[a] Mass-spectrometer sensitivity variations throughout the period of these analyses did not exceed ±10 percent. However, because krypton (Kr) was split between 2 gas fractions, the uncertainty in its abundance is ±20 percent. Isotopic ratios and abundances have been corrected for blanks and multiplier discrimination. All extraction blanks were less than 15 percent, with the exception of those that are indicated.

[b] Extraction blanks greater than or equal to 15 percent.

the two previous Apollo missions. Elemental-abundance ratios and the isotopic ratios of helium (He), neon (Ne), and argon (Ar) for the four rock samples are also similar to those of solar-wind-implanted gases found in previously returned samples.

Rock samples 14063, 14066, 14305, and 14321 contain noble gases that have resulted from radiogenic decay of ^{40}Ar and ^{4}He and from cosmic-ray interactions. Also found are small variable quantities of solar-wind gases that are probably present in the lunar dust that contaminates the surfaces of the small samples (average weight, 13 mg). By combining values for the radiogenic ^{40}Ar concentrations in these rocks with values for the K contents (tables 5–III and 5–IV), gas-retention ages that range from 2.8 to 3.8 billion yr result. Sample heterogeneity and possible gas loss make these values uncertain and, at best, they represent only lower limits to the crystallization ages of the rocks. However, unlike the analyses of the fragmental rocks returned from the Apollo 11 and 12 sites, analyses of these rocks by the use of the ^{40}Ar/^{39}Ar method should result in accurate ages. If the formation of fragmental rocks of low solar-wind-gas content involved the incorporation and degassing of surface fine material, this formation must have occurred more than 3 billion yr ago for radiogenic ^{40}Ar to have accumulated. Alternatively, the material incorporated into the four rocks has never contained a significant solar-wind component.

Calculated concentrations of some spallation-produced noble-gas isotopes that were found in these samples are listed in table 5–VIII (after ref. 5–10). Also listed are some approximate surface-exposure ages calculated from these spallation isotopes. Rock samples 14063, 14066, 14321, and 14305 contain similar concentrations of each spallation nuclide. Exposure ages generally fall in the range of 10 to 20 million yr, with reasonable agreement among the ages calculated on the basis of different spallation isotopes. These four rocks may have actually had the same exposure time, with the age variations resulting from differing shielding conditions and analytical uncertainties. One could speculate that the formation of those rocks dates the occurrence of a single cratering event, possibly the formation of Cone Crater. The age of these rocks are considerably lower than typical ages of 40 to 500 million yr for rocks returned from the Apollo 11 and 12 sites. Rock samples 14047 and 14049 and soil sample 14259 exhibit spallation concentrations approximately an order of magnitude greater than those of the Apollo 11 and 12 samples; but, because of the presence of solar-wind gases, these abundances are highly uncertain.

The isotopic compositions of krypton (Kr) and xenon (e) that were measured in several of the samples are listed in table 5–IX. In samples 14259, 14049, 14047, and 14301, the relative isotopic abundances are similar to those determinations of solar-wind gases in previously returned samples, with the addition of a spallation component. For samples 14259, 14049, 14047,

TABLE 5–VIII. *Cosmic-Ray Spallation-Produced Noble Gases* [a] *and Exposure Ages* [b]

Sample no.	Concentration, × 10⁻⁸ cm³/g (at STP)				Exposure age, × 10⁶ yr		
	^{3}He	^{21}Ne	^{38}Ar	^{126}Xe	^{21}Ne	^{38}Ar	^{126}Xe
14063,5................	....	1.6	1.0	0.001	10	9	
14063,5 (fragments)......	[c]37	2.0	1.3	.003	13	10	
14066,2................	[c]19	1.9	1.7	.002	11	14	9
14321,13...............	[c]20	2.7	2.5	.0017	16	24	22
14305,9................	[c]17	2.5	1.7	.0021	14	19	11
14047,2................	....	[d]33	....	....	....	....	
14049,3................	....	[d]50	....	....	....	....	
14259,10...............	....	[d]26	....	....	....	[d]170	

[a] Concentrations of spallation isotopes have been calculated from the data listed in table 5–VII.

[b] Ages are calculated from the chemical data produced tion rates given in ref. 5–10.

[c] Maximum value.

[d] Approximate value.

TABLE 5–IX. Measured Krypton and Xenon Ratios [a]

Sample no.	Measurement accuracy	Kr/84Kr, for i =					Xe/132Xe, for i =							
		78	80	82	83	86	124	126	128	129	130	131	134	136
14259,10	Value	0.0089	0.0466	0.2097	0.2092	0.3025	0.0121	0.0179	0.1063	1.091	0.1770	0.8808	0.3655	0.2941
	Uncertainty, ±	.0005	.0007	.0011	.0018	.0015	.0004	.0005	.0012	.007	.0018	.0043	.0019	.0015
14047,2	Value	.028	.0464	.2105	.2113	.3039	.0127	.0230	.1121	1.071	.1809	.9041	.3642	.2945
	Uncertainty, ±	.001	.003	.0011	.0020	.0016	.0012	.0004	.0010	.011	.0012	.0056	.0025	.0018
14049,3	Value	.0113	.0511	.2167	.2204	.3043	.0181	.0291	.1216	1.075	.1866	.9251	.3641	.2949
	Uncertainty, ±	.0001	.0004	.0010	.0014	.0029	.0003	.0004	.001	.007	.0015	.0044	.0019	.0017
14301,7	Value	.055	.1959	.2133	.2072	.3004	.0133	.0216	.1059	1.058	.1697	.8824	.4048	.3447
	Uncertainty, ±	.005	.0041	.0021	.0063	.0021	.0016	.0013	.0051	.008	.0038	.0063	.0031	.0050
14321,13	Value	.084	.086	.252	.259	.301	.031	.051	.163	1.102	.206	.945	.389	.325
	Uncertainty, ±	.013	.011	.009	.008	.008	.010	.009	.017	.031	.016	.021	.011	.012

[a] All values have been discrimination corrected; uncertainties given are 1σ statistical fluctuations.

and 14301, a solar-wind component was subtracted that was identical to that reduced by Eberhardt et al. (ref. 5–11) from Apollo 11 soils. The resulting spallation-isotope spectra are generally similar to those determined previously from lunar rocks. Xenon-131 exhibits a high relative yield, as was the case for previously returned lunar samples; and, for sample 14259, the ^{129}Xe and ^{132}Xe yields are also enhanced. Although the uncertainties in these spectra are large, definite differences appear to exist in the various spallation Kr and Xe spectra. Spallation spectra are even more uncertain for the low-gas rocks because of necessary fission and atmospheric Xe corrections, and only one such rock (sample 14321) is listed in table 5–III. Rock sample 14301,7 shows large concentrations of fission-produced ^{134}Xe and ^{136}Xe (1.5×10^{-10} cm^3 excess ^{136}Xe/g). If sample 14301,7 has the characteristic measured U concentration of 3.5 ppm (table 5–V), this amount of excess ^{136}Xe is considerably more than the amount that could be produced by the spontaneous fission of ^{238}U in 4.5 billion yr. However, with the possible exception of sample 14301, there appears to be no need to invoke extinct radionuclides to explain the observed abundances of ^{129}Xe and ^{136}Xe.

Biology

No viable organism has been found in the lunar material, and no evidence exists of fossil material. Direct observations involved light microscopy with white light, ultraviolet light, and phase-contrast techniques. A wide variety of biological systems is now undergoing tests with lunar material to determine the possible existence of any toxicity, microbial replication, or pathogenicity. Histological studies are being made to determine whether any evidence of pathogenicity exists. Other activities involve extensive in vitro study of the lunar samples.

Summary

The major findings of the preliminary examination of the lunar samples returned from the Apollo 14 site are as follows.

(1) The Fra Mauro samples may be contrasted with the samples returned from the Mare Tranquillitatis and the Oceanus Procellarum in that

approximately half the Apollo 11 samples consist of basaltic rocks, and all but three Apollo 12 rocks are basaltic, whereas in the samples returned from the Apollo 14 site, only two of the 33 rocks that weigh over 50 g have basaltic textures. The samples from the Fra Mauro site consist largely of fragmental rocks that contain clasts of diverse lithologies and histories. Generally, the rocks differ modally from previously returned rocks in that they contain orthopyroxene and more plagioclase.

(2) The samples differ chemically from previously returned lunar rocks and from the closest meteorite and terrestrial analogs. The previously returned material closest to their composition is the KREEP component (norite or mottled-gray fragments) (ref. 5–6) in the coarse fines from the Apollo 12 site, the dark portion of rock 12013 (ref. 5–7), and Luny Rock 1 returned from the Apollo 11 site (ref. 5–8). The Apollo 14 material is richer in Ti, Fe, Mg, and Si than the material at the Surveyor 7 site, which is the only lunar highlands material directly analyzed.[2] The rocks from the Apollo 14 site have much lower Fe, Ti, Mn, Cr, and Sc concentrations and higher Si, Al, Zr, K, Ba, Rb, Sr, Na, Li, and La concentrations than the lunar mare basalts; are much richer in K and U than are the mare basalts; but have K/U ratios that remain characteristically lunar, although lower than those of Apollo 11 and 12 samples. Uranium, Th, and some other incompatible elements are significantly enriched (relative to most terrestrial basalts) in the material from the Apollo 14 site.

(3) The chemical composition of the soil closely resembles that of the fragmental rocks, except that some elements (K, La, Yb, and Ba) are depleted in the soil (relative to the average rock composition). Most rocks and soils closely resemble each other chemically.

(4) Rocks display characteristic surface features of lunar material (impact microcraters, rounding, etc.) and display shock effects similar to those observed in rocks and soil returned from the Apollo 11 and 12 sites. The rocks show no

[2] A. L. Turkevich: Comparison of the Analytical Results From the Surveyor and Apollo Lunar Missions. Proceedings of the Apollo 12 Lunar Science Conference, Houston, Tex., Jan. 11–14, 1971. To be published in Geochim. et Cosmochim. Acta.

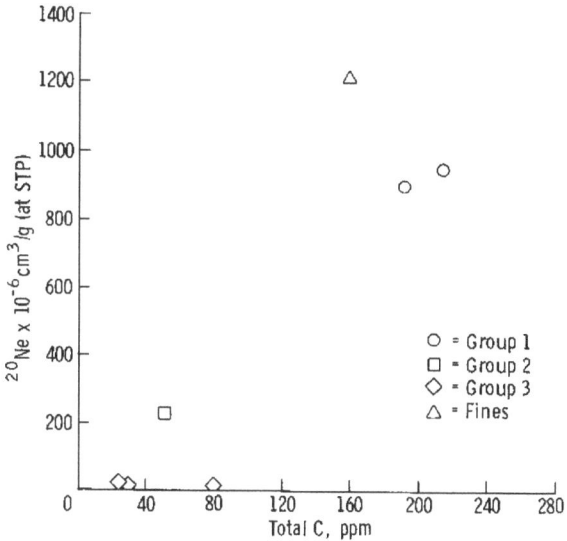

FIGURE 5–12.—The solar-wind-derived ^{20}Ne as a function of C content. Samples are classified according to group; it should be noted that C appears to follow Ne and that some correlation with rock type exists.

evidence of exposure to water, and the metallic-iron content suggests that they were formed and have remained in an oxygen-poor environment, as were the Apollo 11 and 12 materials.

(5) The concentration of solar-wind-implanted material in the soil is large, as was the case for previously returned fines; however, unlike previously examined fragmental rocks, fragmental rocks from the Apollo 14 site possess solar-wind-implanted contents that range from approximately the concentration in the fines to essentially zero, with most of the rocks investigated falling toward either extreme of this range. A positive correlation appears to exist between the solar-wind components (C and ^{20}Ne) and the friability of fragmental rocks (fig. 5–12).

(6) Carbon contents lie within the range of C contents found in samples from the Apollo 11 and 12 sites.

(7) The surface-exposure times estimated for four fragmental rocks are approximately an order of magnitude less than the typical exposure times for rocks from the Apollo 11 and 12 sites and fall in the range of 10 to 20 million yr.

(8) A much broader range of soil mechanics properties was encountered at the Apollo 14 site than has been previously observed at the Apollo and Surveyor landing sites. At different points

along the traverse, lesser cohesion, coarser grain size, and greater resistance to penetration than at the Apollo 11 and 12 sites were found. These variations are indicative of a very complex heterogeneous deposit. The soils are more poorly sorted, but the range of grain size is similar to those of the Apollo 11 and 12 soils.

(9) No evidence of biological material has been found in any lunar material to date.

The Apollo 14 rocks exhibit characteristics that are compatible with their hypothesized derivation as ejecta from the Imbrium Basin. The rocks are largely fragmental and show pronounced shock effects, and the composition of most samples is distinctly different from that of basaltic rocks from lunar maria. The crystallinity observed in any of the fragmental rocks is compatible with the characteristics that would result from a single very large impact even in which annealing took place within a thick hot ejecta blanket.

Fragmental rocks within fragmental rocks and the apparently complex histories of nearly all the returned rocks deserve serious study. Numerous possibilities exist to explain the complex histories of these rocks: (1) excavation by the Imbrian event of preexisting brecciated material, including ejecta from the prior Mare Serenatatis cratering event; (2) formation of breccia within breccia during the Imbrian event; and (3) local cratering events superimposed on the Imbrian event.

If the rocks at the Apollo 14 landing site were excavated by the Imbrian event, then they represent a sample of the premare lunar surface to a depth of some tens of kilometers. The presence of microscopically visible exsolution in clinopyroxenes and orthopyroxenes indicates that some of the rocks from which the fragmental rocks were derived cooled more slowly and thus presumably originated at greater depths than the mare basalts, which typically exhibit only submicroscopic pyroxene exsolution.

The range of compositions and textures of clasts within the fragmental rocks suggests that their source area was complex. Nevertheless, the major- and trace-element abundances in the fragmental rocks and soils that were analyzed are so similar that extensive mixing must have occurred. Thus, it can be inferred that the bulk composition of the source area was enriched in incompatible elements and depleted in Fe relative to the mare basalts. The mineralogical and chemical similarity between the Fra Mauro material, on the one hand, and the fragments with a high abundance of KREEP component (or norite), the dark portion of rock sample 12013 from the Apollo 12 site and Luny Rock 1 from the Apollo 11 site, on the other hand, suggests that material of this composition is widespread on the lunar surface. Because this material is even further removed in composition from average solar or chondritic abundances than are the mare basalts, a profound premare lunar differentiation history that produced a lunar crust is indicated. This differentiation could have occurred during the accretion of the Moon with the near-surface accumulation of material that was rich in incompatible elements or in a process of crust formation that involved the fractionation of a considerable volume of the Moon early in the lunar history.

The basaltic rocks from the maria could not have been derived from a partial melting of material of Fra Mauro composition (and the reverse is also true). Therefore, it would appear that both the mare and nonmare areas of the Moon are not representative of the bulk composition of the Moon but may represent partial melting products of the lunar interior. The extent and variability of the rocks formed in the early lunar crust cannot be determined without samples from other upland areas, and no compelling reason yet exists to assume that the Fra Mauro material is representative of the lunar highlands.

The extremely low abundances of C and the virtual absence of H_2O in lunar rocks are further examples of the depletion of volatile elements that have been observed in other lunar samples and also indicate that the probability of finding indigenous life on the Moon is increasingly slight. The preliminary examination has shown that the samples from the Apollo 14 site are complex and clearly distinct from previously returned lunar rocks and from the closest meteoritic and terrestrial analogs. The samples should present a considerable challenge to lunar scientists as they attempt to increase man's knowledge of lunar history. Their study should elucidate a period of lunar evolution for which no similar terrestrial record is available.

References

5-1. GILBERT, G. K.: The Moon's Face. Bull. Phil. Soc. Wash., vol. 12, no. 241, 1893.

5-2. EGGLETON, R. E.: Preliminary Geology of the Riphaeus Quadrangle of the Moon and Definition of the Fra Mauro Formation. Astrogeol. Studies, Ann. Prog. Rept., Aug. 1962–July 1963. Pt. A, U.S. Geol. Survey open-file Rept., 1964, pp. 46–63.

5-3. ANON.: Preliminary Examination of Lunar Samples From Apollo 11. Science, vol. 165, no. 3899, Sept. 19, 1969, pp. 1211–1227.

5-4. ANON.: Preliminary Examination of Lunar Samples From Apollo 12. Science, vol. 167, no. 3923, Mar. 6, 1970, pp. 1325–1339.

5-5. GAULT, DONALD E.; QUAIDE, WILLIAM L.; AND OBERBECK, VERNE R.: Impact Cratering Mechanics and Structures. Shock Metamorphism of Natural Materials, Bevan M. French and Nicholas M. Short, eds., Mono Book Corp. (Baltimore), 1968, pp. 87–99.

5-6. HUBBARD, NORMAN J.; MEYER, CHARLES, JR.; GAST, PAUL W.; AND WIESMANN, HENRY: The Composition and Derivation of Apollo 12 Soils. Earth Planet. Sci. Lett., vol. 10, no. 3, Feb. 1971, pp. 341–350.

5-7. DRAKE, M. J.; McCALLUM, I. S.; McKAY, G. A.; AND WEILL, D. F.: Mineralogy and Petrology of Apollo 12 Sample No. 12013: A Progress Report. Earth Planet. Sci. Lett., vol. 9, no. 103, 1970, pp. 103–123.

5-8. ALBEE, A. L.; AND CHODOS, A. A.: Microprobe Investigations on the Apollo 11 Samples. Proc. Apollo 11 Lunar Sci. Conf., Geochim. Cosmochim. Acta Suppl. 1, vol. 1, A. A. Levinson, ed., Pergamon Press, 1970, pp. 135–157.

5-9. MOORE, CARLETON B.; LEWIS, CHARLES F.; GIBSON, EVERETT K.; AND NICHIPORUK, WALTER: Total Carbon and Nitrogen Abundances in Lunar Samples. Science, vol. 167, no. 3918, Jan. 30, 1970, pp. 495–497.

5-10. BOGARD, D. D.; FUNKHOUSER, J. G.; SCHAEFFER, O. A.; AND ZAHRINGER, J.: Noble Gas Abundances in Lunar Material—Cosmic-Ray Spallation Products From the Sea of Tranquility and the Ocean of Storms. J. Geophys. Res., vol. 76, no. 11, Apr. 1971, pp. 2757–2779.

5-11. EBERHARDT, P.; GEISS, J.; GRAF, H.; GROEGLER, N.; ET AL.: Trapped Solar Wind Noble Gases, Exposure Age and K/Ar Age in Apollo 11 Lunar Fine Material. Proc. Apollo 11 Lunar Sci. Conf., Geochim. Cosmochim. Acta Suppl. 1, vol. 1, A. A. Levinson, ed., Pergamon Press, 1970, pp. 1037–1070.

ACKNOWLEDGMENTS

The people who contributed directly to obtaining the data and to the preparation of this report are D. H. Anderson, NASA Manned Spacecraft Center (MSC); M. N. Bass, MSC; A. Dean Bennett, Brown & Root-Northrop (BRN); D. D. Bogard, MSC; Robin Brett, MSC; L. G. Bromwell, Massachusetts Institute of Technology (MIT); Patrick Butler, Jr., MSC; W. D. Carrier III, MSC; R. S. Clark, MSC; M. B. Duke, MSC; Paul W. Gast, MSC; E. K. Gibson, Jr., MSC; W. R. Hart, BRN; G. H. Heiken, MSC; W. C. Hirsch, BRN; Friedrich Hörz, MSC; E. D. Jackson, U.S. Geological Survey (USGS); Pratt H. Johnson, BRN; J. E. Keith, MSC; C. F. Lewis, Arizona State University; John F. Lindsay, MSC; J. Roger Martin, BRN; William C. Melson, U.S. National Museum; Edgar D. Mitchell, MSC; Carleton B. Moore, Arizona State University; D. A. Morrison, MSC; Welden B. Nance, BRN; William C. Phinney, MSC; A. M. Reid, MSC; M. A. Reynolds, MSC; K. A. Richardson, MSC; W. I. Ridley, MSC; E. Schonfeld, MSC; Alan B. Shepard, MSC; R. L. Sutton, USGS; N. J. Trask, USGS; Jeff Warner, MSC; R. B. Wilkin, BRN; H. G. Wilshire, USGS; and D. R. Wones, MIT.

The members of the Lunar Sample Preliminary Examination Team wish to acknowledge the technical assistance of the NASA Manned Spacecraft Center, Brown & Root-Northrop, and Lockheed Electronics Corp. staffs, which include Travis J. Allen, John O. Annexstad, R. Bell, Linda Bennett, Mike C. Brabham, E. P. Carranza, L. E. Cornitius, C. Cucksee, James B. Dorsey, Adrian Eaton, Paul Gilmore, Paul Graf, George M. Greene, D. W. Hutchinson, Robert W. Irvin, D. Jezek, Stuart W. Johnson, Carl E. Lee, E. Allen Locke, T. M. McPherson, David Mann, Maureen Mitchell, David R. Moore, J. L. Nix, W. A. Parkan, Davis S. Pettus, Clifford M. Polo, W. R. Portenier, G. R. Primeaux, James Ramsay, Sandra Richards, M. K. Robbins, J. Schwartzback, Jr., John Siggin, Louis A. Simms, K. L. Suit, Nancy Trent, Nelson L. Turner, Linda Tyler, and D. R. White.

6. Passive Seismic Experiment

Gary V. Latham,[a][†] *Maurice Ewing,*[a] *Frank Press,*[b] *George Sutton,*[c]
James Dorman,[a] *Yosio Nakamura,*[d] *Nafi Toksoz,*[b] *Fred Duennebier,*[c]
and David Lammlein[a]

The purpose of the passive seismic experiment (PSE) is to detect vibrations of the lunar surface and to use these data to determine the internal structure, physical state, and tectonic activity of the Moon. Sources of seismic energy may be internal (moonquakes) or external (meteoroid impacts and manmade impacts). A secondary objective of the experiment is the determination of the number of the masses of meteoroids that strike the lunar surface. The instrument is also capable of measuring tilts of the lunar surface and changes in gravity that occur at the PSE location.

Since deployment and activation of the PSE on February 5, 1971, the instrument has operated as planned, except as noted in the following subsection. The sensor was installed west-northwest from the lunar module (LM) 178 m from the nearest LM footpad. With the successful deployment and operation of the Apollo 14 PSE and the continued operation of the Apollo 12 PSE 181 km to the west, a major milestone in the geophysical exploration of the Moon has been achieved. For the first time, geophysical measurements can be made on the lunar surface simultaneously at two widely separated stations. These two stations represent the beginning of the network of observatories required for determining the locations of natural events.

Seismic signals from 79 events believed to be of natural origin were recorded by the three long-period (LP) seismometers at the Apollo 14 site during the 44-day period following the LM ascent.

Of these, at least two were moonquakes that originated in the active region (A$_1$ zone), previously identified from recordings at the Apollo 12 site, and 12 others were possibly moonquakes. The moonquakes occurred most frequently near the time of closest approach between the Earth and the Moon (perigee), suggesting that the moonquakes are triggered by tidal stresses. Thirty-one of the events were also recorded at the Apollo 12 site. All events detected by the Apollo 12 PSE were detected by the Apollo 14 PSE.

The greater rate of detection of seismic events by the Apollo 14 PSE is believed to be a consequence of the thick layer of unconsolidated material that blankets the region (Fra Mauro Formation plus overlying regolith) and amplifies seismic motion at the surface. The surface-amplification effect may amplify noise as much as seismic signals; but, on the Moon, natural noise is so low that it has not been appreciably above the threshold of any instrument yet deployed.

The A$_1$ zone is estimated to be nearly equidistant from the Apollo 12 and 14 sites and at a range of approximately 600 to 700 km from them. Except for the absence of the surface-reflected phases usually seen on the records from deep earthquakes, the seismic data suggest that A$_1$ events, and perhaps all moonquakes, are deep. If this theory is verified by future data, the result will have fundamental implications relative to the present state of the lunar interior.

Many of the events recorded by the Apollo 14 PSE were meteoroid impacts. Measurements of the amplitudes permit a tentative estimate for the mass distribution of meteoroids that collide with the lunar surface:

[a] Lamont-Doherty Geological Observatory.
[b] Massachusetts Institute of Technology.
[c] University of Hawaii.
[d] General Dynamics.
[†] Principal investigator.

$$\log N = -2.32 - \log M \qquad (6\text{--}1)$$

where N (given in $km^{-2} \cdot yr^{-1}$) is the cumulative number of meteoroids with mass equal to or greater than M (given in grams). One-half of the recorded impacts occur at ranges less than 1300 km (nearest 13 percent of the lunar surface). Twenty-one percent of the recorded impacts occur on the hemisphere opposite the Apollo 14 site. Thus, it appears that impacts are being detected at very great ranges on the lunar surface.

One of the largest natural impacts detected during this report period may have generated a gas cloud that was detected at the Apollo 14 site by the suprathermal ion detector experiment (SIDE) sensor approximately 37 min after the beginning of the seismic signal. If the gas cloud was indeed the result of the impact that generated the seismic signal, the meteoroid struck the lunar surface at a distance of approximately 2000 km and had a mass of 10 to 20 kg.

Seismic signals were recorded from two man-made impacts during the Apollo 14 mission: The SIVB stage of the Saturn launch vehicle and the LM ascent stage. The LM-impact signal was the first event of precisely known location and time recorded by two instruments on the lunar surface. Data from these impacts combined with data from the impacts accomplished during the Apollo 12 and 13 missions, and the lunar-surface magnetometer results are the main sources of information on the internal structure of the Moon. The general characteristics of the recorded seismic signals suggest that the outer shell of the Moon, to depths of 50 to 100 km, is highly heterogeneous in the region of the Apollo 12 and 14 sites. The heterogeneity of the outer zone results in intensive scattering of seismic waves and greatly complicates the recorded signals.

The nature of the zone in which scattering of seismic waves occurs is not precisely known, but it is evident that the presence of craters must contribute, at least in part, to the general complexity of the zone. The scattering zone appears, however, to be too thick to be accounted for solely by visible craters. Nor is it known whether the entire Moon is mantled by such a layer. It is possible—in fact, probable—that the lunar highlands will reveal structures quite different from the maria in which seismic instruments have thus far been located. The lunar body beneath the scattering zone appears to transmit seismic energy with extremely high efficiency. This observation does not necessarily imply that the rock at these depths (greater than 50 to 100 km) was never fractured; it may simply be the depth at which rock, conditioned by pressure, begins to behave more nearly as a homogeneous elastic solid.

The impact data suggest that long-range propagation of seismic waves, where most of the path would be located beneath the scattering zone, may be much more efficient than thought previously. It now appears likely, for example, that the LM impact anywhere on the Moon would have been recorded at the Apollo 14 site. If so, many of the recorded meteoroid impacts may occur at great ranges from the Apollo 12 and 14 sites. No clear evidence for the presence of a major boundary within the upper 25 km of the Moon has yet emerged, although present data are too limited to preclude this possibility.

Signals were recorded from astronaut activities at all points along their traverse and from the LM ascent. The velocities of the signals from the LM ascent agree with velocities determined for the upper 100 m of lunar material from the active seismic experiment (ASE). The velocity of the upper few meters of unconsolidated material (regolith) is 104 m/sec. This value is quite close to the regolith velocity measured at the Apollo 12 site (108 m/sec).

Instrument Description and Performance

A seismometer consists simply of a mass, free to move in one direction, that is suspended by means of a spring (or a combination of springs and hinges) from a framework. The suspended mass is provided with damping to suppress vibrations at the natural frequency of the system. The framework rests on the surface, the motions of which are to be studied, and moves with the surface. The suspended mass tends to remain fixed in space because of its own inertia while the frame moves around the mass. The resulting relative motion between the mass and the framework can be recorded and used to calculate original ground motion if the instrumental constants are known.

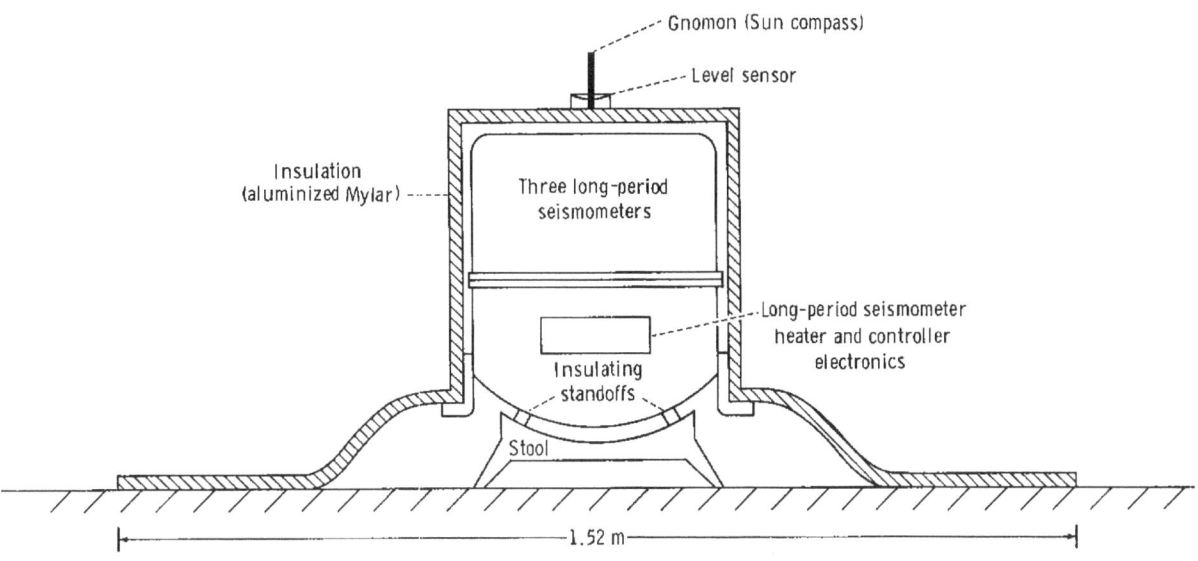

FIGURE 6–1.—Schematic diagram of PSE.

The Apollo 14 PSE consists of two main subsystems: the sensor unit and the electronics module. The sensor, shown schematically in figure 6–1, contains three matched LP seismometers (with resonant periods of 15 sec) alined orthogonally to measure one vertical (Z) and two horizontal (X and Y) components of surface motion. The sensor also includes a single-axis short-period (SP) seismometer (with a resonant period of 1 sec) sensitive to vertical motion at higher frequencies.

The instrument is constructed principally of beryllium and weighs 11.5 kg, including the electronics module and thermal insulation. Without insulation, the sensor is 23 cm in diameter and 29 cm high. The total power drain varies between 4.3 and 7.4 W.

Instrument temperature control is provided by a 6-W heater, a proportional controller, and an insulating wrapping of aluminized Mylar. The insulating shroud is spread over the local surface to reduce temperature variations of the surface material. In this way, it is expected that thermally induced tilts of the local surface will be reduced to acceptable levels.

The LP seismometer will detect vibrations of the lunar surface in the frequency range from 0.004 to 2 Hz. The SP seismometer covers the band from 0.05 to 20 Hz. The LP seismometers can detect ground motions as small as 0.3 nm at maximum sensitivity; the SP seismometer can detect ground motions of 0.3 nm at 1 Hz.

The LP horizontal-component (LPX and LPY) seismometers are very sensitive to tilt and must be leveled to high accuracy. In the Apollo system, the seismometers are leveled by means of a two-axis, motor-driven gimbal. A third motor adjusts the LP vertical-component (LPZ) seismometer in the vertical direction. Motor operation is controlled by command.

Calibration of the complete system is accomplished by applying an accurate increment or step of current to the coil of each of the four seismometers by transmission of a command from Earth. The current step is equivalent to a known step of ground acceleration.

A caging system is provided to secure all critical elements of the instrument against damage during the transport and deployment phases of the Apollo mission. In the present design, a pneumatic system is used in which pressurized bellows expand to clamp fragile parts in place. Uncaging is performed on command by piercing the connecting line by means of a small explosive device.

The seismometer system is controlled from Earth by a set of 15 commands that govern functions such as speed and direction of leveling motors, and instrument gain and calibration. The

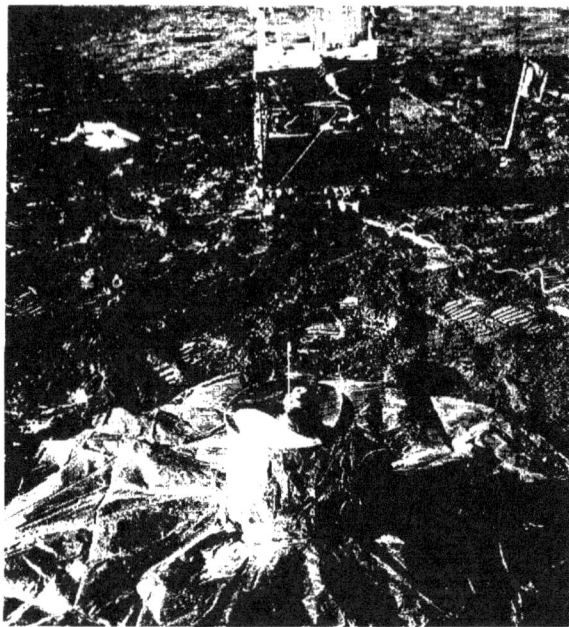

FIGURE 6-2.—Seismometer after deployment on the lunar surface (AS14-67-9363).

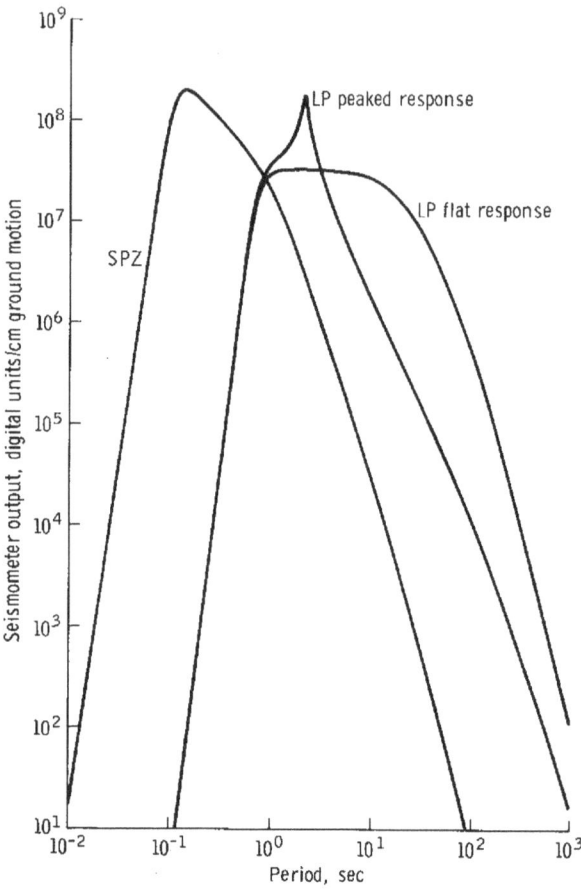

FIGURE 6-3.—Response curves for the LP and SPZ seismometers. The ordinate scale is in digital units (DU) per centimeter ground motion amplitude. A DU is the signal variation that corresponds to a change in the least significant bit of the 10-bit digital data word.

seismometer is shown fully deployed on the lunar surface in figure 6-2.

Two modes of operation of the LP seismometers are possible: the flat-response mode and the peaked-response mode. In the flat-response mode, the seismometers have natural periods of 15 sec. In the peaked-response mode, the seismometers act as underdamped pendulums with natural periods of 2.2 sec. Maximum sensitivity is increased by a factor of 6 in the peaked-response mode, but sensitivity to low-frequency signals is reduced. The response curves for both modes are shown in figure 6-3.

The PSE was deployed 3 m north of the central station. No difficulty was experienced in deploying the experiment. Since initial activation of the PSE, all elements have operated as planned with two exceptions—instability of the LPZ seismometer in the flat-response mode, and intermittent response of the Y-axis of the gimbal leveling system.

In the normal mode of operation (flat-response mode), the LPZ seismometer is unstable for reasons as yet undetermined. This problem has been circumvented by removing the feedback filter and operating all three LP seismometers in the

peaked-response mode. The SPZ seismometer has operated as planned.

Operation of the gimbal motor that levels the LPY seismometer has shown intermittent malfunction. The motor drive has not responded to command on several occasions. In these cases, the reserve-power status has indicated that no power was being supplied to the motor. The power-control circuit of the leveling motor is presently considered to be the most likely cause of this problem. Motor actuation has been achieved in all cases by repeating the motor drive command.

Maximum instrument temperature exceeds the design set point (125° F) by about 5° F near lunar noon. The desired thermal control is maintained during lunar night.

As was recorded at the Apollo 12 site, episodes of seismic disturbances are observed on the LP seismometers throughout the lunar day. These disturbances are most intense near times of terminator passage and are believed to be due to thermal contraction and expansion of the Mylar thermal shroud that blankets the sensor.

Results

This experiment is a continuation of observations made during the Apollo 11, 12, and 13 missions (refs. 6–1 to 6–6).

Preascent Period

Before the LM ascent, many signals corresponding to various astronaut activities within the LM

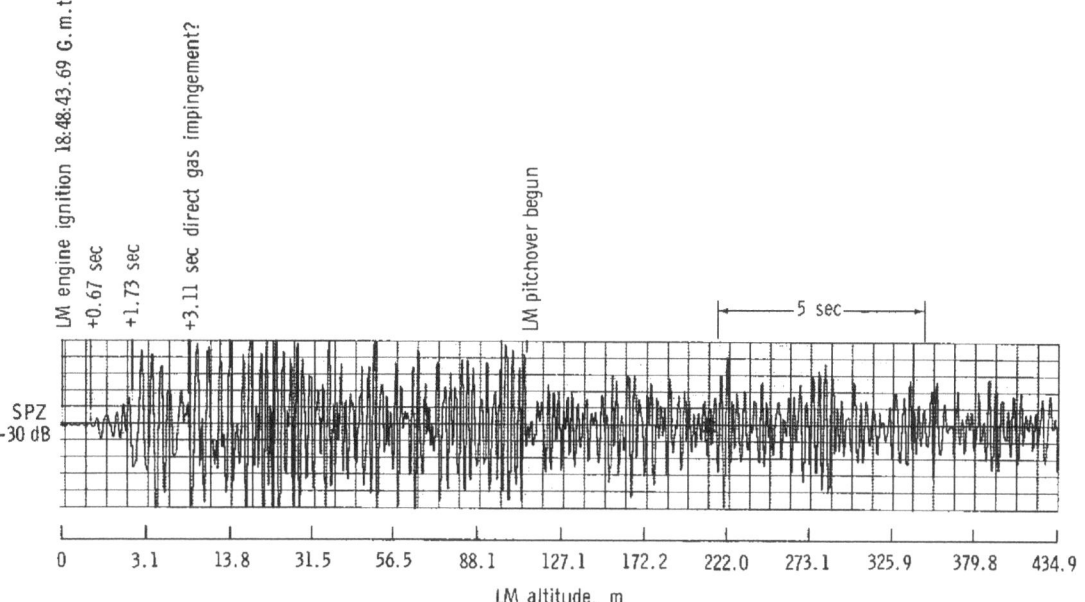

FIGURE 6–4.—Signal recorded by the SPZ seismometer from the liftoff of the Apollo 14 LM ascent stage. The LM rose vertically to an altitude of 100 m and then pitched over to a nearly horizontal attitude.

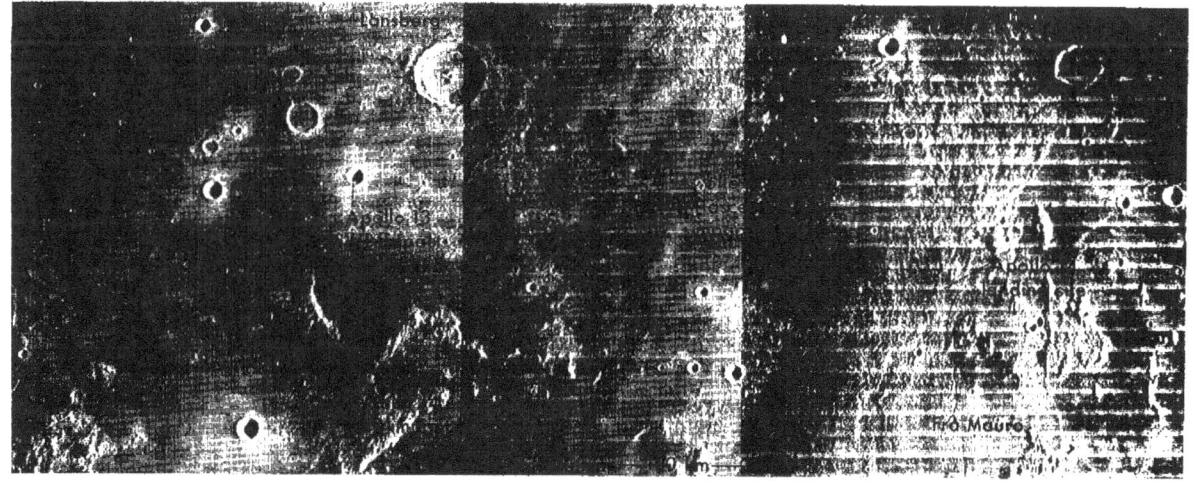

FIGURE 6–5.—Locations of ALSEP stations and LM and SIVB impacts.

and on the surface were recorded, primarily on the SPZ seismometer. The astronauts' footfalls were detected at all points along their traverse (maximum range approximately 1.4 km). A signal of particular interest during this period was generated by the thrust of the LM ascent engine. The signal began 0.67 sec after the burn began and lasted about 4.5 min. As shown in figure 6–4, two secondary arrivals can be recognized in the early part of the signal. These arrivals follow the first arrival by 1.06 and 2.44 sec. By comparison with the traveltime of seismic waves measured by the active seismic experiment using the thumper source, the first arrival was interpreted as a wave refracted along the surface of a higher velocity material (possibly the Fra Mauro Formation) located at a depth of 8.5 m (sec. 7). The second arrival had an apparent velocity of 104 m/sec and corresponded to a wave traveling through the top (regolith) layer. This value is remarkably close to the velocity of 108 m/sec measured from the LM ascent at the Apollo 12 site. The third arrival was probably generated by direct impingement of exhaust gas on the PSE. Following this abrupt increase in signal amplitude, the signal remained at a high level for approximately 8.2 sec and then decreased abruptly. The decrease in signal amplitude corresponded in time to the LM pitch maneuver. Following the pitch maneuver, exhaust gases flowed parallel to the surface and no longer impinged on the surface in the near vicinity of the PSE.

Signals From Impacts of the SIVB and LM Ascent Stages

Signals from two manmade impacts were recorded as part of the Apollo 14 mission: The SIVB stage and the LM ascent stage. The SIVB impact preceded emplacement of the Apollo 14 Apollo lunar-surface experiments package (ALSEP) station and was therefore recorded at the Apollo 12 site only. The LM impact point was nominal and was recorded at both the Apollo 12 and 14 sites. The SIVB impact point was closer to the Apollo 12 site than planned (172 km instead of 300 km) and, though well recorded, did not provide the new information desired. However, data from these impacts have aided greatly

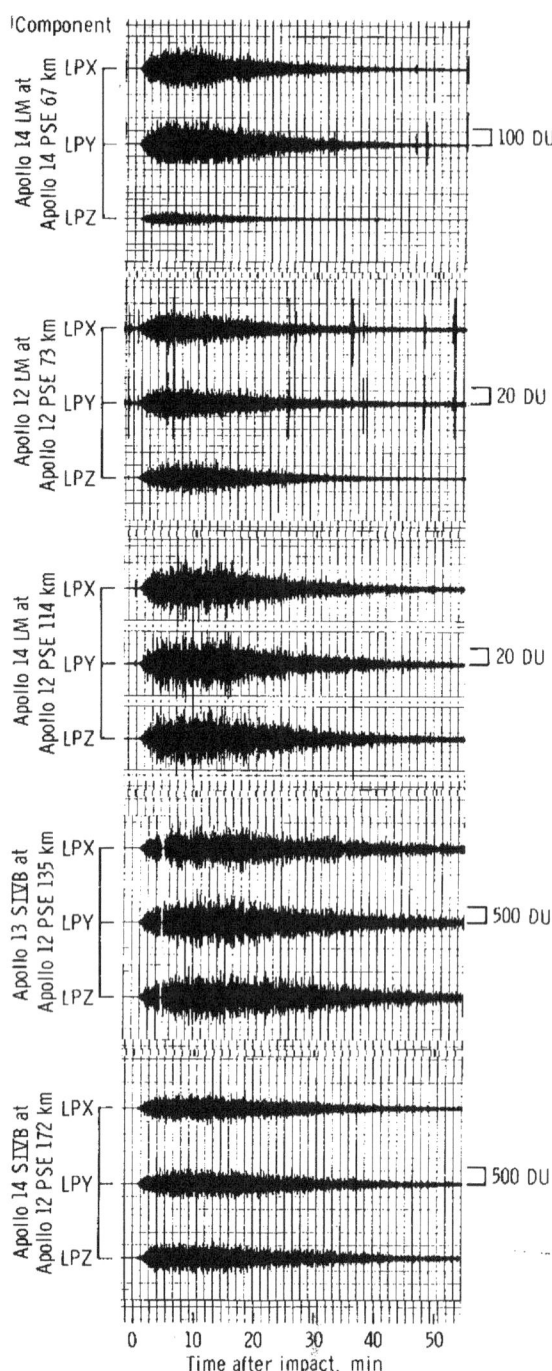

FIGURE 6–6.—Compressed time-scale records of the seismic signals received from the LM and SIVB impacts of the Apollo 12, 13, and 14 missions, as recorded by the PSE LP seismometers at the Apollo 12 and 14 sites. The LPZ is the vertical-component seismometer; the LPX and LPY are the horizontal-component seismometers. The DU is explained in figure 6–3.

TABLE 6–I. *Manmade Impact Parameters*

Impact parameters	Impacting vehicle			
	Apollo 12 LM	Apollo 14 LM	Apollo 13 SIVB	Apollo 14 SIVB
Day, G.m.t.........................	Nov. 20, 1969	Feb. 7, 1971	Apr. 15, 1970	Feb. 4, 1971
Range time,[a] G.m.t., hr:min:sec......	22:17:17.7	00:45:25.7	01:09:41.0	07:40:55.4
Real time, G.m.t., hr:min:sec........	22:17:16.4	00:45:24.4	01:09:39.7	07:40:54.2
Velocity, km/sec....................	1.68	1.68	2.58	2.54
Mass, kg...........................	2383	2303	13 925	14 016
Kinetic energy, ergs.................	3.36×10^{16}	3.25×10^{16}	4.63×10^{16}	5.54×10^{16}
Angle from horizontal, deg..........	3.7	3.6	76	69
Heading, deg.......................	306	282	78	103

[a] Range time is the time signal of associated event was received on Earth.

TABLE 6–II. *Distances From Impact Points to Seismic Stations*

Impacting vehicle	Distance, km, from impact point to—	
	Apollo 12 site	Apollo 14 site
Apollo 12 LM......	73	
Apollo 14 LM......	114	67
Apollo 13 SIVB....	135	
Apollo 14 SIVB....	172	

TABLE 6–III. *Coordinates of Seismic Stations and Impact Points*

[From the Lunar Planning Chart (LOC–2), scale 1:2,500,000, ed. 1, July 1969]

Location	Coordinates, deg	
	South	West
Apollo 12 site.............	3.04	23.42
Apollo 14 site.............	3.65	17.48
Apollo 12 LM-impact point..	3.94	21.20
Apollo 14 LM-impact point..	3.42	19.67
Apollo 13 SIVB-impact point.	2.75	27.86
Apollo 14 SIVB-impact point.	8.09	26.02

in understanding lunar structure and seismic-wave transmission.

The locations of the Apollo 12 and 14 impacts and of all impacts accomplished to date are shown in figure 6–5. Pertinent parameters for the impacts and the distances from the seismic stations

are given in tables 6–I to 6–III. The impact signals are shown in compressed time scales in figure 6–6.

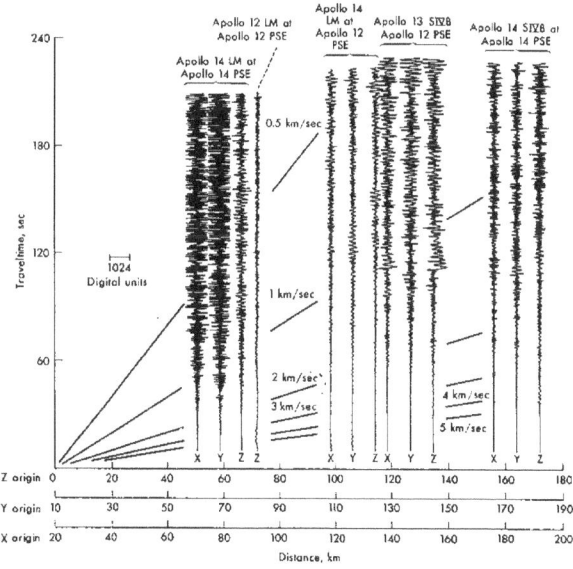

FIGURE 6–7.—Record section of LP seismograms of artificial impacts recorded at the Apollo 12 and 14 sites. Impact and station parameters are given in tables 6–I to 6–III. The Z-component of each event is plotted at the source-receiver distance on the range scale. The X- and Y-components of the Apollo 12 LM impact are not plotted because they would overlap nearby traces. The seismometers were operated at full gain on the Moon, all at peaked response except for flat response for the Apollo 12 LM recording. Length of bar corresponds to 1024 DU for SIVB recordings. The playout system magnification for the LM-impact traces is 10 times that of the SIVB impact traces.

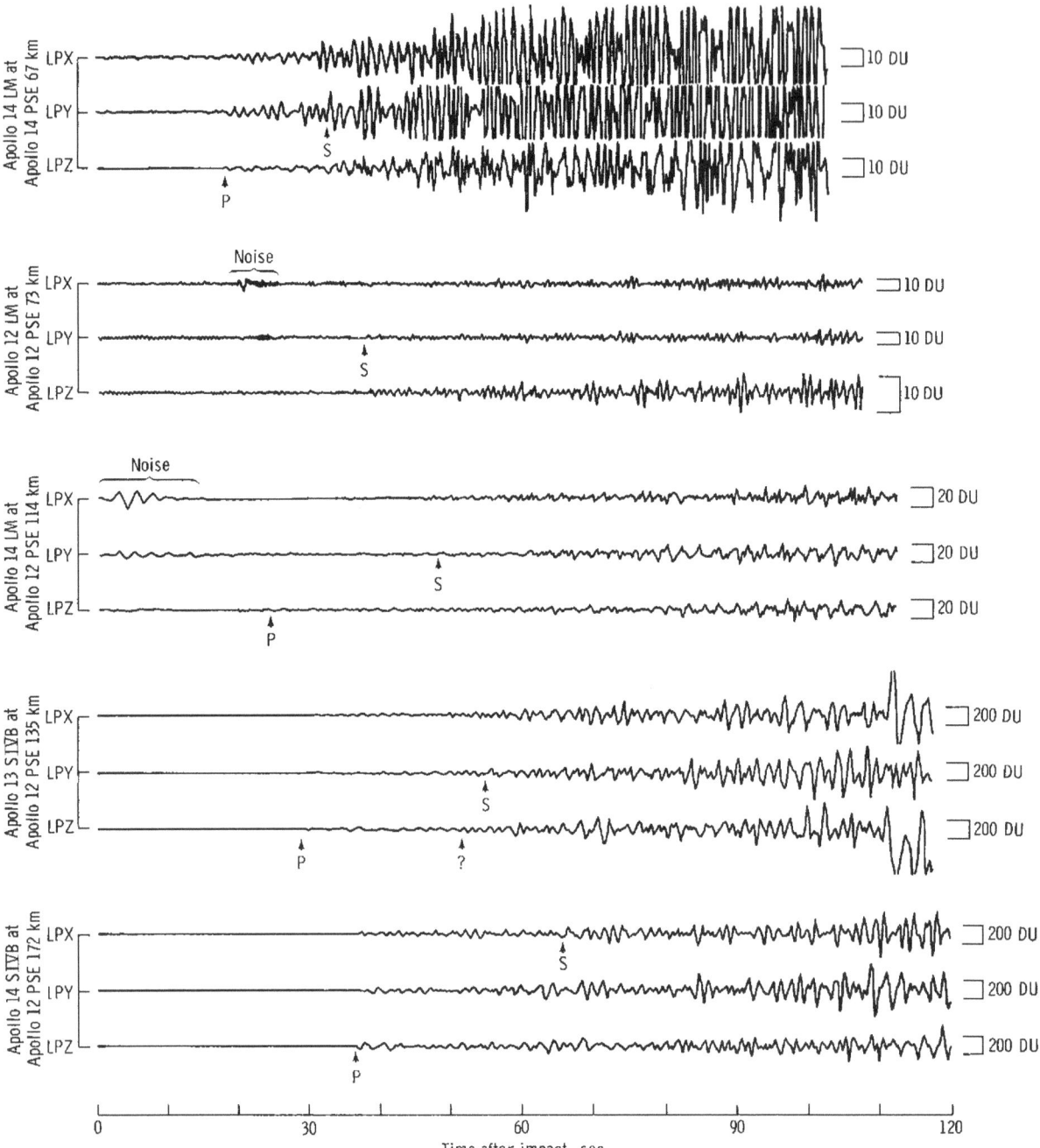

FIGURE 6-8.—Records of the seismic signals received from the LM and SIVB impacts of the Apollo 12, 13, and 14 missions, as recorded by the PSE at the Apollo 12 and 14 sites. The LPZ is the vertical-component seismometer; the LPX and LPY are the horizontal-component seismometers. The earliest detectable signal is identified as the direct compressional-wave (P-wave) arrival, except for the record from the Apollo 12 LM impact where the signal is too weak to identify the first arrival with certainty. A later arrival, tentatively identified as the shear wave (S-wave), is also indicated in each record. The DU is explained in figure 6-3.

The characteristics of the new impact signals are very similar to those of the previous impacts. The signals are prolonged with gradual increase and decrease in signal intensity and with little correlation between any two components of ground motion. Signals corresponding to the arrival of various types of seismic waves can be identified in the early parts of the records, but they are less distinct than in normal earthquake seismic records. These characteristics are believed to result from intensive scattering of the seismic waves, implying that the outer shell of the Moon in the region of the Apollo 12 and 14 sites may be highly heterogeneous, possibly to depths as great as 50 to 100 km. Seismic-wave transmission and inferred lunar structure are discussed in greater detail in the following paragraphs.

The beginnings of the impact signals are shown on expanded time scales in figures 6–7 and 6–8. The seismic records in figure 6–7 are spaced on a

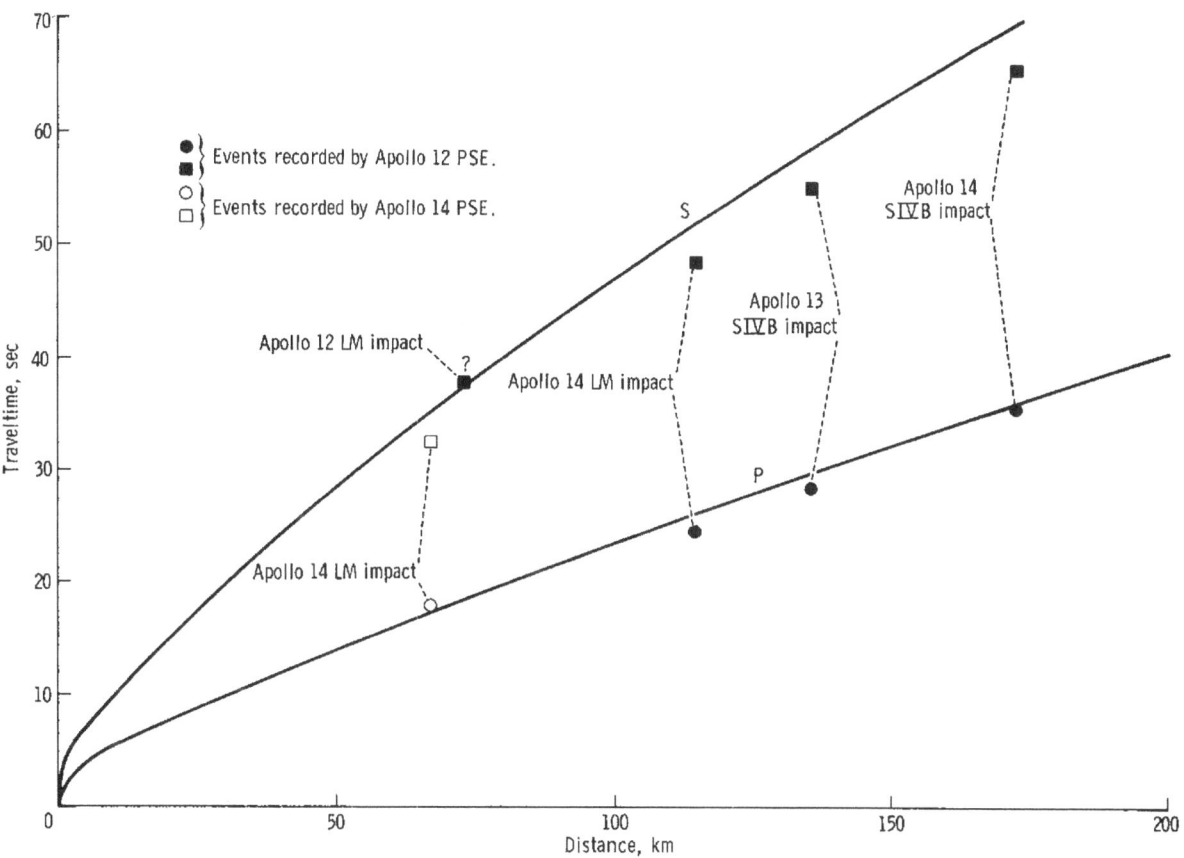

FIGURE 6–9.—Traveltimes of seismic waves, assumed to be the direct compressional and shear waves (P- and S-waves, respectively), recorded from the LM and SIVB impacts of the Apollo 12, 13, and 14 missions. The phases are identified on the records shown in figure 6–8. The beginning of the wavetrain from the Apollo 12 LM impact is so emergent that the first arrival (P-wave) cannot be identified with certainty. The phase identified as a possible shear wave (S-wave) is also uncertain. The solid lines are theoretical traveltime curves for P- and S-waves for the assumed lunar structure shown in figure 6–20. The P-wave data fit the theoretical traveltime curve well, indicating that the assumed lunar model for the variation of compression-wave velocity with depth is a good approximation to depths of at least 25 km. The observed shear-wave traveltimes, if correctly identified, are shorter than those predicted by the assumed model. Thus, the shear velocities are somewhat higher in the upper 30 km of the Moon than those given in the model.

distance scale representing the source-receiver ranges ("record section" format) and are included primarily to illustrate the complexity of the signals and the difficulty of relating, in detail, the waveforms of one to another. The times of phases corresponding to the arrival of various types of seismic waves are indicated in figure 6–8. Corresponding traveltimes for these phases are shown in figure 6–9. Traveltimes derived for compressional waves (P-waves) and shear waves from high-pressure measurements on returned lunar samples are also shown in figure 6–9 for comparison. (This subject is covered in detail in "Discussion.")

Data From the LP Seismometers

Data presently being received at the Apollo 14 site on the LPX, LPY, and LPZ seismometers are similar in their main features to those recorded at the Apollo 12 site, but some significant and useful differences exist. As at the Apollo 12 site, ambient seismic noise appears to be below the threshold of instrument sensitivity, emphasizing again the extreme quietness of the lunar surface. Events believed to be of natural origin, moonquakes or meteorite impacts, are also recorded, as at the Apollo 12 site, though such recordings are more frequent at the Apollo 14 site. The gross characteristics are very similar to those of events recorded at the Apollo 12 site and similar in many respects to the characteristics of the artificial-impact signals described previously. The principal features are long durations measured in tens of minutes, an emergent beginning, a rise or buildup time of approximately 2 to 15 min, and much slower decay or tailing off of signal strength. As at the Apollo 12 site, moonquakes can probably be distinguished from impact events in most cases by several criteria, including the presence of an early abruptly beginning phase (H-phase) that occurs on the horizontal components only. Conspicuously, the horizontal amplitudes at the Apollo 14 site are in the ratio of approximately 3:1 to the vertical amplitude. At the Apollo 12 site, all components are about equally strong, though minor variations from event to event have been recorded by both instruments. For signals recorded by both instruments, the total ground-motion amplitudes recorded by the Apollo 14 PSE are larger than those of the Apollo 12 PSE by an average factor of approximately 2.

All LP events that have been recorded by the Apollo 12 PSE since deployment of the Apollo 14 PSE have been recorded simultaneously by the Apollo 14 PSE. The reverse is not true. This interesting relationship appears to have a useful application in estimating the distance of the seismic sources as explained in the following paragraphs. While simultaneous observation at two sites of signals from the same event is not sufficient for locating the event precisely, many other methods prove useful in characterizing the events and evaluating the comparative geologic structures of the two sites.

Data From the SPZ Seismometer

Hundreds of signals with a great variety of shapes and sizes have been recorded on the SPZ seismometer during the first 45 days of operation of the Apollo 14 PSE. The general level of recorded activity gradually subsided through the first lunar night following the initial activation of the PSE and increased abruptly at sunrise. Most of the events are attributed to venting of gases and thermoelastic "popping" within the LM descent stage. The PSE is located approximately 178 m from the nearest footpad of the LM. Similar signals at even higher levels of activity were observed during the operation of the Apollo 11 PSE. The higher level of activity at the Apollo 11 site is explained by the smaller separation between the LM and the PSE in the Apollo 11 deployment (16.8 m). (The Apollo 12 PSE was deployed 130 m from the nearest LM footpad.)

One family of signals that occur near sunrise and sunset exhibits a sequence of pulses early in the wavetrain; these pulses are identical to those recorded during the LM ascent. This family of signals is almost certainly generated by impulsive sources within the LM descent stage.

The variation of SPZ activity, whether LM-generated or not, is clearly related to the solar cycle at the Apollo 14 site. Thus, possible sources of the activity are thermal effects on the LM, the PSE thermal shroud, the cable connecting the sensor to the central station, other ALSEP instruments, and the lunar soil and nearby rocks.

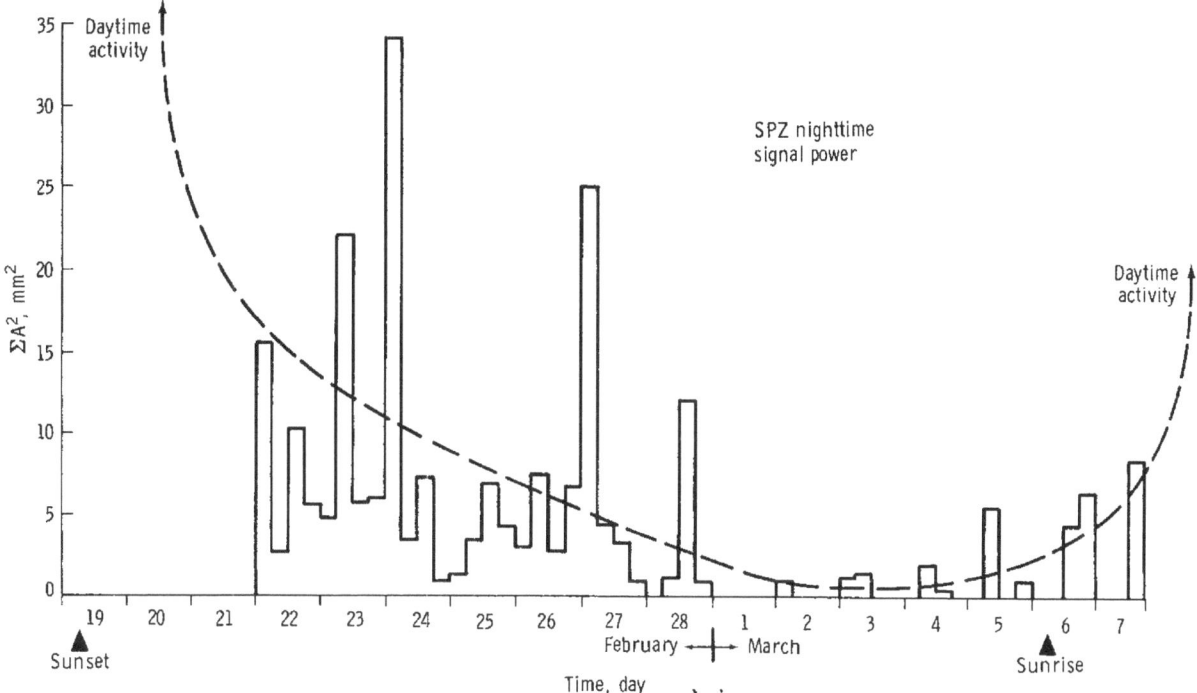

FIGURE 6–10.—Relative SPZ signal power during the first night after deployment, as a function of time. For each event, amplitude A is maximum peak-to-peak signal not exceeded more than 10 percent of the time on seismograms where 0.5 mm is 1 DU. The height of each bar is the sum of squares of the maximum trace amplitude of events recorded during each 6-hr interval. The average level of activity is shown by the dotted line. This graph begins almost 3 days after the preceding sunset, which was at approximately 06:00 G.m.t. on February 19. Activity before, during, and shortly after the sunset was too intense to be represented on the vertical scale of this graph, as was activity following the sunrise of 08:12 G.m.t. on March 6. The apparent nighttime decay constant is several days, probably too long to represent cooling of the LM. However, the cooling time constant of rocks and soil may be of the magnitude indicated in this figure.

The seismic activity recorded on the SPZ seismometer during the first lunar night is shown in figure 6–10. The level of activity clearly decreased throughout the lunar night. The LM descent stage would certainly reach thermal equilibrium at very low temperatures soon after sunset, probably within 1 day. Thus, the continuing activity that was observed raises the question of possible natural sources of seismic signals. Meteoroid impacts at close range would be likely sources for some of the signals. A cumulative-frequency curve of the amplitudes of nighttime SPZ events is shown in figure 6–11. The slope of this curve is in approximate agreement with the predicted slope of the flux (ref. 6–7) for meteoroid impacts in the small-mass range (cometary particles). However,

the frequency of meteoroid impacts would not be expected to show a gradual decrease during the lunar night unless a monthly variation in meteoroid flux is postulated. Thermal cracking of rocks of relatively recent exposure is perhaps a more likely process that might have the observed time characteristic. It is expected that more definite remarks can be made concerning the sources of signals recorded on the SPZ seismometer as the contribution from the LM descent stage decreases during succeeding lunations.

Natural Events

The general discussion of SP and LP data leads to a more detailed consideration of the data on

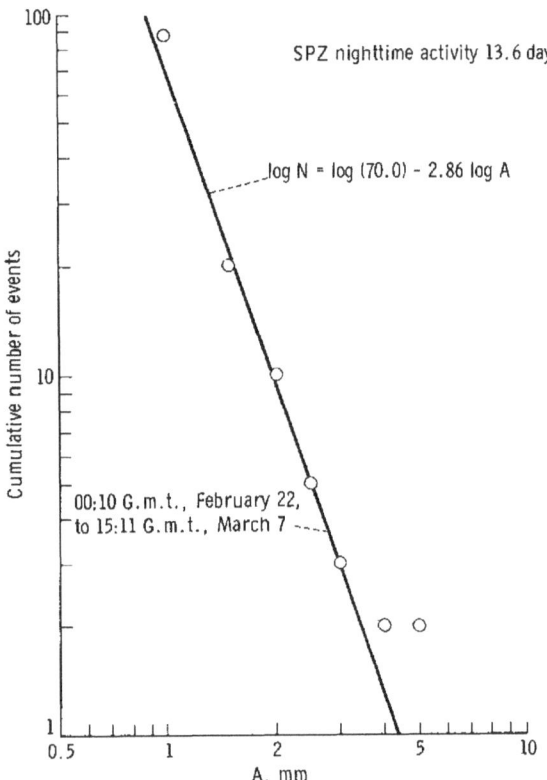

SPZ nighttime activity 13.6 days

$\log N = \log (70.0) - 2.86 \log A$

00:10 G.m.t., February 22, to 15:11 G.m.t., March 7

A, mm

FIGURE 6–11.—Cumulative amplitude distribution of A values of same events represented in figure 6–10. The slope of this curve is appropriate for sources resulting from impacts of cometary particles, but the time variation of intensity as shown in figure 6–10 is not consistent with that hypothesis.

discrete events of natural origin. The interpretation of these data rests on principles developed from study of artificial-impact data, described previously, where the known time, place, and energy input of the seismic source permit the use of a wider range of analytical methods. By contrast, the problems of determining the basic parameters of a natural event, such as distance to source, time of origin, and source nature and strength, are difficult because many techniques used in terrestrial seismology are not applicable under lunar conditions. These conditions include not only the sparsity of seismic stations but also the strikingly different seismic-wave transmission characteristics of the Moon. However, substantial progress is being made and some significant steps are represented in this report. Estimates of source

distance based on rise time of the signal are reliable to perhaps ±20 percent out to a range of 150 km. An estimate of the law of amplitude falloff with a distance can now be placed on a fairly reliable basis, permitting estimates of source energy and providing an indirect method of range estimation beyond 150 km. It can now be estimated with some reliability that the Apollo 15 LM impact near Hadley Rille 1000 km to the north will be detected by the Apollo 14 PSE. Details of these arguments and their application to natural events observed are discussed in the following paragraphs.

Two sources of natural seismic signals are expected: moonquakes and meteoroid impacts. Learning to distinguish them by differences in signal characteristics is obviously one of the most important problems in lunar seismology and is somewhat equivalent to the problem of distinguishing between earthquakes and explosions on Earth. From analysis of data recorded during the first 9 months of operation of the Apollo 12 PSE, the most promising methods for distinguishing between these source mechanisms revealed thus far are given in the following paragraphs.

Waveform matching. Detailed comparison of the seismic signals has revealed that some signals match each other in nearly every detail of the record for the entire duration of the signal. Three methods have been used for identification of matching signals: (1) visual comparison of the waveforms, (2) visual comparison of energy plots (smoothed plots of squared signal amplitude as a function of time), and (3) computed coherence functions between signals. Through this combination of methods, a total of nine sets of matching signals (71 signals) has been identified from among the 208 signals recorded during the first 9 months of operation of the Apollo 12 PSE. These matching signals will be designated category A signals for purposes of this discussion. The largest subset of category A signals (A_1) contains 26 events and accounts for about 80 percent of the total seismic energy from matching events recorded by the Apollo 12 PSE. At least two of the seismic events recorded during the first month of operation of the Apollo 14 PSE fall into subset A_1. The records for these events are shown on an expanded time scale in figure 6–12 along with the

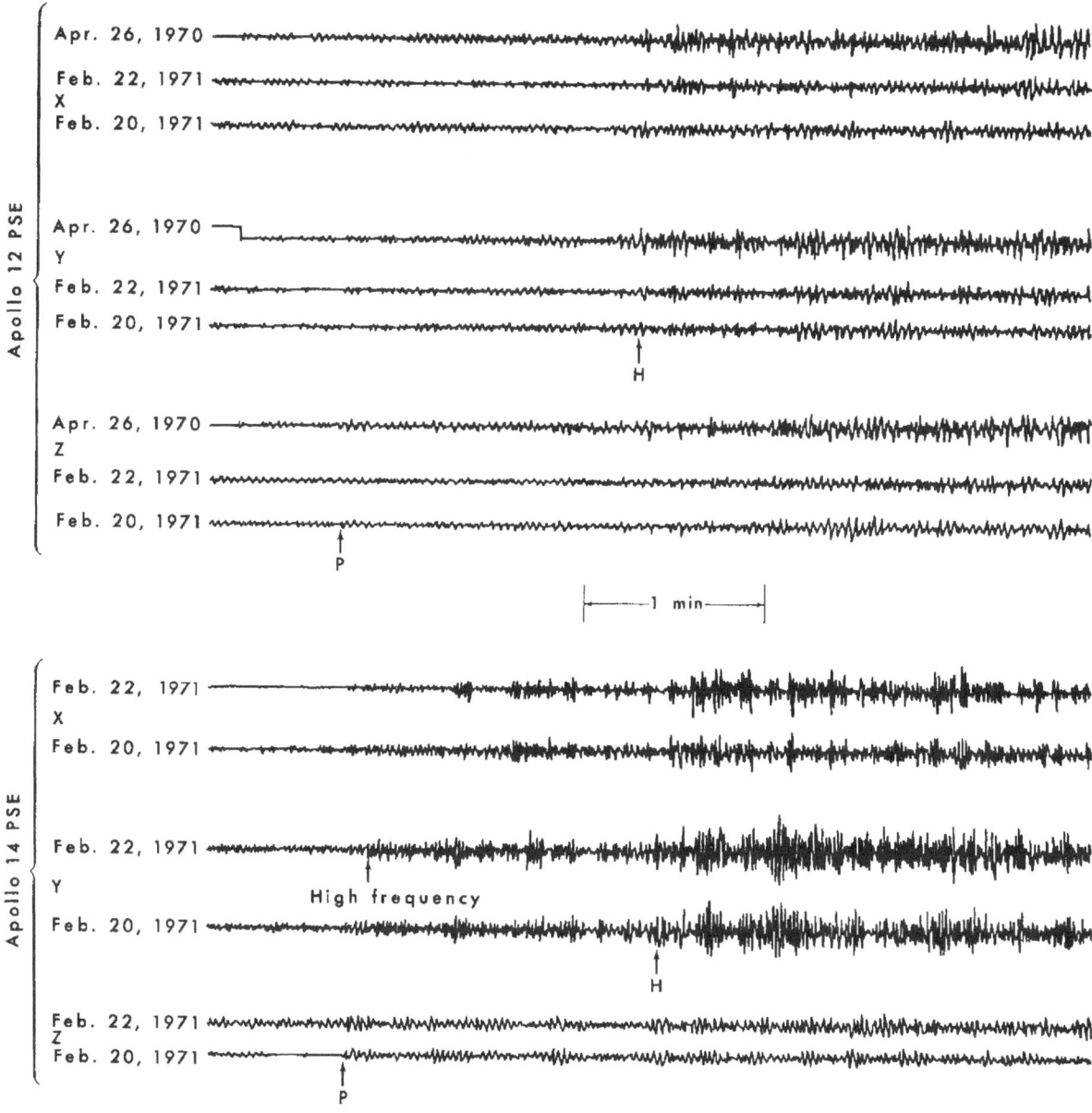

FIGURE 6–12.—Records of two moonquakes detected at the Apollo 12 and 14 sites on February 20 and 22. A moonquake recorded at the Apollo 12 site on April 26, 1970, before the deployment of the Apollo 14 PSE, is included for comparison. These three moonquakes originated within the zone of greatest activity (A_1 zone) detected thus far. The most prominent phases are tentatively interpreted as the direct compressional wave (P-wave) and shear wave (H-wave) arrivals. A high-frequency train of waves begins on the horizontal components at the Apollo 14 site about 8 sec after the P-wave. A similar signal is not evident at the Apollo 12 site.

record from an event recorded by the Apollo 12 PSE before the deployment of the Apollo 14 PSE. The identity between these signals for any one of the components of ground motion is evident. The

lack of similarity between the three components of motion is equally striking. A second set of signals, designated category C, has waveform characteristics similar to those of the artificial impacts, and these signals do not match each other.

Spectral characteristics. The seismic signals can be classified broadly into two categories on the basis of spectra: low-frequency signals and high-frequency signals. If, for example, the ratio of the signal power at 1 Hz to the signal power at 0.45 Hz is plotted as a function of the signal rise time, as shown in figure 6–13, the signals identified as either category A or C fall into two distinct groups. The spectra of category C signals (and the LM and SIVB signals) are richer in higher frequencies than those of the category A signals.

The H-phase. The most prominent phases identifiable in the category A signals are designated P- and H-phase (fig. 6–12). The P-phase is the first detectable signal and is considered to be the direct compressional wave from the source; however, weaker signals may precede those presently being detected. The H-phase is a prominent arrival on the LPX and LPY seismometers of the identified subset A_1 events, beginning about 100 sec after the P-phase. However, the indefinite beginning of the H-phase does not permit a meaningful

estimate of the difference in the arrival time at the two instruments. The H-phase is not observed in the category C of the artificial-impact signals.

Rise time. The time interval from the beginning of a seismic signal to the time of maximum signal amplitude (rise time) is variable, and is generally much shorter for category A events than for category C events. For category A events, the rise time is relatively independent of frequency in narrowband-filtered playouts; whereas category C events and artificial impacts typically have frequency-dependent rise times.

Time of occurrence. With few exceptions, the matching signals (category A) occur at monthly intervals near the time of perigee, with a secondary peak in activity near the time of maximum separation (apogee). The occurrence of category C events does not appear to correlate with the monthly orbital cycle.

Histograms showing the relative seismic energy detected per day at the Apollo 12 and 14 sites during this report period are shown in figure 6–14. These graphs are derived by dividing the events into daily intervals and summing the maximum signal amplitudes squared for signals in each interval. The shaded portions of the bars represent the seismic energy from category A events. These events are found to occur in a time interval within 7 days before and 4 days after the perigee of February 25 and within 6 days before the perigee of March 26. The secondary peak in moonquake activity near apogee is not apparent in this short sample of data.

Events that produce virtually identical seismic signals must have a common point of origin. Meteoroid impacts can be eliminated as a possible source because of the very low probability that they could be concentrated at the same point on the lunar surface and would occur only in association with apogee and perigee. The clear relationship between the occurrence of the category A (matching) events and the occurrence of apogee and perigee suggests strongly that these events are moonquakes induced by tidal strains that reach maximum values at apogee and perigee. The identification of 11 subcategories of category A events suggests that there are at least 11 distinct focuses for the repeating moonquakes detected by the Apollo 12 and 14 seismic instruments. The

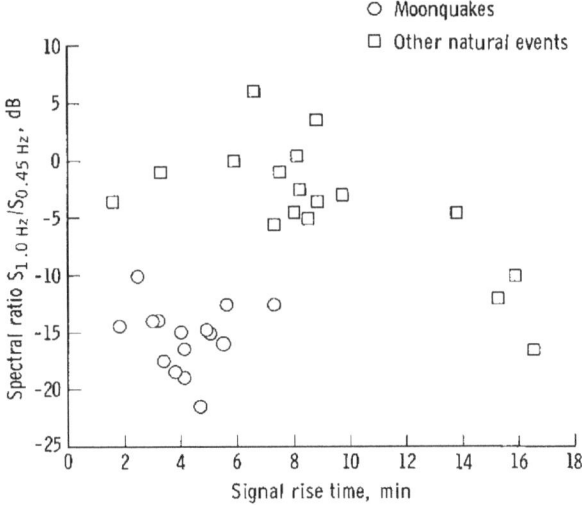

FIGURE 6–13.—Spectral ratio plotted as a function of signal rise time of natural events detected by the Apollo 12 LP seismometers. The rise times are measured from the beginning of the signal to the approximate peak of the signal envelopes of unfiltered seismograms. The ratios of the spectral amplitudes at 1.0 and 0.45 Hz are not corrected for instrument response. To correct for the instrument response, 16.5 dB must be added to this scale. The separate groupings of moonquakes and events believed to be meteoroid impacts are noteworthy.

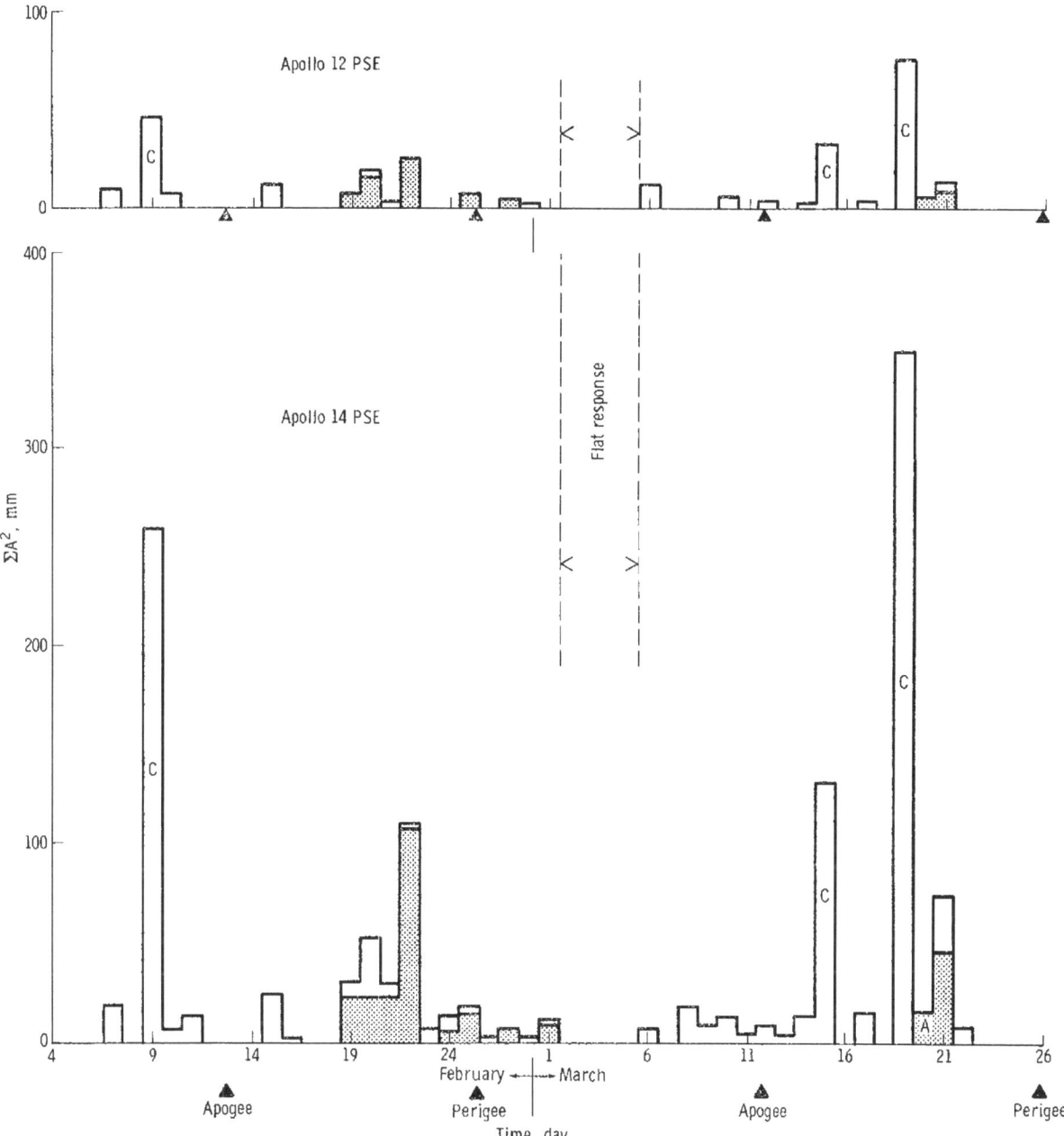

FIGURE 6–14.—Relative signal power observed on the LP components of both seismometers as a function of time. For each station, A^2 for an event is derived from the amplitudes of the X-, Y- and Z-components by $A^2 = X^2 + Y^2 + Z^2$. The height of each bar is the sum of A^2 for all events during a 1-day interval. Amplitudes for X, Y, and Z were each measured by the method used for the SPZ amplitude in figure 6–10. For these components, 0.5 mm is 1 DU. The shaded portions of the bars represent events having moonquake characteristics. The white portions represent events having impact characteristics but may contain some unidentified moonquake contributions. The long bars of February 9 and of March 15 and 19 are each dominated by a single large event with impact characteristics. The bars for February 22 are dominated by a category A_1 event recorded at both sites. Several events very weakly recorded during flat-response operation may have made a significant contribution to these data if recorded with peaked response.

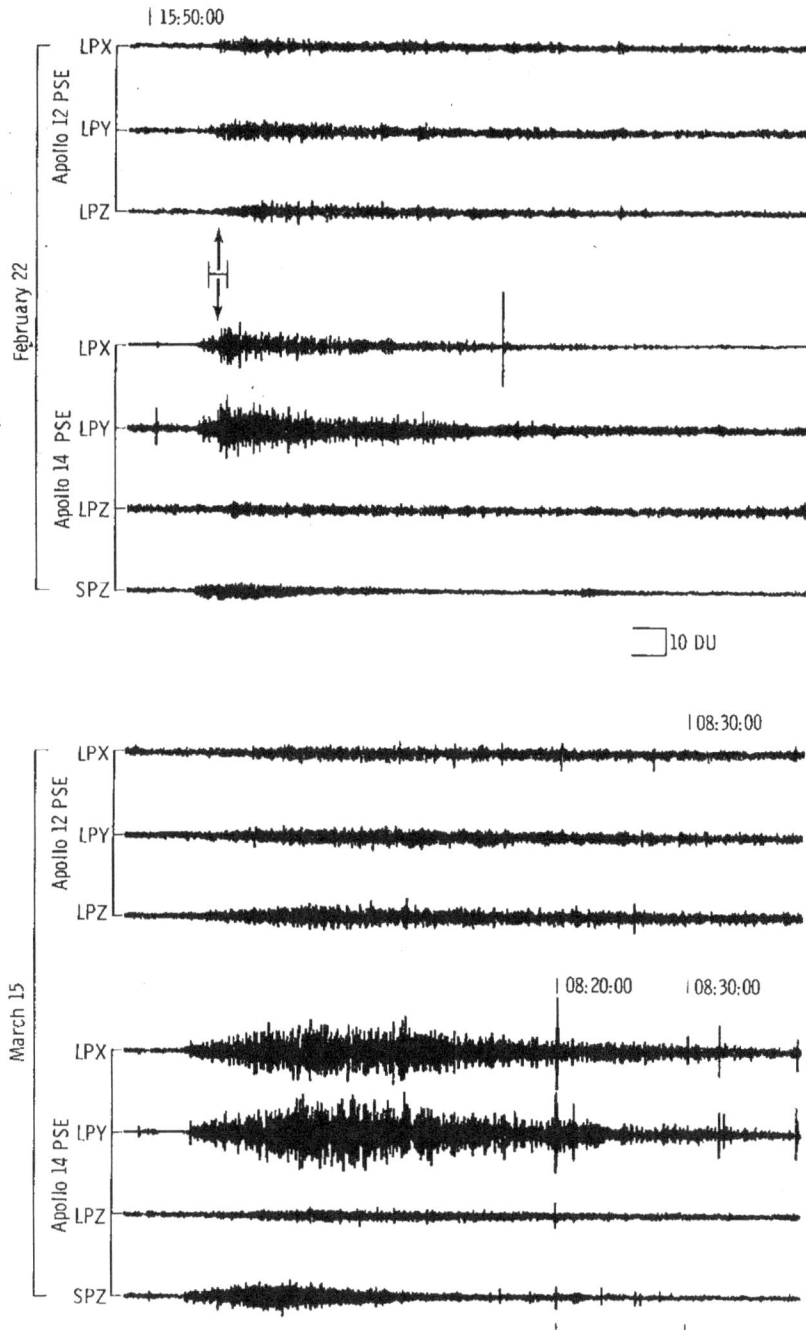

FIGURE 6–15.—Compressed time-scale playouts of two of the natural events observed at both sites. The event on February 22 is a category A or matching event (moonquake) and that on March 15 is a category C event (meteoroid impact). The difference in general shape of the signal envelopes should be noted. The matching events have a relatively abrupt beginning of P- and H-phases, in contrast to the gradual signal buildup of category C events. The nearest perigee occurred on February 25.

similarities between the LM and SIVB impact signals and category C signals strongly suggest that these are generated by meteoroid impacts.

In all, 79 events have been identified from the records of the LP seismometers at the Apollo 14 site for the first 44 days of operation. In the results presented here, six of the events have been omitted from consideration because they were recorded while the instrument was temporarily in the flat-response mode and were of very low amplitudes. Thus, the principal data concern 73 events recorded at the Apollo 14 site over a period of 40

days of operation with the instrument in the peaked-response mode. Twenty-seven of the 73 events recorded at the Apollo 14 site were also recorded at the Apollo 12 site. All events recorded at the Apollo 12 site were recorded at the Apollo 14 site. Based upon the criteria previously discussed, 14 of the 73 events are possible moonquakes (category A events) and 17 are meteoroid impacts (category C events). The remaining events are too small to be classified without further analysis.

Compressed time-scale playouts of some LP

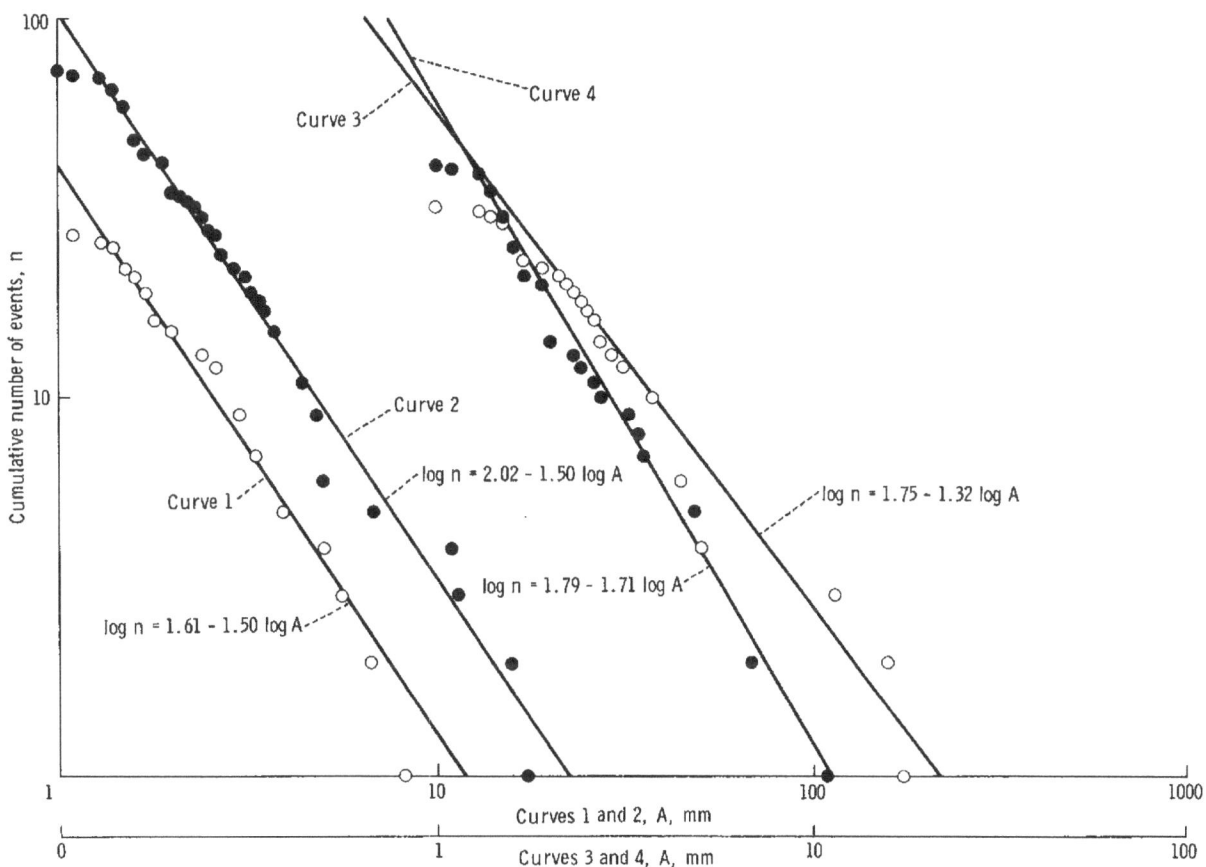

FIGURE 6–16.—Cumulative curves for all LP events at the Apollo 12 and 14 sites; n is the number of events with amplitude equal to or greater than A. The A values are the same ones used in figure 6–14. Curve 1 contains all data for the Apollo 12 PSE; curve 2, all data for the Apollo 14 PSE. Curves 3 and 4 divide the Apollo 14 PSE data into groups recorded during nonperigee and perigee intervals: curve 3, nonperigee (Feb. 7 to 18 and Mar. 6 to 19); curve 4, perigee (Feb. 19 to Mar. 1 and Mar. 20 to 22). This places all identified moonquakes on curve 4 and all others, including most impacts, on curve 3. (See fig. 6–14.) Lines of identical slope fit curves 1 and 2 as expected, because the same distribution of events was observed at both sites. However, the difference in slopes of curves 3 and 4 represents a fundamental difference in the size-distribution laws of moonquakes and impacts.

events are shown in figure 6–15. The playouts are useful for observing the shape of the signal envelope and for measuring the rise time and maximum peak-to-peak signal strength, quantities that are used in further analysis. The H-phase is conspicuous on the LPX and LPY compressed playouts.

The data for each station have been plotted in cumulative amplitude curves in figure 6–16. The ordinate of the highest points on curves 1 and 2 indicates the total number of events recorded at each station: 27 and 73 for the Apollo 12 and 14 sites, respectively. The slope of such a cumulative curve is known (ref. 6–8) to depend only on the distribution of source sizes and to be independent of the law governing amplitude falloff with distance, independent of local variations in propagation effects or instrument sensitivity, and independent of any geographic bias in source locations. Therefore, as expected, the distribution for data from the Apollo 12 and 14 instruments has essentially the same slope because the same distribution of events was recorded by the two instruments. The horizontal separation of curves 1 and 2 indicates that events of equal frequency are approximately 1.9 times larger in amplitude on the Apollo 14 PSE than the Apollo 12 PSE. This should also be the average amplitude ratio for an event observed simultaneously by both instruments. Correspondingly, events of the same amplitude are 2.8 times more frequent on the Apollo 14 PSE than on the Apollo 12 PSE. Because no geographic bias of event locations is known to exist, these differences must result primarily from differences in instrument sensitivity. Indeed, a geographic bias of the impact component of the data would be highly improbable, though quite possible, for the moonquake component. The difference in instrument sensitivity is thought not to be due to instrument differences, but to differences in the ground structure and in near-station propagation effects.

In curves 3 and 4 of figure 6–16, a perigee component (primarily moonquakes) and a nonperigee component (primarily meteoroid impacts) of the data of curve 2 (the Apollo 14 PSE only) are plotted separately. The two groups can be seen to have frequency-amplitude distributions differing significantly in slope, which bears out the initial hypothesis that the two groups of events represent different processes: meteoroid impacts and moonquakes.

The seismograms recorded by the Apollo 14 PSE differ from those recorded by the Apollo 12 PSE, for the same event, in several important respects: (1) As mentioned previously, the average amplitudes at the Apollo 14 site are larger than those at the Apollo 12 site by a factor of 1.9. (2) The peak amplitudes recorded by the Apollo 12 PSE are about the same on all LP seismometers, whereas, at the Apollo 14 site, the amplitudes recorded on the horizontal components are about 3 times larger than on the vertical component. (3) A high-frequency train (1 Hz), beginning approximately 8 sec after the P-phase for the subset A_1 events, is present on the LPX and LPY seismograms recorded at the Apollo 14 site.

These differences in signal characteristics between the two instruments were also noted for the Apollo 14 LM impact, except for the impulsive beginning of the high-frequency train observed in the moonquake records. The beginning of this train cannot be distinguished in the impact signal. As discussed previously, these differences appear to be explained by the presence of a thick layer of weak material at the Apollo 14 site.

The abrupt beginning of the high-frequency train may be explained by the arrival of shear waves, converted from compressional waves at an interface below the instrument. Assuming the velocity-depth relationship shown in figure 6–17, the interface would have to be at a depth of approximately 25 km to give the 8-sec delay between P-phase and the onset of the high-frequency train.

Discussion

Seismic-Wave Propagation and Lunar Structure

The characteristics of lunar seismic signals previously described (duration of signal and lack of coherence between components) suggest scattering of the seismic waves. If the structure through which the observed signals traveled contains structural or compositional irregularities with dimensions of the order of 1 wavelength (from less than 100 m to 10 km), waves will be scattered in all

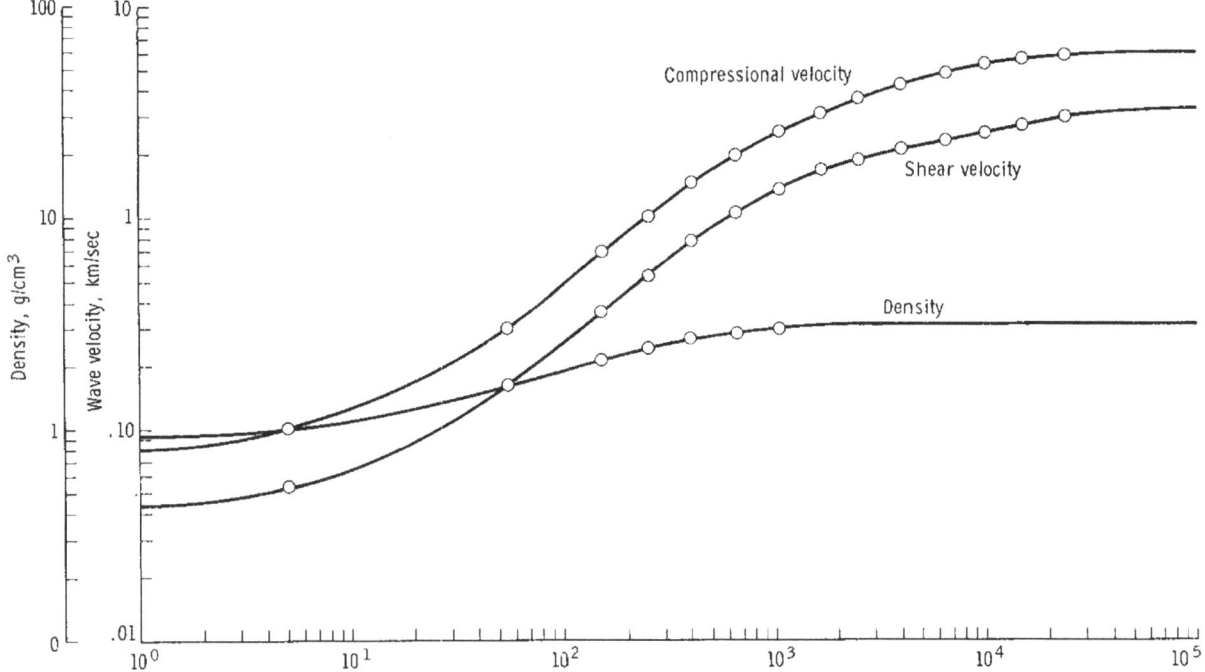

FIGURE 6–17.—Smoothed variation of compressional- and shear-wave velocity and density with depth for the outer shell of the Moon. This model corresponds approximately to results of the ASE at very shallow depth and to laboratory rock measurements and PSE data at greater depth. A fixed value, 0.30, of Poisson's ratio was taken for material shallower than 3 km, and shear velocities were calculated from laboratory compressional velocity results to obtain shear velocity values consistent with landing-site data. Strong variation of velocities, especially near the surface, makes it necessary to represent them on logarithmic scales.

directions. The arrivals at a given instrument will then be the summation of waves from many directions and will prolong the signal. Constructive and destructive interference between these arrivals will reduce correlation between components of ground motion.

Scattering of seismic waves can be treated by the methods derived from the kinetic theory of gases to explain gaseous diffusion or heat flow in a solid. It is expected that a strong velocity gradient in the outer shell of the Moon forms a surface waveguide, and therefore the two-dimensional diffusion equation with dissipation was used to solve for an instantaneous point source of energy, equivalent to an impact.

$$E = \frac{E_0}{\pi \xi t} \, \exp \left(\frac{-r^2}{\xi t} - \frac{\omega t}{Q} \right) \qquad (6\text{--}2)$$

where

E = seismic energy density measured at distance r from the source

E_0 = seismic energy introduced at the source

$\xi = 2V\mu$ = diffusivity of the medium

V = average velocity of seismic waves

μ = equivalent mean free path for seismic waves in the medium

t = time

ω = angular frequency of the signal

$1/Q$ = dissipation constant for the lunar material, where $2\pi/Q$ is defined as the fractional loss of elastic energy per cycle of vibration of the system

The degree of scattering of seismic waves in the medium is inversely proportional to ξ; that is, the larger the diffusivity, the smaller the amount of scattering.

An accurate fit to the envelopes of the narrow-band-filtered signals from the LM and SIVB impacts can be obtained with proper selection of the coefficients ξ and Q in equation (6–2). Values of ξ required to fit the signal envelopes at 1 and 0.45 Hz are shown in figure 6–18. The rise times of the impact signals are shown in figure 6–19. From figure 6–18, it is evident that the apparent diffusivity is a function of both distance and frequency; that is, increasing with increasing range at a given frequency.

From figure 6–19, the rise times of the signal envelopes appear to reach a maximum value and to remain at this value, or even possibly decrease, for signals from more distant events. Both the maximum rise time and the range at which it occurs are dependent upon the signal frequency. Thus, the rise time of the signal envelope apparently cannot be used as a measure of source distance beyond some maximum range, probably 150 to 200 km. Each curve shown in figure 6–19

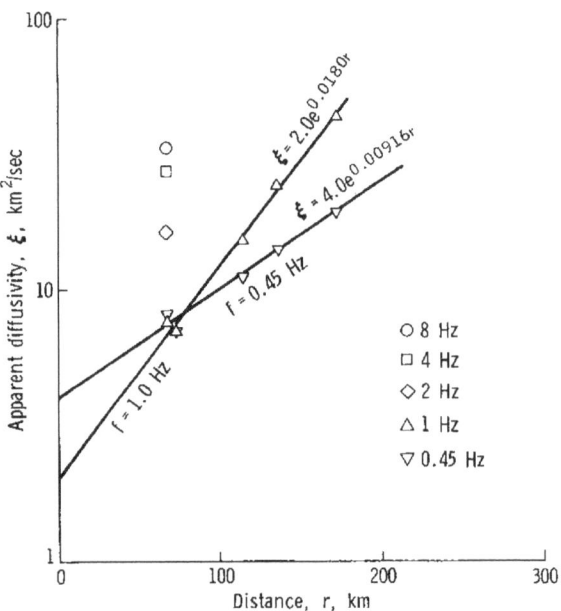

FIGURE 6–18.—Apparent diffusivity of LM- and SIVB-impact signals obtained by matching envelopes of narrowband-filtered LPZ (for 0.45 and 1 Hz) and SPZ (for 2, 4, and 8 Hz) seismograms with theoretical envelopes given by equation (6–2). The difference in variation of apparent diffusivity with distance at 1.0 and 0.45 Hz should be noted.

was computed by substituting into equation (6–2) values of ξ and Q derived from figure 6–18.

A model for seismic-wave propagation and the structure of the outer shell of the Moon (fig. 6–20) that appears to account for the properties of the signal envelopes previously described is one in which waves at shallow depths are intensely scattered. Wave scattering must decrease rapidly with increasing depth, or decrease abruptly at some depth; that is, the lunar material must behave more nearly as an ideal seismic-wave conductor with increasing depth. According to this model, the primary surface waves and body waves generated by an impact and secondary surface waves generated by scattering of primary surface waves will predominate in the recorded seismic signal at near ranges. At far ranges, most of the seismic energy received will have traveled as body waves through the less disturbed zone at depth. Energy will "leak" into this zone by scattering of the primary waves in the source region. Near the receiving seismometer, some conversion of body waves back to surface waves will occur as the waves propagate through the scattering zone near the surface. Thus, the apparent diffusivity increases with increasing range as the signals that have traveled at greater depths contribute increasingly to the recorded signal. Also, the envelope rise time will not increase appreciably at a range greater than that at which most of the propagation has occurred in the relatively undisturbed material below. These ranges are estimated to be about 70 and 150 km for signal frequencies of 1.0 and 0.45 Hz, respectively. Thus, the outer shell of the Moon may be heterogeneous on a scale of 1 wavelength (0.1 to 10 km) to depths approaching 50 to 100 km.

The precise nature of the heterogeneity present in the scattering zone cannot be determined from present data; however, it is evident that craters and cratering processes must contribute significantly to the general complexity of this zone. The presence of craters, in an otherwise homogeneous medium, has been shown to introduce appreciable scattering of surface waves (ref. 6–9). However, the thickness inferred for this zone appears to be too great to be explained entirely by visible craters.

Alternatively, it can be assumed that seismic velocity increases with depth as a result of self-compaction in a fragmented outer layer many

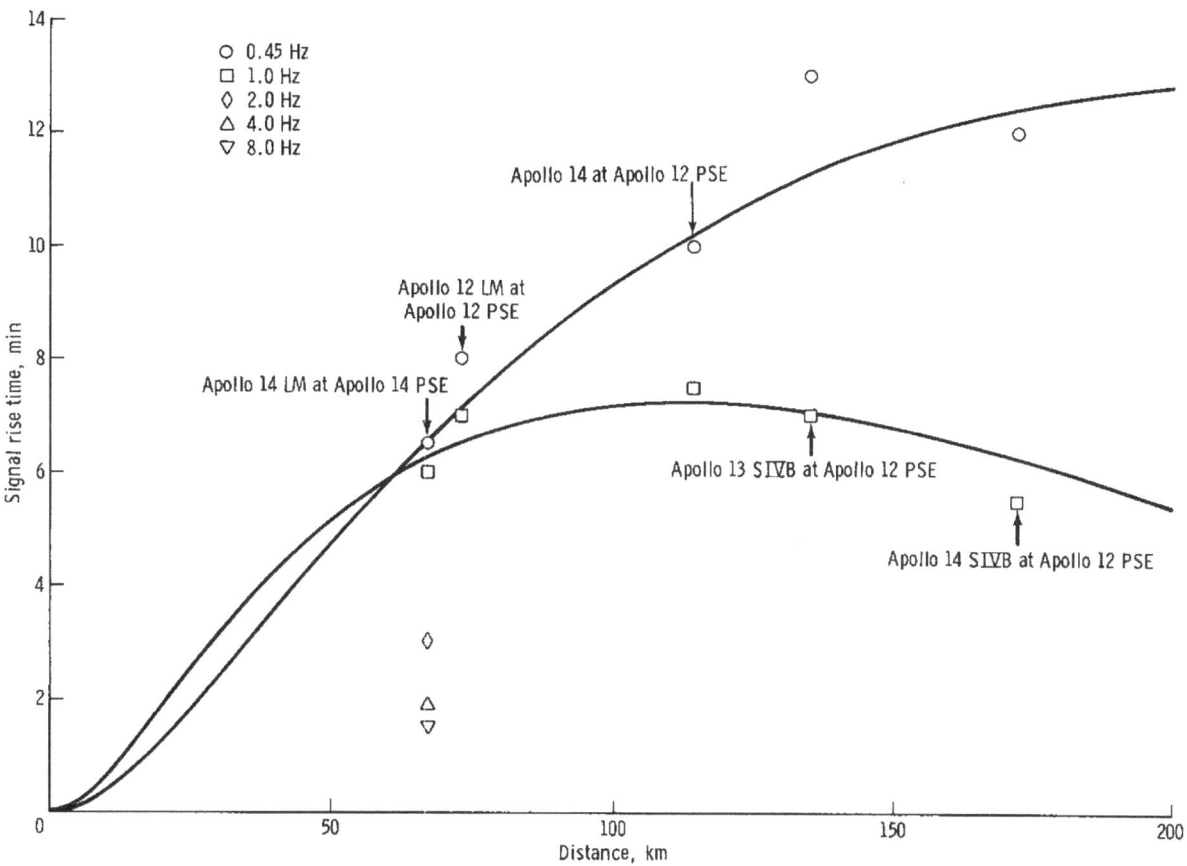

FIGURE 6–19.—Rise times of LM- and SIVB-impact signals measured on narrowband-filtered LPZ and SPZ seismograms. The curves represent rise times of the theoretical signal envelopes given by equation (6–2) for the diffusivities given by the straight lines shown in figure 6–18.

kilometers thick. Even without a downward increase in block size, this is equivalent to a downward increase in diffusivity because diffusivity is proportional to the product of the mean free path and wave velocity.

The efficiency of transmission of seismic energy in the lunar structure is extremely high. Attenuation of elastic energy in a vibrating system is frequently specified by the quality factor Q where $2\pi/Q$ is the fractional loss of elastic energy per cycle of vibration. Thus, a high Q implies low attenuation. The value of Q is normally found to be independent of frequency in terrestrial seismic experiments. From empirical fitting of signal-diffusion envelopes based on equation (6–2), the Q of the structure in the region of the Apollo 12

and 14 sites is approximately 2000 at 0.45 Hz and 3000 at 1.0 Hz; that is, the measured Q for the lunar case increases with increasing signal frequency. When measured from the lunar seismic signals from a single instrument, Q results from a combination of leakage of energy into the deep lunar interior and the intrinsic energy dissipation within the lunar material. If attenuation of detectable seismic signals introduced by downward loss of seismic energy is relatively independent of frequency, then the measured Q will increase with increasing frequency, as is observed, and the intrinsic energy dissipation within the lunar material must be higher than the measured values.

The values of Q measured from the lunar seismic signals are an order of magnitude higher than

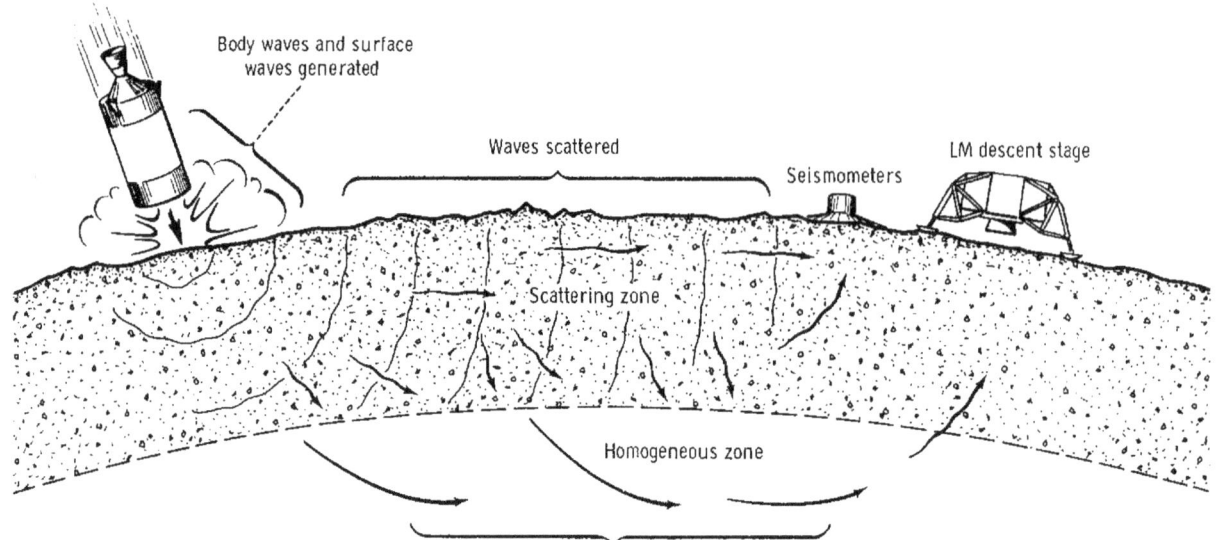

FIGURE 6-20.—Schematic diagram illustrating seismic-wave transmission in the Moon as described in the text. According to the proposed model, seismic waves generated at or near the surface are scattered intensively in a surface zone (scattering zone) with maximum thickness of 50 to 100 km. As the primary waves travel outward from the source, some of the seismic energy is scattered downward and "leaks" into the interior as compressional waves and shear waves. Seismic waves travel through the interior (homogeneous zone) with little or no scattering and with very high efficiency, returning to the surface through the scattering zone.

those measured in surficial layers on Earth. Low absorption of seismic-wave energy in the lunar material may be explained by the nearly complete absence of volatiles. In the absence of volatiles, grain or fragments may be locked securely together (possibly cold welded), thus preventing frictional losses introduced by slippage at points of contact for the small strains associated with the passage of a seismic wave. Also, frictional losses accompanying small motions of interstitial fluids induced by passage of seismic waves must be absent. Some experimental evidence in support of this suggestion has been given (ref. 6-10).

Additional detailed information on the lunar structure in the vicinity of the Apollo 12 and 14 sites can be gained from the velocities of compressional and shear waves observed from the LM and SIVB impacts. As shown in figure 6-9, the traveltimes of seismic waves from these impacts are in close agreement with those predicted from laboratory measurements of the pressure dependence of the elastic properties of returned lunar samples.

Compressional-wave data were determined for a porous igneous rock (ref. 6-11), and shear-wave data were determined for a sample of similar rock (ref. 6-12). The variations of compressional- and shear-wave velocities with depth, as derived from these measurements, are modified (fig. 6-17) at shallow depths to be consistent with the direct measurements obtained from the LM ascent and the ASE (sec. 7). Given the lack of traveltime data for distances between the local landing-site measurements and the Apollo 14 LM impact of 67 km, the curves of figure 6-17 are poorly determined at intermediate depths of a few kilometers. Recognizing these obvious limitations, the velocity-depth functions of figure 6-17 can be accepted as the best estimates obtainable from the present data.

The simplest hypothesis consistent with this model is that there is loosely consolidated material near the surface with a strong velocity increase from self-compaction to depths of at least 25 km (the greatest depth of penetration of compressional waves received from the SIVB impact). No evidence exists for a major horizontal discontinuity in this zone that might be equivalent to the crust-mantle interface on Earth, although present data are not sufficient to rule out the possibility of significant layering. Seismic-wave velocities at shallow depth are too low for homogeneous igneous rock. A compressional-wave velocity of 6 km/sec is attained at depths somewhere between 6 and 20 km. In this depth range, the elastic properties of the material, conditioned by pressure, approach those of homogeneous rock. This structure may have formed from individual fragments during the final stages of accretion of the Moon; or it may be the remnant of a solidified primitive igneous crust that was pulverized by meteoroid impact. By either of these hypotheses, the outer layer may have been modified by episodes of maria filling that left a basaltic surface that has since been modified by continuing meteoroid impacts.

The much higher seismic amplitude recorded by the Apollo 14 PSE from the LM impact is thought to reflect primarily the shallow structure near the site. A very prominent 1-Hz signal is present on horizontal-component seismograms that is not present on the vertical component nor on any component at the Apollo 12 site. The spectra for the vertical component at the Apollo 14 site peak at about 1.8 Hz, whereas the 1-Hz signal component accounts for horizontal signal envelopes that are much larger than for the vertical component (fig. 6-6). The ratio of horizontal-to-vertical particle motion taken from the signal spectra near the time of maximum signal intensity is shown in figure 6-21. The amplification of horizontal motion near 1 Hz is obvious from this figure.

The most obvious structural difference that may account for the observed seismic differences between the Apollo 12 and 14 sites is the presence at the Apollo 14 site of a surface layer of unconsolidated material of very low velocity overlying material of higher velocity that is probably the Fra Mauro Formation. The thickness of the surface layer as determined from the ASE results is approximately 8.5 m (sec. 7). This layer probably thickens toward the ridges to the east and west of the Apollo 14 site. Seismic measurements at the Apollo 12 site do not indicate the presence of a

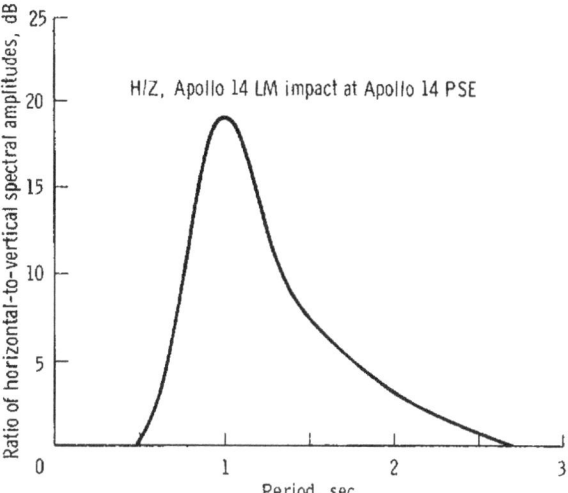

FIGURE 6–21.—Ratio of the spectral amplitudes of horizontal ground motion H (average of the two horizontal components) to vertical ground motion Z for the Apollo 14 LM-impact signal recorded at the Apollo 14 site. A 3-min data sample, taken near the time of maximum signal intensity, was used in the calculation of the spectra. The ratio plot shows that a strong 1-Hz signal, present on the horizontal components, is either weak or entirely absent in the vertical component of ground motion.

surface layer with a well-defined lower boundary at the Apollo 12 landing site.

Amplification of seismic waves by near-surface, low-velocity layers is a well-known phenomenon. Because the 1-Hz signal can be seen on the horizontal components only, the wave motion is assumed to be horizontally polarized shear waves or steeply incident vertically polarized shear waves. The frequency of the recorded signal will be given approximately by $f = \beta/4H$, where f is the signal frequency, β is the shear-wave velocity in the top layer, and H is the thickness of the layer. Using shear-wave velocities shown in figure 6-17, $\beta = 50$ m/sec. Thus, for $f = 1.0$ Hz, H is 12.5 m, which is near the value obtained from the Apollo 14 ASE.

Thus, the presence of the low-velocity surficial layer at the Apollo 14 site appears to account for the observed amplification of horizontal motion. The same spectral peaks and associated horizontal amplification are observed in the signals from natural events at the Apollo 14 site and appear to explain why twice as many events are detected at the Apollo 14 site as at the Apollo 12 site.

Important elements of a lunar seismic-transmission model may be drawn from the data previously discussed. From equation (6-2), it may be shown that, neglecting losses ($Q \rightarrow \infty$), the peak amplitude of a wave diffusing in the surface-scattering layer varies with range as $1/r$. Regarding the lunar interior, arguments based on petrology and bulk density lead to models in which variations of wave velocity with depth are quite small (ref. 6-13). Indeed, from the traveltime data and rock experiments previously discussed, it appears that, within a 20-km depth, the seismic velocity rises to approximately 6 km/sec, which is near the maximum value expected at any depth. Thus, even though surface values are low, the zone of strong gradients is thin, and wave amplitudes in the interior of the Moon may closely follow the $1/r$ law that applies to a medium with constant velocity. Through this combination of effects, the $1/r$ spreading law seems reasonable for arrivals observed both at near and far ranges. According to these observations, amplitudes fall off less rapidly with distance in the Moon than for waves transmitted through the crust and mantle of the Earth.

Tests of the $1/r$ spreading relation were made using (1) observations of the Apollo 14 LM-impact event at the Apollo 12 and 14 sites and (2) observations of the Apollo 13 and 14 SIVB-impact events at the Apollo 12 site. Each comparison involved use of r and signal amplitudes A measured on the various recordings. In the first instance, the same event (Apollo 14 LM impact) was recorded by both instruments; so a correction for the average difference in amplitude sensitivity of the instruments was used ($A_{14} = 2A_{12}$, as obtained from fig. 6-16). In the second instance, two events (the Apollo 13 and 14 SIVB impacts) were compared on records of the Apollo 12 PSE; and a correction was made (according to the equation given in a following portion of this section entitled "Meteoroid Flux") for the slight difference in kinetic energy of the two impacting vehicles. In each case, the inverse relationship, $Ar =$ constant, held within the estimated precision of the peak-amplitude measurements (5 to 10 percent). The ranges covered by these tests, 67

to 172 km, are more critical for behavior in the near range, or diffusion regime, than for the far range for which spherical spreading is postulated. The more distant impacts of the Apollo 15 vehicles will be an important further test of present assumptions. The amplitude-spreading relation is extremely important in providing a basis for estimating meteoroid flux from observed impact events, as explained in a later portion of this section.

Location and Focal Mechanism of Moonquakes

One of the most important objectives of the present analysis is the location of the zone in which moonquakes originate. Moonquakes recorded at the Apollo 12 site during the first year of operation are believed to have originated at no less than 11 different focuses, although some of these zones may be quite close to one another. A single focus (A_1 zone), however, accounts for nearly 80 percent of the seismic energy from moonquakes recorded at the Apollo 12 site. At least two moonquakes from the A_1 zone (fig. 6-12) were recorded by the Apollo 12 and 14 instruments during the first perigee period following activation of the Apollo 14 PSE. The first detectable signals from these events arrived at the Apollo 12 site 0.9 sec earlier than at the Apollo 14 site. Thus, the A_1 zone is nearly equidistant from the two Apollo sites (slightly closer to the Apollo 12 site). The time interval between the P- and H-phases is between 91 and 98 sec at the Apollo 12 site and between 102 and 107 sec at Apollo 14 site. These ranges in the P- and H-phase time interval represent the uncertainty in identification of the onset of the H-phase. The computed distance of the source from each site depends upon the velocities assumed for the propagation of each of these two phases.

As previously discussed, the rise time of the signal envelopes is a measure of the distance between the seismometers and the source at ranges less than some maximum value (approximately 200 km). However, plotting rise time at one station as a function of rise time at the other for events believed to be moonquakes (category A events) and meteoroid impacts (category C events) results in the graph in figure 6-22 in

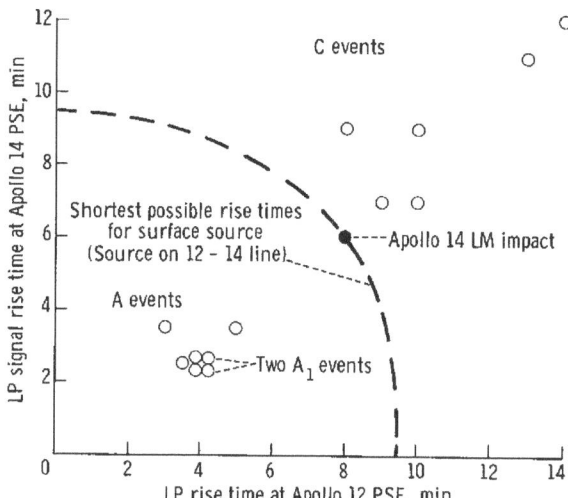

FIGURE 6–22.—Signal rise time at both sites measured on the LP unfiltered seismograms. The dashed line indicates approximate rise times when a surface source is located on the line joining the two sites. The intercepts at 9.5 min are based upon extrapolation of the rise times of the LM- and SIVB-impact signals to a range of 181 km (the distance between the Apollo 12 and 14 sites). Events falling inside the line, including two matching A₁ events, are, therefore, not likely to have surface sources. All events falling outside the line are category C events consistent with the hypothesis that these are impacts on the lunar surface.

which the points for these two categories are found to fall into groups. A point corresponding to the Apollo 14 LM impact falls into the impact grouping as expected. Because the LM impact occurred nearly along the line between the two stations, the shortest possible rise time for surface-focus events would fall along the dashed line shown in figure 6–22. The moonquakes obviously have rise times shorter than this minimum value. A possible interpretation of these data is that moonquakes do not originate near the surface, but at appreciable depth. Other factors that suggest this possibility are (1) the interval between the P- and H-phases is generally less complicated than the early parts of the signals from known impacts; (2) the P-wave amplitudes are larger for the moonquake signals, relative to the maximum amplitudes in the wavetrain, than for impacts; and (3) the high-frequency train begins abruptly approximately 8 sec after the P-phase in the moonquake signals recorded at the Apollo 14 site, suggesting conver-

sion from P-waves to shear waves (S-waves) at a depth of approximately 25 km below the site.

If the H-phase is interpreted as the direct S-wave arrival, the abrupt beginning indicates travel mostly below the scattering layer. Using the H- and P-phase interval of 100 sec and velocities appropriate for those depths (fig. 6–17), the source of the moonquakes is found to be approximately 600 to 700 km from the Apollo 12 and 14 sites. Although the evidence is certainly not conclusive, it is suggested that the moonquakes originate at significant depth within the Moon. Maximum depths of approximately 700 km are possible. The source of strain energy released as moonquakes is not known; but, if significant depth of focus for the moonquakes is verified by future data, it will have profound implications concerning the dynamics of the lunar interior.

Two pieces of evidence that bear on the question of the focal mechanism for moonquakes are the polarities of the seismic signals and the cumulative distribution of moonquake amplitudes. With few exceptions, the polarities of signals belonging to a set of matching events are identical. This situation implies that the source mechanism is progressive and not one that periodically reverses direction. This observation suggests a secular accumulation of strain triggered periodically by tidal stresses. It is possible, of course, that detectable movements in one direction are compensated by many small, undetectable movements in the opposite direction.

The cumulative distribution curve for signal amplitudes recorded for moonquakes at the Apollo 14 site is shown in figure 6–16, curve 4. The slope of this curve, normally referred to as the b value in terrestrial seismic experiments, is 1.7. The corresponding value for moonquakes detected over a 9-month period at the Apollo 12 site is 1.6. The b values measured for tectonic earthquakes are normally close to 1, whereas quakes associated with volcanic activity (B-type volcanic events) have b values near 2. Volcanic earthquakes are presumably associated with the movements of magma under pressure beneath the surface.

The b-values for seismic signals generated by heating and cooling of small samples of brick and sintered perlite were measured as approximately 2 (ref. 6–14). This calculation indicates that high *b* values may be associated with seismic energy release generated by thermal stresses, whereas *b* values near 1 are associated with mechanical stress in the rocks. If this argument is valid for the lunar stress mechanisms, one possibility is that moonquakes are generated by thermal stresses. However, because the correlation between maximum moonquake activity and perigee has been demonstrated, it is evident that tidal stresses must also play an important role. Because depths as great as 700 km have been indicated by other considerations, it is possible that tidal pumping of magma at great depth may cause local thermal gradients that induce the moonquakes. To account for almost exact repetition of these events, one can imagine that cracks, propagating slowly by tidal stress cycles, are being invaded by magma.

High *b* values are also measured in laboratory tests when two surfaces are rubbed together under high contact pressure. The situation might be analogous to the shuffling of large blocks at depth by tidal stresses. These suggestions are obviously of the most tentative nature, but they serve as hypotheses against which future data will be tested.

Meteoroid Flux

With $1/r$ as a reasonable approximation for the range dependence of lunar seismic signal amplitudes at all ranges, a relationship between meteoroid kinetic energy and seismic energy generated at the source can be expressed as

$$MV^2 \propto (Ar)^{2\beta} \qquad (6\text{-}3)$$

where M and V are the meteoroid mass and velocity, respectively, and A is the peak-to-peak trace amplitude (mm) recorded at distance r (km) from the impact. The parameter β allows for a nonlinear dependence between seismic energy generated by an impact and the energy of the impacting body. Thus, if $\beta = 1$, seismic energy is directly proportional to the impact kinetic energy, and $\beta < 1$ implies that the proportion of impact kinetic energy converted to seismic energy increases with the magnitude of the impact. This might be expected, for example, if the conversion efficiency increases for larger impacts that penetrate more deeply into more compacted material beneath the surface. It has been found

that $2\beta = 1.32$ from the distribution of meteorite impact amplitudes, as explained as follows.

Assuming a constant meteoroid velocity, equation (6-3) becomes

$$M = K(Ar)^{2\beta} \qquad (6\text{-}4)$$

where K is a constant of proportionality that can be evaluated using data from an impact of known parameters. Using the SIVB impact data, from the Apollo 13 and 14 missions and an assumed meteoroid velocity of 20 km/sec, $\log K = 1.86$ is obtained. The same value of $\log K$ was obtained from the data on both SIVB impacts (tables 6-I and 6-II). This result tends to verify the assumed $1/r$ spreading law because range and signal amplitude are the only factors in equation (6-4) that differ significantly for these two impacts.

Assuming a slope of -1 (ref. 6-7), the cumulative meteoroid flux distribution is of the form

$$\log N = -B - \log M \qquad (6\text{-}5)$$

where N is the number of meteoroids of mass M or greater that strike a surface of unit area per year, and B is an arbitrary constant.

Substituting equation (6-4) into equation (6-5) and integrating over the lunar surface, the cumulative number n of lunar impacts that produce signal amplitudes $\geq A$, can be expressed as

$$\log n \ (\text{yr}^{-1}) = -B - \log K - 2\beta \log A + \log \int_0^{2a} r^{-2\beta} 2\pi r \ dr \qquad (6\text{-}6)$$

where a is the radius of the Moon and r is the chord distance between the impact point and the seismic site. The quantity 2β in equation (6-6) is seen to be the slope of the amplitude distribution curve for impact events. It is read directly from figure 6-16, curve 3, as $2\beta = 1.32$. From this same curve, $\log n = 1.75$ at $A = 1$ mm for the 26-day period monitored or $\log n$ for 1 yr = 2.90. Taking $a = 1738$ km for the radius of the Moon, the definite integral in equation (6-6) has a value of 2.33×10^3. Substituting these values into equation (6-6), $B = 2.32$ is obtained. Thus, the meteoroid flux distribution obtained from seismic measurements is

$$\log N = -2.32 - \log M \qquad (6\text{-}1)$$

where N is given in $km^{-2} \cdot yr^{-1}$, and M is given in g. The cumulative flux distribution (ref. 6–7), reduced by a factor of 2 to account for the reduced lunar gravity and Earth shielding, is

$$\log N = -1.03 - \log M \qquad (6-7)$$

where N is given in $km^{-2} \cdot yr^{-1}$, and M is given in g. Thus, the distribution of meteoroid impacts on the lunar surface derived from seismic measurements is a factor of 20 smaller than the flux estimate of reference 6–7 for meteoroids in the kilogram-mass range. This result must be carefully examined but the method is believed to be a promising one.

It can also be shown from equation (6–6) that 50 percent of the recorded impacts occur at ranges of less than about 1300 km (median chord distance), or the nearest 13 percent of the lunar surfaces; the remaining 50 percent occur at ranges greater than 1300 km or the most distant 87 percent of the Moon. Also, 21 percent of the impacts detected must occur on the far half of the Moon relative to the Apollo 14 site. Thus, the instruments appear to be recording meteoroid impacts from very great ranges on the lunar surface. According to the present calibration values, a meteoroid of mass 1 kg or greater falling on the far side of the Moon could be detected above the threshold of the Apollo 14 PSE.

Radius of Detectability for Impacts

During the nonperigee periods (figs. 6–14 and 6–16), 15 events were recorded by the Apollo 12 PSE and an additional 17, or a total of 32, were recorded by the Apollo 14 PSE. The Apollo 14 PSE events included all those detected at the Apollo 12 site. The overlap relationship of these data suggests an approximate statistical method for estimating a minimum value for the ranges to which events are detectable from the two sites. Assuming that the nonperigee events are randomly distributed meteoroid impacts, then the ratio of the areas from which impacts are detected at the two sites (A_{12}/A_{14}) must approximately equal the ratio of the number of recorded impacts (N_{12}/N_{14}). Therefore,

$$\frac{A_{12}}{A_{14}} = \frac{R_{12}^2}{R_{14}^2} = \frac{N_{12}}{N_{14}} = \frac{15}{32} = 0.47$$

or $\qquad (6-8)$

$$\frac{R_{12}}{R_{14}} = 0.69$$

where R_{12} and R_{14} are the radii of the areas of perceptibility. Thus, impacts of equal energy will be detected at greater range from the Apollo 14 site than from the Apollo 12 site by a factor of 1.45. Also, because all events detected at the Apollo 12 station are also recorded at the Apollo 14 site, the area of perceptibility for the Apollo 12 PSE is completely enclosed by the area of perceptibility for the Apollo 14 PSE. Thus,

$$R_{12} + 181 \text{ km} \leq R_{14} \qquad (6-9)$$

because the sites are 181 km apart. From equations (6–8) and (6–9), $R_{14} \geq 580$ km. This method is weak for ranges large in comparison with the separation of the two stations, but it tends to confirm that events are being detected from ranges much larger than can be estimated from the rise-time measurements, and it is consistent with the $1/r$ spreading law for amplitudes of seismic signals, which is favored for reasons previously given.

Correlation Between Seismic Events and Measurements of Neutral and Ionized Gas

No definite correlation between the occurrence of seismic events and the detection of increased gas pressure by the lunar atmosphere detector has been noted during this report period. However, detailed comparison of records has not been possible during the preliminary analytic period represented by this report.

Ionized gas particles have been recorded from the LM and SIVB impacts by both the charged-particle lunar environment experiment and the SIDE. The apparent velocity of the leading edge of the expanding gas cloud varies between 1 and 2 km/sec (secs. 8 and 10). The variations in velocity possibly result from interaction between the gas cloud and the solar wind. Ionized gas particles were also recorded by the SIDE sensors at both the Apollo 12 and 14 sites on March 19, beginning approximately 36 min after the beginning of the largest natural impact signal recorded (at 18:20 G.m.t.) during this report period (sec. 8). Assuming a common source for the increased ion concentration and the seismic signal and a velocity of 1 km/sec for the leading edge of the gas cloud, the impact must have occurred at a range of approximately 2000 km from the Apollo 14 site to explain the difference in arrival time of seismic signal and ionized gas particles. This observation strengthens the belief that meteoroid impacts are being recorded at very great ranges. Assuming an impact velocity of 20 km/sec, the mass of the meteoroid is estimated, from comparison with signals from the SIVB impacts, to be approximately 10 to 20 kg.

Conclusions

Seismic events of natural origin were recorded by the seismometers of the Apollo 14 PSE at an average rate of 1.8/day during the 45-day period covered by this report. This quantity is more than twice the number recorded by the Apollo 12 PSE during the same interval. The greater number of recorded events by the Apollo 14 PSE is believed to be a consequence of the thick layer of unconsolidated material that blankets the region (Fra Mauro Formation plus overlying regolith). Amplification of seismic-wave motion at the surface of such a layer is a well-known phenomenon.

Hundreds of events were detected by the SP seismometer at the Apollo 14 site. Most of these events are attributed to thermoelastic stress relief and venting of gases within the LM descent stage. Many of the events, however, are believed to be of natural origin, because the occurrence of events persists at a reduced rate throughout the lunar night. Local meteoroid impacts are expected to be the main source of these signals, although other source mechanisms, such as thermal fracturing of rocks, are possible.

The natural seismic events are moonquakes and meteoroid impacts. Of the total number of recorded events, 14 have been tentatively identified as moonquakes. The moonquakes occur at monthly intervals near times of perigee. The repeating moonquakes detected at the Apollo 12 and 14 sites are believed to occur at not less than 11 different locations. However, a single focal zone accounts for 80 percent of the total seismic energy detected. At least two of the moonquakes observed by the Apollo 14 PSE during the first perigee crossing were generated in this zone. The

active zone appears to be 600 to 700 km from the Apollo 12 and 14 sites and may be deep within the Moon, though the precise location has not been determined. If, however, appreciable depth of focus is verified by future data, it will be of fundamental importance relative to the present state of the lunar interior.

Cumulative strain at each location is inferred. Thus, the moonquakes appear to be releasing internal strain of unknown origin, the release being triggered by tidal stresses.

A simple wave-propagation model for the entire Moon is now available. The velocity of compressional waves increases from 100 m/sec or less near the surface to 6 km/sec depths of 6 and 20 km. This velocity structure results in a strong surface waveguide. Natural events and artificial impacts produce seismic reverberations of unusually long duration; this occurrence can be explained by intensive scattering of waves in the outer shell of the Moon. The thickness of the scattering zone may be as great as 50 to 100 km. The presence of craters alone is sufficient to explain near-surface scattering, but the scattering zone appears to be too thick to be accounted for by visible cratering alone. The nature of the heterogeneity that results in scattering below the cratered zone is unknown. Extremely low seismic-energy absorption, even near the surface, may be a result of the nearly complete absence of volatiles in the Moon. Material below the zone of scattering behaves more nearly as an ideal seismic-wave conductor, and variation of seismic-wave velocity with depth is small at depths below 20 km. Consequently, seismic-wave amplitudes vary approximately as the inverse of range (chord distance to both near and far ranges. This falloff of amplitude with distance is less rapid than in the Earth, a factor that, along with the very low noise level of the Moon, favors the detection of distant weak events. Accordingly, it appears that the impact of the LM ascent stage, or even a meteoroid of mass as small as 1 kg, can be detected at the Apollo 14 site from any point on the surface of the Moon. The Moon appears to be a more efficient "sounding board" for meteoroid impacts than had previously been suspected. Artificial impacts of the Apollo 15 mission will be a good test of this hypothesis.

Based upon the distribution of the seismic signals received from events believed to be meteoroid impacts, the following tentative estimate for the distribution of meteoroids in the kilogram-mass range that collide with the lunar surface has been deduced: $\log N = -2.32 - \log M$, where N is the cumulative number of meteoroids (in $km^{-2} \cdot yr^{-1}$) of mass equal to or greater than M (in g). This flux is smaller by a factor of 20 than that given previously (ref. 6-7).

Based upon the present wave-propagation model, one-half of the recorded impacts occur at ranges less than 1300 km (nearest 13 percent of the lunar surface), and the remaining impacts occur at ranges greater than 1300 km (87 percent of the lunar surface). Twenty-one percent of the detected impacts occur on the far half of the Moon. Thus, it would appear that impacts are being detected from the entire lunar surface. Approximately one meteoroid impact per year of kinetic energy equal to the SIVB impact is predicted from the present data.

References

6-1. LATHAM, G.; EWING, M.; PRESS, F.; SUTTON, G.; ET AL.: Passive Seismic Experiment. Sec. 6 of Apollo 11 Preliminary Science Report. NASA SP-214, 1969.

6-2. LATHAM, G.; EWING, M.; PRESS, F.; SUTTON, G.; ET AL.: Apollo 11 Passive Seismic Experiment. Science, vol. 167, no. 3918, Jan. 30, 1970, pp. 455-467.

6-3. LATHAM, G.; EWING, M.; PRESS, F.; SUTTON, G.; ET AL.: Apollo 11 Passive Seismic Experiment. Vol. III of Proc. Apollo 11 Lunar Sci. Conf. Geochim. Cosmochim. Acta Supp. 1, 1970, pp. 2309-2320.

6-4. LATHAM, G.; EWING, M.; PRESS, F.; SUTTON, G.; ET AL.: Passive Seismic Experiment. Sec. 3 of Apollo Preliminary Science Report. NASA SP-235, 1970.

6-5. LATHAM, G.; EWING, M.; PRESS, F.; SUTTON, G.; ET AL.: Seismic Data from Man-Made Impacts on the Moon. Science, vol. 170, no. 3958, Nov. 6, 1970, pp. 620-626.

6-6. EWING, M.; LATHAM, G.; PRESS, F.; SUTTON, G.; ET AL.: Seismology of the Moon and Implications on Internal Structure, Origin, and Evolution. Highlights of Astronomy, D. Reidel Pub. Co. (Dordrecht, Holland), 1971.

6-7. HAWKINS, G. S.: The Meteor Population. Research Rept. 3. NASA CR-51365, 1963.

6-8. ISACKS, B.; AND OLIVER, J.: Seismic Waves With Frequencies From 1 to 100 Cycles per Second Recorded in a Deep Mine in Northern New Jersey. Bull. Seism. Soc. Amer., vol. 54, no. 6, 1964, pp. 1941-1979.

6-9. STEG, R. G.; AND KLEMENS, P. G.: Scattering of Rayleigh Waves by Surface Irregularities. Phys. Rev. Lett., vol. 24, no. 8, Feb. 23, 1970, pp. 381-383.

6-10. PANDIT, B. I.; AND TOZER, D. C.: Anomalous Propagation of Elastic Energy Within the Moon. Nature, vol. 226, no. 5243, Apr. 25, 1970, p. 335.

6-11. SCHREIBER, E.; ANDERSON, O.; SOGA, N.; WARREN, N.; AND SCHOLZ, C.: Sound Velocity and Compressibility for Lunar Rocks 17 and 46 and for Glass Spheres From the Lunar Soil. Science, vol. 167, no. 3918, Jan. 30, 1970, pp. 732-734.

6-12. KANAMORI, H.; NUR, A.; CHUNG, D.; WONES, D.; AND SIMMINS, G.: Elastic Wave Velocities of Lunar Samples at High Pressures and Their Geophysical Implications. Science, vol. 167, no. 3918, Jan. 30, 1970, pp. 726-727.

6-13. NAKAMURA, YOSIO; AND LATHAM, GARY V.: Internal Constitution of the Moon: Is the Lunar Interior Chemically Homogeneous? J. Geophys. Res., vol. 74, no. 15, July 1969, pp. 3771-3780.

6-14. WARREN, NICHOLAS V.; AND LATHAM, GARY V.: An Experimental Study of Thermally Induced Microfracturing and Its Relation to Volcanic Seismicity. J. Geophys. Res., vol. 75, no. 23, Aug. 1970, pp. 4455-4464.

7. Active Seismic Experiment

Robert L. Kovach,[a][†] Joel S. Watkins,[b] and Tom Landers[a]

The purpose of the active seismic experiment (ASE) is to generate and monitor seismic waves in the lunar near surface and to use these data to study the internal structure of the Moon to a depth of approximately 460 m (1500 ft). Two seismic energy sources are used: an astronaut-activated thumper device containing 21 small explosive initiators and a mortar package containing four high-explosive grenades. The grenades are rocket launched by command from Earth and are designed to impact at ranges of about 150, 300,

[a] Stanford University.
[b] University of North Carolina.
[†] Principal investigator.

900, and 1500 m (about 500, 1000, 3000, and 5000 ft). A secondary objective of the experiment is to monitor high-frequency seismic activity during periodic listening modes.

Analysis to date of the seismic signals generated by the astronaut-activated thumper has revealed important information concerning the near-surface structure of the Moon. Two compressional wave (P-wave) seismic velocities were measured at the Fra Mauro site. The near-surface material possesses a seismic-wave velocity of 104 m/sec (340 ft/sec). Underlying this surficial layer at a depth of 8.5 m (28 ft), the lunar material has a velocity of 299 m/sec (980 ft/sec). The measured thickness of the upper unconsolidated debris

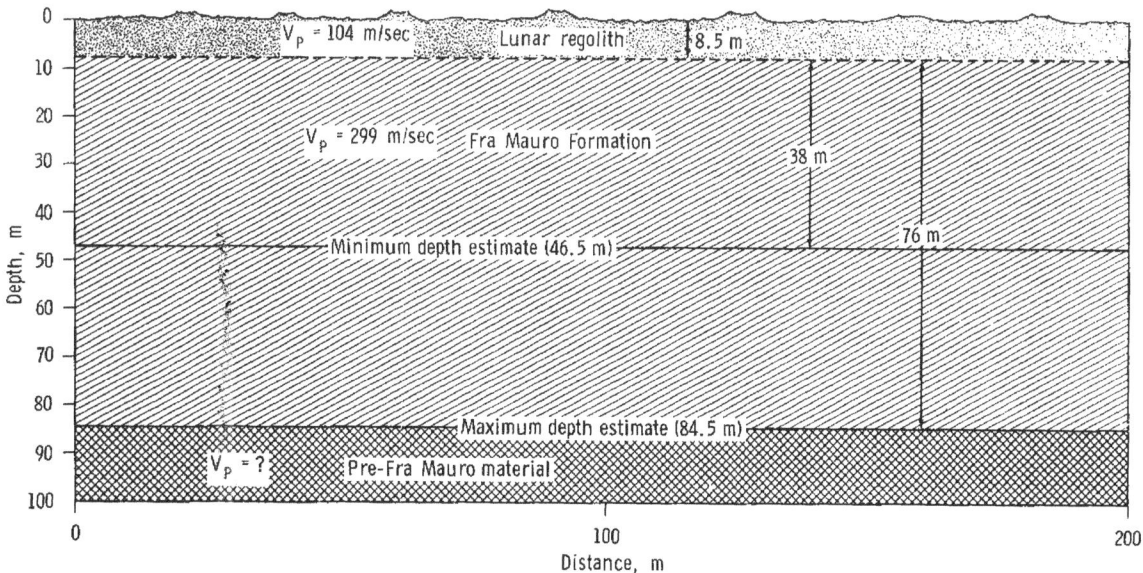

FIGURE 7-1.—Seismic cross section at Fra Mauro landing site. (V_P = seismic-wave velocity.)

layer is in good agreement with geological estimates of the thickness of the regolith at this site.

By combining the seismic-refraction results from the ASE with the lunar module (LM) ascent seismic data recorded by the Apollo 14 passive seismic experiment (PSE), estimates of the thickness of the underlying material can be made (fig. 7–1). These estimates range from 38 to 76 m (124 to 250 ft) and may be indicative of the thickness of the Fra Mauro Formation at this particular site. More definitive conclusions must await the seismic results from the four grenade firings.

Interesting signals, similar to some events recorded by the PSE, have also been recorded during the intermittent passive listening periods of the ASE. Analysis of these signals together with similar data from the PSE may shed light on the origin of these signals.

Instrument Description and Performance

The ASE consists of a thumper and geophones, a mortar package assembly, electronics within the Apollo lunar-surface experiments package (ALSEP) central station, and interconnecting cabling. The components of the ASE are shown schematically in figure 7–2.

The astronaut-activated thumper is a short staff (fig. 7–3) used to detonate small explosive charges—single bridgewire Apollo standard initiators. Twenty-one initiators are mounted per-

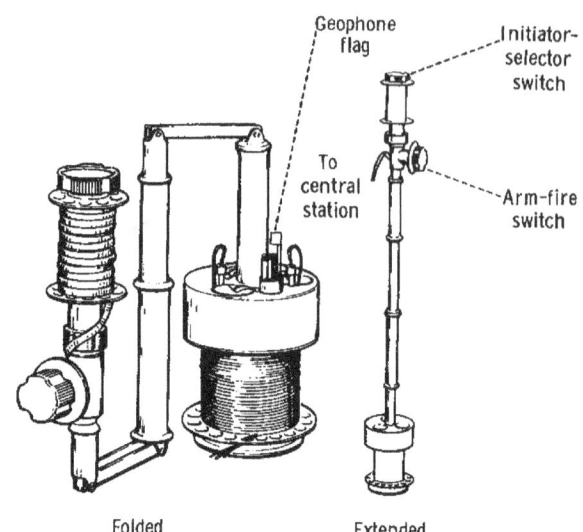

FIGURE 7–3.—Schematic diagram of the thumper in the folded and extended positions.

pendicular to the base plate at the lower end of the staff. A pressure switch in the base plate detects the instant of initiation. An arm-fire switch and an initiator-selector switch are located at the upper end of the staff. A cable connects the thumper to the central station to transmit real-time event data. The thumper also stores the three geophones and connecting cables until deployment on the lunar surface. In figure 7–4, the lunar module pilot (LMP) is shown beginning to unwind the

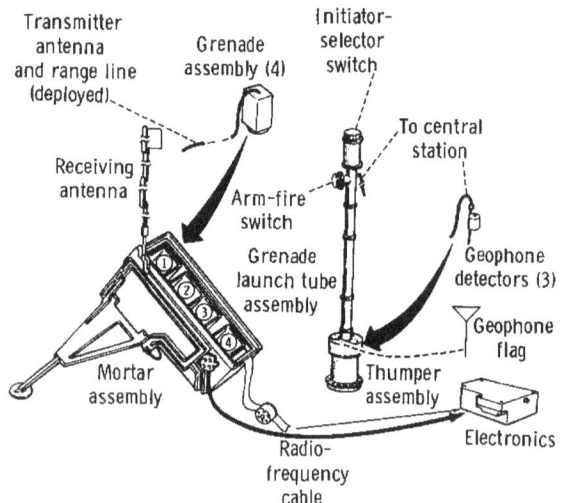

FIGURE 7–2.—Schematic diagram of the ASE.

FIGURE 7–4.—Enlargement of 16-mm sequence camera photograph showing the LMP with hand-held thumper (S–71–19509).

geophone line from the thumper on the Moon just before activation of the ASE at 17:59 G.m.t. on February 6, 1971.

The three identical geophones are miniature seismometers of the moving coil-magnet type. The coil is the inertial mass suspended by springs in the magnetic field. Above the natural resonant frequency of the geophones (7.5 Hz), the output is proportional to ground velocity. The geophones are deployed at 3-, 49-, and 94-m (10-, 160-, and 310-ft) intervals in a linear array from the central station and connected to it by cables.

A three-channel amplifier and log compressor condition the geophone signals before conversion into a digital format for telemetering to Earth. The low signal-to-noise ratios expected and the lack of knowledge as to the character of the expected waveforms made it desirable to widen the frequency response as much as possible within the constraints of the digital sampling frequency of 500 Hz. Because signal levels were expected to be distributed throughout the system dynamic range, a logarithmic compression scheme was selected to give signal resolution as some constant fraction of

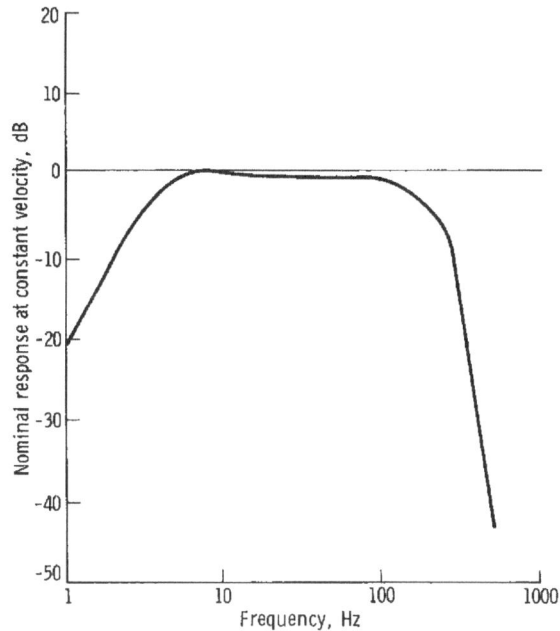

FIGURE 7–5.—Frequency response of the ASE.

signal amplitude. The system deployed on the Moon has the properties listed in table 7–I, and

TABLE 7–I. *Apollo 14 ASE Characteristics*

Component characteristics	Channel no.		
	1	*2*	*3*
Geophones:			
Generator constant, V/m/sec	250.4	243.3	241.9
Frequency, Hz	7.32	7.22	7.58
Resistance, ohm	6065	6157	6182
Amplifiers:			
Noise level, μV rms at input	0.300	0.325	0.272
Dynamic range, rms signal to rms noise in dB	86.8	86.5	87.5
Gain (at 10 Hz and $V_{input} = 0.005$ V rms)	666.7	666.7	675.7
Log compressor (compression accuracy for temperature range 15° to 50° C):			
Positive signal error, percent	3.79	4.71	2.00
Negative signal error, percent	2.07	1.32	3.33
System:			
Signal-to-noise ratio (rms signal to rms noise in dB for a 1-nm peak-to-peak signal at 10 Hz)	33.6	33.1	32.9
Calibrator accuracy:			
Generator constant, percent error	4.21	9.70	6.40
Natural frequency, percent error	3.28	4.99	8.58

TABLE 7–II. *Apollo 14 ASE Grenade Parameters*

Parameter	Grenade no.			
	1	2	3	4
Range, m...............................	1 500	900	300	150
Mass, g.................................	1 236	1 020	810	719
High-explosive-charge mass, g.............	454	272	136	45.4
Rocket-motor mean peak thrust, N........	20 460	11 450	7005	5337
Mean velocity, m/sec....................	49	40	23	16
Lunar flight time, sec...................	44	32	19	13
Rocket-motor-propellant mass, g..........	47	31	16.8	11.5
Quantity of propellant pellets, number.....	2 435	1 596	648	550
Launch angle, deg.......................	45	45	45	45
Rocket-motor thrust duration, msec.......	7.0	8.2	12.5	12.5

the nominal frequency response shown in figure 7–5.

The mortar assembly comprises a mortar box, a grenade-launch-tube assembly, and interconnecting cables. To provide an optimum launch angle for the grenades, the mortar package is deployed at an angle approximately 45° to the lunar surface. A two-axis inclinometer provides pitch- and roll-angle (deviation from the vertical) information on the mortar package. The mortar box is a rectangular fiber-glass-and-magnesium construction in which is mounted the grenade-launch-tube assembly containing four grenades.

Each grenade is attached to a range line, which is a thin-stranded cable wound around the outside of the launch tube. Two fine copper wires are looped around each range line. The first loop is spaced so that it will break when the grenade is approximately 0.4 m (16 in.) from the launch tube. A second loop is spaced to break when the range line has deployed an additional 8 m (25 ft) from the first breakwire. Breaking the loops starts and stops a range-gate pulse to establish a time interval for the determination of the initial grenade velocity.

The four grenades are similar but differ in the amount of propellant and high explosive (table 7–II). Each grenade possesses a square cross section with a thin fiber-glass casing. The casing contains the rocket motor, safe slide plate, high-explosive charge, ignition and detonation devices, thermal battery, and a 30-MHz transmitter. The

range line is attached to the transmitter to serve as a half-wave end-feed antenna.

In operation, an arm command from ground control applies a pulse to charge condensers in the mortar box and grenade; a fire command discharges the condenser through an initiator, which ignites the rocket motor. When the grenade leaves the tube, a spring-ejected safe slide is removed, activating a microswitch in the grenade.

A thermal battery and the electronics in the grenade make up the firing circuit. The microswitch discharges a condenser across a thermal match to activate the thermal battery, which in turn powers the transmitter and produces a capacitor charge for the detonator. At impact, an omnidirectional impact switch closes, discharging the capacitor into the detonator to ignite the explosive. The explosion terminates radiofrequency transmission as an indication of detonation time. The critical parameters measured are the detonation time, time of flight, initial velocity, and launch angle. Because of the ballistic trajectory followed by the grenades in the lunar vacuum, the necessary data are available to determine grenade range. The planned mortar mode of operation for the ASE is shown in figure 7–6.

Because some of the geophone parameters might drift on the lunar surface, a calibrator circuit is provided in the system to measure these parameters to within 10 percent of the preflight values. The damping resistance across the geophone is altered to underdamp the geophone, and

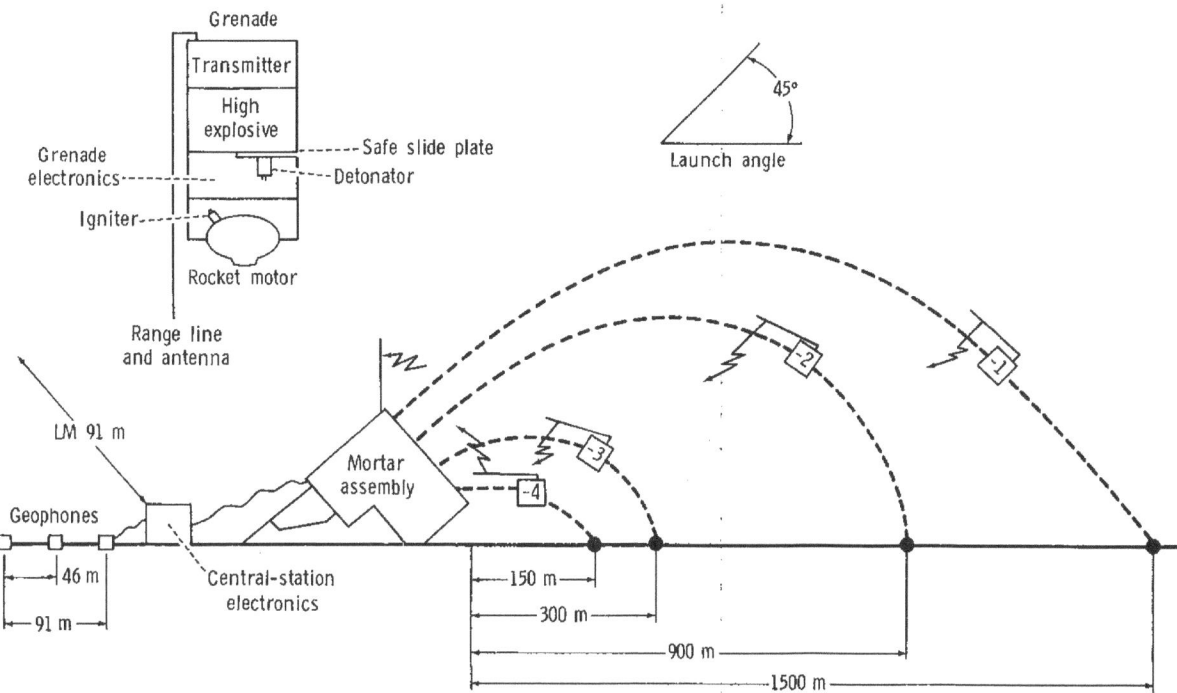

FIGURE 7–6.—Schematic diagram showing the mortar mode of operation for the ASE.

current is introduced into the geophone coil to react with the magnetic field of the geophone, producing a force on the geophone coil. This force moves the coil and, with an underdamped geophone, the signal from the geophone is a logarithmically decaying sinusoidal signal. A typical calibration pulse recorded on the three geophones on February 12 is shown in figure 7–7. Analysis of similar calibration pulses transmitted between thumper operations on the Moon demon-

strated close agreement of the natural frequency and generator constant of the geophones with measured preflight values.

The ASE system is controlled from Earth by a number of commands that control such functions as switching to the high-bit data rate and firing the grenades from the mortar box assembly. Further technical details of the ASE, particularly of the electronics, can be found in reference 7–1.

Thermal Control

The ASE electronics are part of the ALSEP central station and do not require separate thermal control. The ASE mortar package assembly was designed to be maintained between —60° and 85° C. Thermal control of the mortar package assembly is accomplished with multilayer aluminized Mylar insulation used in conjunction with small heaters. Heater operation begins automatically at a temperature of about —17° C when the ASE is in the standby mode of operation. Four temperatures are monitored in this mode: central-electronics temperature, grenade-launch-tube-assembly temperature, mortar-package tempera-

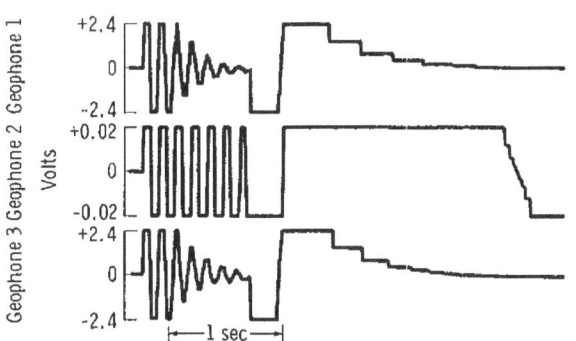

FIGURE 7–7.—Calibration pulse recorded on February 12 during passive listening mode.

ture, and temperature of the geophone closest to the ALSEP.

Deployment

At the Apollo 14 site, the three geophones are alined in a southerly direction from the ALSEP central station (fig. 3–1 in sec. 3). No difficulty was experienced in deploying and arming the

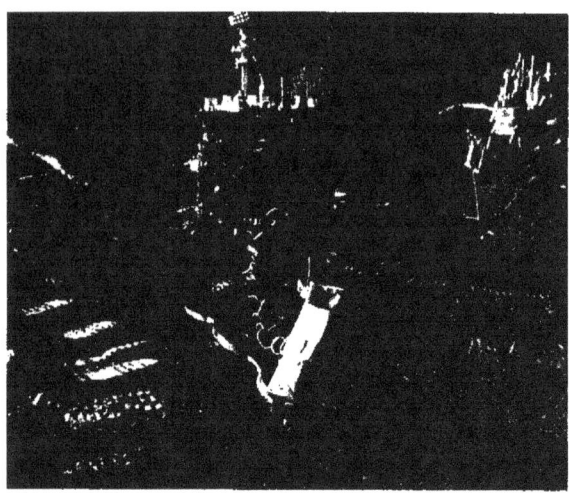

FIGURE 7–8.—Mortar assembly deployed on the lunar surface (AS14–67–9361).

mortar package. The mortar package is positioned to fire the four grenades in a northerly direction in alinement with the geophone line. The mortar package assembly is shown in figure 7–8 as deployed on the lunar surface. A view looking downrange from the mortar package is shown in figure 7–9. The geophone line appears in the right foreground.

For convenience, the geophone closest to the ALSEP is designated "geophone 1" and the most distant is designated "geophone 3." No difficulty was experienced in implanting the geophones and maintaining them vertically in the lunar soil. Before beginning thumper operations at geophone 3, it was noted that the alinement flag at geophone 2 had fallen over. Because of time constraints, thumper operations were begun at geophone 3 before returning to check the flag and geophone at the middle position.

Thumper operations were begun at 18:09 G.m.t. and continued until 18:37 G.m.t. Thumper firings were begun with shot 1 at geophone 3 and continued at 4.6-m (15-ft) intervals along the geophone line to shot 21 at geophone 1. In figure 7–10, the LMP is shown firing the thumper along the geophone line on the lunar surface. The thumper failed to fire after several attempts at

FIGURE 7–9.—Photograph looking downrange from mortar assembly on lunar surface. Geophone line is deployed in right foreground (AS14–67–9377).

FIGURE 7–10.—The LMP firing thumper along geophone line on the lunar surface (AS14–67–9374).

several initiator positions, and several firing positions were skipped to gain extravehicular activity (EVA) time. Successful thumper shots were recorded at positions 1 (located at geophone 3); 2, 3, 4, 7, 11 (located at geophone 2); 12, 13, 17, 18, 19, 20, and 21 (located at geophone 1).

Upon reaching position 11, at the middle geophone, the LMP observed that this geophone had pulled out of the ground, apparently because of the effects of set or elastic memory in the cable. After repositioning the geophone, he resumed thumping operations. Even though geophone 2 was resting on its side during the first five thumper firings, usable seismic data were recorded. The net result of tipping a vertical-component geophone off vertical is to translate the mass and effectively increase the natural frequency. Analysis of the calibration pulse sent before beginning thumping operations showed that the effective natural frequency of geophone 2 had increased from 7.5 to 13.4 Hz. The total time spent on thumping operations was 28 min, within allowable EVA constraints, and valuable lunar seismic data were obtained.

Several thumper shots were attempted while the commander was moving on the lunar surface near the ALSEP central station. Unfortunately, his movements generated seismic energy that was recorded by the highly sensitive geophones. As a result, his movements had to be restricted during the remaining thumper operations. However, it still may be possible to conduct thumper operations on the Moon and allow the second astronaut to move about, provided he is sufficiently far removed from the central station and geophone line.

Description of Recorded Seismic Signals

Thumper Mode

During thumper operations on the lunar surface, the LMP was instructed to stand still for 20 sec before and 5 sec after each firing. Therefore, 5 sec of seismic data were recorded for each thumper firing. The seismic data recorded for thumper shots 18 and 20 are shown in figure 7–11. Characteristically, the seismic signals produced by thumper firings within 9 m (30 ft) of a geophone

have extremely impulsive beginnings and saturate the dynamic range of the amplifier for about 0.5 sec. The predominant frequency of these signals range from 27 to 29 Hz.

As the distance between the thumper firings and the geophones is increased, the seismic signals possess more emergent beginnings. In figure 7–12, a record section is alined in time to the same instant of firing for thumper shots 18 to 21 as recorded at geophone 2. The wave trains build up to a maximum amplitude within the first 0.25 to 0.5 sec from onset of signal and then gradually decrease in amplitude. This effect can best be seen in figure 7–11 by comparing the seismic signals recorded at 14, 32, and 41 m (45, 105, and 135 ft). Little difficulty exists in picking the onset of the seismic signals out to a distance of 46 m (150 ft); but, at greater distances, uncertainty arises in determining the beginning of the seismic wave arrival because the signals are much weaker. The peak amplitude of the recorded signals typically decreases by a factor of approximately 60 in 61 m. More refined data-analysis techniques applied to the recorded seismic signals from the thumper mode of operation are underway.

Passive Listening Mode

The ASE is also capable of operating in a passive listening mode and is commanded into this high-bit-rate mode for a 30-min period each week. Several interesting signals have been recorded in this mode of operation. Two of these events, recorded on February 19, are shown in figure 7–13. The signals have the largest amplitude on a single geophone, although the signal is definitely discernible above the ambient noise level, but greatly reduced in amplitude on the other geophone channels.

The signal recorded on geophone 1 beginning at about 15:29:41 G.m.t. has a predominant frequency of approximately 36 Hz, whereas the signal recorded by geophone 2 at approximately 15:38:46 G.m.t. has a frequency of approximately 47 Hz. Both of these signals have impulsive beginnings and relatively short durations of 6 to 10 sec. Maximum amplitudes occur at the beginning of the wave train.

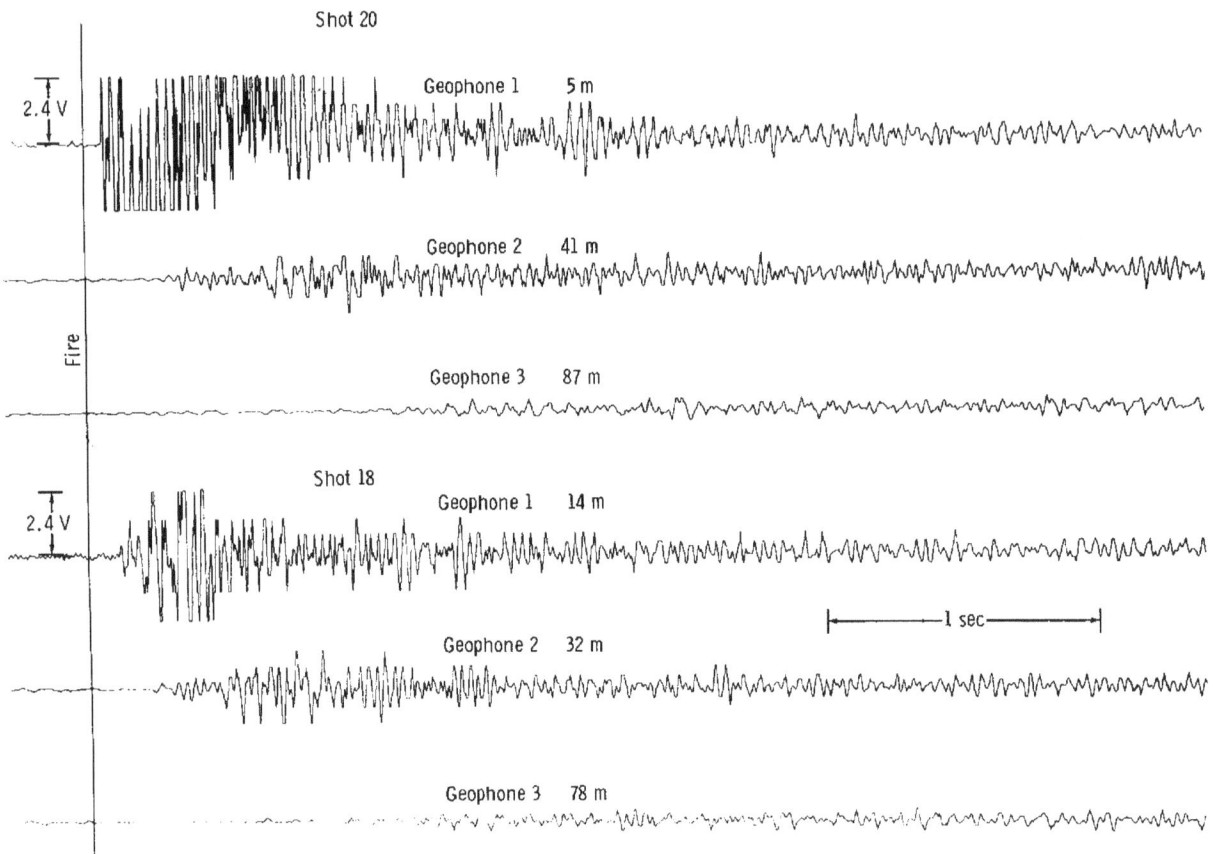

FIGURE 7–11.—Seismic signals produced by thumper firings 18 and 20 on the lunar surface.

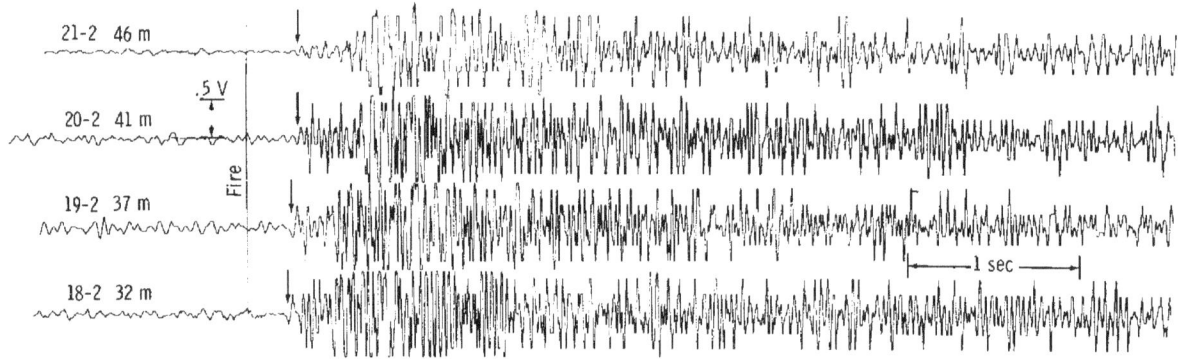

FIGURE 7–12.—Seismic signals produced by thumper firings 18 to 21 as recorded at geophone 2. The traces are alined to the same relative instant of the firing of the thumper. The small arrows point to the onset of the seismic signal.

One can note a gross similarity to the seismic signals produced by thumper firings close to a geophone, although the amplitude of these signals is approximately 100 to 200 times smaller than the thumper-generated signals. These events are also strikingly similar to the type X events (impulsive beginnings and relatively short durations—normally less than 10 sec) recorded during the

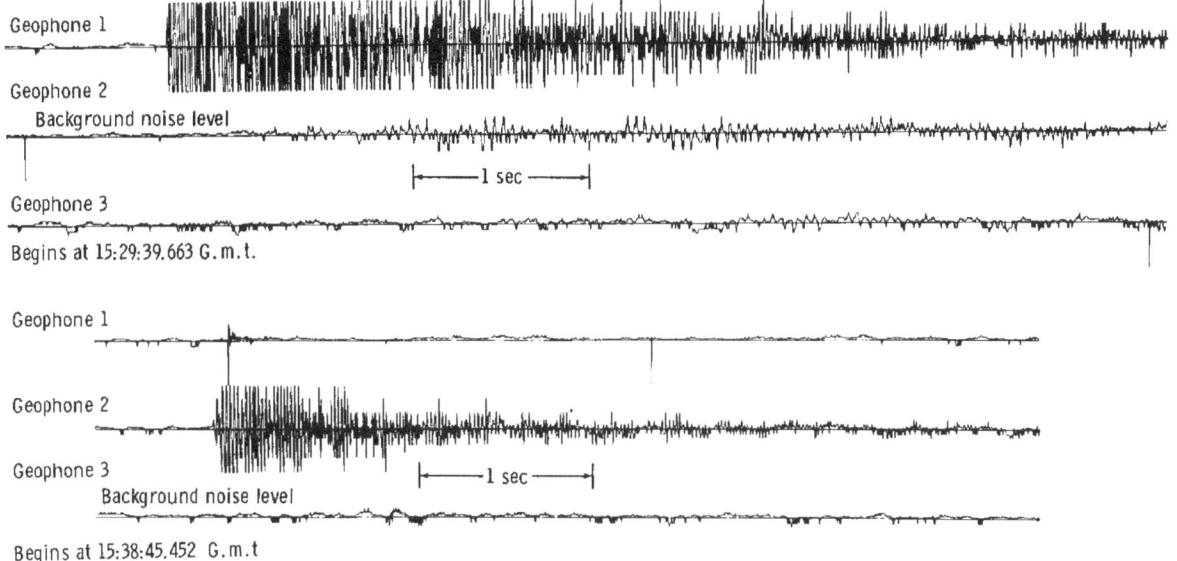

FIGURE 7–13.—Signals recorded during the passive listening mode on February 19; all plots clipped at 0.01 V.

Apollo 11 mission (ref. 7–2) that were speculated to be produced by direct micrometeoroid impacts on the PSE. Continuing work is being directed toward detailed analysis of these events.

Discussion

A preliminary interpretation of the traveltime/distance data (fig. 7–14) obtained from the thumper firings is rewardingly consistent in that

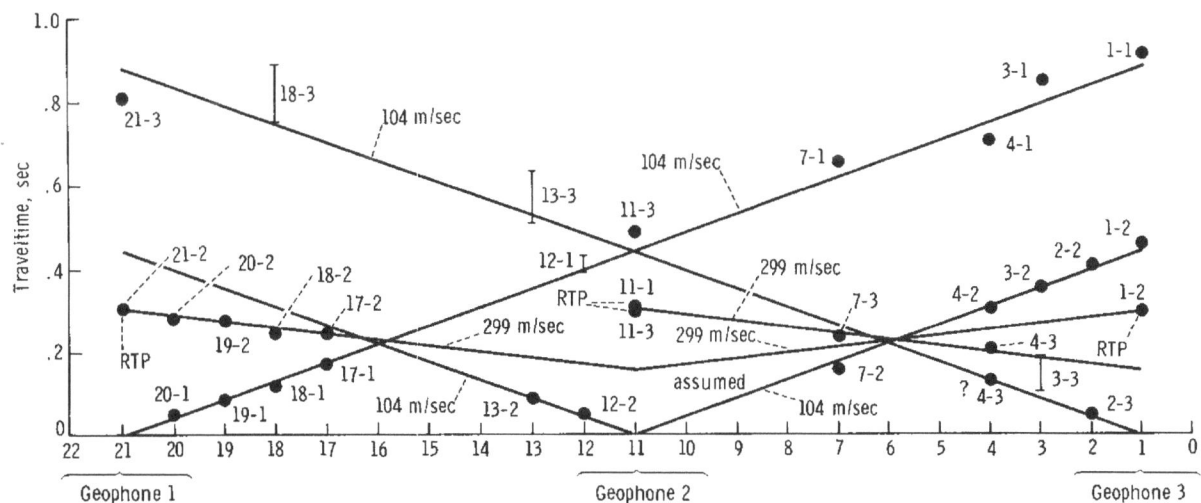

FIGURE 7–14.—Seismic arrivals from the thumper firings plotted on a traveltime/distance graph. The data points are shown as black circles; the first number refers to the thumper firing, the second number to the geophone on which the data were recorded. Reverse tie points are indicated as RTP; distance between thumper shot locations is 4.6 m.

very good agreement exists in reverse tie points (RTP); that is, the principle of seismic reciprocity states that traveltimes must be identical when the positions of a geophone and a shot (thumper firing) are interchanged. For example, the traveltime for thumper shot 21 to geophone 2 and thumper shot 11 to geophone 1 should be identical. The near identity of reverse tie points 21–2 and 11–1 and of 11–3 and 1–2 lends strength to the data interpretation. Agreement between reverse tie points 1–1 and 21–3 is somewhat poorer, but it should be remembered that the more distant thumper firings produced much weaker seismic signals for which it is difficult to determine unambiguously an initial onset. An example of traveltimes that cannot be precisely determined is 18–3, and the range of possible traveltimes is shown by the line.

Two P-wave velocities are evident in the traveltime data. A direct arrival is observed with a P-wave velocity of 104 m/sec together with a faster arrival possessing a velocity of 299 m/sec. No apparent variation exists in P-wave velocities across the section sampled as is evidenced by the conformance of seismic velocities measured along the geophone line.

The depth to the 299-m/sec refracting horizon is 8.5 m. It is proposed that this thin upper layer possessing a seismic velocity of 104 m/sec represents the fragmental veneer of unconsolidated particulate debris—the lunar regolith that covers the surface at the Fra Mauro site. If the discontinuity between the 104-m/sec and the 299-m/sec material is accepted as the base of the regolith, the thickness of the regolith is very similar to that estimated solely on geological evidence. By photographic studies of the depth at which blocky floors appear in fresh craters, it has been inferred that the fragmental, surficial layer that overlies the more consolidated or semiconsolidated substrate at the Fra Mauro site ranges in thickness from 5 to 12 m (ref. 7–3).

From measurement of the elapsed time between engine ignition and signal arrival at the Apollo 12 PSE for the reaction control system test firings and for the LM ascent, the compressional velocity of the lunar-surface material at the Apollo 12 site was determined to be approximately 108 m/sec

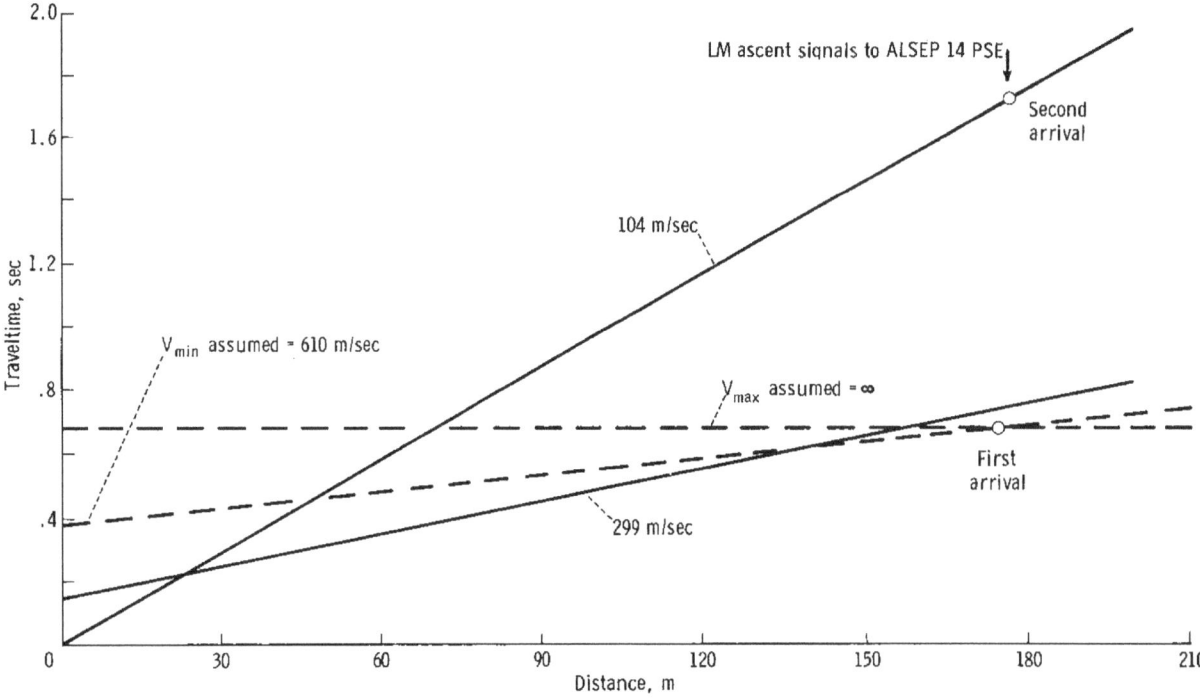

FIGURE 7–15.—First and second arrivals from the LM ascent as recorded by the PSE compared to the extrapolated traveltime/distance data derived from the thumper firings.

(ref. 7–4). This value is in exceedingly close agreement with that measured for the surface material at the Apollo 14 Fra Mauro site and is also consistent with estimates derived on the basis of mechanical properties measured by Surveyor (ref. 7–5). Therefore, it can be argued that the fragmental and comminuted layer that covers much of the lunar surface, although locally variable in thickness, possesses remarkably similar seismic or acoustic properties.

One other piece of evidence lends support to this hypothesis. Seismic signals were also generated by the LM ascent and recorded by the Apollo 14 PSE at a distance of 178 m (584 ft). These data are shown in figure 7–15 compared with the extrapolated traveltime/distance curves derived from the thumper firings of the ASE. The observed traveltime for the second arrival is nearly identical to that predicted for a direct seismic wave propagating with a velocity of 104 m/sec. A first arrival is observed with a traveltime somewhat faster than that predicted by a refraction from the top of the 299-m/sec horizon, suggesting that a material with a faster intrinsic compressional-wave velocity lies beneath the 299-m/sec material.

If the assumption is made that this underlying material possesses an infinite compressional-wave velocity V and that the traveltime curve behaves in the manner shown in figure 7–15, a maximum estimate to the thickness of the 299-m/sec material can be derived. Similarly, if only a modest increase in seismic velocity, such as to 610 m/sec (2000 ft/sec), in the underlying material is proposed, then a minimum estimate is obtained. These assumptions lead to a minimum thickness estimate of 38 m and a maximum thickness estimate of 76 m.

At this writing, it is somewhat premature to speculate what the 299-m/sec material represents other than to comment that this velocity is similar to that measured in situ in blocky basalt flows near Flagstaff, Ariz., or in blocky pumice deposits such as found at the Southern Coulee of the Mono Craters, California (ref. 7–6). However, no inference to a specific rock type should be made because a wide range of velocities is often determined for similar rocks, and similar velocities are often measured for widely different rock types.

Nevertheless, it is interesting to point out that the thickness estimate of 38 m to 76 m for this material is not in disagreement with the postulated thickness of 100 m or so for the Fra Mauro Formation (ref. 7–3).

The relatively low compressional wave velocities that were measured by the ASE argue against the presence of substantial amounts of permafrost in the lunar near surface at this particular site. Measured velocities in permafrost vary greatly—depending on such factors as lithology, porosity, and degree of interstitial freezing—but typically range from 2438 to 4572 m/sec (8000 to 15 000 ft/sec) (ref. 7–7).

It has also been proposed (ref. 7–8) that the unconsolidated surface debris layer (the lunar regolith) together with a shattered crystalline layer forms a surface low-velocity zone to "trap" seismic surface waves effectively; this proposal helps to explain the prolonged reverberations recorded by the Apollo 12 PSE after the impact of the Apollo 12 LM and the Apollo 13 SIVB. However, the assumed working model consisted of a 30-m-thick, surface, low-velocity layer overlying crystalline material that has an intrinsic seismic velocity some 20 times greater than that of the surface-debris layer. To date, the results of the ASE argue against this hypothesis on a moonwide basis. Further details concerning the deeper structure of the lunar near surface of the Fra Mauro site must await the results from the ASE grenade firings.

References

7–1. McAllister, Bruce D.; Kerr, James; Zimmer, John; Kovach, Robert L.; and Watkins, Joel: A Seismic Refraction System for Lunar Use. IEEE Trans. Geosci. Electron., vol. GE–7, no. 3, July 1969, pp. 164–171.

7–2. Latham, G. V.; Ewing, M.; Press, F.; et al.: Passive Seismic Experiment. Sec. 6 of Apollo 11 Preliminary Science Report. NASA SP–214, 1969.

7–3. Offield, T. W.: Geologic Map of the Fra Mauro Site—Apollo 13, Scale 1:5000. USGS map, 1970.

7–4. Latham, G. V.; Ewing, M.; Press, F.; et al.: Passive Seismic Experiment. Sec. 3 of Apollo 12

Preliminary Science Report. NASA SP–235, 1970.

7–5. SUTTON, GEORGE H.; AND DUENNEBIER, FREDERICK: Elastic Properties of the Lunar Surface From Surveyor Spacecraft Data. J. Geophys. Res., vol. 75, no. 35, Dec. 1970, pp. 7439–7444.

7–6. WATKINS, J. S.: Annual Report, Investigation of In Situ Physical Properties of Surface and Sub-surface Site Materials by Engineering Geo-

physical Techniques. NASA Contract T–25091 (G), July 1966.

7–7. BARNES, D. F.: Geophysical Methods for De-lineating Permafrost. Proc. Int. Conf. Perma-frost, NAS–NRC pub. 1287, 1965, pp. 349–355.

7–8. LATHAM, G. V.; EWING, M.; DORMAN, J.; PRESS, F.; ET AL.: Seismic Data From Man-Made Impacts on the Moon. Science, vol. 170, no. 3958, Nov. 1970, pp. 620–626.

8. Suprathermal Ion Detector Experiment
(Lunar Ionosphere Detector)

H. K. Hills[a] and J. W. Freeman, Jr.[a][†]

The suprathermal ion detector experiment (SIDE), part of the Apollo lunar-surface experiments package (ALSEP), is designed to achieve the following experimental objectives.

(1) Provide information on the energy and mass spectra of the positive ions close to the lunar surface that result from solar-ultraviolet or solar-wind ionization of gases from any of the following sources: residual primordial atmosphere of heavy gases, sporadic outgassing such as volcanic activity, evaporation of solar-wind gases accreted on the lunar surface, and exhaust gases from the lunar module (LM) descent and ascent engines and the astronauts' portable life-support equipment

(2) Measure the flux and energy spectra of positive ions in the magnetotail and magnetosheath during those periods when the Moon passes through the magnetic tail of the Earth

(3) Provide data on the plasma interaction between the solar wind and the Moon

(4) Determine a preliminary value for the electric potential of the lunar surface

Instrument

The SIDE instrument is basically identical to that flown on the Apollo 12 mission and described in reference 8–1. However, the Apollo 14 instrument is completely described herein.

Description

The SIDE consists of two positive-ion detectors.

[a] Rice University.
[†] Principal investigator.

The first of these, the mass analyzer detector, is provided with a crossed electric- and magnetic-field (or Wein) velocity filter and a curved-plate electrostatic energy-per-unit-charge filter in tandem in the ion flightpath. The requirement that the detected ion must pass through both filters

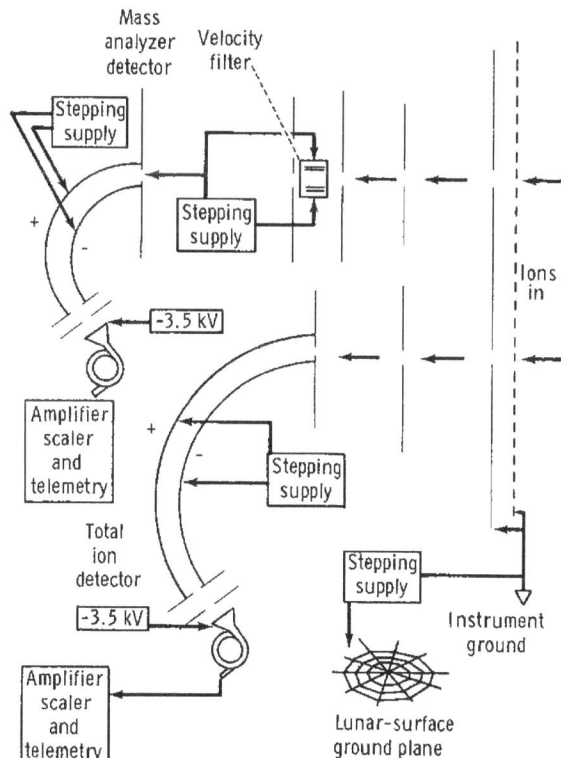

FIGURE 8–1.—Schematic diagram of the SIDE.

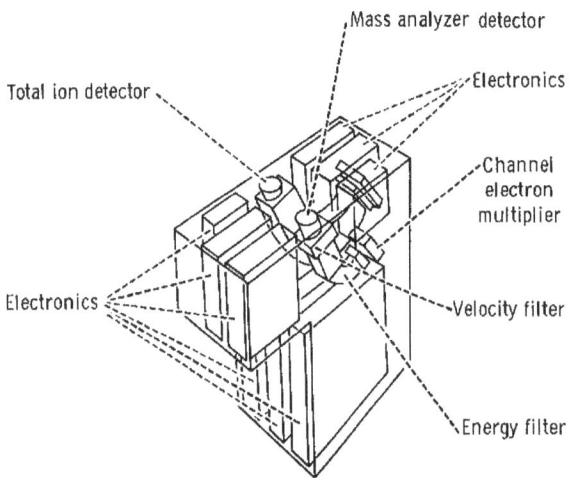

FIGURE 8-2.—Cutaway drawing showing the interior of the SIDE.

allows a determination of the mass-per-unit charge. The ion sensor itself is a channel electron multiplier operated as an ion counter that yields saturated pulses for each input ion. The second detector, the total ion detector, uses only a curved-plate electrostatic energy-per-unit-charge filter. Again, the ion sensor itself is a channel electron multiplier operated as an ion counter. Both channel electron multipliers are biased with the input ends at —3.5 kV, thereby providing a postanalysis acceleration to boost the positive-ion energies to yield high detection efficiencies. The general detector concept is illustrated in figure 8-1; figure 8-2 is a cutaway drawing that illustrates the location of the filter elements and the channel electron multipliers.

A primary objective of the experiment is to provide a measurement of the approximate mass-per-unit-charge spectrum of the positive ions near the lunar surface as a function of energy for ions from approximately 50 eV down to near-thermal energies. Therefore, the mass analyzer detector measures mass spectra at six energy levels: 48.6, 16.2, 5.4, 1.8, 0.6, and 0.2 eV. However, for the Apollo 14 instrument, dependable laboratory calibrations of the mass analyzer detector were achieved only at the two highest energy levels. The total ion detector measures the differential positive-ion energy spectrum (regardless of mass) from 3500 eV down to 10 eV in 20 energy steps. For

the Apollo 14 mass analyzer detector, the range of the mass spectrum covered is approximately 6 to 750 atomic mass units (amu). Twenty mass channels span this range. The relative width for each mass channel $\Delta M/M$ is approximately 0.2 near the lower masses. In principle, the flux of ions with masses less than 6 amu/Q can be obtained by subtracting the integrated mass-spectrum flux obtained with the mass analyzer detector from the total ion flux, at the same energy, obtained with the total ion detector.

To compensate for a possible large (tens of volts) lunar-surface electric potential, a wire screen is deployed on the lunar surface beneath the SIDE. This screen is connected to one side of a stepped voltage supply, the other side of which is connected to the internal ground of the detector and to a grounded grid mounted immediately above the instrument and in front of the ion entrance apertures (fig. 8-1). The stepped voltage is advanced only after a complete energy and mass scan of the mass analyzer detector (i.e., every 2.58 min). The voltage supply is programed to step through the following voltages: 0, 0.6, 1.2, 1.8, 2.4, 3.6, 5.4, 7.8, 10.2, 16.2, 19.8, 27.6, 0, —0.6, —1.2, —1.8, —2.4, —3.6, —5.4, —7.8, —10.2, —16.2, —19.8, and —27.6. This stepped supply and the ground screen may function in either of two ways. If the lunar-surface potential is large and positive, the stepped supply, when on the appropriate step, may counteract the effect of the lunar-surface potential, thereby allowing low-energy ions to reach the instrument with their intrinsic energies. However, if the lunar-surface potential is near zero, then on those voltage steps that match or nearly match the energy level of the mass analyzer detector or the total ion detector (1.2, 5.4, etc.), thermal ions may be accelerated into the SIDE at energies optimum for detection. The success of this method depends on the Debye length and on the extent to which the ground-screen potential approximates that of the lunar surface. It is not yet possible to assess either of these two factors; however, the data from the Apollo 12 instrument indicate that the ground-screen voltage does influence the instrument response to the incoming ions.

The SIDE is shown deployed on the lunar surface in figure 8-3. The experiment is deployed

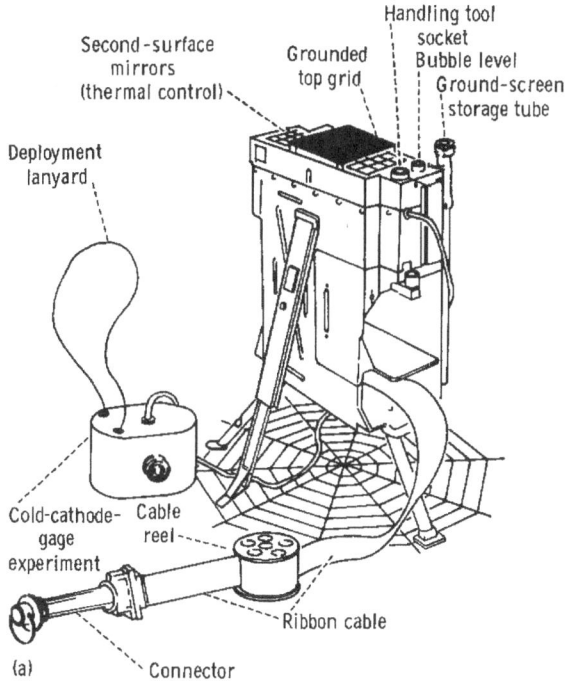

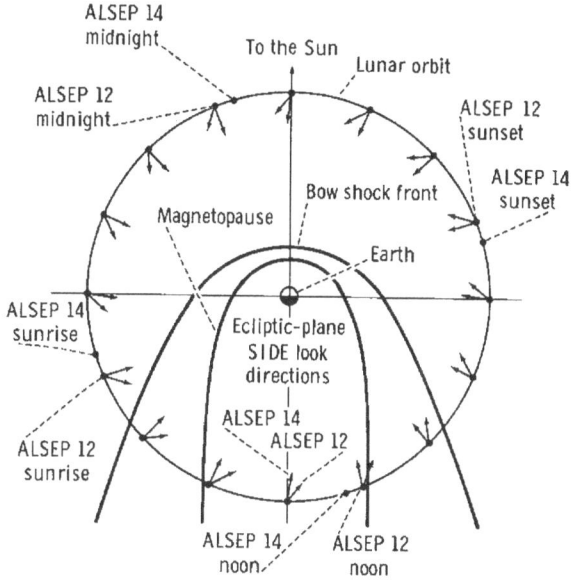

FIGURE 8–4.—The look directions of the Apollo 12 and 14 SIDE instruments at various points along the lunar orbit. (The diameter of the Earth is not drawn to scale.)

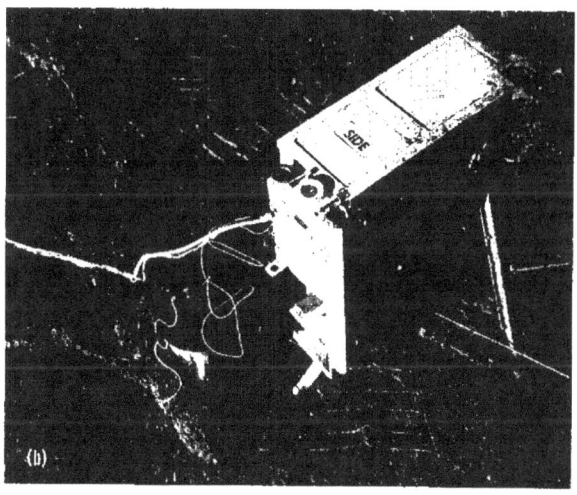

FIGURE 8–3.—The SIDE as deployed on the Moon. (a) External diagram. (b) NASA photograph AS14–63–9371.

approximately 16 m (50 ft) from the ALSEP central station in a southeasterly direction. The top surface stands 0.5 m (20 in.) above the lunar surface. The sensor look directions include the ecliptic plane, and the look axes are canted 15° from the local vertical and to the east. The look directions of both the Apollo 12 and 14 instru-

ments are shown in figure 8–4 in an Earth-Sun coordinate system at various points along the lunar orbit. The field of view of each sensor is roughly a square solid angle 6° on a side. The sensitivities of the total ion detector and mass analyzer detector are approximately 5×10^{17} and 10^{17} counts/sec/A of entering ion flux, respectively.

Performance

At the time of preparation of this report, the operation of the SIDE and the associated cold-cathode-gage-experiment electronics continued to be excellent; and all temperatures and voltages were nominal. At initial turn-on of the instrument, an illegal digital logic condition appeared, but it was cleared with the transmission of a command. The high voltages within the instrument have been commanded off for the periods of higher instrument temperature (up to ~83° C), centered on local noon, to allow the instrument to outgas without danger to the electronics. The high voltages were not operated with the instrument temperature above 25° C on the lunar day of deployment. They were operated the following lunar day up to 45° C in the morning, and from 55° C down in the afternoon, with no problems. Present plans are to increase the operating temperature limit by 10° C each successive lunar day until full-time operation is reached. The background counting rates have been quite low, well under 1 count/sec even at an internal temperature of 55° C.

Results

Magnetosheath of the Earth

The look directions (fig. 8–4) of the Apollo 12 and 14 SIDE instruments allow a two-point study of the angular distribution of the energetic ions

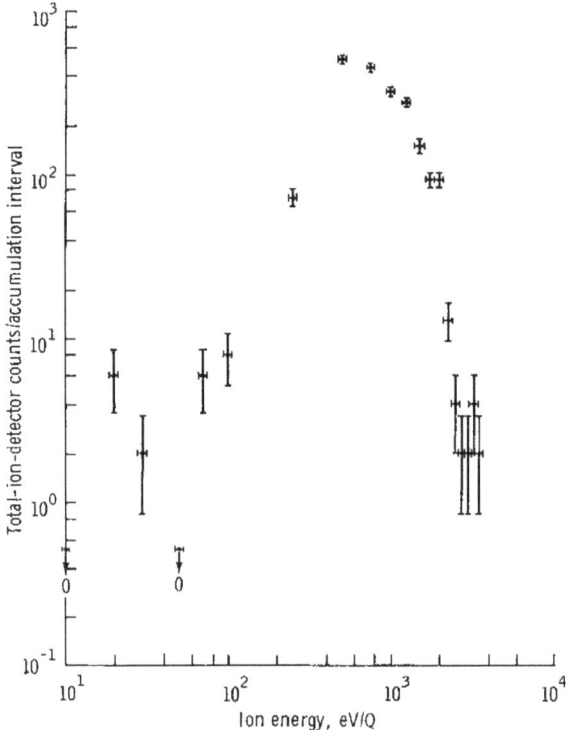

FIGURE 8–5.—Typical energy spectrum observed (17:21: 59 to 17:22:22 G.m.t., March 7) by the total ion detector in the magnetosheath region approximately 1.5 days after sunrise. The horizontal bars show the energy bandwidths of the instrument and the vertical bars denote the standard deviation of the counting rate. The accumulation interval is 1.13 sec of the 1.2 sec/ frame. The high counting rates in the 500- to 1500-eV channels, with low rates in the 10- to 100-eV channels, should be noted.

observed streaming along the magnetosheath toward the magnetospheric tail of the Earth. The Apollo 12 SIDE looks upstream, approximately parallel to the magnetosheath, as the Moon leaves the magnetosphere, and records intense fluxes of energetic ions there (ref. 8–2); the Apollo 14 SIDE looks in a direction 36° closer to the Earth (i.e., about 36° from parallel to the magnetosheath) although it has not yet operated in this magnetosheath region. When the Moon enters the magnetospheric tail, the situation is reversed. In this instance the Apollo 12 SIDE records extremely low fluxes (or none) of magnetosheath ions while looking in a direction about 55° from parallel to the magnetosheath, whereas the Apollo 14 instrument looks upstream in a direction approximately 25° from parallel to the magnetosheath and observes energetic ions streaming down the magnetosheath. A typical spectrum observed in this region is shown in figure 8–5. The fluxes observed are considerably lower than those recorded by the Apollo 12 SIDE on the other side of the magnetosphere; this occurrence is attributed primarily to the difference in look directions relative to the magnetosheath. A long-term study is expected to yield information on the symmetry of the structure of the magnetosheath.

Mass Spectra

The ion energy spectrum displayed in figure 8–5 is typical of those observed more or less steadily on March 7, 1971, for a period of more than 12 hr. However, there were numerous instances of the temporary appearance of an enhanced flux of 50- to 70-eV ions, which resulted in an energy spectrum such as that shown in figure 8–6, observed minutes after the spectrum in figure 8–5. Because mass analysis is carried out by the SIDE for ions of energy up to 48.6 eV, the mass analyzer detector can record these 50-eV ions if the mass-per-unit charge is greater than 6. The mass spectrum shown in figure 8–7 was recorded shortly before the energy spectrum of figure 8–6 and is remarkable in the high response in channel 5 relative to the other channels. Masses 17 to 24 are recorded in channel 5, so the conclusive identification of these ions is not possible. Further investigation is necessary to see if these observations are repeated in later lunar cycles, if the ion

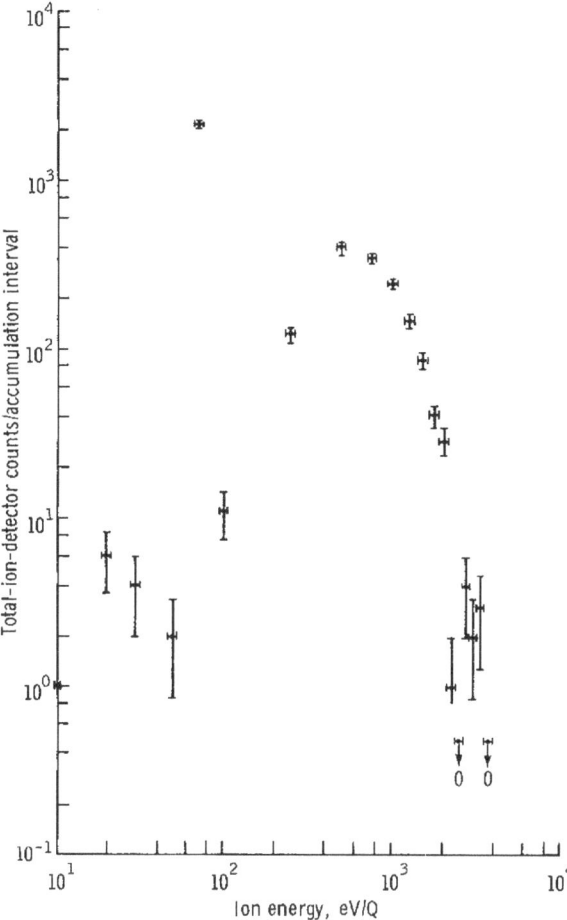

FIGURE 8–6.—Energy spectrum observed (17:25:46 to 17:26:09 G.m.t., March 7) by the total ion detector less than 4 min after that shown in figure 8–5. The two spectra are almost identical except for the appearance of the high counting rate in the 70-eV channel. This low-energy flux varies substantially in time, sometimes completely disappearing or reappearing within the 24 sec required to complete one spectrum. The energy also varies between 50 and 70 eV.

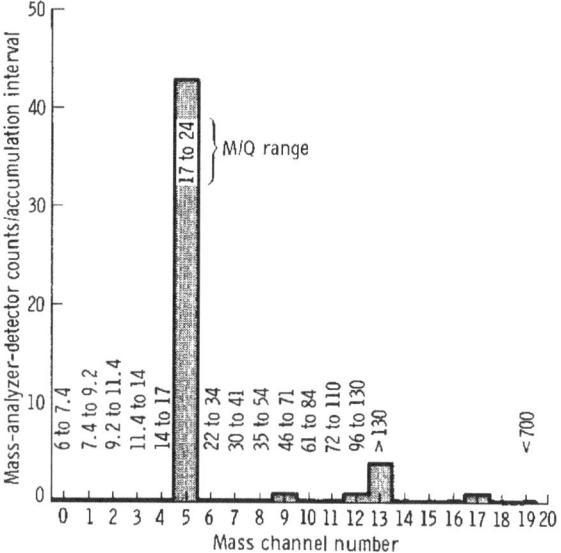

FIGURE 8–7.—Mass spectrum observed (17:16:00 G.m.t., March 7) at 48.6 eV from an ion energy spectrum of the type shown in figure 8–6. The high response in channel 5 should be noted. The mass-per-unit charge M/Q range is indicated for each channel. The accumulation interval is 1.13 sec, the same as for the total ion detector.

intensities decay, or if there is a periodicity to the occurrences. Such information is needed to evaluate the possibilities that these ions originate from natural lunar processes, from the terrestrial magnetosphere, from contamination caused by landing activities, from venting in the LM descent stage, or from some other source.

It is of interest to note that venting of the LM cabin before the second period of extravehicular activity (EVA) yielded a convenient calibration example of the mass analyzer detector. The mass spectrum recorded by the SIDE at 48.6 eV as a result of the ionization of the vented oxygen atmosphere is shown in figure 8–8(a). The corresponding total-ion-detector energy spectrum (fig. 8–8(b)) features a strong peak at 50 eV that had not been present previously and that persisted for only a short time. All the mass spectra observed to date with the Apollo 14 SIDE show statistically significant mass peaks only in the 48.6- and 16.2-eV energy levels. This result indicates that thermal ions in the lunar ionosphere must be promptly accelerated to suprathermal energies.

Correlation With Apollo 12 SIDE

The successful deployment and operation of the Apollo 14 SIDE in addition to the earlier Apollo 12 SIDE established a pair of essentially identical, simultaneously operating instruments separated by approximately 6° in longitude and less than 1° in latitude. The look directions, however, are 36° apart, as illustrated in figure 8–4. These two instruments make possible the investigation of the dimensions of the observed ion phenomena to

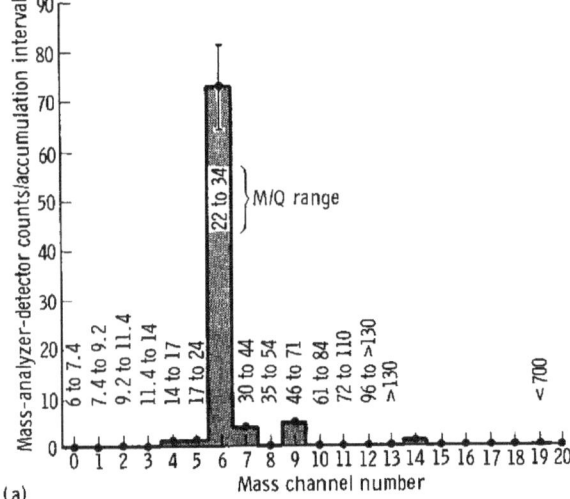

(a)

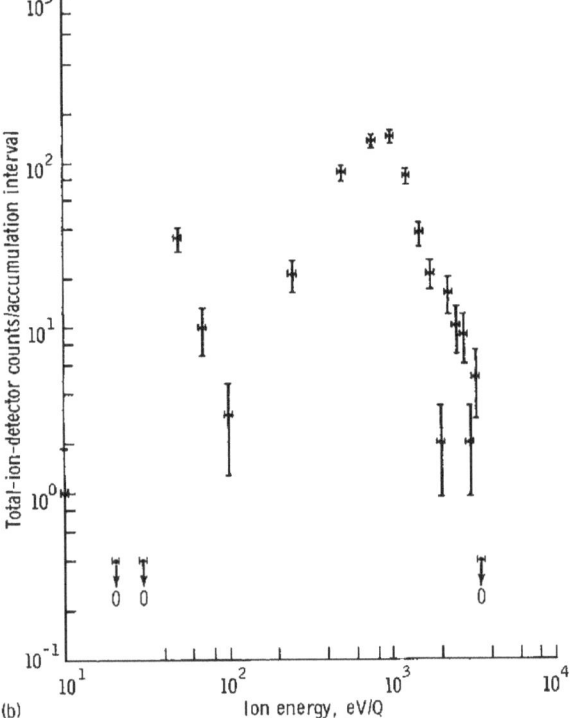

(b)

FIGURE 8-8.—Spectra recorded as a result of the ionization of the vented LM oxygen atmosphere at the beginning of the second period of EVA, February 6. (a) Mass-analysis-detector spectrum recorded 08:38:25 to 08:38:48 G.m.t., at 48.6 eV. (b) Total-ion-detector energy spectrum recorded 08:38:26 to 08:38:49 G.m.t.

ascertain whether the events are local or global in character and, if local, the apparent speed and direction of motion of the ion clouds.

Energetic ion events of the type observed in the nighttime and previously attributed to protons coming from the bow shock front of the Earth have been recorded simultaneously by the Apollo 12 and 14 instruments, further supporting the contention that these events were not of local origin, because they were recorded simultaneously at detectors approximately 180 km apart.

A striking example of the apparent movement of an ion cloud was observed on March 19, when an ion event was detected by the Apollo 14 SIDE starting at 18:57 G.m.t., and by the Apollo 12 SIDE starting at 19:02 G.m.t. The total-ion-detector records for the two instruments are shown on the same time base in figure 8–9. The similar magnitudes of the fluxes observed, the similar time durations of the major parts of the event, and the high ion energies recorded (2 to 3.5 keV, which is very unusual for this part of the orbit, based on previous Apollo 12 data) all indicate that this event is due to a rather well-defined cloud of ions that moved westward passing over the Apollo 14 site, then over the Apollo 12 site. The most intense portion of the cloud took about 3 min to pass one site, which implies an east-west dimension of approximately 130 km, based on the apparent travel velocity of ~0.72 km/sec between the two sites.

The Apollo 12 and 14 passive seismic experiments both recorded a relatively large seismic disturbance that began at 18:21 G.m.t. and continued until well after the ion observations.[1] This occurrence would be the first observed case of an ion event associated with a natural impact event, if indeed these two events are related. The long time delay between the onset of the seismic disturbance and that of the energetic ions casts some doubt on this hypothesis, because the ions observed from previous vehicle impacts arrived within seconds of the seismic signals. However, a strong dependence on distance or on direction or both to the impact may exist. The character and intensity of the ion clouds observed are quite different from any events recorded in this region during the previous 16 months of the Apollo 12 SIDE data, which lends support to indications that this event is related to the seismic impact. No mass

[1] G. V. Latham, private communication, Mar. 20, 1971.

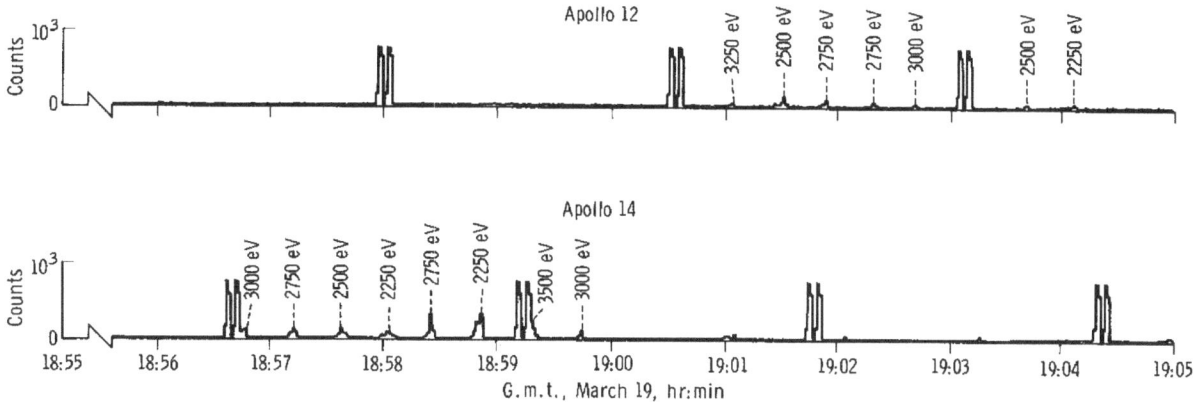

FIGURE 8–9.—Total-ion-detector data from the real-time records of the Apollo 12 and 14 instruments. The well-defined burst of ions, recorded first at the Apollo 14, then at the Apollo 12, SIDE should be noted. This event occurred approximately 36 min after the beginning of a long seismic disturbance. The high counting rates at 2.58-min intervals are due to calibration signals.

analysis during this event was possible because the ion energies are far above the range of the mass analyzer detector. For the present, the relationship between the ion cloud and the seismic disturbance is considered to be an open question.

Apollo 14 Activities Observed by the Apollo 12 SIDE

Three major events of the Apollo 14 mission were clearly recorded by the Apollo 12 SIDE: the SIVB impact, the LM ascent-stage flight nearly over the Apollo 12 site, and the LM ascent-stage impact.

The two impacts (the SIVB and the LM ascent stage) did not produce as intense or as complex events as did the impact of the Apollo 13 SIVB [2] or the Apollo 12 LM (ref. 8–1), but they were nevertheless unambiguously recorded.

The nominal flightpath of the LM as it left the Fra Mauro site was, at closest approach, approximately 22 km north of the Apollo 12 site at an altitude of approximately 15 km. Ions from the LM were recorded by the Apollo 12 SIDE for a period of 100 sec starting 1 min after the LM had passed by at the minimum slant range of approxi-

mately 27 km. The time delay is considered to be due to expansion of the neutral gas for some time before its ionization and subsequent acceleration. The observations of the SIDE are shown in figure 8–10. The lower histogram gives the sum of the counting rates recorded throughout the event in the 50-, 70-, and 100-eV channels by the total ion detector, while the mass-analyzer-detector data are given on an expanded scale for the 20-channel spectrum nearest the maximum of the event. This mass spectrum was the only one during the event that indicated fluxes significantly above background and was the only one taken at an energy (48.6 eV) at which the total ion detector indicated high ion fluxes. The 48.6-eV spectra taken before and after this one (at 2.58-min intervals) both indicated only background rates. This situation was to be expected, because the total ion detector recorded no ions in the 50-eV range at those times. The observations indicate that the flux of 50-eV ions reached the maximum at or very near the time of the scan of the 48.6-eV mass spectrum. If the peak counting rate recorded in the total-ion-detector 50-eV channel is taken to be a good indication of the flux 11 frames later, then the comparison of the total-ion-detector and the mass-analyzer-detector data indicates that the ions observed were predominantly heavy ions, rather than protons. The mass spectrum indicated appears to differ substantially from that expected from the

[2] J. W. Freeman, Jr., H. K. Hills, and M. A. Fenner: Some Results From the Apollo 12 Suprathermal Ion Detector. Proc. Apollo 12 Lunar Sci. Conf. (Houston), Jan. 11–14, 1971. To be published in Geochim. Cosmochim. Acta.

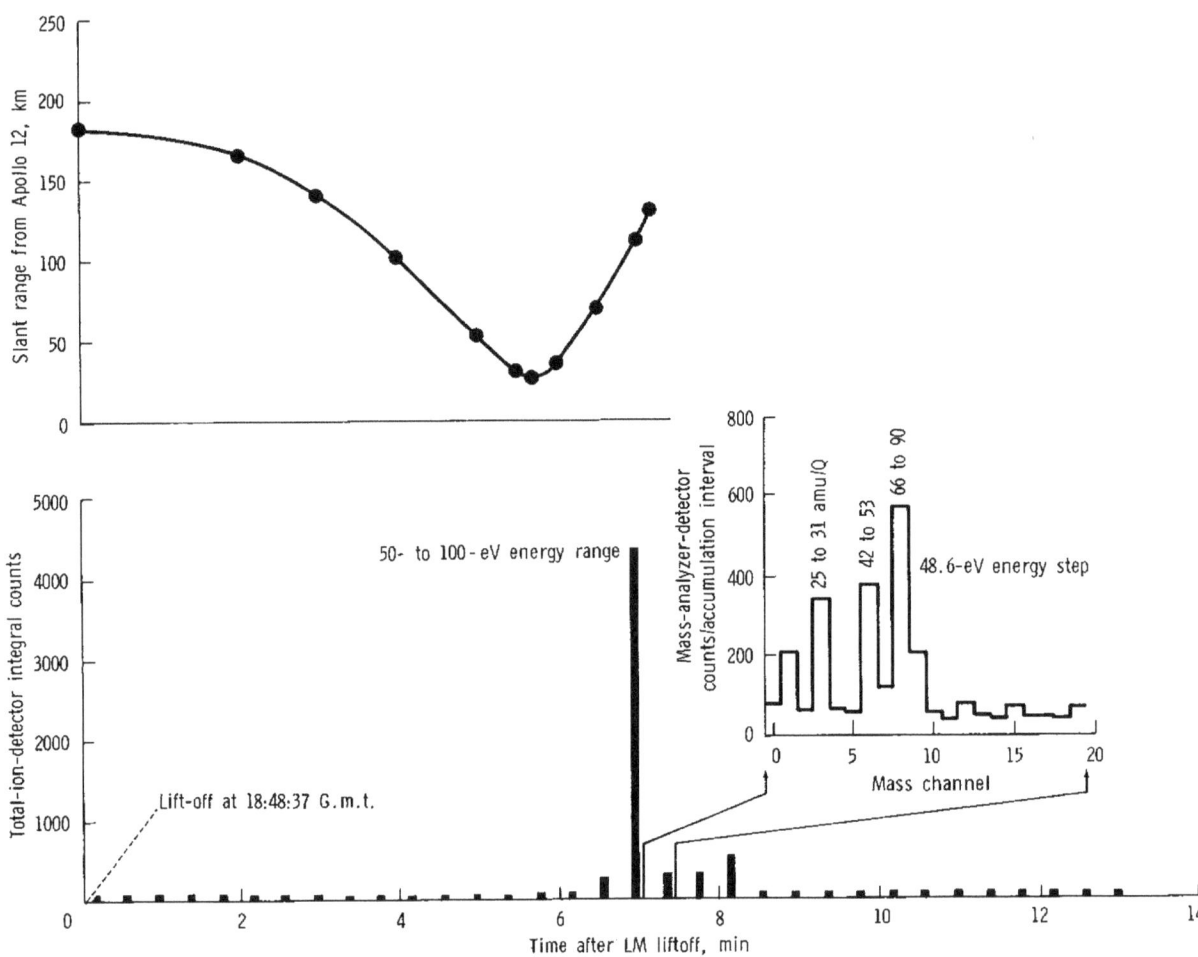

FIGURE 8–10.—Observation of the Apollo 14 LM ascent by the Apollo 12 SIDE on February 6. All total-ion-detector counts accumulated in the adjoining 50-, 70-, and 100-eV channels of each 20-channel spectrum are shown as a function of time after liftoff. Nearly all the counts significantly above background fell in these three channels. The corresponding mass spectrum at 48.6 eV is shown in the inset at the right. The upper curve shows the slant range to the LM (i.e., the line-of-sight distance) from liftoff until orbit insertion.

LM ascent-engine exhaust products (ref. 8–3), particularly in the existence of a high flux of ions (the origin of which is still uncertain) recorded in the 66- to 90-amu/Q mass channel. One possible origin may be unused excess nitrogen tetroxide from the hypergolic-fuel engine. The other mass peaks recorded are compatible with expected exhaust gases from the engine, including water vapor, nitrogen, carbon monoxide, and carbon dioxide.

Summary

The performance of the SIDE has been ex-

cellent. The preliminary data analysis yields the following significant observations.

(1) As the Moon enters the magnetospheric tail of the Earth, ions of 250- to 1000-eV energy have been observed streaming down the magnetosheath toward the tail. The Apollo 12 and 14 instruments may now be used jointly to investigate this phenomenon on both sides of the magnetospheric tail.

(2) Intermittent intense fluxes of 50- to 70-eV ions have been observed about 2 days after sunrise, with masses indicated to be in the 17- to 24-amu/Q range. These are in addition to the steady, more energetic fluxes (noted in the preced-

ing paragraph) observed at the same time. The origin of the ions is undetermined.

(3) Energy and mass spectra were observed during the venting of the oxygen atmosphere in the LM cabin.

(4) Correlation of Apollo 14 SIDE data with Apollo 12 SIDE data allows determination of whether the ion events in question are local or cover a large area, or are due to moving ion clouds.

(5) A large ($\sim$130-km) ion cloud was observed to pass over both the Apollo 14 and 12 sites during a relatively large seismic disturbance. However, it is not certain that this cloud was related to the seismic disturbance.

(6) The ascent flight of the Apollo 14 LM was observed by the Apollo 12 SIDE, both by the total ion detector and the mass analyzer detector. In addition to fluxes of ions of lower mass, a flux of ions was observed in the 66- to 90-amu/Q range, which is heavier than the expected ascent-engine exhaust-gas products.

(7) Ions resulting from the impacts of the Apollo 14 SIVB and the LM ascent stage were recorded by the Apollo 12 SIDE.

References

8-1. FREEMAN, J. W., JR.; BALSIGER, H.; AND HILLS, H. K.: Suprathermal Ion Detector Experiment (Lunar Ionosphere Detector). Sec. 6 of Apollo 12 Preliminary Science Report. NASA SP-235, 1970.

8-2. FREEMAN, J. W., JR.; FENNER, M. A.; HILLS, H. K.; AND BALSIGER, H.: Preliminary Results From Apollo 12 ALSEP Lunar Ionosphere Detector: II. Detection of Ions of Solar Wind Energies. Trans. Amer. Geophys. Union, vol. 51, no. 7, July 1970. p. 590 (rev. abstract).

8-3. ARONOWITZ, LEONARD; KOCH, FRANK; SCANLON, JOSEPH; AND SIDRAN, MIRIAM: Contamination of Lunar Surface Samples by the Lunar Module Exhaust. J. Geophys. Res., vol. 73, no. 10, May 1968, pp. 3231-3238.

ACKNOWLEDGMENTS

The authors gratefully acknowledge the support of those persons who contributed to the success of the SIDE. Particular thanks are extended to Wayne Andrew Smith, Paul Bailey, James Ballentyne, and Alex Frosch of Rice University. Thanks are also due to Martha Fenner, Robert Lindeman, and René Medrano, graduate students in the Space Science Department who assisted with the project. Jürg Meister, a European Space Research Organization/National Aeronautics and Space Administration (ESRO/NASA) international fellow, and Hans Balsiger, a former ESRO/NASA fellow, both contributed to the SIDE project. Time Zero Corp. was the subcontractor for the design and fabrication of the instrument. Personnel of the NASA Manned Spacecraft Center and of Bendix Aerospace Systems Division also provided valuable support. This research has been supported by NASA contract NAS 9-5911.

9. Cold-Cathode-Gage Experiment
(Lunar Atmosphere Detector)

F. S. Johnson,[a]† *D. E. Evans,*[b] *and J. M. Carroll* [a]

Purpose of the Experiment

Although the lunar atmosphere is known to be very tenuous, its existence cannot be doubted. At the very least, the solar wind striking the lunar surface constitutes a source. The atmospheric concentration to be expected depends upon the equilibrium between source and loss mechanisms. The observations of the lunar atmosphere will turn out to be of greatest significance if the dominant source mechanism for the atmosphere is found to be internal (i.e., geochemical in nature) rather than external (from the solar wind) to the Moon.

The dominant loss mechanisms for lunar gases are expected to be thermal escape for particles lighter than neon and escape through ionization for neon and heavier particles. At temperatures encountered on the lunar surface, thermal velocities for the lighter gas particles are such that a significant fraction of the particles have greater-than-escape velocity. The average thermal-escape lifetime for particles on the warmest portion of the Moon is approximately 10^4 sec for helium and 10^{10} sec for neon. Heavier particles require much longer times for escape by thermal motion.

Particles exposed to solar ultraviolet radiation become ionized in approximately 10^7 sec and, once ionized, the particles are accelerated by the electric field associated with the motion of the solar wind. The initial acceleration is at right angles to both the direction of the solar wind and of the embedded magnetic field; then the direction of motion is deviated by the magnetic field so that the ionized particle acquires a velocity equal to the solar-wind-velocity component that is perpendicular to the embedded magnetic field. The time required for this acceleration is approximately the ion gyro period in the embedded magnetic field, and the radii of gyration for most ions are comparable to the lunar radius (or greater). As a consequence of this process, particles in the lunar atmosphere are largely swept away into space within a few hundred seconds (the ion gyro period) after becoming ionized. Thus, the time required for ionization regulates the loss process, which results in lifetimes for particles in the lunar atmosphere on the order of 10^7 sec.

The cold-cathode-gage experiment (CCGE) gives indications of the amount of gas present, but not the gas composition. The CCGE indications can be expressed as concentrations of particles per cubic centimeter or as pressure in torrs, which depends on ambient temperature and concentration. The amount of gas observed can be compared with the expectation resulting from the solar-wind source as an indication of whether other sources of gases are present. Contamination from the lunar module (LM) and the astronaut suits constitutes an additional source, but one that should decrease with time in an identifiable way. However, in the long run, measurements of actual composition of the lunar atmosphere should be made with a mass spectrometer to examine constitutents of particularly great interest and to iden-

[a] University of Texas at Dallas.
[b] NASA Manned Spacecraft Center.
† Principal investigator.

tify and discriminate against known contaminants from the vehicle system.

Instrument Description

The essential sensing element of the CCGE consists of a coaxial electrode arrangement as shown in figure 9–1. The cathode consists of a spool that is surrounded by a cylindrical anode. A magnetic field of approximately 0.090 T (900 G) is applied along the axis, and 4500 V are applied to the anode. A self-sustained electrical discharge develops in the gage in which the electrons remain largely trapped in the magnetic field with enough

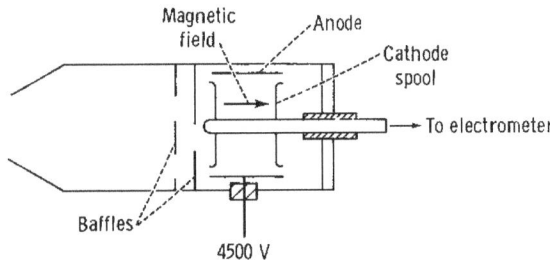

FIGURE 9–1.—Sensor in the CCGE.

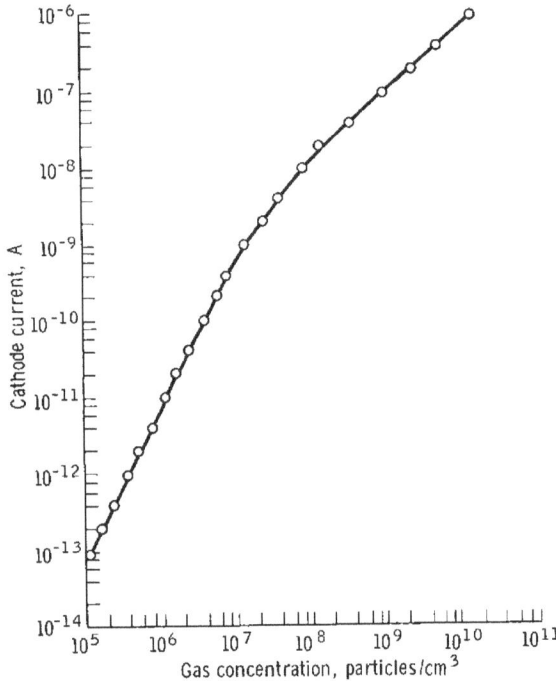

FIGURE 9–2.—Response of the CCGE.

energy to ionize any gas particles that they strike. The current of ions collected at the cathode is a measure of the gas density in the gage.

The response of the CCGE in terms of cathode current as a function of gas concentration is shown in figure 9–2. The CCGE response depends in a rather modest degree on the gas composition; thus, as long as the gas composition remains unknown, a fundamental uncertainty remains in the interpretation of the data. It is usual to express the results in terms of equivalent nitrogen response (i.e., the concentration of nitrogen that would produce the observed response). The true concentration varies from this result by a factor that is usually less than 2.

Instrument temperature was monitored by means of a sensor on the CCGE. Because no temperature control exists, the expected temperature range is approximately 100 to 400 K.

The CCGE was closed with a dust cover that did not constitute a vacuum seal. The cover was removed by command using a squib motor and was then pulled aside by a spring. Because the CCGE was not evacuated, adsorbed gases produced an elevated level of response when the gage was initially turned on. However, the baking of the CCGE on the lunar surface at 400 K for more than a week during each lunar day drove the adsorbed gases out of the gage.

Electronic Circuitry

A description of each of the major CCGE assemblies follows.

Electrometer amplifier. An autoranging, autozeroing electrometer amplifier monitors current outputs from the sensor or from the calibration current generators in the 10^{-13}- to 10^{-6}-A range. The output ranges from −15 mV to −15 V. The output of the electrometer is routed to the analog-to-digital converters for conversion. The electrometer consists of a high-gain, low-leakage differential amplifier with switched high-impedance feedback resistors and an autozeroing network.

The electrometer operates in three automatically selected overlapping ranges: (1) most sensitive, (2) midrange, and (3) least sensitive. Range 1 senses currents from approximately 10^{-13}

to 9.3×10^{-11} A; range 2, currents from approximately 3.3×10^{-12} to 3.2×10^{-9} A; and range 3, currents from approximately 10^{-9} to 9.3×10^{-7} A.

Power supply. The 4500-V power supply consists of a regulator, a converter, a voltage-multiplier network, and the associated feedback network of a low-voltage power supply. The regulator furnishes approximately 24 V for conversion to a 5-kHz squarewave, which is applied to the converter transformer. The output of the converter transformer is applied to a voltage-multiplier network (stacked standard doublers), the output of which is filtered and applied to the CCGE anode.

Deployment

The electronics for the CCGE were contained in the suprathermal ion detector experiment (SIDE), and the command and data-processing systems of the SIDE also served the CCGE. The CCGE was separable from the SIDE package and was connected to it by a cable approximately 1 m (3 ft) long. The experiment was deployed so that the LM descent stage was just outside the CCGE field of view, which looked south toward Fra Mauro. The CCGE is shown deployed on the lunar surface in figure 9-3.

FIGURE 9-3.—The CCGE deployed on the lunar surface (AS14-67-9372).

Experiment Results

The CCGE was turned on at approximately 23:59 G.m.t. on February 5, 1971. A full-scale response was indicated until 01:20 G.m.t. on February 6, at which time the concentration began to decrease.

Both the CCGE 4.5-kV and the SIDE 3.5-kV high-voltage power supplies were turned off at 01:40 G.m.t. on February 6 after a mode change occurred that suggested that arcing might be taking place. The 4.5-kV power supply was activated again at 07:50 G.m.t. on February 6 to observe the LM depressurization before the second period of extravehicular activity (EVA). After observation of this event, the voltage was again turned off at 08:39 G.m.t., and it remained off until just before the LM depressurization for housekeeping functions after the second EVA. The CCGE was operated from 13:13 to 13:16 G.m.t. for this observation. No other CCGE operation was attempted until the start of the first lunar night because of the operational constraints on the SIDE and CCGE high voltages.

A plot of density as a function of time during the depressurization for the second EVA is shown in figure 9-4, and a similar plot during the venting for disposal of the backpacks following the second EVA is shown in figure 9-5. A noticeable

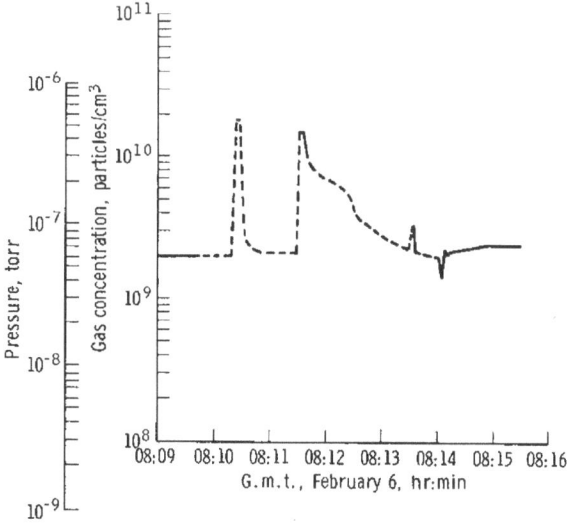

FIGURE 9-4.—Gas concentration detected during depressurization of the LM for the second EVA $T = 295°$ K).

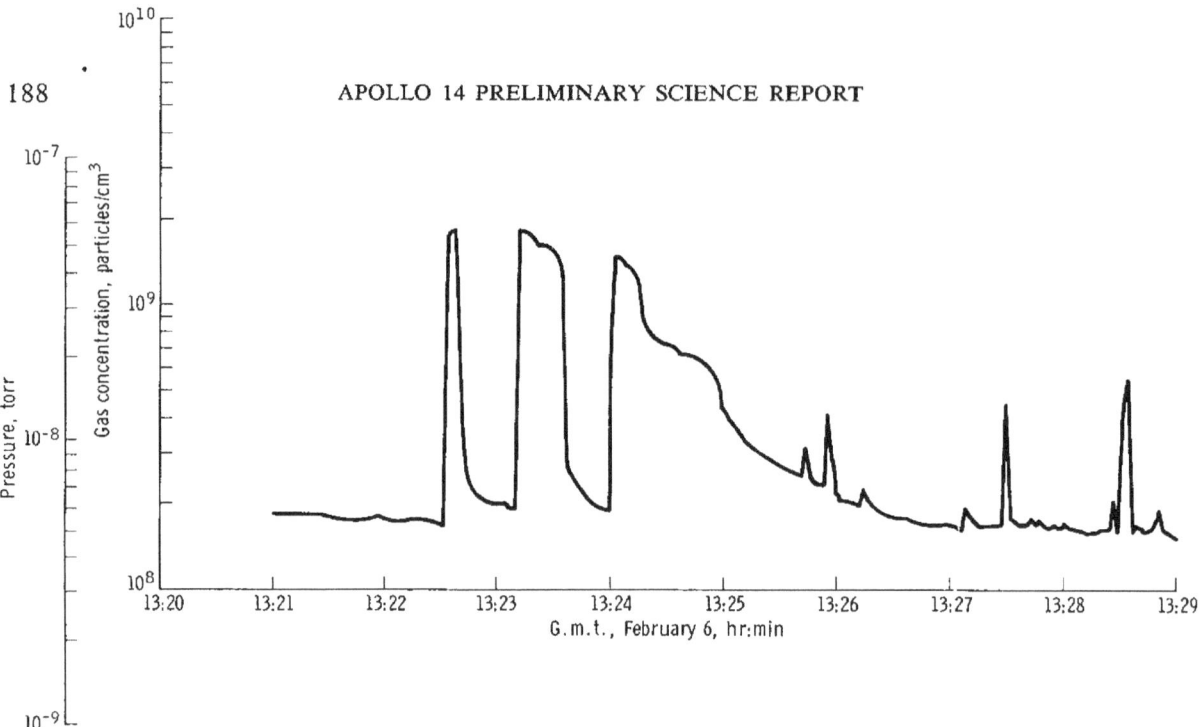

FIGURE 9–5.—Gas concentration detected during depressurization of the LM to discard life-support backpacks and other unneeded equipment ($T = 296°$ K).

similarity is evident in the two plots. In figure 9–5, extra peaks are shown that were apparently caused by the venting being interrupted because of a loose hose on one of the spacesuits. The two peaks at 13:27:30 and 13:28:30 G.m.t. were caused by equipment being thrown from the LM and were observed at the time of occurrence while the recorder and television were watched simultaneously at the NASA Manned Spacecraft Center. No obvious explanation exists for the peaks at 13:26 G.m.t., except that this time was probably when the LM hatch was opened.

When the high voltage was turned on at the start of the first lunar night on February 19, the output indication of the CCGE electrometer was in the low area of range 3. When the CCGE went through the automatic-calibration cycle, the output indication went to a low value in range 1, which is the most sensitive range. This occurrence was improperly interpreted as an out-of-strike condition. This improper interpretation was caused partially by the density of the lunar atmosphere being much lower than had been anticipated and partially by the lack of opportunity to track the density downward from the higher values that had prevailed earlier.

The CCGE was placed in an operational mode that eliminated the calibration cycle and was forced into range 3 by turning the 4.5-kV power supply off and then back on. This method was believed to be the only way to prevent the gage discharge from going out of strike, and the indication in range 3 was considered to be proper. As more data were analyzed, it was determined that the range 1 indication was actually correct and that a zero-drift error that is normally corrected by an autozeroing during the calibration cycle was large enough (for the operational mode that was selected) to prevent the instrument from switching out of range 3. The result of the confusion was a determination that the CCGE had been operating correctly only during part of the time from February 19 to March 5.

Three notable gas events, as shown in figure 9–6, were observed during operation from February 19 to March 5. At 15:30 G.m.t. on February 19, the concentration of the lunar atmosphere was approximately 2×10^5 cm⁻³. At 22:00 G.m.t. on February 20, the CCGE switched into a proper mode of operation in range 2, with an indicated concentration of 2×10^6 cm⁻³. The first event occurred between these times. After the first event, the density decreased steadily until the second event occurred. The time and magnitude of the first event were not determined because of the CCGE being in the improper operational mode

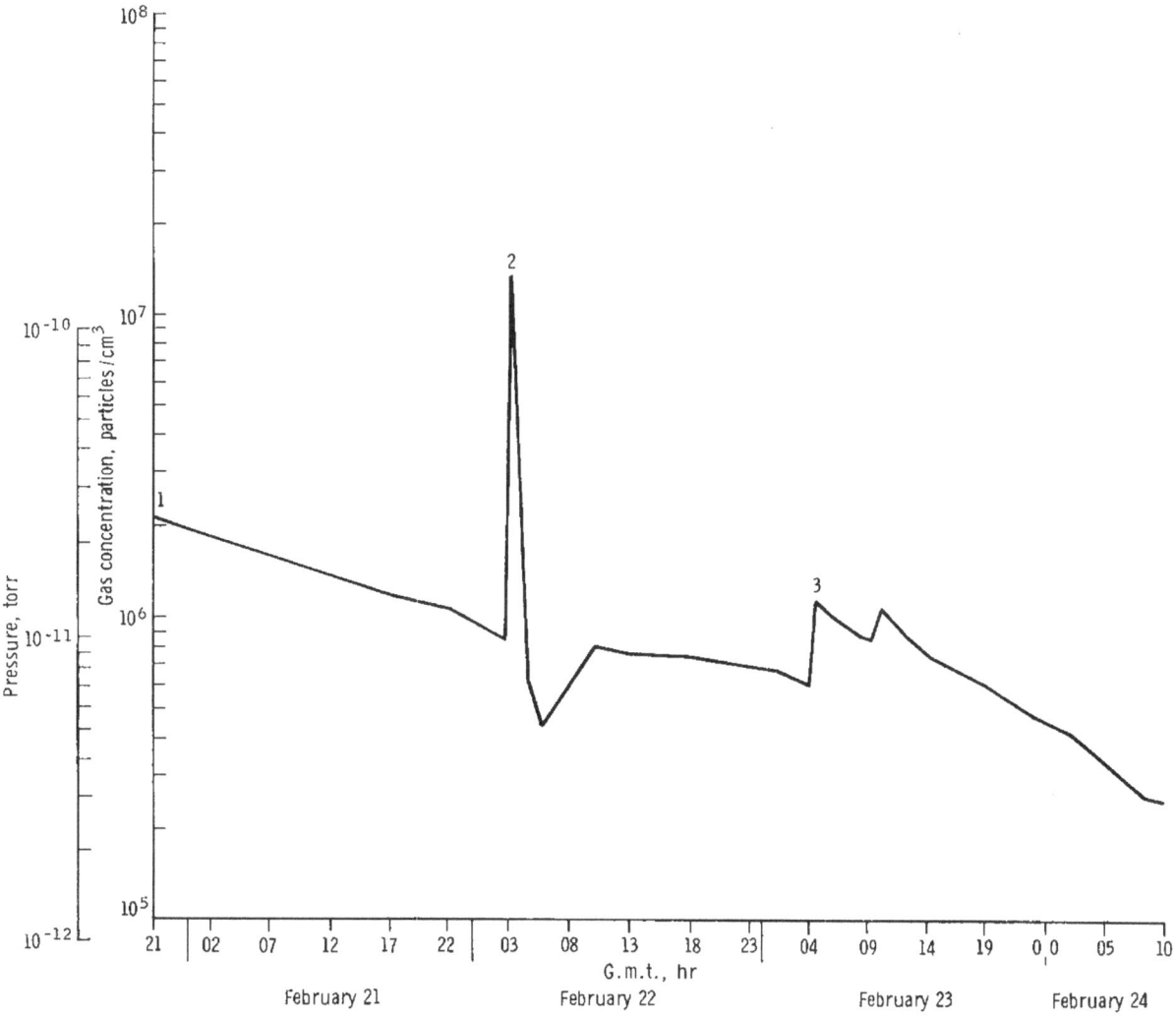

FIGURE 9–6.—Gas concentration detected during the first lunar evening after deployment of the CCGE, showing gas events 1, 2, and 3 ($T = 110°$ K).

at the time of occurrence. The event apparently caused an increase in concentration between one and two orders of magnitude and had a duration of between 89 and 113 hr.

The second and third events occurred at 02:46 G.m.t. on February 22 and 04:36 G.m.t. on February 23, as shown in figure 9–6. The second event was slightly greater than one order of magnitude and had a duration of approximately 1 hr. The rise time of the event was approximately 10 sec, as shown in figure 9–7.

Event 3, which occurred in range 1, had a rise time of approximately 6 min, and the amplitude was approximately doubled (fig. 9–8). As shown

in figure 9–6, event 3 may have been a double event; however, this observation is not definite, and the event will be explored further when the data tapes are available and more data points are obtained. The concentration decreased to a level of approximately 2×10^5 cm^{-3} after event 3 and remained at this level until sunrise at approximately 00:08 G.m.t. on March 6. At sunrise, the concentration increased by two orders of magnitude within 2 min, as shown in figure 9–9. The indication decreased to approximately 2×10^6 cm^{-3} by March 8, when another event occurred and increased the concentration by approximately an order of magnitude to 1.2×10^7 cm^{-3}. This event

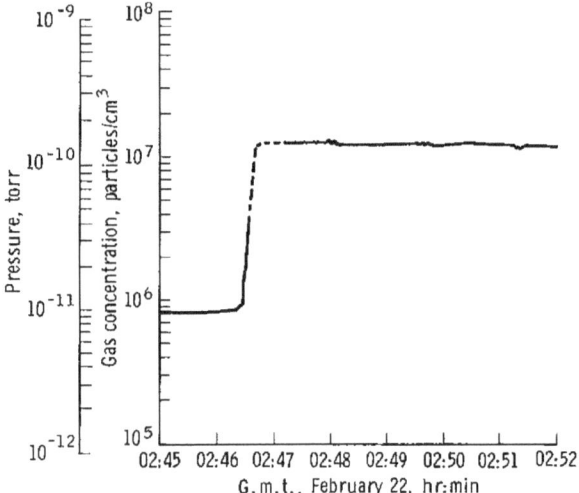

FIGURE 9-7.—Gas concentration during the commencement of gas event 2, showing rise time of approximately 10 sec ($T=110°$ K).

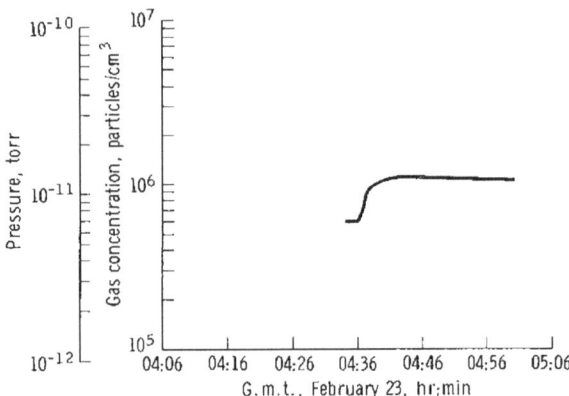

FIGURE 9-8.—Gas concentration during the commencement of gas event 3, showing rise time of approximately 6 min ($T=110°$ K).

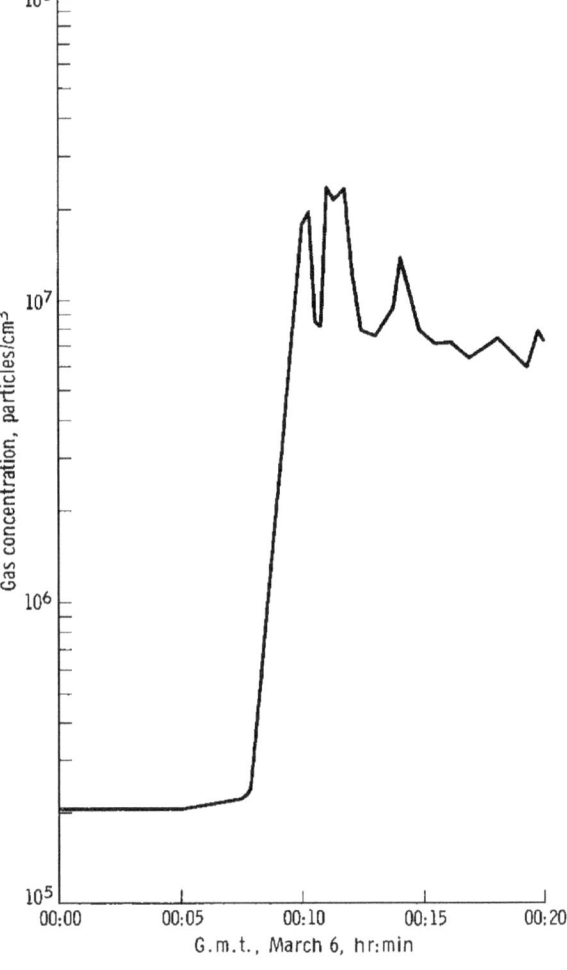

FIGURE 9-9.—Gas concentration during first lunar sunrise after deployment of the CCGE.

lasted approximately 8 hr. The high voltages were turned off, and no further data were available until March 19.

On March 20, another event occurred that caused the CCGE output to change from range 1 to range 3 for a concentration increase of approximately two orders of magnitude. This event lasted approximately 20 min.

In summary, the results to date indicate that the concentration of the lunar atmosphere during lunar nighttime is approximately 2×10^5 cm⁻³, except when outgassing or other gas events occur. In lunar daytime, the concentration after one lunar cycle is no greater than 1×10^7 cm⁻³ and is still influenced by outgassing from the lunar-surface experiments and the LM descent stage. Numerous gas events occur during both day and night operation.

Data analysis at the time of this writing is very preliminary because tapes are not yet available for analysis. Consequently, the results presented herein are very tentative.

ACKNOWLEDGMENTS

The CCGE sensor was supplied by the Norton Co., with the assistance of Frank Torney. The electronics were built by Marshall Laboratories (now Time Zero Corp.) under subcontract to Rice University. The authors are grateful to Dr. John Freeman and his associates at Rice University for their assistance in assuming contractual responsibility for development and production of the gage electronics in connection with the SIDE. Many other individuals at the NASA Manned Spacecraft Center and their contractors have rendered valuable assistance.

10. Charged-Particle Lunar Environment Experiment

Brian J. O'Brien [a][†] *and David L. Reasoner* [b]

Purpose of the Experiment

The primary scientific objective of the charged-particle lunar environment experiment (CPLEE) is to measure the fluxes of charged particles (electrons and ions) with energies ranging from 50 to 50 000 eV that bombard the lunar surface. The following list illustrates the wide variety of phenomena that may be responsible for these particles:

(1) Relatively stable plasma population in the magnetospheric tail including the so-called plasma sheet and neutral sheet (ref. 10–1)

(2) Transient particle fluxes in the magnetospheric tail resulting from such phenomena as geomagnetic substorms and particle-acceleration mechanisms similar to those that produce auroras (ref. 10–2)

(3) Plasma in the transition region between the magnetospheric tail and the shock front

(4) The solar wind, and particles resulting from the interaction between the solar wind and the lunar surface (refs. 10–3 and 10–4)

(5) Solar cosmic rays, those particles thrown into interplanetary space by solar-flare eruptions

(6) Photoelectrons at the lunar surface produced by the interaction of solar photons with the lunar-surface material

(7) "Artificial events"; for example, particles produced by the impact of the lunar module (LM)

[a] University of Sydney.
[b] Rice University.
[†] Principal investigator.

Thus, in one sense the Moon serves as a satellite to carry the CPLEE instrument through various regions of space, and in another sense the CPLEE is a detector of phenomena resulting from the interaction of radiation with the lunar surface.

Summary of Observations

In conjunction with the preceding list of scientific objectives, the following preliminary observations have been noted:

(1) Detection of stable, low-energy photoelectron fluxes at the lunar surface

(2) Observation of plasma clouds produced by the impact of the Apollo 14 LM ascent stage

(3) Observations of rapidly fluctuating, low-energy (50 to 200 eV) electrons in the magnetosheath and magnetospheric tail

(4) Detection of fluxes of medium-energy electrons, with durations of a few minutes to some tens of minutes, deep within the magnetospheric tail

(5) Observation of electron spectra in the magnetospheric tail remarkably similar to electron spectra observed above terrestrial auroras

(6) Observation of rapid time variations (10 sec) in solar-wind fluxes observed in interplanetary space

Theoretical Basis

The objectives of the CPLEE are to measure the proton and electron fluxes at the lunar surface and to study their energy, angular distributions, and time variations. The results of these measurements will provide information on a variety of particle phenomena, important both in themselves and for their relevance to lunar-surface properties.

A category of radiation exists that may periodically envelop the Apollo lunar-surface experiments package (ALSEP) at the times of the full Moon, when it is in the magnetospheric tail of the Earth, which is swept downstream like a comet tail by the solar wind. It has been speculated (ref. 10–2) that the electrons and protons that cause auroras when they plunge into the terrestrial atmosphere are accelerated in the magnetospheric tail. Indeed, it has been shown (ref. 10–5) that the ultimate source of auroral particles is the Sun and furthermore that an almost-continuous replenishment of the magnetospheric-particle population is necessary to sustain the observed auroral fluxes (ref. 10–6). The mechanisms that accelerate these particles to auroral energies are not understood, and simultaneous observations near the Earth and near the Moon are essential for detailed study of their general characteristics and morphology.

The solar wind may occasionally strike the surface of the Moon. The wind is caused by the expansion into interplanetary space of the very hot outer envelope of the Sun. The stream apparently carries energy and perturbations toward the Earth-Moon system; consequently, the solar wind may be the source of energy that leads to such terrestrial phenomena as auroras and Van Allen radiation. For this study, the Moon would serve as an excellent stable observation post in space.

However, apparently the pure interplanetary solar wind does not always hit the lunar surface (ref. 10–3). Because the solar wind is supersonic and because the Moon is sufficiently large to prove an obstacle to the flow of the wind, it is possible that, at times, there is a standing shock front. To date, the only such phenomenon observed is caused by the terrestrial magnetic field, which hollows out a cavity in the solar wind. The detailed physical processes that occur at such shock fronts are not understood fully, and they are of considerable fundamental interest in plasma research. If, occasionally, such a shock front exists near the

Moon, the CPLEE will observe the disordered (or thermalized) fluxes of electrons and protons that share energy on the downstream side of the shock. Apparently, the "shadowing" of the solar wind by the lunar surface, causing a plasma "void" on the dark side, is the most frequently occurring situation (ref. 10–3).

The instrument can also measure the lower energy solar cosmic rays occasionally produced in solar eruptions or flares. To observe these low-energy particles, the experimental packages must be placed beyond the reach of the modifying effects of an atmosphere and a magnetic field such as exists on Earth. The Moon is an excellent platform for such studies because both the atmosphere and magnetic field are so relatively negligible.

The sunlit lunar surface may be a veritable sea of low-energy photoelectrons generated by solar photons striking the surface. If such electrons are present, the CPLEE, with the capability to detect electrons with energies down to 40 eV, will be in an excellent position to study them. Studies of any such photoelectron layer are important in deducing surface properties related to photoemission and in gaining indirect information about lunar-surface electric fields.

Observations of the charged-particle environment of the Moon are of interest also, not merely for its own sake but because such particles affect the lunar environment. They may cause luminescence or coloration effects on the lunar surface. The charged particles may also sweep away a large proportion of the lunar atmosphere. Furthermore, they constitute a very important proportion of the electrical environment, and they may, for example, nullify electrostatic effects that would otherwise occur on the lunar surface.

Equipment

Description of the Instrument

The CPLEE consists of a box supported by four legs. The box contains two similar physical charged-particle analyzers, two different programable high-voltage supplies, twelve 20-bit accumulators, and appropriate conditioning and shifting circuitry. The total weight on Earth is approxi-

FIGURE 10–1.—The CPLEE deployed on the lunar surface. Photograph illustrates the absence of dust contamination and the east-west alinement of the CPLEE.

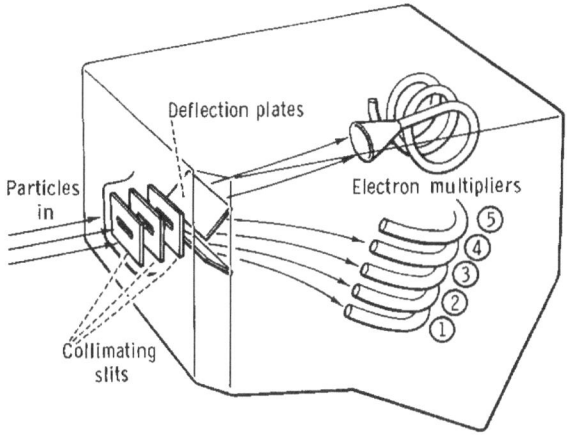

FIGURE 10–2.—Schematic sketch of the CPLEE physical particle analyzer, showing the deflection plates and channel electron-multiplier stack.

FIGURE 10–3.—The CPLEE physical particle analyzer.

mately 2.7 kg (6 lb), and normal power dissipation is 3.0 W rising to approximately 6 W when the lunar-night survival heater is on. The CPLEE is shown deployed on the lunar surface in figure 10–1.

Each physical analyzer contains five C-shaped channel electron multipliers with a nominal aperture of 1 mm each and one helical channel electron multiplier with a nominal aperture of 8 mm. These are shown schematically in figure 10–2 and an actual analyzer is shown in figure 10–3.

The channel electron multiplier is a hollow glass tube, the inside surface of which, when bombarded by charged particles, ultraviolet light, etc., is an emitter of secondary electrons. In the CPLEE, the aperture of each electron multiplier is operated nominally at ground potential (actually at 16 V), while a voltage of 2800 or 3200 V (selected by ground command) is placed on the other (i.e., anode) end. Thus, if an incident particle enters the aperture and secondary electrons are produced, these are accelerated and hit the walls to

generate more secondary electrons, so that a multiplication to an order of 10^7 is achieved by the time the pulse arrives at the anode. After conditioning, pulses from each electron multiplier are accumulated in a register for later readout as described in the following paragraphs.

As shown in figure 10–2, incident particles enter an analyzer through a series of slits and then pass between two deflection plates across which a voltage can be applied. Thus, at a given deflection voltage, the five small-aperture electron multipliers make a five-point measurement of the energy spectrum of charged particles of a given polarity (e.g., electrons), while, simultaneously, the large-aperture electron multiplier makes a single wideband measurement of particles with the opposite polarity (e.g., protons). The advantages of simultaneously measuring particles of opposite polarity and of simultaneous multiple-spectral samples are considerable in studies of rapidly varying particle fluxes.

The CPLEE particle analyzer is quite similar to the switched proton electron Channeltron spectrometer (SPECS) (ref. 10–6) and, in fact, the SPECS instrument was the prototype of the CPLEE anlyzer. The capability of the SPECS, and thus of the basic particle analyzer of the CPLEE, was demonstrated by a series of sounding-

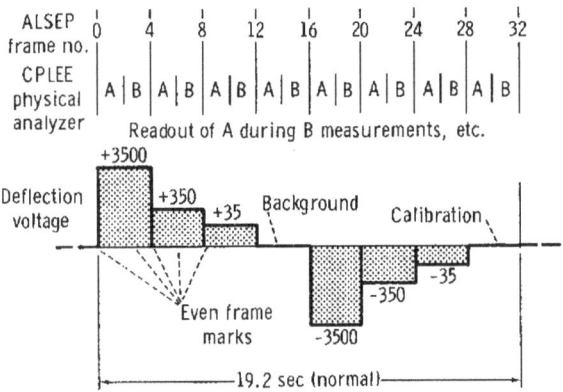

FIGURE 10–4.—Deflection-voltage stepping sequence of the CPLEE in the automatic mode. At the +0 step, the background level is measured and at the −0 step, a test oscillator is injected into the accumulators.

rocket flights in 1967 and 1968 (refs. 10–1 and 10–5) and on the Rice University-Office of Naval Research satellite Aurora 1 (refs. 10–7 and 10–8.

In the CPLEE, the deflection-plate voltage, in the normal mode, is stepped in the sequence shown in figure 10–4. As a consequence, the energy passbands shown in figure 10–5 are sampled. Although data acquired by the six sensors are not transmitted simultaneously, the six sensors are connected to six accumulators for exactly the same time (viz, 1.2 sec) and the contents transferred to shift registers for later sequential transmission.

Two analyzers, A and B, point in the directions shown in figure 10–6. The same deflection voltage is applied to each analyzer simultaneously, with counts from 1.2-sec accumulation time of analyzer A being transmitted while counts from analyzer B are accumulating. Thus, each voltage is normally on for 2.4 sec with the result that the total cycle time is 19.2 sec (fig. 10–4), when allowance is made for two sample times when the deflection voltage is zero. On one of those two occasions, counts are accumulated as usual to measure background or contaminating radiation. On the other occasion, a pulse generator of about 375 kHz is connected to the accumulators to verify operation.

The command link with the ALSEP provides a variety of options for CPLEE operation. Aside from the usual power commands common to all

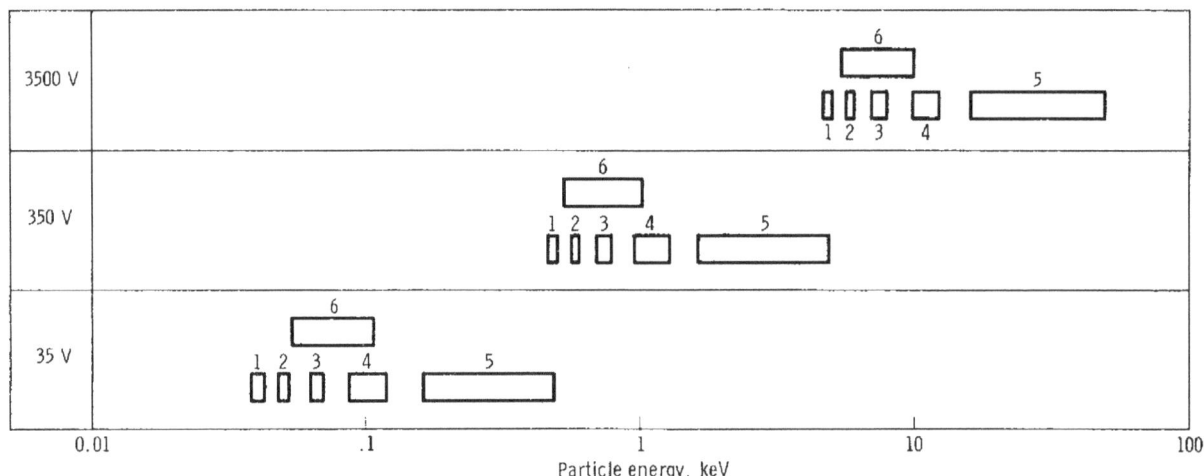

FIGURE 10–5.—Rectangular equivalent energy passbands of the CPLEE.

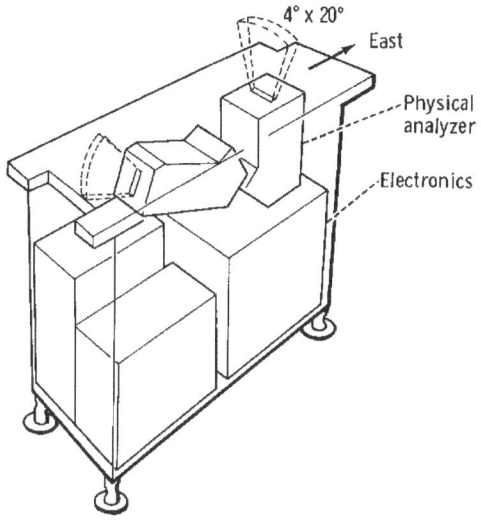

FIGURE 10–6.—Sketch of the CPLEE, showing the fields of view and the look directions of the physical analyzers.

ALSEP experiments, three commands are provided that allow the normal automatic stepping sequence to be modified. The sequence can be stopped and then the deflection plate supply can be manually stepped to any one of the eight possible levels. This is done to study a particular phenomenon (e.g., low-energy electrons) with higher time resolution (2.4 sec). A second set of commands allows the electron-multiplier high-voltage supply to be set at either 2800 or 3200 V. The higher voltage is used in the event the electron-multiplier gains decrease during lunar operations. A third pair of commands allows the normal thermal-control mode to be bypassed in the event of failure of the thermostat, thus offering manual control of the heaters.

The CPLEE apertures are covered with a dust cover to avoid contamination during deployment and, particularly, during LM ascent (ref. 10–9). The dust cover was made doubly useful because a ^{63}Ni radioactive source was placed on the underside over each aperture. Thus, the sensors were proof calibrated on the Moon, and the data compared with measurements made in the same way with the same system when the unit was last calibrated on Earth.

Calibration

Calibration of the CPLEE was extensive and will be described briefly in this report. The major portion of the calibration was performed with an electron gun that fired a large, uniform beam of electrons of adjustable energy levels, monoenergetic to approximately 2 percent. Under the control of a computer, the instrument was tilted at various angles to the beam. The computer stored the count rates of each channel at each angle and electron energy level as well as the beam current measured by use of a Faraday cup. The absolute geometric factors were then computed from the several million accrued measurements. In addition, the ^{63}Ni sources were used as broadband near-isotopic electron sources for standard calibrations.

In practice, the exact passbands were derived, rather than the rectangular equivalent passbands of figure 10–5. However, the finer details, together with information gained from measuring susceptibility to ultraviolet light and to scattered electrons, can be shown to be negligible for this preliminary study.

Deployment

The CPLEE was deployed with no difficulty at approximately 18:00 G.m.t. on February 5, 1971. Leveling to within 2.5° and east-west alinement to within ±2° were to be accomplished with a bubble level and a Sun compass, respectively.

It has since been determined by a careful study of the lunar photographs and a comparison of predicted and actual solar ultraviolet response profiles that the experiment is 1.7° off level, tipped to the east, and 1° away from a perfect east-west alinement. This error is well within the preflight specifications. Furthermore, the photograph (fig. 10–1) shows no visible dust accretion on the exterior surfaces.

Operation of the Experiment

The CPLEE was first commanded on at 19:00 G.m.t., February 5, during the first period of extravehicular activity for a brief functional test of 5-min duration. All data and housekeeping channels were active, and the instrument began operation in the proper initial modes (i.e., automatic sequencer, on; electron-multiplier voltage increase, off; and automatic thermal control, on).

A complete instrument checkout procedure was initiated at 04:00 and continued until 06:10 G.m.t., February 6. During this period, data from the dust-cover beta sources were accumulated and compared with prelaunch calibrations. A partial comparison is shown in a following section of this report. Also during this period, all command functions of the CPLEE were exercised except the forced heater mode and dust-cover removal commands. The instrument responded perfectly to all commands. After the checkout procedure, the CPLEE was commanded to the standby mode to await LM ascent.

Following LM ascent, the CPLEE was commanded on at 19:10 G.m.t. and the dust cover was successfully removed at 19:30 G.m.t., February 6. The CPLEE immediately began returning data on charged-particle fluxes in the magnetosphere.

The instrument temperatures were carefully and continuously monitored for 45 days after deployment. It was found that the temperature range was nominal, with the internal electronics temperature ranging from 58° C at lunar noon to −24° C during lunar night. The total lunar eclipse of February 10 offered an excellent opportunity to determine various thermal parameters and to test the capability of the CPLEE to survive extreme thermal shocks. A plot of the physical-analyzer temperature during the eclipse is shown in figure 10–7. The maximum thermal shock occurred after umbra exit, with a temperature change rate of 25° C/hr. Also from this figure, it is possible to derive a thermal time constant of approximately 1.9 hr. The CPLEE suffered no ill effects from this period of rapid temperature changes.

The command capability of the CPLEE was used extensively during the 45-day real-time support period to optimize scientific return from the instrument. Alternate 1-hr periods of manual operation at the −35-V step and automatic operation have been used to concentrate on rapid temporal variations in low-energy electrons. Similarly, alternate periods of 350-V manual and automatic operation have been used to focus on rapid changes in magnetopause ions and the solar wind. In fact, the manual operation capability and the attendant 2.4-sec sampling interval made possible the detection of phenomena that would have been impossible to detect otherwise because of sampling problems and aliasing. Most of the decisions concerning operational modes were based on viewing the real-time data stream.

To date, the CPLEE has operated continuously with all high voltages on except for one brief period of approximately 15 sec when it was commanded to standby, and then back to on, to restore automatic thermal control at the termination of the first lunar night. No evidence of high-voltage discharge or corona has been observed.

Results

The following paragraphs are a detailed discussion of the scientific phenomena recorded by the CPLEE and listed in the "Summary of Observations." In many cases, these phenomena are quite distinct, and hence each phenomenon, complete with data, discussion, and conclusions, is presented in following portions of this section.

Beta-Source Tests

Abbreviated results of three separate beta-source tests (with the CPLEE subjected to exci-

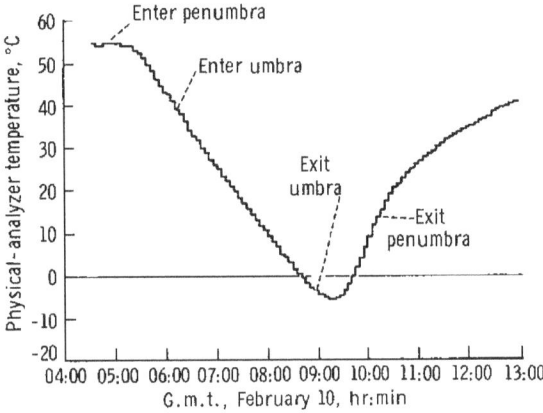

FIGURE 10–7.—Temperature profile of CPLEE during the total lunar eclipse of February 10.

TABLE 10–I. *Beta-Source Tests of the CPLEE*

Calibration and date	Readout on channel—					
	1	2	3	4	5	6
Analyzer A:						
Precalibration, Oct. 24, 1969..................	8.7	22.2	38.8	80.7	165.7	1280.5
Postcalibration, Jan. 20, 1970................	8.2	18.9	38.5	86.6	205.7	1323.0
Postdeployment, Feb. 6, 1971................	10.68	20.5	39.6	82.4	195.9	1259.0
Analyzer B:						
Precalibration, Oct. 24, 1969..................	5.8	12.7	19.8	43.6	113.1	777.7
Postcalibration, Jan. 20, 1970................	4.5	9.1	14.6	34.8	96.6	577.9
Postdeployment, Feb. 6, 1971................	7.68	12.0	17.8	35.4	90.0	763.8

tation by the ^{63}Ni beta sources mounted under the dust cover) are presented in table 10–I. The tests were conducted before the complete laboratory calibration, immediately after the laboratory calibration, and after lunar deployment. These tests span an interval of approximately 15 months.

The counting rates tabulated are for deflection voltages of −3500 V for channels 1 to 5 and 3500 V for channel 6, when the channels were sensitive to electrons with energies between 5 and 50 keV. Variations in analyzer A are not more than 30 percent, with 4 to 6 percent being typical. In analyzer B, there was a general trend of gain loss between the precalibration and postcalibration tests of approximately 20 percent, but there was a partial recovery between the postcalibration test and the postdeployment test. This effect is attributed to the well-known characteristic of temporary electron-multiplier fatigue resulting from exposure to high fluxes (e.g., during the calibration) and later recovery. This phenomenon has been documented (ref. 10–10). These beta-source tests show that no major changes occurred in the electron-multiplier characteristics between calibration and deployment, and they verified the operation of CPLEE. The small variations in gain observed are to be expected and are tolerable.

Photoelectron Fluxes

One of the most stable and persistent features in the CPLEE data is the presence of low-energy electrons when the lunar surface, in the vicinity of the ALSEP, is illuminated by the Sun. It was possible early in the mission to prove that these fluxes were of photoelectric origin, by observing the disappearance of the fluxes during the total lunar eclipse of February 10. The counting rates of channel 6 at 35-V deflection (sensitive to electrons with 50 eV $< E <$ 150 eV) of both analyzers A and B before, during, and after the eclipse are shown in figures 10–8 and 10–9. The

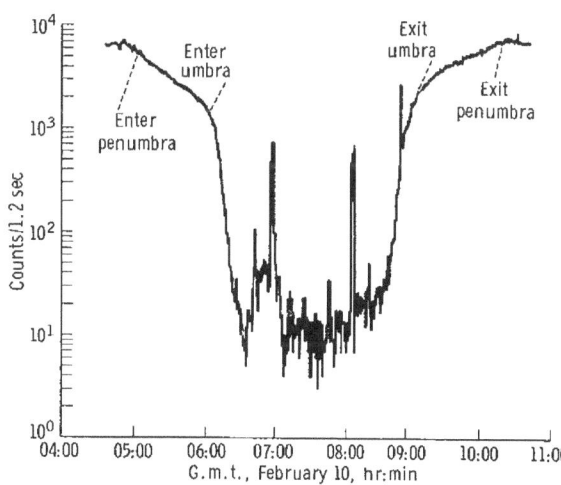

FIGURE 10–8.—Counting rate of channel 6 of analyzer A at 35 V, measuring electrons with energies between 50 and 150 eV for the period including the lunar eclipse.

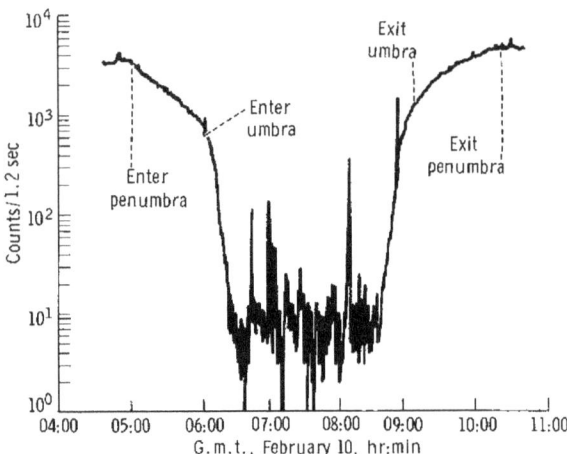

FIGURE 10-9.—Counting rate of channel 6 of analyzer *B* at 35 V, measuring electrons with energies between 50 and 150 eV for the period including the lunar eclipse.

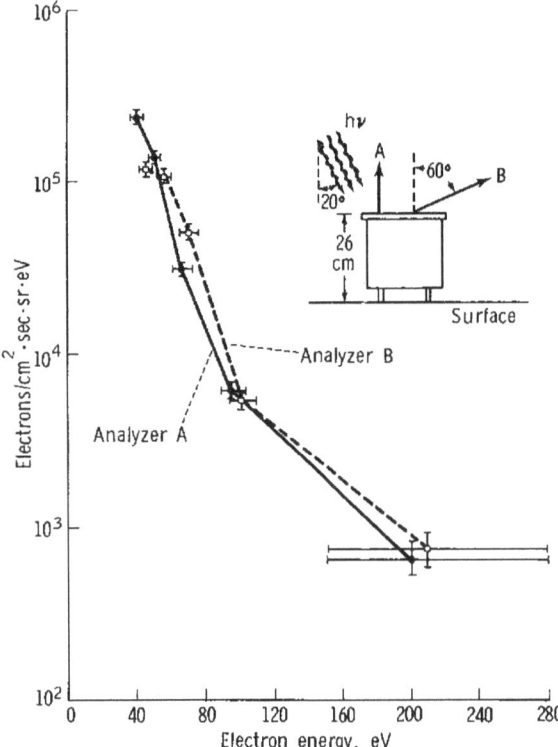

FIGURE 10-10.—Energy spectrum of photoelectrons with energies between 40 and 200 eV. The sketch on the figure shows the geometry of CPLEE relative to the lunar surface and to the direction of solar radiation.

flux is seen to correlate exactly with the presence of illumination; and, during the eclipse, sporadic bursts of electrons, presumably of magnetospheric origin, occur with flux levels that are normally undetectable because of the masking effect of the photoelectrons.

The energy spectrum of these photoelectrons, obtained from channels 1 to 5 at -35 V at a period just before eclipse onset, is shown in figure 10-10 for both analyzers. As would be expected, the spectrum is quite steep, as the CPLEE was observing essentially a high-energy and possible nonthermal tail of an electron distribution with an average energy of approximately 2 eV. In fact, the high-energy tail that was measured was almost certainly nonthermal, because the spectrum between 40 and 100 eV can be represented by an equation of the form

$$j(E) = j_0 \exp \left[\frac{-(E-40)}{14} \right]$$

where $j(E)$ is electron flux in units of electrons/ $cm^2 \cdot sec \cdot sr \cdot eV$ and j_0 is the flux at $E = 40$ eV. Clearly, this does not agree with a simple maxwellian distribution at low energies with $kT \approx 2$ eV. Two possible explanations for this discrepancy exist: (1) Some process is acting to accelerate part of the photoelectron gas; (2) the CPLEE itself is at a positive potential with respect to the surrounding lunar-surface average potential. The second explanation is entirely possible in view of the facts that the CPLEE is well insulated from the lunar surface by fiber-glass legs and that the photoemissive properties of the CPLEE and of the lunar surface are almost certainly different. It is hoped that this question will be resolved with detailed studies of these photoelectron fluxes, especially during periods of terminator crossings and the eclipse.

It should also be noted that, although in one sense the photoelectron fluxes are a contaminant obscuring weak fluxes of magnetospheric origin (fig. 10-8), they are valuable not only because they furnish information of solar-radiation/lunar-surface interactions but also because they furnish a stable "calibration source" for monitoring long-term changes in electron multiplier operating characteristics. To put it another way, the photoelectrons offer a continuing "beta-source" test for monitoring the performance of the instrument.

At certain angles between the Sun line and the analyzer look directions, preflight ultraviolet-rejection tests showed enhanced counting rates because of photoelectrons produced inside the analyzers. The enhanced counts reported here must be due to photoelectrons from the lunar surface because both analyzers A and B recorded comparable fluxes. It would be impossible for a single-point ultraviolet source (e.g., the Sun) to produce such similar counting rates in both analyzers.

LM Impact Event

On February 7, the Apollo LM ascent stage impacted the lunar surface 66 km west of the CPLEE. The terminal mass and velocity were 2303 kg and 1.68 km/sec, respectively, resulting in an impact energy of 3.25×10^{11} J (sec. 6). The LM contained approximately 180 kg of volatile propellants, primarily dimethylhydrazine fuel and nitrogen tetroxide oxidizer. For the purpose of reference and orientation, figure 10–11 is a lunar map showing the location of the impact point relative to the Apollo 12 and 14 ALSEP sites.

The counting rates of channel 6 of analyzer A (measuring ions with energies of 50 to 150 eV/unit charge) and channel 3 of the same analyzer (measuring negative particles with energies of 61 to 68 eV) are shown in figure 10–12 from 00:44:53 to 00:48:55 G.m.t. on February 7.

FIGURE 10–11.—Lunar map showing the locations of the CPLEE and of the Apollo 14 LM ascent-stage impact point.

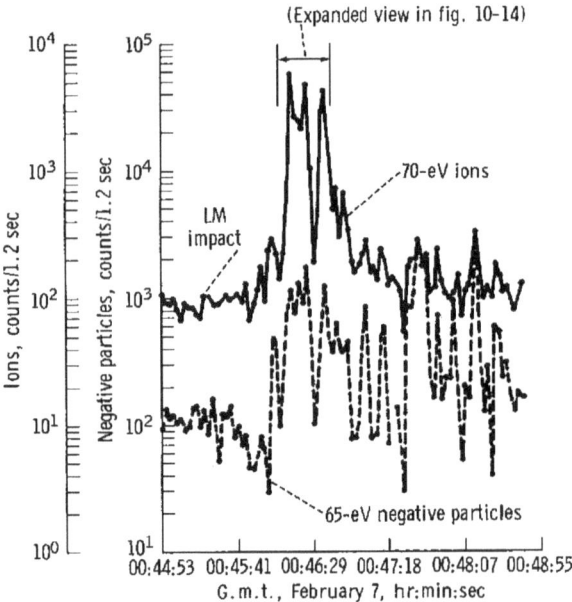

FIGURE 10–12.—Counting rates of channels 3 and 6 of analyzer A at -35 V, measuring 65-eV negative particles and 70-eV ions, respectively, showing the particle fluxes resulting from the LM impact.

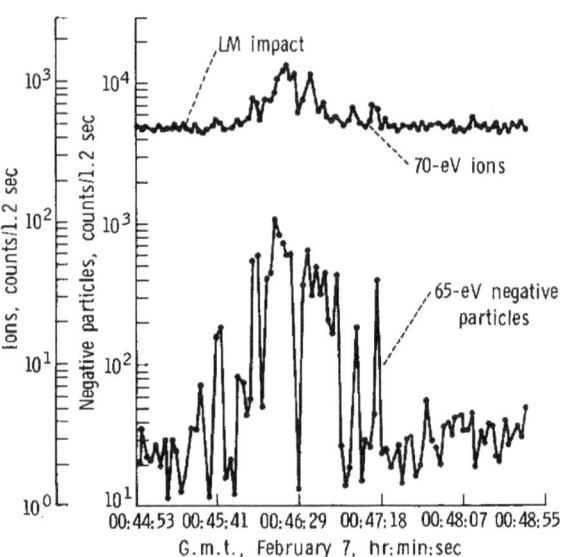

FIGURE 10–13.—Counting rates of channels 3 and 6 of analyzer B at -35 V, measuring 65-eV negative particles and 70-eV ions, respectively, showing the particle fluxes resulting from LM impact.

As can be seen from figure 10–12, the counting rates before and during the LM impact were reasonably constant and, by all indications, were due to the ambient population of low-energy electrons and ions that are present whenever the lunar surface, in the vicinity of the CPLEE, is illuminated. (This conclusion is supported by the observation that these ambient fluxes disappeared entirely during the total lunar eclipse (figs. 10–8 and 10–9) that occurred a few days later on February 10.) The counting rates increased by a factor of about 4 approximately 40 sec after LM impact and then reverted to ambient levels for a few seconds. However, 48 sec after LM impact, the ion electron counting rates increased very rapidly by a factor of up to 40 as the plasma cloud enveloped the CPLEE. A second plasma cloud passed the CPLEE a few seconds later, as shown by the second large peak. On the assumption that the plasma clouds traveled essentially in a linear path between the impact point and the CPLEE, the average velocity calculated was 1.0 km/sec; and the horizontal dimensions were 14 and 7 km for the first and second clouds, respectively.

The same data for analyzer B oriented 60° from the vertical toward the lunar west (i.e., toward the impact point) are shown in figure 10–13. By comparing figures 10–12 and 10–13, it can be noted that the flux enhancements were essentially simultaneous in the two directions, but that the positive-ion flux measured by analyzer A was five times higher than the flux measured by analyzer B. On the other hand, the negative-particle flux measured by analyzer A was only one-third as great as the negative-particle flux measured by analyzer B.

The detailed characteristics of the plasma clouds are shown in figure 10–14, which is an expanded-time-scale plot of the negative-particle fluxes in five energy ranges and the ion flux in a single energy range measured by analyzer A. The plot shows clearly that the negative-particle enhancement was confined to energies less than 100 eV, because the 200-eV flux was essentially constant throughout the event. Furthermore, the spectrum of negative particles during the enhancement is quite different from the background electron spectrum. This point is illustrated further in figure 10–15, which shows the negative-particle spectra for 00:42:33 G.m.t. (before the LM impact) and

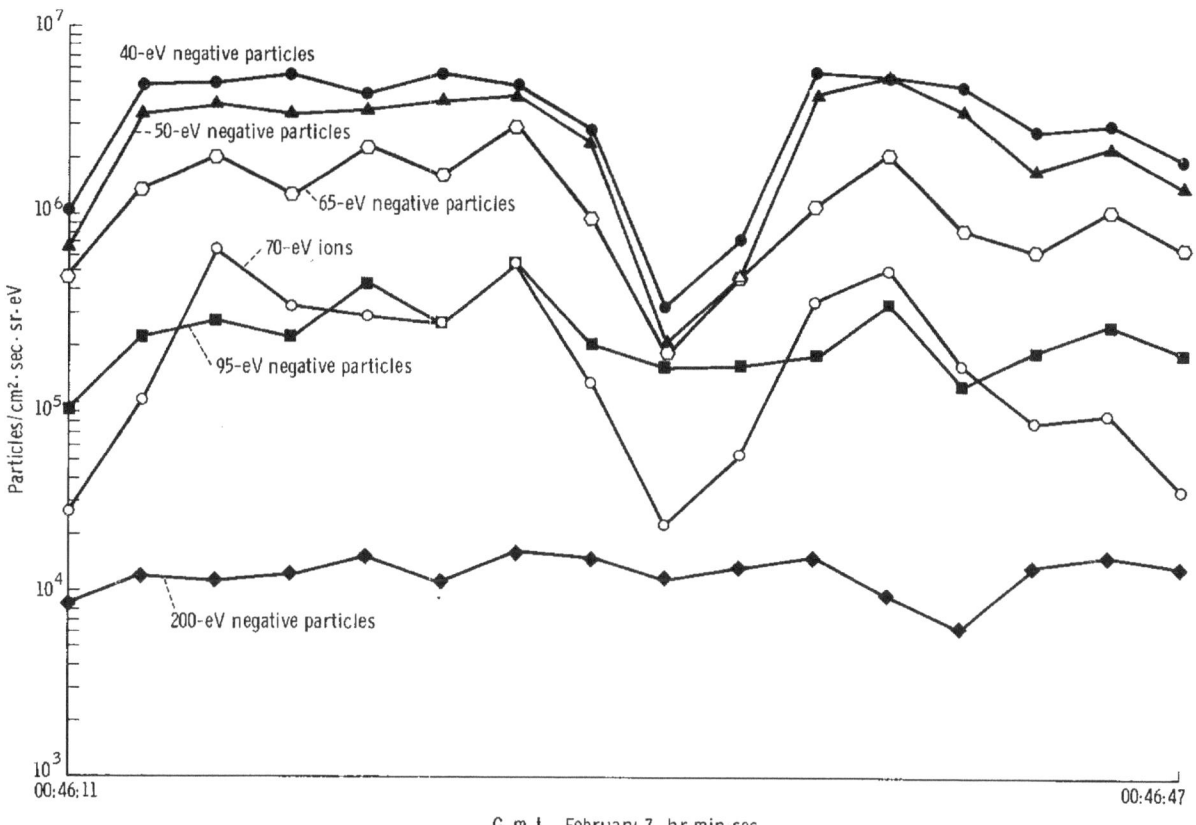

FIGURE 10–14.—Expanded view of the data of figure 10–12, showing details of the two prominent peaks. The fluxes are computed from five negative-particle energy ranges and a single positive-ion energy range.

00:46:19 G.m.t., February 7 (at the peak of the first plasma cloud).

It might well be questioned whether the flux enhancements at 48 and 67 sec after LM impact were actually initiated by the event. In the period of approximately 2 days following the impact event, several rapid enhancements in the low-energy electron fluxes (by a factor of up to 50) were observed. However, these other enhancements were not correlated with positive-ion-flux increase; and, in fact, the event referred to is the only such example of perfectly correlated positive- and negative-particle enhancements recorded to date. In addition, careful monitoring before the LM impact revealed that the fluxes were relatively stable, constant to within a factor of 2 over periods of a few minutes. This observation lends

credence to the belief that this was a valid case of cause and effect.

Further confidence in the interpretation that the flux enhancements were artificially impact produced rather than of natural origin is gained by noting that, although no such plasma clouds have previously been detected resulting from impact events, positive-ion clouds have been detected by the Apollo 12 suprathermal ion detector experiment. These positive-ion clouds were interpreted as resulting from the Apollo 13 and 14 SIVB impacts.[1]

[1] J. W. Freeman, Jr.; H. K. Hills; and M. A. Fenner: Some Results From the Apollo 12 Suprathermal Ion Detector. Proc. Apollo 12 Lunar Sci. Conf. (Houston), Jan. 11–14, 1971. To be published in Geochim. Cosmochim. Acta.

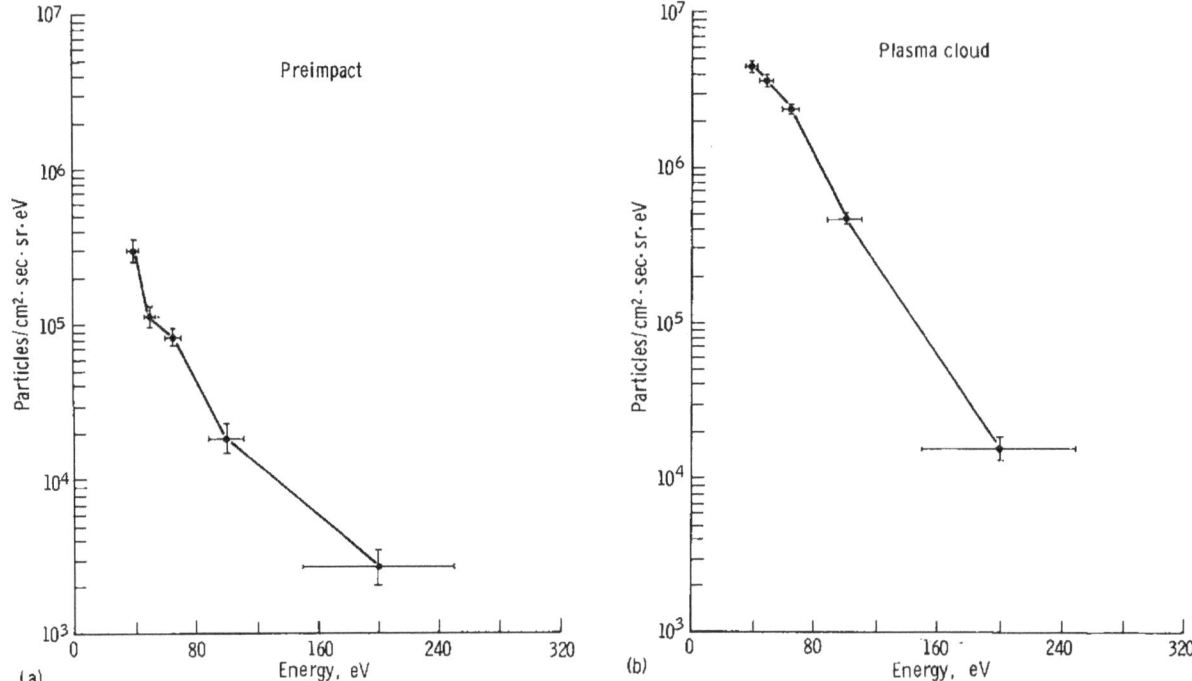

FIGURE 10–15.—Negative-particle spectra measured by analyzer A for two periods. (a) The spectrum a few minutes before LM impact. (b) The spectrum at the height of the first large peak shown in figure 10–12.

Some detailed parameters calculated from the flux enhancements are discussed next. An average cloud velocity of 1 km/sec has been previously noted, and it is of interest to compare this with particle velocities in the cloud. Some assumptions must, of course, be made as to the ion species present. Considering that the most likely source of ions was the LM propellants, an average ion mass of 25 is estimated. If it is assumed that the negative particles detected were electrons and that the positive particles had an average mass of 25, this assumption yields velocities ($E = 50$ eV) of 4000 km/sec and 20 km/sec, respectively. The charged-particle energy densities based upon the ions actually measured are calculated to be 5.6×10^{-10} erg/cm³, assuming the ions were protons, and 28.0×10^{-10} erg/cm³, assuming an average ion mass of 25. These are lower limits, because positive ions were measured in only a single energy range, and an overall ion-energy spectrum is required to make a more exact calculation. The magnetic-field energy density at the lunar surface, based on Apollo 12 lunar-surface-

magnetometer measurements (ref. 10–11) of a steady 35γ field is 50×10^{-10} erg/cm³; hence, the particle energy density appears to be at least comparable to and possibly dominant over the magnetic-field energy density. This conclusion, however, may be modified by the Apollo 14 lunar portable magnetometer observations of lunar-surface fields of 40 to 100 gammas at two distinct sites (sec. 13). It should also be noted that the solar-wind energy density is 80×10^{-10} erg/cm³.

Therefore, the conclusion can be made that the LM impact resulted in the production of two annular plasma clouds that contained negative particles and ions with energies up to 100 eV. The particles traveled across the lunar surface with a velocity of approximately 1 km/sec. No speculation has been made as to the mechanisms responsible for production of these clouds, but the simultaneous arrival of both positive- and negative-charge species is impossible to reconcile with a simple model of photodissociation and ionization and subsequent acceleration by a static electric field.

The fact that the electron and ionic components were detected simultaneously offers a unique problem; because, if one assumes that the particles were energized at the instant of impact, a mechanism must be found that is able to hold the cloud together, in view of the fact that measured ion velocities exceed the cloud velocity by an order of magnitude. This fact in itself argues against ambipolar diffusion. Processes such as charge exchange, scattering, and wave-particle interactions can also be rejected by appealing to considerations based on the size of the clouds (10 km). The only remaining possibility is magnetic confinement, or a process whereby the local magnetic field confines the particles in circular orbits. There is, however, a criticism of the hypothesis that magnetic confinement can explain the observation. The cyclotron radii of the particles must be no greater than the dimension of the plasma cloud. The cyclotron radius of a 50-eV mass-25 ion in a 100γ magnetic field is 50 km, or a factor of approximately 3 to 6 times larger than the inferred cloud dimensions.

It thus appears that a simple model of the particles being energized at the instant of impact is untenable not in itself, but because a mechanism to contain the plasma after energization is not readily evident and, in fact, may not exist. The alternate conclusion is that the impact produced expanding gas clouds, and the particles in these gas clouds were then ionized by any one of several means (e.g., photoionization) and subsequently accelerated by a continuously or erratically active acceleration mechanism. The solar-magnetospheric (SM) coordinates of the CPLEE at the time of impact were $Y_{SM}=34R_E$, $Z_{SM}=21R_E$, and solar elevation angle $=30°$, where R_E is Earth radius. (The solar-magnetospheric coordinate system is based on the Earth-Sun line (X-axis) and the magnetic dipole axis of the Earth. The Z-axis is perpendicular to the Earth-Sun line and in the plane formed by the Earth-Sun line and the Earth magnetic dipole axis; the Y-axis completes the right-handed coordinate system.) Hence, it is highly likely that the solar wind had direct access to the lunar surface at this time. Noting the energy densities of the solar wind and the plasma cloud particles, the solar wind is energetically capable of being the energy source. Whether any such mechanism can work is unknown at this time,

although calculations have indicated that the solar wind can interact with a neutral gas through means other than simple particle-particle collisions (ref. 10–12).

In summary, the impact-event data apparently indicate a situation in which the gas cloud, solar wind, and local magnetic field are all interacting, offering a unique and fascinating problem in plasma physics.

Low-Energy Electron Fluctuations

In addition to the stable low-energy photoelectron population that the CPLEE records whenever the lunar surface in the vicinity of the CPLEE is illuminated, the CPLEE also observes rapidly varying fluxes of low-energy electrons of magnetospheric origin, with intensities large enough to be detected above the photoelectron background. Examples of these fluxes are shown in figure 10–16, wherein the counting rates of channel 3 (65-eV electrons) and channel 5 (200-eV electrons) are plotted for a brief time segment. At approximately 21:20 G.m.t., February 7, the solar magnetospheric coordinates of the CPLEE were $Y_{SM}=24R_E$ and $Z_{SM}=14R_E$, locating the instrument within the magnetospheric tail near the boundary. The instrument was in the manual mode, and hence the individual measurements are 2.4 sec apart. The flux enhancements range up to a factor of 10 above the background level on time scales on the order of a few seconds.

At first glance, the enhancements in the two energy ranges appear to be well correlated, but a closer examination of the figure reveals temporal dispersions in the enhancements. To illustrate this point more clearly, the data for the period 21:20:07 to 21:20:41 G.m.t. have been plotted in a special manner in figure 10–17: a log-log plot of the counting rates in the two energy channels was made with the higher energy channel on the vertical axis and the lower energy channel on the horizontal axis. Each pair of count rates from the two channels is represented by a single point, and a vector is drawn between successive points in the direction of increasing time. On this type of plot, if the enhancements are perfectly correlated, all vectors will lie along a constant slope, the magnitude of which is a function of the relative enhancements. A burst where the higher energy electrons

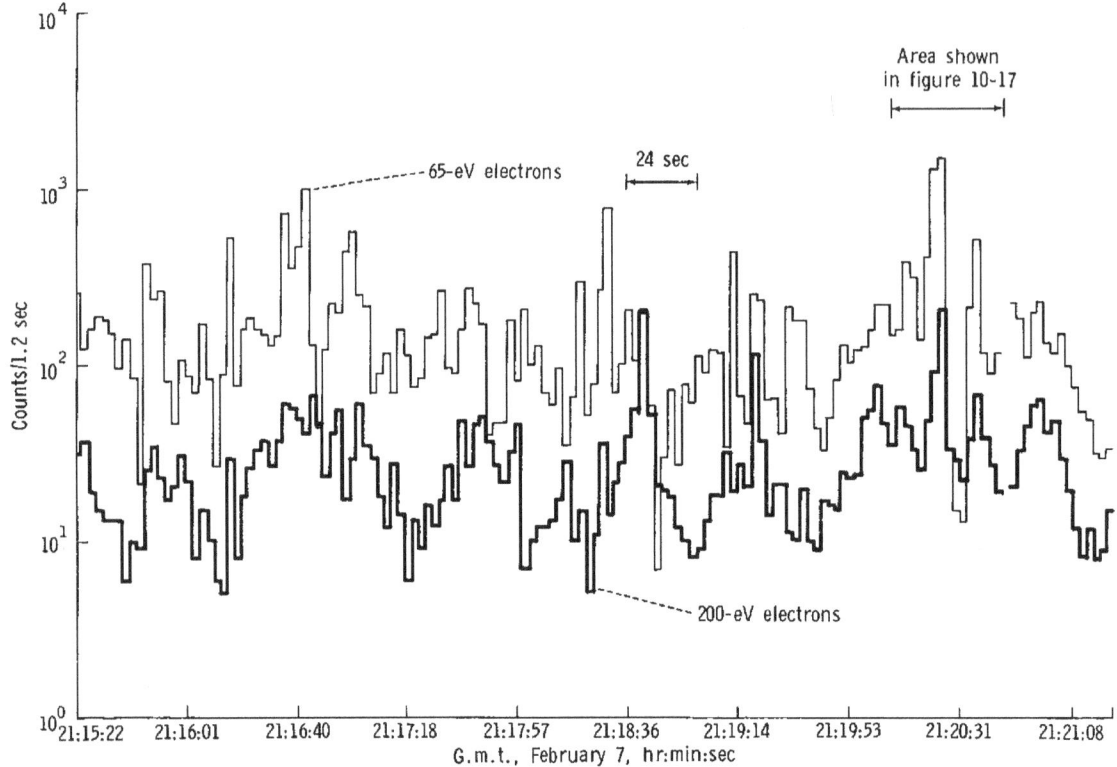

FIGURE 10–16.—Example of rapid variations in magnetospheric low-energy electron fluxes. The data are from channels 3 and 5 of analyzer A at -35 V, measuring 65- and 200-eV electrons, respectively.

lead the lower energy electrons will result in an open figure with the vectors rotating clockwise; likewise, if the higher energy electrons lag the lower energy electrons, the vectors will rotate counterclockwise. An examination of figure 10–17 shows that, in general, for the longest vectors, the constant-slope rule is followed; but that on smaller scales (e.g., points 1 to 5 and 9 to 12), considerable deviations from the constant-slope rule exist. For these events, the vectors rotate clockwise, indicating that the higher energy electrons lead the lower energy electrons.

Although plots such as these are indicative in nature, they do show the general character of the enhancements and suggest that low-energy electrons are being accelerated or modulated by processes relatively near the Moon. An approximate estimate of the distance can be obtained by considering the velocity difference at the two energies and the dispersion times in the enhancements (0 sec to approximately 2 sec), resulting in a maxi-

mum distance of some 20 000 km, or $3R_E$. An extensive cross-correlation analysis will be necessary to refine these calculations, but the preliminary studies indicate the presence of local (with reference to the Moon) processes capable of modulating or accelerating low-energy electron fluxes.

Medium-Energy Electron Event

At approximately 18:30 G.m.t. on March 10, distinct enhancements in medium-energy (approximately 1-keV) electron fluxes were observed in both analyzers A and B. The enhancements ranged up to an order of magnitude above background and lasted from a few minutes up to 2 hr; the entire event lasted approximately 4 hr. The gross temporal features of these enhancements are shown in figure 10–18 by giving the counting rate of channel 6 at 350 V (500- to 1500-eV electrons) from 18:30 to 23:00 G.m.t. The data gaps at 19:30 and 21:30 G.m.t. were due to the fact

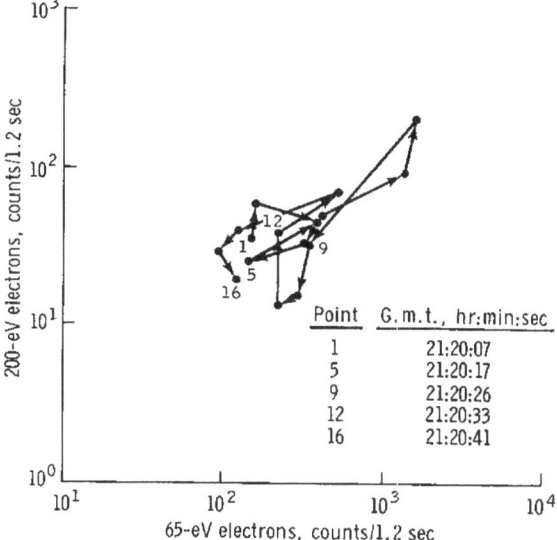

FIGURE 10–17.—Electron correlation analysis of the CPLEE analyzer *A*. A detailed study of a portion of the data of figure 10–16. The counting rate of channel 5 (200-eV electrons) is plotted against the counting rate of channel 3 (65-eV electrons) on a log-log scale. Perfect temporal simultaneity would result in all vectors lying parallel to a line of constant slope. The marked deviations from this rule should be noted.

that the CPLEE was in the manual mode at -35 V at these times, and the data gap at 21:00 was due to a temporary shutdown of the data display system at the NASA Manned Spacecraft Center.

It is seen from the figure that the event is characterized by erratic, relatively short-duration flux enhancements between 18:30 and 21:00 G.m.t.; by a period of stable high fluxes between 21:10 and 22:00 G.m.t.; and by a return to erratic enhancements between 22:00 and 23:00 G.m.t. The magnetic-activity index K_p was 3 or less on March 10, and there were no enhancements in the solar X-ray flux. Thus, this event apparently is characteristic of the quiet-time magnetosphere, and the electrons are truly magnetospheric in origin.

On the basis of the particle measurements alone, it is difficult to resolve the question of whether the enhancements are of a spatial or temporal nature; that is, whether the effects of the CPLEE moving in and out of stable spatial region or regions of flux enhancements are being recorded or whether a large-scale temporal event is being recorded. The cyclotron radius of a 1-keV

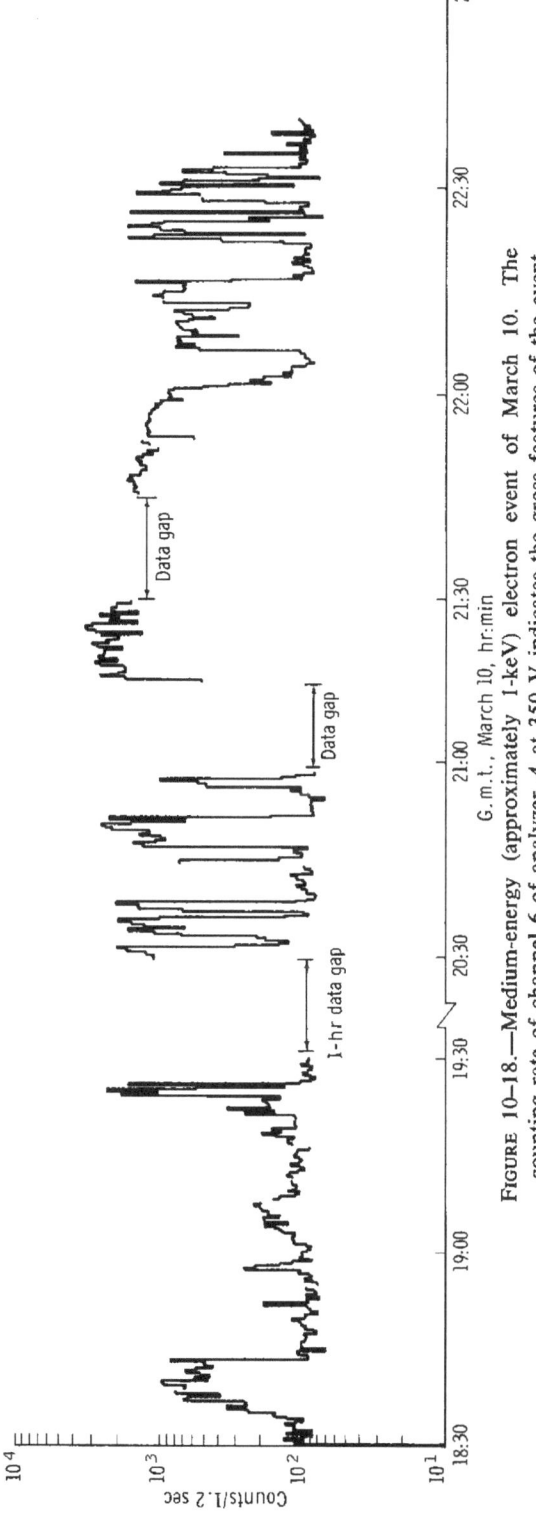

FIGURE 10–18.—Medium-energy (approximately 1-keV) electron event of March 10. The counting rate of channel 6 of analyzer *A* at 350 V indicates the gross features of the event.

electron in a 10γ field, typical of the magnetic tail at lunar distances (ref. 10–13), is 10.6 km, and the Moon moves a distance of approximately 20 km between data samples.

The path of the CPLEE in the solar-magnetospheric Y-Z plane is shown in figure 10–19. Of particular interest is the fact that the event was seen only during the period when Z_{SM} was near the maximum positive excursion of $6R_E$. This is highly suggestive, though certainly not conclusive proof, that the CPLEE was sampling a stable spatial structure located at $Z_{SM} = 6R_E$ and $Y_{SM} = 11R_E$ to $13R_E$. Further indirect evidence is that there is no extended trailing edge in the events. The leading and trailing edges appear equally sharp.

The electron energy spectrum—averaged over the time from 21:45 to 22:00 G.m.t., March 10, the period of the most stable fluxes (fig. 10–18) —is shown in figure 10–20. The photoelectron continuum is the dominant contribution between 40 and 100 eV, but a suggestion of a peak exists in the spectrum of these magnetospheric electrons at 600 eV. Also shown is an upper limit to the background equivalent flux from all other sources at 500 eV, showing the order-of-magnitude enhancement seen in the event. The integrated flux for electrons with energies between 500 and 2000 eV is 4.5 × 10⁶ electrons/cm² · sec · sr.

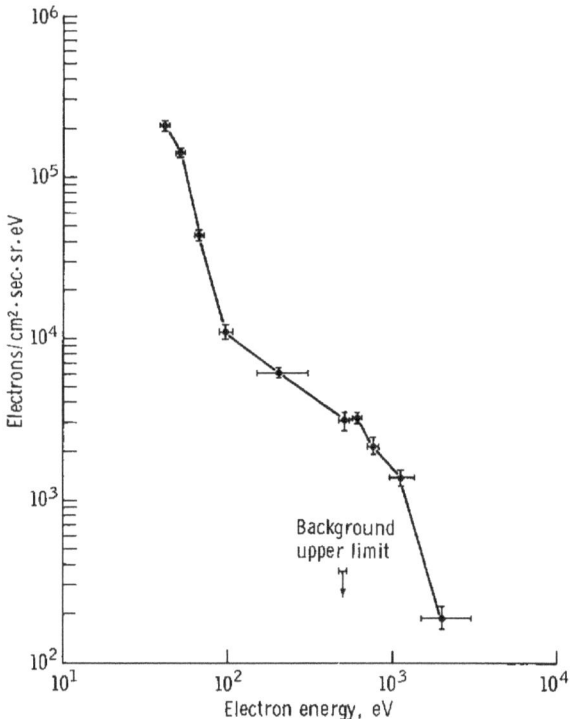

FIGURE 10–20.—Electron energy spectrum measured by analyzer A between 40 and 2000 eV from 21:45 to 22:00 G.m.t., March 10, the period of high, stable flux shown in figure 10–18.

The energy spectrum and total flux of electrons and the temporal history of the event suggest that these data represent an observation of the particles of a thin neutral sheet moving across the site. The difficulty with this interpretation lies in the fact that, at this time, the CPLEE was approximately $6R_E$ away from the theoretical location of the neutral sheet, the Y_{SM} axis (fig. 10–19). Strong indications exist, however, that the solar-magnetospheric coordinate system does not aid in locating the neutral sheet with an error of less than approximately $10R_E$ at lunar distances. The neutral-sheet observations with a magnetometer on board the Interplanetary Monitoring Platform 1 satellite (ref. 10–14) locate the neutral sheet at various times during the period March 22 to May 26, 1964, in the range $-2R_E < Z_{SM} < 5R_E$. Hence, it is plausible that the neutral sheet could have been located at $Z_{SM}=6R_E$ at the time of the CPLEE observation. Further measurements during forthcoming magnetospheric tail passes by the

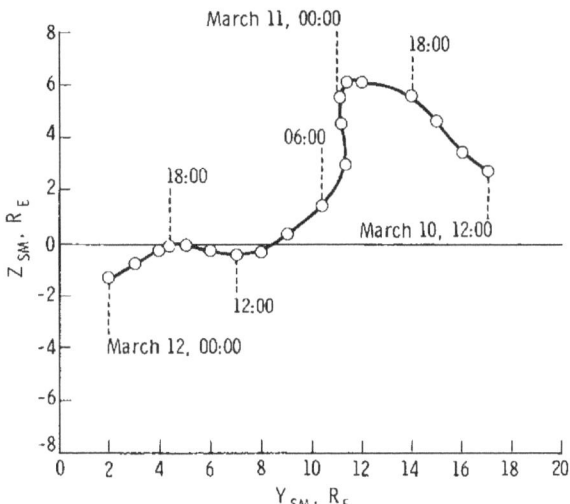

FIGURE 10–19.—Track of the CPLEE in the Y_{SM}-Z_{SM} plane for March 10 to March 12, including the period of the electron event shown in figure 10–18.

CPLEE are needed to effect a definite resolution of this question.

Electron Spectra Similar to Auroral Spectra

On several occasions when the CPLEE was in the magnetospheric tail, short-duration electron enhancements in all ranges of the instrument were observed. These enhancements typically had durations of a few minutes. The energy spectrum of one such enhancement at 23:16 G.m.t., February 7, is shown in figure 10–21. As in most electron spectra observed when the lunar surface is illuminated, the spectrum between 40 and 100 eV is dominated by the photoelectron continuum. However, in the higher energy ranges, a double-peak structure with a low-energy peak in the range 300 to 500 eV and a high-energy peak at 5 to 6 keV can be seen. It is interesting to compare these spectra with spectra observed above a terrestrial aurora. A set of spectra observed above an aurora (measured with a SPECS detector on

board a Javelin sounding rocket) is shown in figure 10–22 from reference 10–1. (It should be recalled that the basic particle detectors of both the SPECS and the CPLEE are very similar.)

The photoelectron continuum is, of course, absent from these auroral spectra; but, aside from that, a remarkable similarity between the electron fluxes recorded by the CPLEE and the auroral electrons is readily evident. The double-peak structure in both spectra, the low-energy peaks in the 100- to 500-eV range, and the high-energy peaks at 5 to 6 keV are particularly noteworthy. The flux levels in the auroral spectrum are within a factor of 5 of the flux levels measured by the CPLEE (fig. 10–21). Furthermore, while particles measured above an aurora tend to be more or less isotropically distributed about the field lines, the magnetic-tail particles observed by the CPLEE were strongly peaked along the field lines. This deduction was made on the basis of the observation that no flux enhancements were seen in analyzer B, and that the angles between the magnetic field and the directions of analyzers A and B were approximately 20° and 80°, respectively. Particles energized near the Earth and subsequently traveling back into the tail would be sharply peaked along field lines at lunar distances according to the first invariant $\sin^2 \alpha/B =$ constant, where α is the particle pitch angle and B is the local magnetic-field magnitude.

Consequently, the process that produces energetic particles above a terrestrial aurora may well result in the appearance of similar particles in the magnetospheric tail. A definite resolution of this question awaits further study of the data and correlation between the CPLEE data and Earth-based measurements of auroral activity. However, this preliminary indication of auroral particles at large distances from the Earth in the magnetospheric tail implies that some auroral-zone magnetic-field lines are linked with field lines stretching far into the tail and, hence, give information on the general topology of the magnetosphere.

Rapid Temporal Solar-Wind Variations

When the Moon crosses from the magnetospheric-tail regions into interplanetary space on the dawn side of the magnetosphere, the CPLEE analyzer B is pointed toward the Sun and, hence,

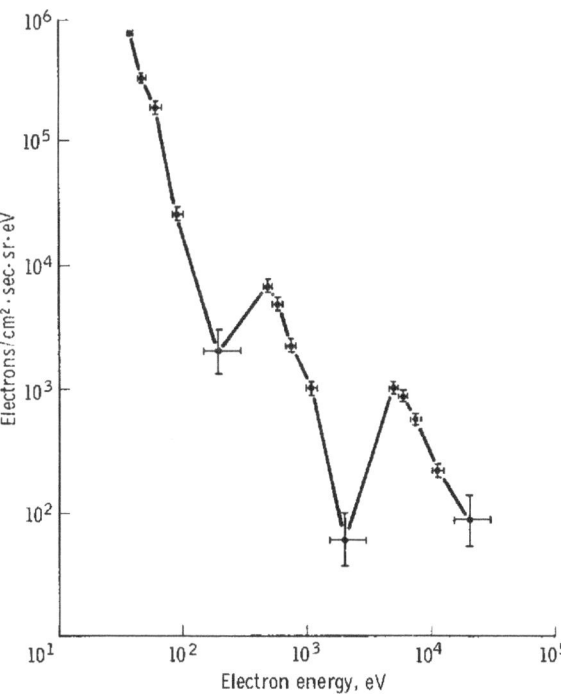

FIGURE 10–21.—Electron energy spectrum of a typical "auroral-electron" event measured by the CPLEE in the magnetospheric tail. Of particular note is the double-peak structure, with a low-energy peak at 300 to 500 eV and a higher energy peak at 5 to 6 keV.

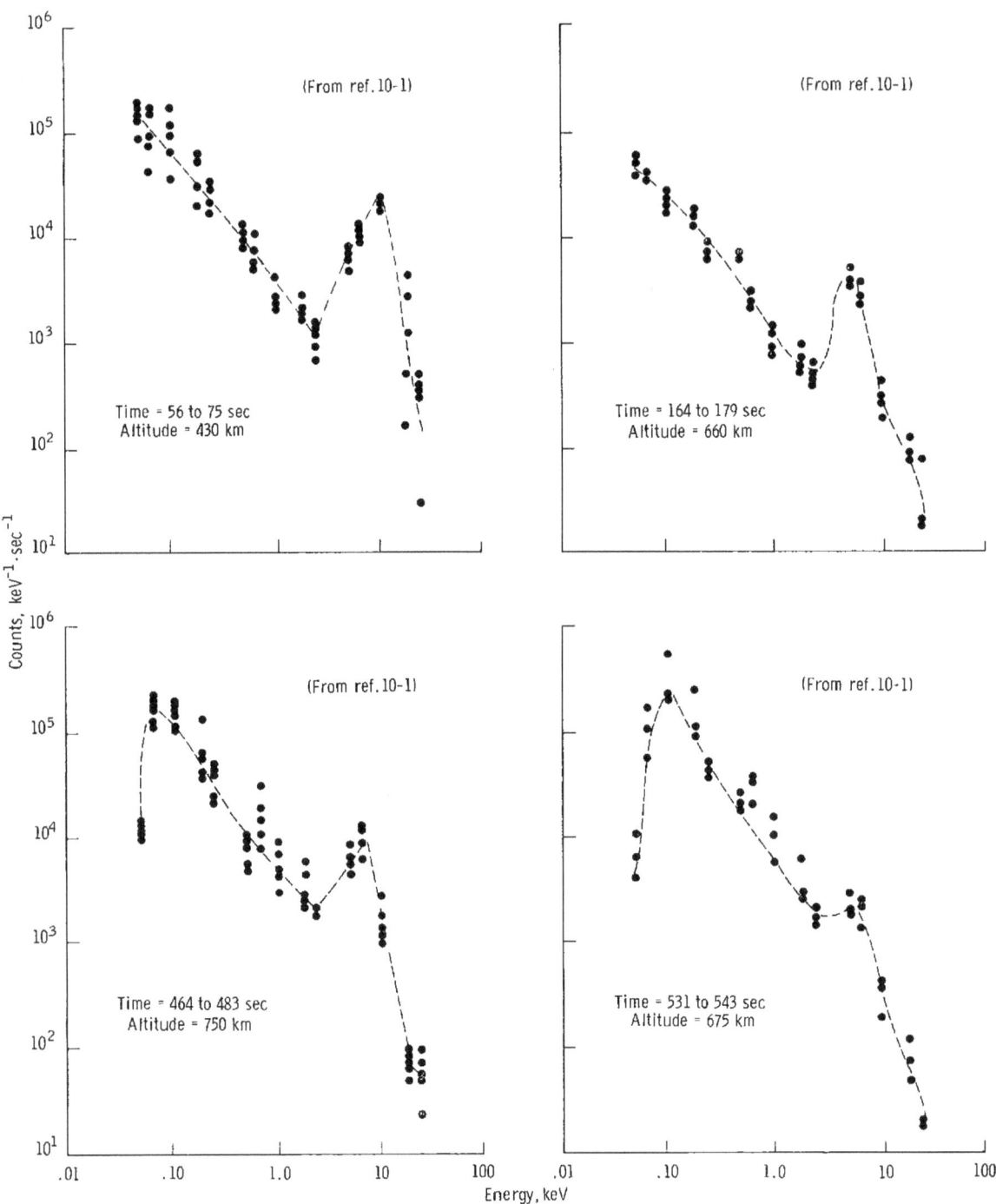

FIGURE 10–22.—Electron spectra measured above a terrestrial aurora by a device similar to the CPLEE on a sounding rocket probe (ref. 10–1). Striking similarities exist between these spectra and the CPLEE magnetospheric-tail electron spectrum shown in figure 10–21.

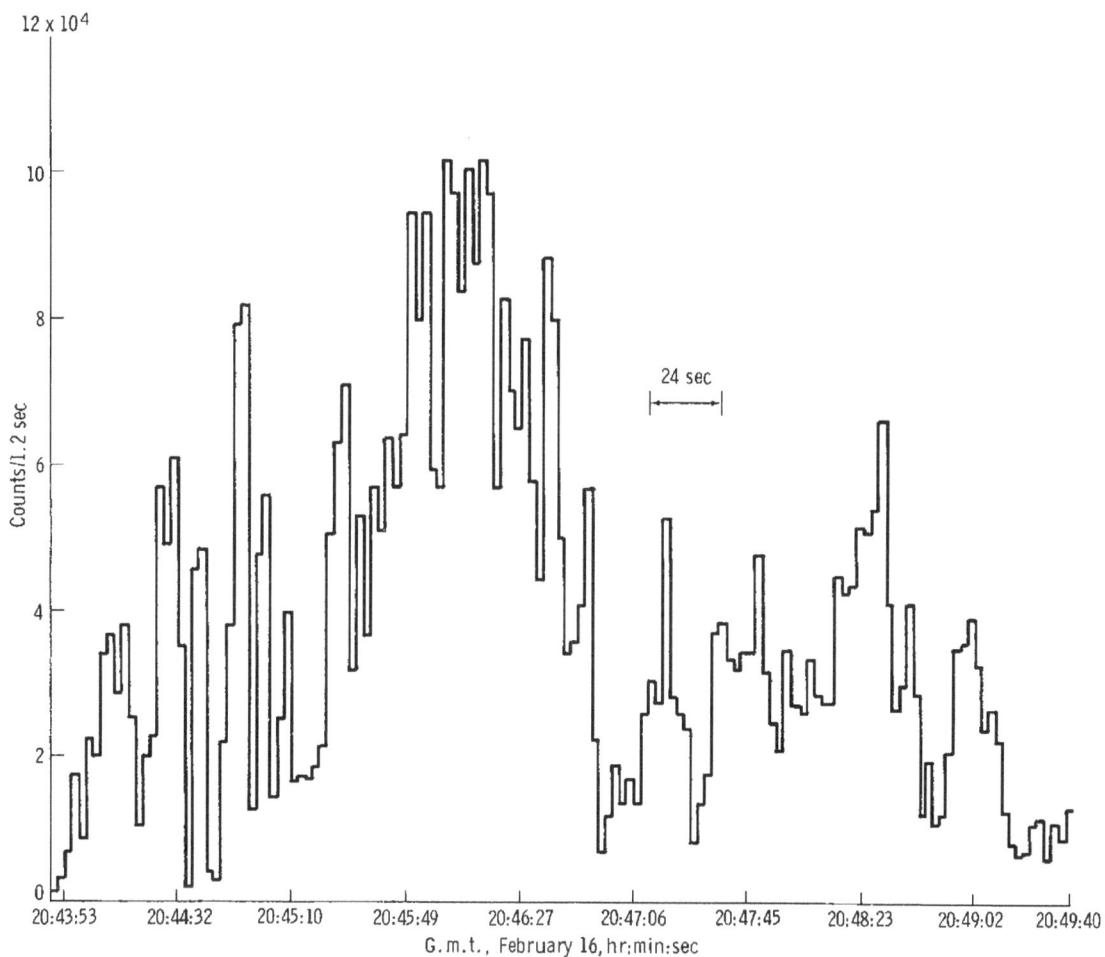

FIGURE 10–23.—An example of rapid temporal variations in solar-wind fluxes. Data from channel 5 of analyzer *B* at 350 V, sensitive to ions with energies between 1.5 and 3 keV, are plotted with a time resolution of 2.4 sec.

is able to detect solar-wind fluxes striking the Moon. Some of the detailed characteristics of solar wind at the lunar surface have been reported previously (ref. 10–4), and the CPLEE measurements appear to be in general agreement with the measurements of average solar-wind parameters by the Apollo 12 solar-wind spectrometer. The unique rapid-sampling capability of the CPLEE has been used to study rapid temporal variations in the solar wind. The sampling interval of the CPLEE (2.4 sec) is compared with those of other experiments designed to measure solar-wind fluxes, notably the Vela 3A and 3B detectors with sampling intervals of 256 sec in reference 10–15

and with the solar-wind spectrometer sampling interval of 28.1 sec in reference 10–4.

An example of rapid solar-wind variations is shown in figure 10–23. At 20:45 G.m.t. on February 16, the solar-magnetospheric coordinates of the CPLEE were $Y_{SM} = -67R_E$ and $Z_{SM} = -32R_E$, placing the instrument well away from the magnetospheric-tail boundary. The angle between the center of the detector field of view and the CPLEE-Sun line was 2°. The CPLEE data showed the counting rate was concentrated in channel 5 at 350-V deflection (the channel sensitive to ions with energies between 1.5 and 3.0 keV, exactly what would be expected if the instrument were viewing the direct solar wind). The ratio of the counting rates of this channel in analyzer B to the corresponding channel in analyzer A was on the order of 1000:1, indicating the extreme directionality of the flux.

Variations in the solar-wind flux of up to a factor of 10 on time scales as short as 5 sec are shown in figure 10–23. A comparison of the cyclotron radius of a 1.5-keV proton in the 5γ interplanetary field (1000 km) with the linear velocity of the Moon (1 km/sec) indicates that these variations are indeed temporal in nature. If the variations were spatial in origin, variations in the flux of a factor of 10 over distances as short as 1/200 of a cyclotron radius would be required. This situation is highly unlikely.

An isolated feature of the data has not been selected, and the indication is that these rapid temporal variations are a persistent feature of the solar-wind flux. Lacking a detailed analysis of the frequency spectra of these variations, their origin can only be speculated upon at this time. It is noted, however, that the observed frequency of variations (approximately 0.2 Hz) is similar to the expected observed frequency of magneto-acoustic waves (low-frequency waves that propagate in a magnetized plasma such as the solar wind), the wavelength of which is on the order of an ion-cyclotron radius, as seen by a stationary observer (approximately 0.5 Hz). This suggests that the variations seen are due to magnetoacoustic waves modulating the particle fluxes, and these waves may be generated at the shock surface between the solar wind and the magnetospheric tail.

Summary

During the first month of operation, the CPLEE has detected particle fluxes at the lunar surface resulting from a wide range of lunar-surface, magnetospheric, and interplanetary phenomena. Preliminary data analysis has revealed the presence of a lunar photoelectron layer, an indication of modulation or acceleration of low-energy electrons in the vicinity of the Moon, penetration of auroral particles to lunar distances in the magnetospheric tail, detection of electron fluxes in the magnetospheric tail possibly associated with the neutral sheet, strong modulations of solar-wind fluxes, and the appearance of ions and electrons or negative ions with energies up to 100 eV associated with the LM impact. Many of these discoveries were possible only because of the rapid sampling capability of the CPLEE and its ability to measure particles of both charge signs over a wide energy and dynamic range, coupled with the real-time data display and command capability of the ALSEP.

These preliminary findings have resulted from analysis of "quick-look" hardcopy data. Other phenomena are apparent in the data, but adequate characterization and description must await detailed computer analysis of the 200 measurements/min being returned by the CPLEE.

References

10–1. WESTERLUND, L. H.: Rocket-Borne Observations of the Auroral Electron Energy Spectra and Their Pitch-Angle Distribution. Ph.D. thesis, Rice Univ., 1968.

10–2. O'BRIEN, B. J.: Interrelations of Energetic Charged Particles in the Magnetosphere. Ch. IV of Solar Terrestrial Physics, Newman and King, eds., Academic Press (London), 1967, pp. 169–211.

10–3. LYON, E. F.; BRIDGE, H. S.; AND BINSAK, J. H.: Explorer 35 Plasma Measurements in the Vicinity of the Moon. J. Geophys. Res., vol. 72, no. 23, Dec. 1967, pp. 6113–6117.

10–4. SNYDER, CONWAY W.; CLAY, DOUGLAS R.; AND NEUGEBAUER, MARCIA: The Solar-Wind Spectrometer Experiment. Sec. 5 of the Apollo 12 Preliminary Science Report. NASA SP–235, 1970.

10–5. REASONER, D. L.; EATHER, R. H.; AND O'BRIEN, B. J.: Detection of Alpha Particles in Auroral Phenomena. J. Geophys. Res., vol. 73, no. 13, July 1968, pp. 4185–4198.

10–6. O'BRIEN, B. J.; ABNEY, F.; BURCH, J.; HARRISON, R.; ET AL.: SPECS, A Versatile Space-Qualified Detector of Charged Particles. Rev. Sci. Instrum., vol. 38, no. 8, Aug. 1967, pp. 1058–1068.

10–7. BURCH, JAMES L.: Low-Energy Electron Fluxes at Latitudes Above the Auroral Zone. J. Geophys. Res., vol. 73, no. 11, June 1968, pp. 3585–3591.

10–8. MAEHLUM, BERNT N.: On the High Latitude, Universal Time Controlled F-Layer. J. Atmos. Terr. Phys., vol. 31, no. 1, Jan. 1969, pp. 531–538.

10-9. O'BRIEN, B. J.; FREDEN, S.; AND BATES, J.: Degradation of Apollo 11 Deployed Instruments Because of Lunar Module Ascent Effects. J. Appl. Phys., vol. 41, no. 11, Oct. 1970, pp. 4538–4541.

10-10. EGIDI, A.; MARCONERO, R.; PIZZELLA, G.; AND SPERLI, F.: Channeltron Fatigue and Efficiency For Protons and Electrons. Rev. Sci. Instrum., vol. 40, no. 1, Jan. 1969, pp. 88–91.

10-11. DYAL, P.; PARKIN, C. W.; AND SONNET, C. P.: Lunar Surface Magnetometer Experiment. Sec. 4 of Apollo 12 Preliminary Science Report. NASA SP-235, 1970.

10-12. LEHNERT, B.: Minimum Temperature and Power Effect of Cosmical Plasmas Interacting With Neutral Gas. Rept. 70-11, Royal Inst. Tech., Div. Plasma Phys., Stockholm, Sweden, 1970.

10-13. NESS, N. F.; BEHANNON, K. W.; SCEARCE, C. S.; AND CANTARANO, S. C.: Early Results From the Magnetic Field Experiment on Lunar Explorer 35. J. Geophys. Res., vol. 72, no. 23, Dec. 1967, pp. 5769–5778.

10-14. SPEISER, T. W.; AND NESS, N. F.: The Neutral Sheet in The Geomagnetic Tail: Its Motion, Equivalent Currents, and Field Line Connection Through It. J. Geophys. Res., vol. 72, no. 1, Jan. 1967, pp. 131–141.

10-15. GOSLING, J. T.; ASHBRIDGE, J. R.; BAME, S. J.; HUNDHAUSEN, A. J.; AND STRONG, I. B.: Satellite Observations of Interplanetary Shock Waves. J. Geophys. Res., vol. 73, no. 1, Jan. 1968, pp. 43–50.

ACKNOWLEDGMENTS

A large number of people contributed to the success of the CPLEE, both directly and indirectly through development of the SPECS and of the calibration equipment. Rice University personnel who made major contributions are Wayne Smith, John Musselwhite, James Ballentyne, John McGarity, David Nystrom, William Porter, Foster Abney, James Burch, and Tad Winiecki.

Bendix personnel who played principal roles were Joe Clayton, Park Curry, Al Robinson, Charles Hocking, William Stanley, George Burton, Lou Paine, Lowell Ferguson, John Ioannau, Jack Dye, Mark Brooks, Charles Flint, and Jerome Pfeiffer.

Numerous NASA personnel played important roles in many phases of the program. Among them must be included Dick Moke, Jack Small, Don Wiseman, Ausley Carraway, J. B. Thomas, and W. K. Stephenson of the NASA Manned Spacecraft Center as well as Dr. J. Naugle, Dr. A. Opp, and E. Davin of NASA Headquarters and A. Spinak of Wallops Station.

This work was supported by NASA contract NAS 9-5884 and analysis was assisted in part by the Science Foundation in Physics at the University of Sydney, Sydney, Australia. Feasibility studies of the CPLEE were supported in part by NASA contracts NAS r-209, NAS 6-1061, and NAS 9-4822.

11. Laser Ranging Retroreflector

J. E. Faller,[a][†] C. O. Alley,[b] P. L. Bender,[c] D. G. Currie,[b] R. H. Dicke,[d]
W. M. Kaula,[e] G. J. F. MacDonald,[f] J. D. Mulholland,[g] H. H. Plotkin,[h]
E. C. Silverberg,[i] and D. T. Wilkinson[d]

Concept of the Experiment

During the Apollo 14 mission, a second laser ranging retroreflector (LRRR) was deployed on the lunar surface. The Apollo 11 and 14 retroreflector packages permit ground-based stations to conduct short-pulse laser ranging to these arrays on the lunar surface. An observation program of several years' duration that results in an extended sequence of high-precision Earth-Moon distance measurements will provide data from which a variety of information about the Earth-Moon system can be derived (refs. 11–1 to 11–7).

An obvious immediate use of these data will be to define more precisely the motion of the Moon in its orbit. Another experimental result will be the measurement of the lunar librations—the irregular motions of the Moon about its center. Most of the apparent librations are caused by the ellipticity of the lunar orbit and the inclination of lunar axis of rotation, but residual motions are present because the mass of the Moon is not evenly distributed. The Apollo 11 retroreflector site and the Fra Mauro Apollo 14 site, though both near the equator, are well separated from one another and consequently will yield high-quality informa-

tion concerning the librations of the Moon in longitude. The ability to separate the librations in longitude from the center of mass motion of the Moon will be of significant value in the analysis of the lunar-range results.

It has not yet been possible to conduct a full calculation of the lunar librations with the accuracy needed for the laser ranging experiment. A third U.S. retroreflector is to be deployed as a part of the Apollo 15 mission in the area near Rima Hadley. The three Apollo arrays, well separated in longitude and latitude, will permit a complete geometrical separation of the lunar librations. Already, the existence of two arrays should make it possible to see a phase shift in the 3-yr physical librations, which Eckhardt has recently pointed out (ref. 11–8) should exist unless the Q of the Moon is very high.

Another major objective of this experiment is to learn more about the Earth. Current theories suggest that the surface of the Earth is subdivided into a number of large plates that move with respect to one another. These movements are believed to explain continental drift. As an example, the Pacific plate is thought to be moving toward Japan at the rate of about 10 cm/yr. After observation stations are established in Hawaii and Japan, the lunar-distance measurements will give the longitudes of these stations with such high accuracy that this expected motion should be observable within 2 or 3 yr.

Data obtained from the lunar-distance measurements will also determine the position of the North Pole with an accuracy of approximately 15 cm,

[a] Wesleyan University.
[b] University of Maryland.
[c] Joint Institute for Laboratory Astrophysics.
[d] Princeton University.
[e] University of California at Los Angeles.
[f] Council on Environmental Quality.
[g] University of Texas at Austin.
[h] NASA Goddard Space Flight Center.
[i] University of Texas, McDonald Observatory.
[†] Principal investigator.

which is five or 10 times more accurate than that presently known by current methods. The position of the pole moves around the surface of the Earth in a rather complicated manner. It may travel nearly 70 m along a rather elliptical path during any year. The excitation mechanism for this polar wobble is still much in debate. It cannot be conclusively stated whether the mechanism is atmospheric mass shifts, variations in the coupling of the core and the mantle, or mass shifts in the crust. The last hypothesis has been suggested by a correlation of observed polar shifts with major earthquakes, and hence better measurements may lead to a more complete understanding of earthquake phenomena.

Lunar-distance measurements will also permit more accurate determinations of the Earth rotational rate than has previously been possible. And, finally, the sensitivity afforded by the presence of these retroreflecting arrays on the lunar surface will make it possible to use the Moon again as a testing ground for gravitational theories. Many observers are interested in discovering whether the tensor theory of gravity is sufficient or whether a scalar component is necessary, as has been suggested. A definitive test of the hypotheses may be obtained by monitoring the motion of the Moon. Additionally, the possibility exists of seeing some very small but important effects in the motion of the Moon that are predicted by the general theory of relativity.

Properties of the LRRR

The Apollo 14 LRRR (figs. 11–1 and 11–2) is a wholly passive device containing an array of 100 small, fused-silica corner cubes, each 3.8 cm in diameter. The Apollo 14 LRRR was deployed during the first period of extravehicular activity approximately 30 m west of the central station; thus, the array was placed approximately 200 m west of the lunar module (LM). Leveling and alinement to point the normal-to-the-array face toward the center of the Earth libration pattern was accomplished with no difficulty (fig. 11–3). Each corner cube in the array has the property of reflecting light parallel to the incident direction; that is, a light beam incident on a corner cube is internally reflected in sequence

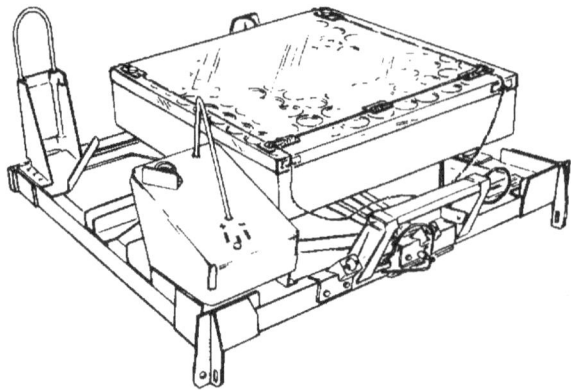

FIGURE 11–1.—The LRRR in a stowed configuration.

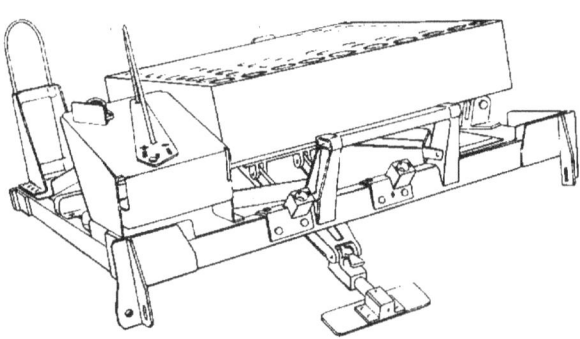

FIGURE 11–2.—The LRRR in a deployed configuration.

FIGURE 11–3.—Photograph of the deployed Apollo 14 array.

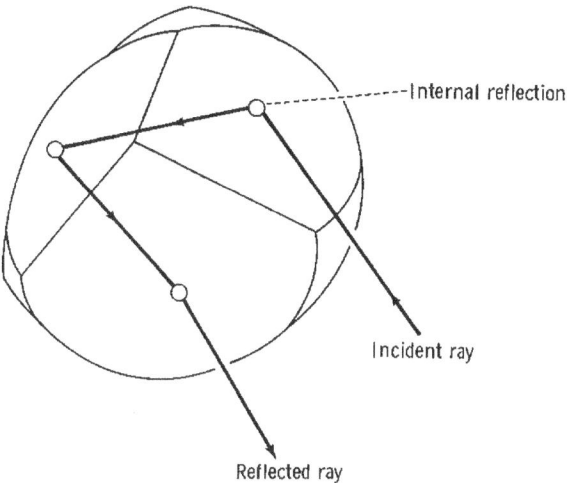

FIGURE 11–4.—Typical laser-ray path in retroreflector.

from the three back faces and then returned along a path parallel to the incident beam (fig. 11–4). This parallelism between the reflected and incident beams insures that the reflected laser pulse will return to the vicinity of origin on the Earth.

The Apollo 14 LRRR is almost identical to the Apollo 11 array placed on the Moon in July 1969 (ref. 11–9). The basic array design is a result of the need to meet and simultaneously satisfy many different and sometimes conflicting requirements. In an ideal environment, the choice would be relatively simple because, for a given geometry and allowable weight (payload), the return signal is maximized by making a single diffraction-limited retroreflector as large as weight restrictions and fabrication techniques will permit. Two aspects of the practical problem, however, vitiate this conclusion: (1) a displacement occurs in the returned laser beam because of the relative velocity between the Moon and the laser transmitter (velocity aberration) and (2) a wide lunar-temperature variation occurs from full Moon to new Moon, as well as the fact that the retroreflector is exposed essentially half the time to an energy input from direct sunlight. The velocity aberration that displaces the center of the returned diffraction pattern between 1.5 and 2 km limits the diameter of the diffraction-limited retroreflector that can be used to approximately 12 cm unless two telescopes spatially separated from one another are used, one for transmitting and the other for receiving. In

the situation in which laser light is transmitted and received at the same location, analysis shows that the loss in efficiency that results from using a large number of smaller diameter corners is almost exactly compensated as a result of the increased diffraction spreading of each corner. This has the effect of placing the transmitter-receiver site higher up the side of the returned diffraction pattern. These two effects result in essentially the same optical efficiency for a given payload weight for corners ranging from approximately 3.8 to 12 cm in diameter. With the use of a corner smaller than 3.8 cm, an overall loss in efficiency is experienced because further diffraction spreading is ineffective; at that size, the single transmitting and receiving site is already, for all practical purposes, at the center or peak of the returned diffraction pattern. The observations that dictated the choice of 3.8 cm as the diameter of the corners were as follows: (1) using this smallest still-efficient size made it possible to minimize the thermal gradients that would distort the individual cube-corner diffraction patterns and (2) with this size corner, one could expect to achieve essentially diffraction-limited performance throughout the lunar day as well as during lunar night. The temperature gradients in the individual corner cubes are further minimized by recessing each reflector by half its diameter in a circular socket (fig. 11–5). Furthermore, each reflector is tab mounted between two Teflon rings to afford all possible thermal insulation. The mechanical mounting structure serves

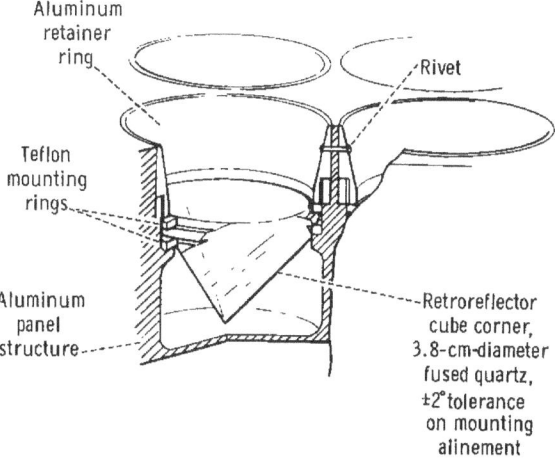

FIGURE 11–5.—Retroreflector mounting.

also to provide passive thermal control by means of surface properties. A transparent polyester cover assembly protected the array from dust during storage, transportation, handling, and flight. The Apollo 14 crew removed this cover at the time of deployment.

The Apollo 14 LRRR differs from the earlier Apollo 11 design in only two main aspects:

(1) The array cavity design was changed to increase the half-angle taper from 1.5° to 6° to decrease the obscuration and thereby increase the array optical efficiency approximately 20 to 30 percent for off-axis Earth positions.

(2) The supporting pallet is lighter and somewhat simpler in design.

Successful range measurements to the Apollo 14 array were first made from the McDonald Observatory of the University of Texas on February 5, 1971, the day on which the LRRR was deployed by the crew. Ranging subsequent to LM liftoff indicated that no serious degradation of the retroreflectors has occurred as a result of the ascent-stage engine burn. Signal strengths compare favorably with the levels obtained over the past year from the Apollo 11 array.

Ground-Station Operation

At present, range measurement to the two retroreflector packages at nearly all phases of the Moon are being conducted at the McDonald Observatory with NASA support. Return signals from the Apollo 11 array have also been obtained by the Pic du Midi Observatory in France and the Air Force Cambridge Research Laboratories Lunar Laser Observatory near Tucson, Ariz. Recent reports suggest that the ranging group in Japan has also had some initial success. It is hoped that several other lunar ranging stations will be in operation within the next year or two, including stations in Hawaii, Russia, and the Southern Hemisphere.

A line drawing of the laser ranging station at the McDonald Observatory is shown in figure 11–6. A schematic drawing of the telescope matching optics and guider is shown in figure 11–7. The present observation program at the McDonald Observatory consists of three observing periods on most nights when the weather permits,

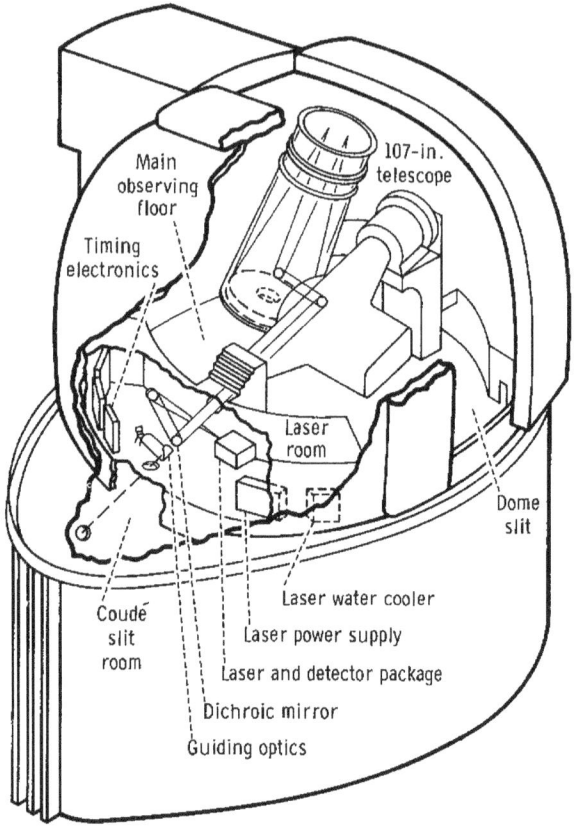

FIGURE 11–6.—Cutaway drawing of McDonald Observatory 107-in. telescope.

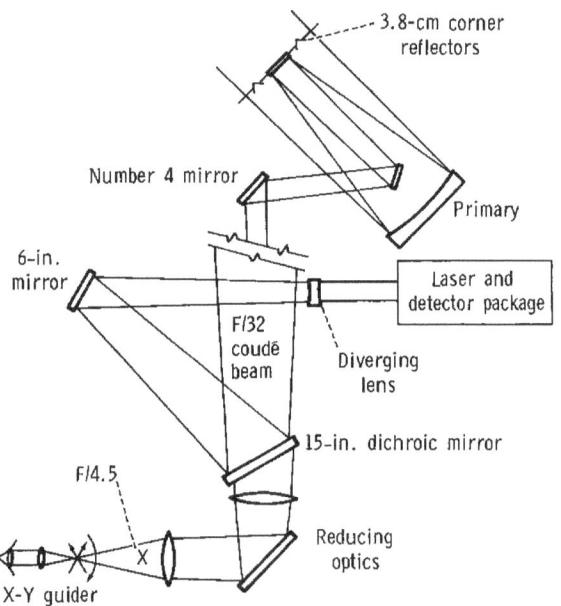

FIGURE 11–7.—Telescope matching optics and guider.

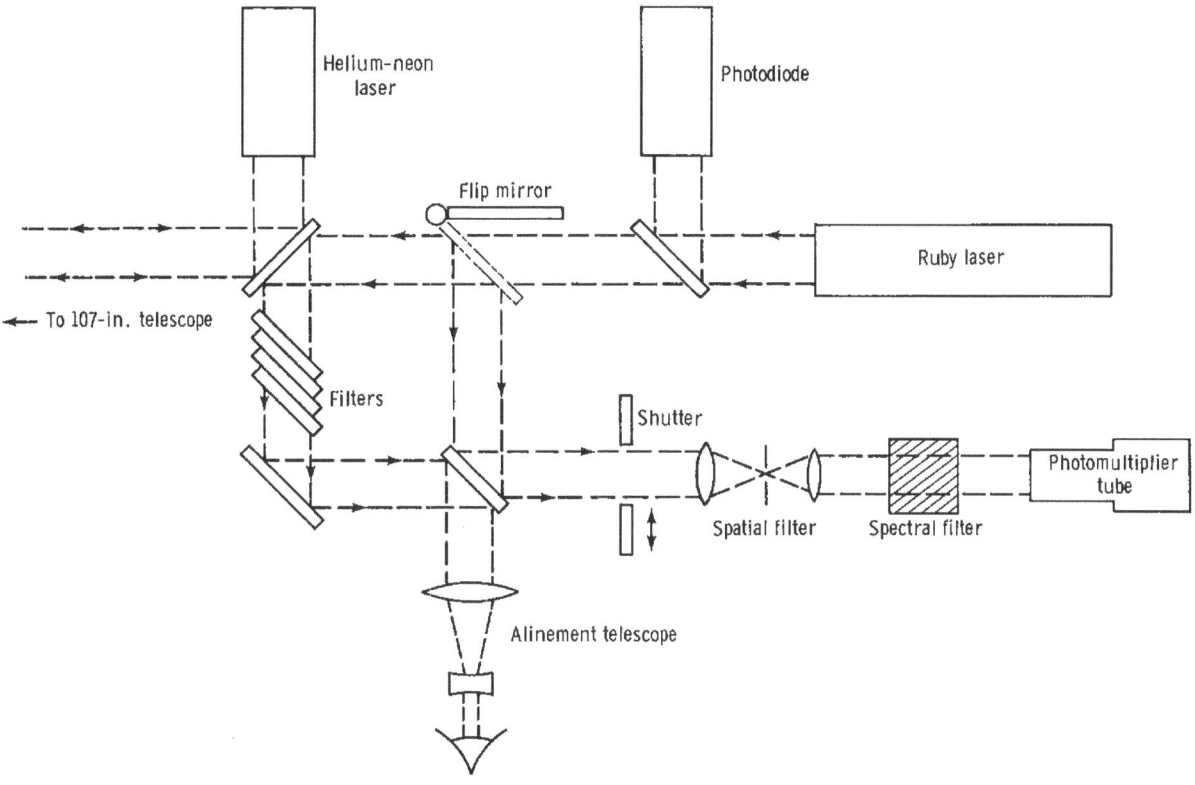

FIGURE 11–8.—Lunar ranging detector package.

except for a period of 5 days around the new Moon. One observing period is near the time of meridian transit for the Moon, and the others are 3 or 4 hr earlier and later. Several runs of approximately 50 shots each are normally fired during each observing period.

During the past year, considerable effort has been successfully exerted in the areas of equipment and calibration procedures.

The ruby laser system being used at present gives 3–J pulses with a repetition rate of one pulse every 3 sec. The total pulse length between the 10-percent intensity points is 4 nsec. The root-mean-square variation in the observed transit time, because of the laser pulse length and the jitter in the photomultiplier receiving the returned signal, is 2 nsec. The present overall accuracy of the measured transit time is ±1 nsec (equivalent to an error of approximately ±30 cm in distance measurements). Improvement to less than 1 nsec is expected with further refinements in the calibration procedures. Recent progress in lasers per-

mits the use of pulses with a 0.2-nsec length. This narrow pulse capability should prove extremely valuable in achieving the scientific goals of the experiment. The present detector-package arrangement is shown diagrammatically in figure 11–8.

With data from two or more well-located observing stations, the lunar range can be corrected accurately for the effects of polar motion and fluctuations in the Earth rotational rate. The uncertainty in range, as a result of the atmosphere in general, will be less than 1 cm for zenith angles of up to 70°.

Summary

The placing of the Apollo 11 retroreflector array on the Moon in July 1969, together with successful deployment of the Apollo 14 array, has resulted in a dramatic change in man's ability to measure the Earth-Moon distance. As in many astronomical and geophysical measurements, full scientific results will be obtained only after many years of monitoring variations in the lunar distance. Experience to date gives every indication that both arrays will continue to function toward achieving the scientific ends of the experiment by providing primary benchmarks on the lunar surface for years to come.

References

11-1. ALLEY, C. O.; BENDER, P. L.; DICKE, R. H.; FALLER, J. E.; ET AL.: Optical Radar Using a Corner Reflector on the Moon. J. Geophys. Res., vol. 70, no. 9, May 1, 1965, pp. 2267–2269.

11-2. ALLEY, C. O.; AND BENDER, P. L.: Information Obtainable From Laser Range Measurements to a Laser Corner Reflector. Continental Drift, Secular Motion of the Pole, and Rotation of the Earth. Symp. 32 IAU, William Markowitz and B. Guinot, eds., D. Reidel Pub. Co. (Dordrecht, Holland), 1968, pp. 86–90.

11-3. ALLEY, C. O.; BENDER, P. L.; CURRIE, D. G.; DICKE, R. H.; AND FALLER, J. E.: Some Implications for Physics and Geophysics of Laser Range Measurements From Earth to a Lunar Retroreflector. The Application of Modern Physics to the Earth and Planetary Interiors. Proc. NATO Adv. Study Inst., S. K. Runcorn, ed., Wiley-Interscience (London and New York), 1969, pp. 523–530.

11-4. FALLER, JAMES; WINER, IRVIN; CARRION, WALTER; JOHNSON, THOMAS S.; ET AL.: Laser Beam Directed at the Lunar Retroreflector Array: Observations of the First Returns. Science, vol. 166, no. 3901, Oct. 3, 1969, pp. 99–102.

11-5. ALLEY, C. O.; CHANG, R. F.; CURRIE, D. G.; MULLENDORE, J.; ET AL.: Apollo 11 Laser Ranging Retroreflector: Initial Measurements From the McDonald Observatory. Science, vol. 167, no. 3917, Jan. 23, 1970, pp. 368–370.

11-6. ALLEY, C. O.; CHANG, R. F.; CURRIE, D. G.; POULTNEY, S. K.; ET AL.: Laser Ranging Retroreflector: Continuing Measurements and Expected Results. Science, vol. 167, no. 3918, Jan. 30, 1970, pp. 458–460.

11-7. FALLER, JAMES E.; AND WAMPLER, E. JOSEPH: The Lunar Laser Reflector. Sci. Amer., vol. 222, no. 3, Mar. 1970, pp. 38–50.

11-8. ECKHARDT, DONALD H.; AND DIETER, KENNETH: A Nonlinear Analysis of the Moon's Physical Libration in Longitude. The Moon, vol. 2, no. 3, Feb. 1971, pp. 309–319.

11-9. ALLEY, C. O.; BENDER, P. L.; CHANG, R. F.; CURRIE, D. G.; ET AL.: Laser Ranging Retroreflector. Sec. 7 of Apollo 11 Preliminary Science Report. NASA SP–214, 1969.

12. The Solar-Wind Composition Experiment

J. Geiss,[a][†] *F. Buehler,*[a] *H. Cerutti,*[a]
P. Eberhardt,[a] and *J. Meister* [a]

Measurements of the relative ion abundances in the solar wind give information on the dynamics of the solar corona and constitute an important method for investigating elemental and isotopic abundances in the outer convective zone of the Sun. Furthermore, solar-wind-abundance data are essential for a detailed interpretation of the trapped gases in meteorites and lunar material. Based on these investigations, the evolution of the lunar surface and a possible transient lunar atmosphere can be studied. Noble-gas studies of the solar wind may also help in tracing the evolution of the terrestrial atmosphere.

It has been known for several years that the helium/hydrogen (He/H) ratio in the solar wind is highly variable and ranges from less than 0.01 to 0.25, with an average of approximately 0.04 (refs. 12–1 to 12–5). During periods of low solar-wind-ion temperature, the elements oxygen, silicon, and iron have been measured by means of the high-resolution electrostatic analyzers on board the Vela satellites and, in some cases, even ^{3}He has been detected (refs. 12–6 and 12–7). In the Apollo program, a different technique is used for studying elemental and isotopic abundances in the solar wind.

During the Apollo 11, 12, and 14 missions, aluminum foils were deployed at the lunar surface and used as targets for collecting solar-wind ions. The foils were returned to Earth, and the implanted solar-wind particles are being analyzed in the laboratory. The Apollo 11 and 12 solar-wind composition (SWC) experiments have so far yielded absolute solar-wind fluxes of ^{4}He, ^{3}He, neon-20 (^{20}Ne), and ^{22}Ne (refs. 12–8 and 12–9). In the case of the Apollo 12 experiment, an approximate figure for the abundance of ^{21}Ne was also obtained. In this report, preliminary results of the first analyses on sections of the Apollo 14 foil are presented.

Proton detectors and magnetometers have been used to establish that the Moon behaves like a passive obstacle to the solar wind, and no evidence of a lunar bow shock has been found (refs. 12–10 to 12–13).[1] Thus, during the normal lunar day, the solar-wind particles strike the lunar surface with essentially unchanged direction and energy, except perhaps in a few places where local magnetic fields are unusually high (ref. 12–14). It was, in fact, shown from the Apollo 11 and 12 SWC experiment data that He reaches the lunar surface in an undisturbed, highly directional flow (refs. 12–8 and 12–9). The Apollo 12 solar-wind spectrometer has recorded the plasma flow that arrives at the Apollo 12 lunar-surface experiments package site as a function of phase of the lunar day. The result is that, most of the time, the plasma flux is not affected by the proximity of the surface of the Moon.[1] Thus, the expectation is that experiments deployed on the lunar surface will yield solar-wind-abundance data that are valid for the undisturbed solar wind.

[a] Physikalisches Institut, University of Bern.
[†] Principal investigator.

[1] D. R. Clay, M. Neugebauer, and C. W. Snyder: Solar Wind Observations on the Lunar Surface With the Apollo 12 ALSEP. Proc. Apollo 12 Lunar Sci. Conf. (Houston), Jan. 11–14, 1971. To be published in Geochim. Cosmochim. Acta.

Principle of the Experiment

An aluminum foil 30 cm wide and approximately 140 cm long, with an area of approximately 4000 cm², was exposed to the solar wind at the lunar surface by the Apollo 14 crew on February 5, 1971, at 15:16 G.m.t. The foil was positioned perpendicular to the solar rays in the azimuthal direction (fig. 12–1), exposed for 21 hr, and returned to Earth. Laboratory experiments have determined that solar-wind ions arriving with an energy of approximately 1 keV/nucleon penetrate approximately 10^{-5} cm into the foil (ref. 12–15), and a large and well-known fraction is firmly trapped (refs. 12–16 and 12–17). In the laboratory, the returned foil is analyzed for trapped solar-wind noble-gas atoms. Parts of the foil are melted in ultra-high-vacuum systems, and the noble-gas atoms of solar-wind origin thus released are analyzed with mass spectrometers for elemental abundance and isotopic composition. Further details of the principle and the procedures of this experiment have been discussed elsewhere (refs. 12–16, 12–18, and 12–19).

In addition to the solar-wind investigations, a search will be made for radon (Rn) emanating from the Moon by using a small portion of the Apollo 14 SWC foil. When Rn decays in the very thin lunar atmosphere, the recoil energy of the daughter nuclides is sufficient that any such atom striking the foil becomes firmly trapped. In the foil, the Rn decay product lead-210 will be searched for and, if found, its concentration will be determined to derive the ambient Rn concentration in the lunar atmosphere.

Instrumentation and Lunar-Surface Operation

The experiment hardware was similar to the hardware used on the Apollo 11 and 12 missions (ref. 12–19). The experiment consisted of a metallic telescopic pole approximately 4 cm in diameter and 38 cm in length when collapsed. In the stowed position, the foil was enclosed in the tubing and rolled up on a spring-driven reel. The instrument weighed 430 g. When extended on the lunar surface, the pole was approximately 1.5 m long and a 30- by 125-cm foil area was exposed. Only the foil assembly was recovered at the end of the lunar-exposure period; it was rolled on the spring-driven reel and returned to Earth. The instrument is shown deployed on the lunar surface at the Apollo 14 landing site in figure 12–1. For the Apollo 14 SWC instrument, the reel handle was color coded to give the exact angular position during exposure of the reel and the portion of foil rolled around it. Detailed analyses of this portion of the foil are intended to yield the angular distribution of the arriving solar-wind ions. After evaluation of a number of Apollo 14 photographs, it was concluded that the foil was standing vertically (within a few degrees) at the lunar surface.

After retrieval, the return unit was placed in a special Teflon bag and returned to Earth in the interim stowage assembly. In the Lunar Receiving Laboratory (LRL) quarantine area, the foil was removed from the Teflon bag and found to be in good condition. No dust was detected on the foil with the unaided eye. The foil and reel were subsequently placed in a container and stored in the LRL crew reception area during the quarantine period.

Preliminary Results

The Apollo 14 foil was made available for analysis after the lifting of the quarantine in early

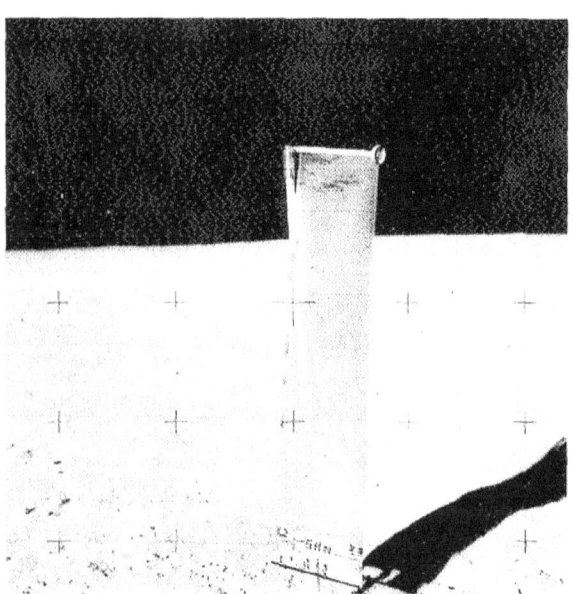

FIGURE 12–1.—Apollo 14 SWC experiment deployed on the lunar surface (AS14–64–9199).

TABLE 12–I. *First Preliminary Results From the Analyses of the Foil From the Apollo 14 SWC Experiment*

Sample no.	Area, cm^2	^{4}He concentration, $\times 10^{10}$ atoms/cm^2	^{4}He/^{3}He	^{4}He/^{20}Ne	^{20}Ne/^{22}Ne	^{20}Ne/^{36}Ar
4–2................	9.6	27.6	2360	460	13.2	34
4–3 [a]...............	10.0	26.8	2240	520	14.0	33
4–4 [a]...............	10.0	26.5	2280	500	13.5	38
4–5................	20.7	25.9	2240	480	13.6	37
5–1................	10.0	24.5	2300	480	13.9	40

[a] Oxide layer removed on back side of foil.

April. For the initial analyses, five small pieces from the upper part of the foil were decontaminated by means of the ultrasonic treatments that had proven effective during the Apollo 11 and 12 SWC experiment analyses. The results of these first measurements are presented in table 12–I. The data given should be considered as preliminary, because recalibration of the gas standards and reevaluation of the mass-spectrometer charts could lead to somewhat different final results. Along with these measurements on the flight foil, numerous pieces that had been cut from the Apollo 14 foil before flight for the purpose of noble-gas blank measurements were analyzed. The blanks that had been determined in this way were subtracted from the noble-gas concentrations measured in the pieces of the flight foil, and the solar-wind-particle concentrations presented in table 12–I were obtained. The foil blanks for He and Ne were 0.1 and 7 percent, respectively, relative to the solar-wind-particle content.

The oxide layer on the back side was removed from two of the investigated pieces of the Apollo 14 foil before analysis. The expectation was that this procedure would reduce a possible residual-dust contamination. Examinations of contaminated portions of the Apollo 12 foil have shown that, after ultrasonic treatment, 50 to 80 percent of the residual dust was located on the back side of the foil.

For the He and Ne isotopes, the results from the five foil pieces are in good agreement. This agreement is further evidence that the He and Ne data given in table 12–I are not appreciably affected by a residual-dust contamination. Further

analyses will be conducted to substantiate this conclusion. In particular, measurements on a shielded portion of the flight foil and on the section of the foil that had been exposed on the back side of the reel (facing away from the Sun) will be used to demonstrate the absence of significant lunar-dust contamination.

Neon-21 has been detected in the five foil pieces listed in table 12–I. Within the limits of error, the ^{21}Ne data agree with the results obtained from the Apollo 12 SWC experiment (ref. 12–9). By using larger portions of the foil, it is expected that the ^{21}Ne abundance can be determined from the Apollo 14 SWC experiment with good accuracy. To obtain the ^{4}He/^{20}Ne ratio in the solar wind, the data given in table 12–I must be corrected for the difference in the He- and Ne-trapping efficiencies. A preliminary ^{4}He/^{20}Ne value of 550 has been obtained.

The argon (Ar) content of the solar wind has been determined for the first time. For ^{36}Ar and ^{38}Ar, the foil blank is actually higher than the solar-wind-particle content. However, in this case, the blank can be uniquely determined for each analyzed foil piece from the ^{40}Ar content. Virtually all the ^{40}Ar that was detected in the foil is of atmospheric origin, because the ^{40}Ar/^{36}Ar ratio of 295 in terrestrial Ar is much larger than the solar-wind ^{40}Ar/^{36}Ar ratio, which is estimated to be smaller than unity (refs. 12–20 and 12–21). The ^{40}Ar/^{36}Ar ratios that were actually measured in the foil pieces range from 240 to 260 (i.e., between 10 and 20 percent of the ^{36}Ar was of solar-wind origin). The solar-wind ^{38}Ar concentrations obtained after blank correction are used to calculate the $^{20}Ne/^{36}Ar$ ratios given in table 12–I. Argon-38 has also been detected in the five foil pieces that have been analyzed. Values between 4.2 and 5.9 for the ^{36}Ar–^{38}Ar ratio were obtained. This ratio will be determined more accurately by analysis of the Ar contained in larger pieces of the foil.

The $^{20}Ne/^{36}Ar$ ratios in the five foil pieces agree within ±10 percent. This variation corresponds approximately to the analytical errors. It should be noted in particular that the Ar concentration is not reduced by the removal of the oxide layer on the back side of the foil. On the basis of this observation, it is estimated that a possible residual-dust contamination has not greatly affected the five $^{20}Ne/^{36}Ar$ ratios given in table 12–I. By taking the average of the measured $^{20}Ne/^{36}Ar$ ratios for the five foil pieces, an estimated value of 37^{+10}_{-5} is obtained for the $^{20}Ne/^{36}Ar$ ratio in the solar wind during the Apollo 14 foil exposure. This value is much higher than the $^{20}Ne/^{36}Ar$ ratios (between 5 and 10) found in unseparated lunar dust. In ilmenite samples that have been separated from the Apollo 11 lunar-fines material, $^{20}Ne/^{36}Ar$ ratios between 25 and 33 have been found (ref. 12–21). The ilmenite values are fairly close to the values obtained from the Apollo 14 SWC experiment and indicate that the composition of the trapped solar-wind particles in ilmenite is much less affected by diffusion or other processes than is the composition in the bulk material.

The $^{20}Ne/^{36}Ar$ ratio of 37 is considerably higher than solar-corona values and cosmic-abundance estimates. Measurements on forbidden lines and ultraviolet analysis (ref. 12–22) indicate that a $^{20}Ne/^{36}Ar$ ratio of 3 would be expected for the solar corona. Cameron (ref. 12–23) estimates a $^{20}Ne/^{36}Ar$ abundance ratio of 11 for the solar system, and the earlier Suess-Urey abundance compilation predicted a higher value of 67 for the $^{20}Ne/^{36}Ar$ ratio (ref. 12–24). The results from the Apollo 14 SWC experiment seem to indicate that an intermediate value for the $^{20}Ne/^{36}Ar$ ratio might be appropriate. However, in the solar-wind-acceleration process, fractionation between ^{20}Ne and ^{36}Ar might occur. Both ^{20}Ne and the much heavier ^{36}Ar are most likely eightfold charged in the solar wind. Separation processes caused by quasi-static electromagnetic fields or dynamical friction could thus deplete ^{36}Ar in the solar wind.[a]

The average ^4He flux during the Apollo 14 exposure period can be calculated by using the data given in table 12–I. The trapping probabilities of the foil for noble-gas ions depend only slightly on energy in the general solar-wind-velocity region. For He with a velocity of approximately 300 km/sec, the trapping probability is 89±2 percent for normal incidence and approximately 9 percent less for an incidence angle of 65°.

The angular distribution and the average angle of incidence on the Apollo 14 foil have not yet been determined. Thus, for the purpose of this report, the average angle of incidence is estimated. The average solar elevation during the foil exposure was 19°. By taking into account the effects of aberration and corotation, an angle of incidence on the foil of 68° is obtained for the undisturbed solar wind. At the present time, it is not known whether the Moon had already passed through the shockfront of the Earth during the Apollo 14 foil exposure. If such were the case, the angle of incidence would be further lowered by 5° to 7° (ref. 12–25). The assumption is therefore made for this report that the average angle of incidence of the solar-wind particles on the foil was approximately 65°. With this assumption, the ^4He flux during the Apollo 14 SWC foil exposure can be calculated and is given in table 12–II, together with the ^4He fluxes previously determined for the Apollo 11 and 12 exposure periods (refs. 12–8 and 12–9).

The flux obtained for the Apollo 14 exposure period is definitely lower than the flux determined during the Apollo 12 mission. However, the He/Ne and $^4He/^3He$ ratios are closer to the results of the Apollo 12 SWC experiment than to those of the Apollo 11 mission. It will be interesting to compare the Apollo 14 SWC experiment data with proton fluxes measured simultaneously by the Apollo 12 solar-wind spectrometer or by instrumentation on unmanned spacecraft to determine whether the flux for all ion species was generally low during the Apollo 14 exposure time, or whether the heavier ions were all depleted relative to H by a similar factor.

[a] J. Geiss: On Elemental and Isotopic Composition of the Solar Wind. Proc. Asilomar Conf. Solar Wind, Mar. 1971.

TABLE 12–II. *Comparison Between the Preliminary Average ^{4}He Flux Obtained From the Apollo 14 SWC Experiment With Solar-Wind Fluxes Obtained From the Apollo 11 and 12 SWC Experiments*

Mission	Exposure date	Time of exposure initiation, G.m.t., hr:min	Exposure duration, hr:min	Average solar-wind ^{4}He flux, $\times 10^6$ cm^{-2} sec^{-1}
Apollo 11	July 21, 1969	03:35	01:17	6.2±1.2
Apollo 12	Nov. 19, 1969	12:35	18:42	8.1±1.0
Apollo 14	Feb. 5, 1971	15:16	21:00	4.2±0.8

The ^{4}He/^{3}He ratio that was obtained from the Apollo 14 SWC experiment is again significantly lower than the ratios found in ilmenite separated from lunar fine material (ref. 12–21) and in the returned Surveyor 3 material (ref. 12–26). It appears that the ^{4}He/^{3}He ratio varies, even when averaged over times of one or several years, and even a secular change in the ^{4}He/^{3}He ratio in the outer convective zone of the Sun cannot be excluded.

References

12-1. SNYDER, CONWAY W.; AND NEUGEBAUER, MARCIA: Interplanetary Solar-Wind Measurements by Mariner II. Space Res., vol. 4, 1964, pp. 89–113.

12-2. WOLFE, J. H.; SILVA, R. W.; MCKIBBIN, D. D.; AND MASON, R. H.: The Compositional, Anisotropic, and Nonradial Flow Characteristics of the Solar Wind. J. Geophys. Res., vol. 71, no. 13, July 1966, pp. 3329–3335.

12-3. HUNDHAUSEN, A. J.; ASBRIDGE, J. R.; BAME, S. J.; GILBERT, H. E.; AND STRONG, I. B.: Vela 3 Satellite Observations of Solar Wind Ions: A Preliminary Report. J. Geophys. Res., vol. 72, no. 1, Jan. 1967, pp. 87–100.

12-4. OGILVIE, K. W.; BURLAGA, L. F.; AND WILKERSON, T. D.: Plasma Observations on Explorer 34. J. Geophys. Res., vol. 73, no. 21, Nov. 1968, pp. 6809–6824.

12-5. ROBBINS, D. E.; HUNDHAUSEN, A. J.; AND BAME, S. J.: Helium in the Solar Wind. J. Geophys. Res., vol. 75, no. 7, Mar. 1970, pp. 1178–1187.

12-6. BAME, S. J.; HUNDHAUSEN, A. J.; ASBRIDGE, J. R.; AND STRONG, I. B.: Solar Wind Ion Composition. Phys. Rev. Lett., vol. 20, no. 8, Feb. 19, 1968, pp. 393–395.

12-7. BAME, S. J.; ASBRIDGE, J. R.; HUNDHAUSEN, A. J.; AND MONTGOMERY, MICHAEL D.: Solar Wind Ions: ^{56}Fe^{+8} to ^{56}Fe^{+12}, ^{28}Si^{+7}, ^{28}Si^{+8}, ^{28}Si^{+9}, and ^{16}O^{+6}. J. Geophys. Res., vol. 75, no. 31, Nov. 1970, pp. 6360–6365.

12-8. BUHLER, F.; EBERHARDT, P.; GEISS, J.; MEISTER, J.; AND SIGNER, P.: Apollo 11 Solar Wind Composition Experiment: First Results. Science, vol. 166, no. 3912, Dec. 19, 1969, pp. 1502–1503.

12-9. GEISS, J.; EBERHARDT, P.; BUHLER, F.; MEISTER, J.; AND SIGNER, P.: Apollo 11 and 12 Solar Wind Composition Experiments: Fluxes of He and Ne Isotopes. J. Geophys. Res., vol. 75, no. 31, Nov. 1970, pp. 5972–5979.

12-10. LYON, E. F.; BRIDGE, H. S.; AND BINSACK, J. H.: Explorer 35 Plasma Measurements in the Vicinity of the Moon. J. Geophys. Res., vol. 72, no. 29, Dec. 1967, pp. 6113–6117.

12-11. NESS, N. F.; BEHANNON, K. W.; SCEARCE, C. S.; AND CANTARANO, S. C.: Early Results From the Magnetic Field Experiment on Lunar Explorer 35. J. Geophys. Res., vol. 72, no. 23, Dec. 1967, pp. 5769–5778.

12-12. SISCOE, G. L.; LYON, E. F.; BINSACK, J. H.; AND BRIDGE, H. S.: Experimental Evidence for a Detached Lunar Compression Wave. J. Geophys. Res., vol. 74, no. 1, Jan. 1969, pp. 59–69.

12-13. FREEMAN, J. W., JR.; HILLS, H. K.; AND BALSIGER, H.: Preliminary Results From the Apollo 12 ALSEP Lunar Ionosphere Detector I. General Results. Trans. Amer. Geophys. Union, vol. 51, no. 4, Apr. 1970, p. 407.

12-14. MIHALOV, J. D.; SONETT, C. P.; BINSACK, J. H.; AND MOUTSOULAS, M. D.: Possible Fossil Lunar Magnetism Inferred From Satellite Data. Science, vol. 171, no. 3974, Mar. 5, 1971, pp. 892–895.

12-15. DAVIES, J. A.; BROWN, F.; AND MCCARGO, M.: Range of Xe^{133} and Ar^{41} Ions of Kiloelectron Volt Energies in Aluminum. Can. J. Phys., vol. 41, no. 6, June 1963, pp. 829–843.

12-16. BUHLER, F.; GEISS, J.; MEISTER, J.; EBERHARDT, P.; ET AL.: Trapping of the Solar Wind in Solids. Earth Planet. Sci. Lett., vol. 1, 1966, pp. 249–255.

12-17. MEISTER, J.: Ein Experiment zur Bestimmung der Zusammensetzung und der Isotopenverhältnisse des Sonnenwindes: Einfangverhalten von Aluminium für niederenergetische Edelgasionen. Ph.D. thesis, Univ. of Bern, 1969.

12–18. SIGNER, PETER; EBERHARDT, PETER; AND GEISS, JOHANNES: Possible Determination of the Solar Wind Composition. J. Geophys. Res., vol. 70, no. 9, May 1965, pp. 2243–2244.

12–19. GEISS, J.; EBERHARDT, P.; SEGNER, P.; BUHLER, F.; AND MEISTER, J.: The Solar-Wind Composition Experiment. Sec. 8 of Apollo 11 Preliminary Science Report. NASA SP–214, 1969.

12–20. HEYMAN, D.; YANIV, A.; ADAMS, J. A. S.; AND FRYER, G. E.: Inert Gases in Lunar Samples. Science, vol. 167, no. 3918, Jan. 30, 1970, pp. 555–558.

12–21. EBERHARDT, P.; GEISS, J.; GRAF, H.; GROGLER, N.; ET AL.: Trapped Solar Wind Noble Gases, Exposure Age and K/Ar Age in Apollo 11 Lunar Fine Material. Proc. Apollo 11 Lunar Sci. Conf., Geochim Cosmochim. Acta Supp. 1, vol. 2, A. A. Levinson, ed., Pergamon Press, Inc., 1970, pp. 1037–1070.

12–22. POTTASCH, S. R.: On the Abundances in the Solar Corona. Origin and Distribution of the Elements, L. H. Ahrens, ed., Pergamon Press, Inc., 1968.

12–23. CAMERON, A. G. W.: A New Table of Abundances of the Elements in the Solar System. Origin and Distribution of the Elements, L. H. Ahrens, ed., Pergamon Press, Inc., 1968, pp. 125–143.

12–24. SUESS, HANS E.; AND UREY, HAROLD C.: Abundances of the Elements. Rev. Mod. Phys., vol. 28, no. 1, Jan. 1956, pp. 53–74.

12–25. GEISS, JOHANNES; HIRT, PETER; AND LEUTWYLER, HEINRICH: On Acceleration and Motion of Ions in Corona and Solar Wind. Solar Phys., vol. 12, 1970, pp. 458–483.

12–26. BUHLER, F.; EBERHARDT, P.; GEISS, J.; AND SCHWARZMULLER, J.: Trapped Solar Wind Helium and Neon in Surveyor 3 Material. Earth Planet. Sci. Lett., vol. 10, no. 3, Feb. 1971, pp. 297–306.

ACKNOWLEDGMENTS

Hardware construction and foil analyses were supported by the University of Bern and the Swiss National Science Foundation.

13. Lunar Portable Magnetometer Experiment

P. Dyal,[a]† C. W. Parkin,[a] [b] C. P. Sonett,[a] R. L. DuBois,[c] and G. Simmons[d]

The Apollo 14 lunar portable magnetometer (LPM) (fig. 13–1) was used to measure the steady magnetic field at different sites in the Fra Mauro region. The LPM recorded steady magnetic fields of 103 ± 5 gammas and 43 ± 6 gammas at two sites separated by 1.12 km. These measurements showed that the unexpectedly high 38-gamma steady field measured [1] at the Apollo 12 site 180 km away (ref. 13–1) was not unique. Indeed, these measurements and studies of lunar samples [2,3] (refs. 13–2 to 13–5) and lunar-orbiting Explorer 35 data (ref. 13–6) indicate that much of the lunar-surface material was magnetized at some prior time in lunar history. These data can be used to gain information concerning present magnetic and structural properties of the local region as well as the thermal and magnetic histories of the area.

Background and Theory of the Experiment

Magnetic fields are believed to permeate most of space, and forces associated with these fields are very important on a cosmological scale. In this solar system, strong magnetic fields associated with sunspot regions extend far enough to affect radio reception and auroral activity on the Earth. The dipolar field of the Earth acts as a shield against the hot solar plasma by deflecting it around the terrestrial sphere. Magnetic-field measurements on the surface of the Earth have a wide range of application (e.g., in navigation; paleomagnetic studies of the geological past, such as sea-floor spreading; and surveys of subsurface ore bodies).

The magnitudes of magnetic fields in the universe vary widely. Measurements range from ap-

[a] NASA Ames Research Center.

[b] National Research Council postdoctoral associate.

[c] Oklahoma University.

[d] NASA Manned Spacecraft Center and Massachusetts Institute of Technology.

† Principal investigator.

[1] P. Dyal and C. W. Parkin: The Apollo 12 Magnetometer Experiment: Internal Lunar Properties From Transient and Steady Magnetic Field Measurements. Proc. Apollo 12 Lunar Sci. Conf. (Houston), Jan. 11–14, 1971. To be published in Geochim. Cosmochim. Acta.

[2] G. W. Pearce, D. W. Strangway, and E. E. Larson: Magnetism of Two Apollo 12 Igneous Rocks. Proc. Apollo 12 Lunar Sci. Conf. (Houston), Jan. 11–14, 1971. To be published in Geochim. Cosmochim. Acta.

[3] C. E. Helsley: Evidence for an Ancient Lunar Magnetic Field. Proc. Apollo 12 Lunar Sci. Conf. (Houston), Jan. 11–14, 1971. To be published in Geochim. Cosmochim. Acta.

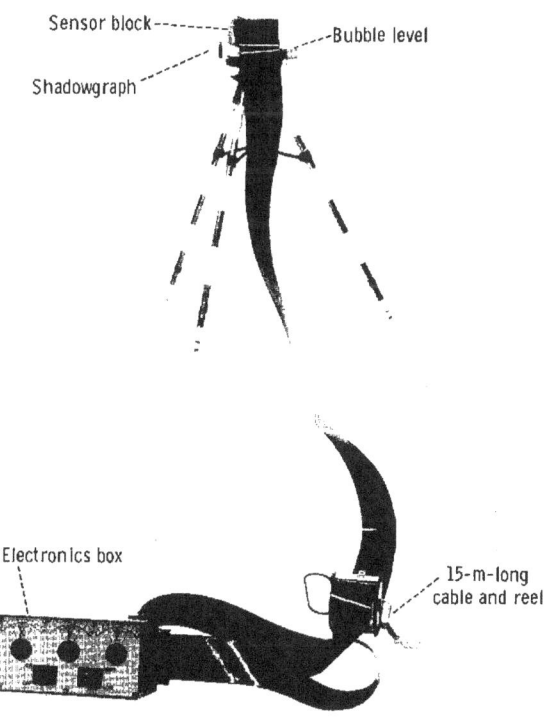

FIGURE 13–1.—The LPM deployed in the laboratory.

proximately 10^{12} gammas for a white dwarf star in the constellation Draconis (ref. 13–7) to approximately 10^3 gammas inside sunspots (ref. 13–8) to 5 gammas in interplanetary space near the Earth (ref. 13–9). The Earth surface field near the Equator is approximately 30 000 gammas, and the fields measured at three lunar-surface sites average about 60 gammas. The sources of these fields are associated with the intrinsic physical properties of materials or moving electrical charges.

Magnetic fields associated with the Moon have been found to be much more important to lunar studies than had been anticipated before the Apollo lunar-landing missions. The inductive response of the lunar interior to time-dependent solar-magnetic fields transported past the Moon permits the present electromagnetic and thermal properties of the lunar interior to be studied (refs. 13–10 and 13–11). Present and future Apollo missions will yield information about the past extrinsic and intrinsic magnetic fields and help clarify major events in lunar history. Past extrinsic fields will have been influential during lunar formation, and the study of the resultant remanence should yield information about ancient solar and terrestrial fields. Intrinsic fields should yield information concerning the history of lunar internal temperature, rotation, volcanism, and tectonic processes.

During the last decade, many investigators (refs. 13–12 to 13–16) conducted experiments to determine the magnetic field associated with the Moon. These experiments, conducted aboard lunar-orbiting and flyby spacecraft, yielded no direct measurement of an intrinsic magnetic field. One of the lunar-orbiting experiments, Explorer 35, aided directly in the LPM experiment. During the second period of astronaut extravehicular activity (EVA), the Explorer 35 recorded the time-independent ambient magnetic field caused by external sources (fig. 13–2). The Explorer 35 measurements, made simultaneously with the LPM measurements, were later vectorially subtracted from the LPM data.

Before the Apollo 14 mission, the measurements made by the Explorer 35 and other lunar-orbiting and flyby spacecraft were an aid in establishing a maximum limit of 4 gammas for an assumed intrinsic global magnetic field. It was, therefore,

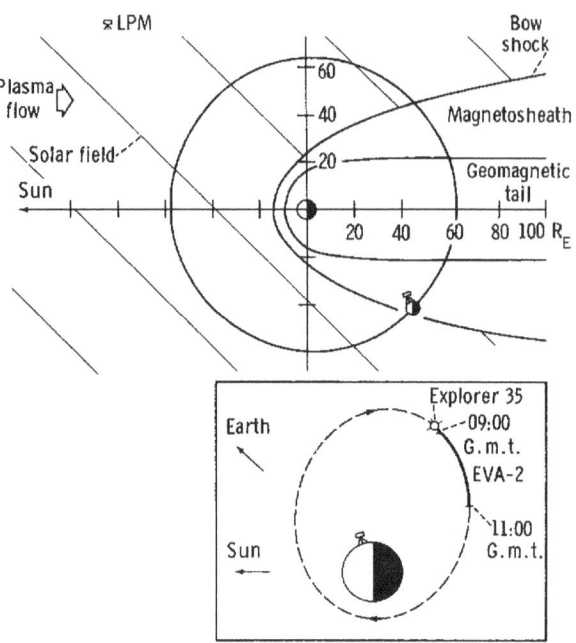

FIGURE 13–2.—The lunar orbit projected onto the ecliptic plane, showing the position of the Apollo 14 LPM on February 6, 1971, during the time of LPM magnetic-field measurements. The inset shows the orbit and location of the Explorer 35 satellite at the time of the surface measurements.

surprising to measure a 38-gamma steady field at the Apollo 12 landing site. This was the first direct surface measurement of a magnetic field intrinsic to an extraterrestrial planetary-sized body.

The magnetic field is probably due to magnetized material that acquired remanence during a lunar epoch involving an inducing field much larger ($\gtrsim 10^3$ gammas) than presently exists at the Moon. This conclusion is consistent with the high remanent magnetization found in lunar samples.

Discovery of the unexpectedly high steady field at the Apollo 12 site resulted in the concept of the Apollo 14 LPM experiment. This instrument was designed to make multiple measurements of the steady (time-independent) magnetic field during the astronauts' traverse at their landing site. The entire project from inception to data return required slightly more than 1 yr. The effort was rewarded by the measurement at Fra Mauro of a 103-gamma field—approximately 25 times greater than had been predicted only 15 months before.

FIGURE 13–3.—Apollo 14 LPM on the lunar surface. (a) On the MET. (b) Deployed at site A during measurement of the 103-gamma magnetic field.

Equipment Description

The LPM was designed as a totally self-contained, portable experiment package. During transit from the Earth to the Moon, the LPM was stowed and attached to the scientific equipment bay of the lunar module (LM); and, on the lunar surface, the LPM was carried on the modularized equipment transporter (MET) (fig. 13–3(a)). In the deployed configuration (fig. 13–3(b)), three orthogonal fluxgate sensors are mounted on top of a tripod. This sensor-equipped tripod is connected by a 15-m-long ribbon cable to the electronics box that contains a battery pack, electronics package, and three milliammeters. (A list of pertinent LPM characteristics is given in table 13–I.)

Fluxgate Sensor

The fluxgate sensor, shown schematically in figure 13–4, records the vector components of the magnetic field. Three fluxgate sensors (refs. 13–17 and 13–18) are orthogonally mounted in the sensor block shown in figure 13–1. Each sensor weighs 18 g and uses 15 mW of power during operation. A sensor consists of a flattened toroidal core of Permalloy that is driven to saturation by a square wave at a frequency of $f_0 = 7250$ Hz. This constant-voltage square wave drives the core to saturation during alternate half-cycles and modulates the permeability at twice the drive frequency. The voltage induced in the sense windings

TABLE 13–I. *Apollo 14 LPM Characteristics*

Parameter	Value
Ranges, gamma............	0 to ±100
	0 to ±50
Resolutions, gamma........	±1.0, ±0.5
Frequency response, Hz.....	dc to 0.01
Battery:	
Power, W.............	1.5
Life, hr..............	60
Mass, kg.................	4.58
Stowage size, cm..........	55.9×15.3×14.3
Operating temperature, °C...	−30 to +60°
Angular response..........	Proportional to cosine of angle between sensor axis and magnetic field

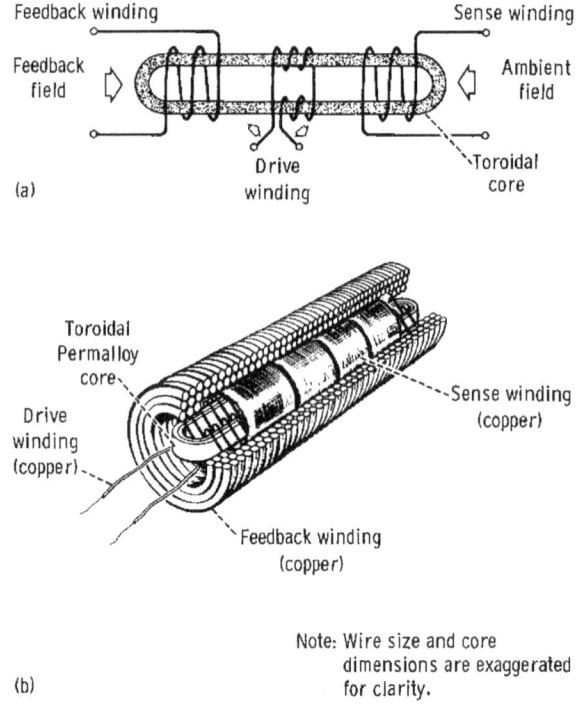

FIGURE 13-4.—Fluxgate sensor. (a) Functional schematic. (b) Racetrack tape 0.0005 in. thick and 0.0625 in. wide.

is equal to the time rate of change of the net flux contained in the area enclosed by the sense winding. This net flux is the superposition of the flux from the drive winding and the ambient magnetic field. The signal generated in the sense winding at the second harmonic of the drive signal will be amplitude modulated at a magnitude proportional to the ambient magnetic field. The phase of this second harmonic signal, with respect to the drive waveform, indicates the polarity of the magnetic field. The sensor electronics amplify and filter the $2f_0$ sense-winding signal and synchronously demodulate it to derive a voltage proportional to the ambient magnetic field. After demodulation, the resulting signal is amplified and used to drive the feedback winding to null out the ambient field within the sensor. Operating at null increases thermal stability by making the circuit independent of core-permeability variations with temperature.

The sensor block, mounted on the top of a tripod, is positioned 75 cm above the lunar surface. The tripod assembly consists of a latching device to hold the sensor block, a bubble level with 1° annular rings, and a shadowgraph with 3° markings used to aline the device along the Moon-Sun line.

Electronics Subsystem

The magnetometer electronics are self-contained with a set of mercury cells for power and three milliammeters for visual readout of the magnetic-field components. A block diagram of the instrument is shown in figure 13-5. The sensors are driven into saturation by a 7.25-kHz square wave generated by a frequency-divided Colpitts oscillator operating at 29 kHz. The oscillator also provides a 14.5-kHz pulse for demodulation of the second harmonic signal from the sense windings. The amplifier-demodulator is a narrowband amplifier with 80-dB gain at 14.5 kHz. The amplifier output is synchronously demodulated, producing a direct-current (dc) output voltage proportional to the amplitude of the ambient magnetic field. This demodulated output is used to drive the feedback winding of the sensor to operate the sensor at null conditions. The demodulated output from each channel also drives a panel meter. Each

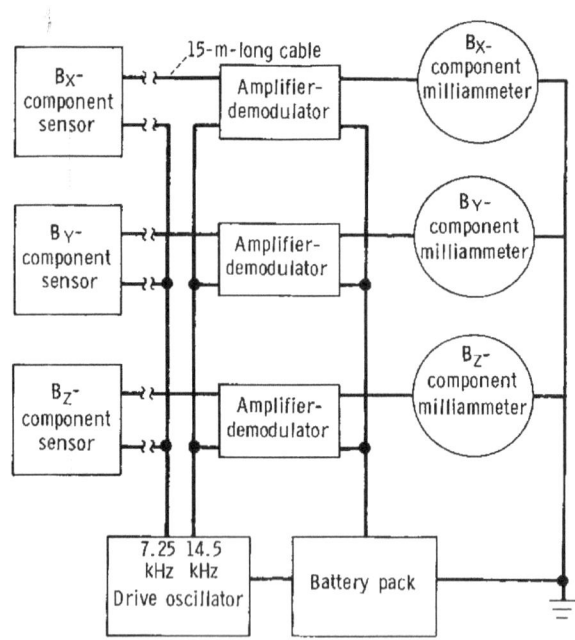

FIGURE 13-5.—Functional block diagram of the LPM electronics.

meter has a 1-mA full-scale movement and associated scaling resistors and capacitors that determine the frequency response of the instrument. This frequency response is chosen so that the instrument can follow changes in steady-field values between measurements at different sites but still filter out high-frequency fluctuations in the ambient solar-wind field. The instrument has an overall low-pass filter frequency response with a 3-dB point at 0.05 Hz. The meters also have a range switch connected to the scaling resistors to permit full-scale deflection for either ±50 gammas or ±100 gammas with reading-resolution capabilities of ±1 and ±2 gammas, respectively.

Exterior surfaces of the instrument were designed so that the temperatures of all components would be maintained between 0° and 50° C for Sun-elevation angles between 7° and 30°. The desired effective optical absorptance and emittance of the instrument were achieved by coating the surface areas with appropriate ratios of electrodeposited gold and white thermal paint. The temperature of the electronics box is measured by visually observing the darkening of temperature-sensitive decals that monitor temperatures in increments between 100° and 290° F.

Instrument Operation

Crew operations are crucial to the execution of this experiment. Both crewmembers remove the LPM mounting pallet from the LM and install the magnetometer on the MET. The MET is then transported a minimum of 100 m from the LM to eliminate the LM as an artificial field source. A measurement sequence is conducted as follows. Leaving the electronics box on the MET, the astronaut deploys the sensor-tripod assembly and levels and alines the sensor block a minimum of 11 m from the MET. He returns to the MET, sets the range switch at either 50 or 100 gammas, then reads the meters in sequence and verbally relays the data to Earth. After all readings are taken, the astronaut reels up the cable and stows the sensor-tripod assembly.

At the first site only, two sets of additional readings are taken with the sensor block rotated first 180° about a horizontal axis, then 180° about a vertical axis. These additional readings allow determination of a zero offset for each axis.

Sensor offsets of vector magnetometers are inherently sensitive to environmental change; thus, a surface measurement of offsets is required. Gain and linearity, however, are relatively insensitive to environmental change.

Results

Magnetic-field measurements were conducted at two sites by the LM pilot during the second period of EVA on February 6. The first measurement was made approximately 170 m from the LM at site A (fig. 13-6), and the second was

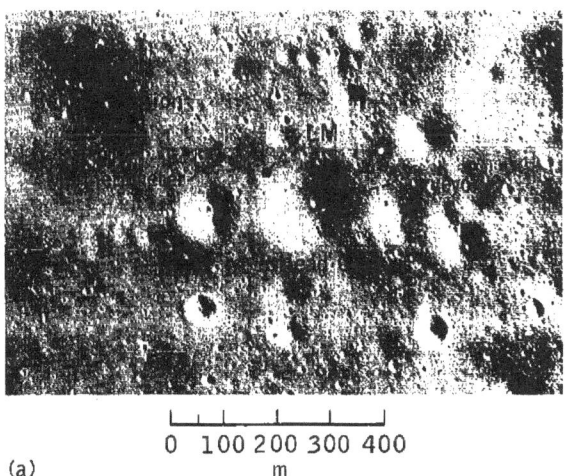

(a)

0 100 200 300 400
m

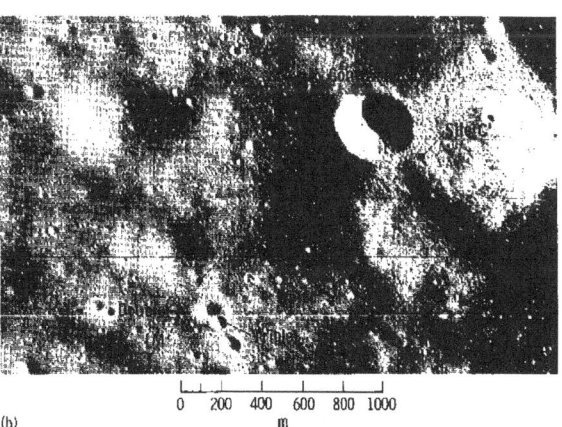

(b)

0 200 400 600 800 1000
m

FIGURE 13-6.—Lunar Orbiter photographs showing the Apollo 12 and 14 landing sites and positions where steady-field measurements were made. (a) Apollo 12 site. (b) Apollo 14 site.

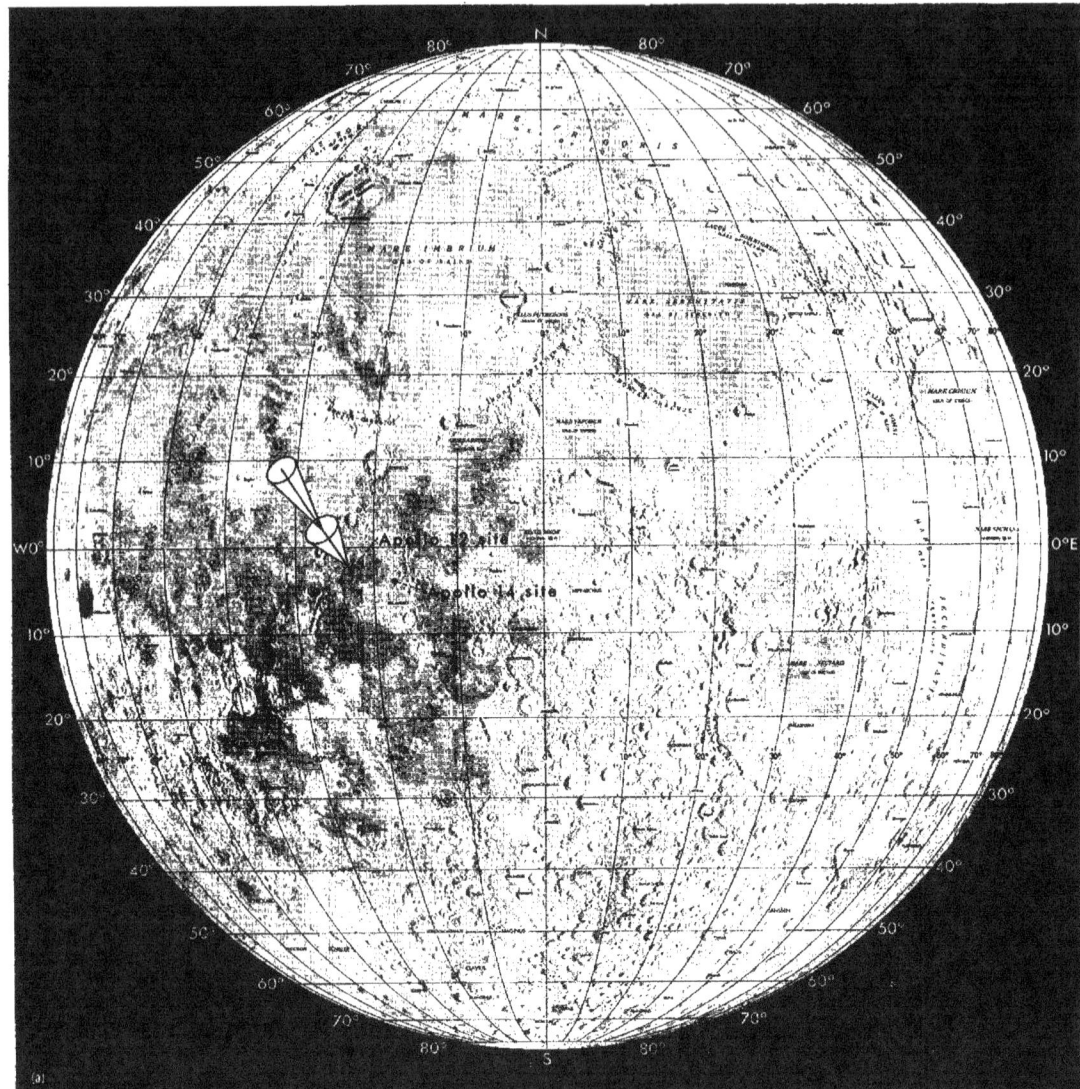

FIGURE 13–7.—Orientation of lunar magnetic-field measurements. (a) Relative location of the Apollo 14 and 12 landing sites.

made near the rim of Cone Crater at site C′. The vector magnetic-field components measured at each site are as follows. At site A, the total LPM-measured components are $B_X = -90$ gammas, $B_Y = 45$ gammas, and $B_Z = -30$ gammas. At site C′, components are $B_X = -11$ gammas, $B_Y = -28$ gammas, and $B_Z = -28$ gammas. These components are listed in a local coordinate system with the origin on the surface; the B_X-component is positive toward the zenith, and the B_Y-

and B_Z-components are positive along the surface to the east and north, respectively.

The two Apollo 14 field measurements are each the vector sum of intrinsic lunar fields and the extralunar (solar or terrestrial or both) fields. The Explorer 35 magnetometer orbiting the Moon, as shown in figure 13–2, simultaneously makes measurements of the extralunar fields. To determine the intrinsic steady field at the two surface sites, it is necessary to transform these extralunar

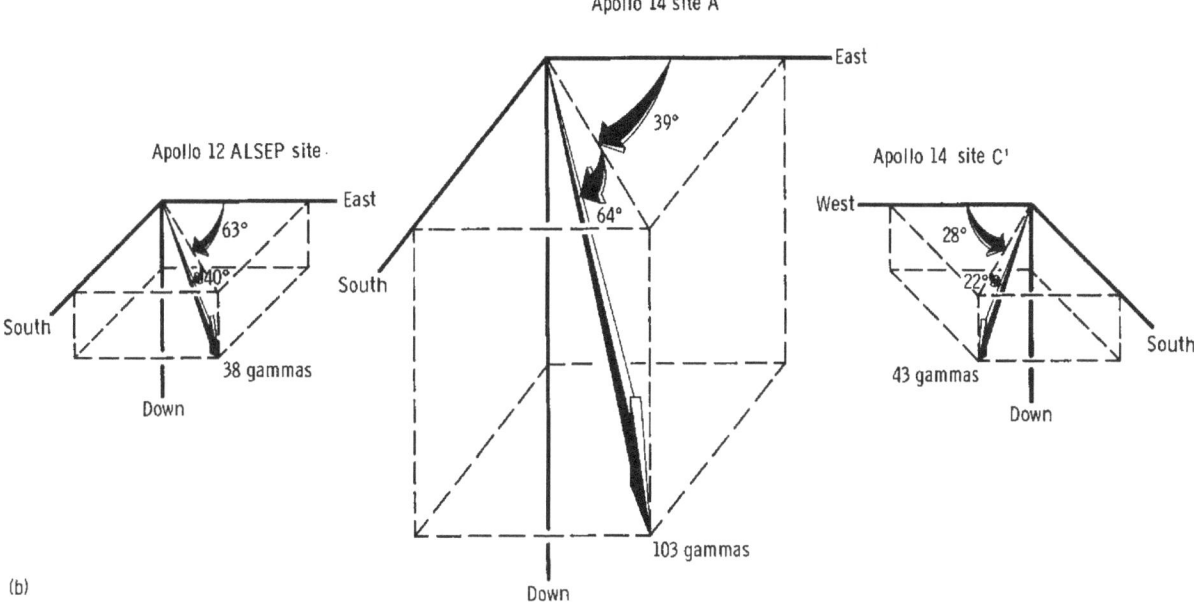

FIGURE 13–7.—Continued—(b) The vector fields for each of the three point measurements.

fields to the Apollo 14 surface coordinate system and vectorially subtract them from the LPM measurements.

In addition to the external fields measured by Explorer 35, there is a time-dependent inductive field generated in the Moon that is proportional to the amplitude and frequency of the Explorer 35 extralunar field (ref. 13–11) and that is not completely filtered out by the LPM. This inductive field must also be subtracted from the Apollo 14 measurements; it is determined by examination of past simultaneous Apollo 12 and Explorer 35 data measured in the same lunar-orbital position (fig.

13–2). The Apollo 12 lunar-surface magnetometer, which would have directly monitored these time-dependent inductive fields, was not operating at the time of the Apollo 14 mission.

The intrinsic lunar steady fields at the Apollo 14 site, which can be calculated by vectorial subtraction of extralunar and induction fields, are listed in table 13–II and are illustrated in figure 13–7. Errors listed in table 13–II include uncertainties in sensor orientation, instrument-temperature measurements, and inductive-field determination.

A magnetic-field gradient can be calculated

TABLE 13–II. *Magnetic-Field Measurements at Apollo 14 and 12 Sites*

Site	Coordinates, deg	Field magnitude, gammas	Magnetic-field components, gammas		
			Up	East	North
Apollo 14:	3.7° S, 17.5° W				
Site A............	(See fig. 3.1 in sec. 3)	103±5	−93±4	+38±5	−24±5
Site C'..........	(See fig. 3.1 in sec. 3)	43±6	−15±4	−36±5	−19±8
Apollo 12 site........	3.2° S, 23.4° W	38±3	−24.4±2.0	+13.0±1.8	−25.6±0.8

from the two Apollo 14 site measurements. Sites A and C' are separated by 1.12 km (fig. 13–6(b)), and the linear gradient is calculated to be 54 ± 7 gammas/km. This value is less than the gradient upper limit determined for the Apollo 12 site 180 km away. At the Apollo 12 site, the steady field is measured to be 38 ± 3 gammas and the gradient upper limit is 133 gammas/km (ref. 13–1). These values are shown for reference in figure 13–7 and table 13–II.

Discussion

The two magnetic-field measurements at the Apollo 14 site have shown that the unexpectedly high field measured at the Apollo 12 site is not a feature confined to one site on the Moon. These three magnetometer measurement sites, separated by distances of from 1 to 180 km, and the high remanence found in the samples[4,5] returned from the Apollo 11 and 12 sites (refs. 13–2 to 13–5), separated by 1400 km, experimentally show that the Moon has been magnetized in these widely separated regions. It is quite possible that thousands of local surface regions of magnetized concentrations (magcons) may exist (ref. 13–19), and reexamination of the Explorer 35 magnetometer and plasma data indicates that several magnetized nonmare areas exist on both the near and

[4] G. W. Pearce, D. W. Strangway, and E. E. Larson: Magnetism of Two Apollo 12 Igneous Rocks. Proc. Apollo 12 Lunar Sci. Conf. (Houston), Jan. 11–14, 1971. To be published in Geochim. Cosmochim. Acta.

[5] C. E. Helsley: Evidence for an Ancient Lunar Magnetic Field. Proc. Apollo 12 Lunar Sci. Conf. (Houston), Jan. 11–14, 1971. To be published in Geochim Cosmochim. Acta.

far sides of the Moon (ref. 13–6). These analyses give evidence that much of the lunar-surface material has been magnetized—perhaps even a crustal shell around the entire Moon.

No obvious mechanism for such large-scale magnetization of surface materials exists at present. Magnetization of the Apollo 11 and 12 samples would have required an external field greater than 10^3 gammas (ref. 13–5). Ambient fields of this magnitude have not been measured in space near the Moon; the largest measured so far are transient fields of a magnitude of approximately 10^2 gammas; these transient fields last only a few minutes (ref. 13–20).

Sources of the Steady Field

The similarities between the Apollo 12 and 14 field measurements (viz, all vectors are pointed down and toward the south and have magnitudes that correspond to within a factor of 3) suggest that the two Apollo 14 sites and possibly the Apollo 12 site are located above a near-surface slab of material that was uniformly magnetized at one time. Subsequently, the magnetization in the slab was altered by local processes, such as tectonic activity or fracturing and shock demagnetization from meteorite impacts. This latter process is graphically illustrated in figure 13–8.

The Apollo 12 and 14 steady magnetic fields could also originate in surface or subsurface dipolar sources, such as meteoroid fragments or ore bodies. Properties of such a source assumed for the Apollo 12 field have been discussed in a previous work (ref. 13–1); a similar analysis will be performed for the Apollo 14 fields.

Numerous other source models exist that could

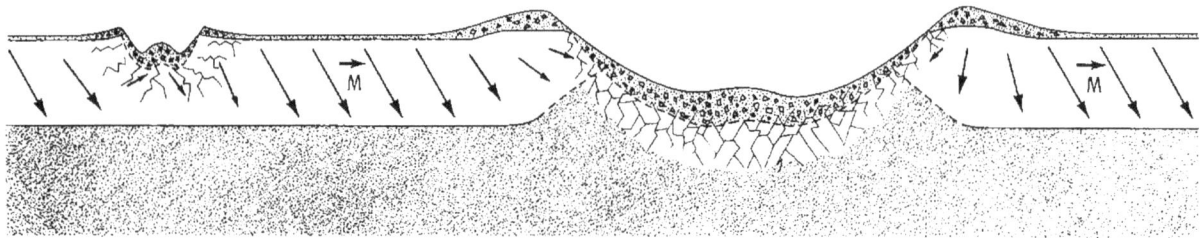

FIGURE 13–8.—Conceptual diagram showing disruption of a previously uniformly magnetized subsurface layer by meteoroid impact. The vectors **M** represent the direction and magnitude of remanent magnetization in the layer.

be postulated for the Apollo 12 and 14 fields. One possibility is that the region was subjected to a uniform magnetic field but that various materials with differing coercivities were magnetized to different strengths. Another model might involve a slow variation in the direction of the ambient field, causing regions that passed through the Curie temperature (ref. 13–21) at different times to be magnetized in different directions.

Models for Magnetizing Lunar Steady-Field Sources

The discovery of magnetic sources on the lunar surface has given strong evidence that, at one time in the lunar past, ambient fields much stronger than at present existed over much or all of the lunar surface. Perhaps the Moon has taken a magnetic snapshot of an early evolutionary phase of the solar system.

Possible origins of the ancient ambient field could have been external to the Moon (Sun or Earth), internal to the Moon (dynamo or thermoelectric currents), or due to induction (eddy cur-

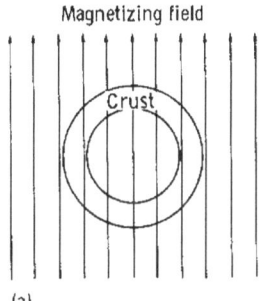
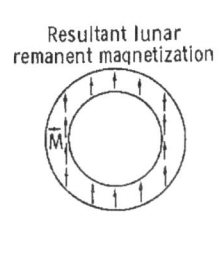

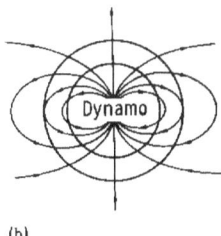
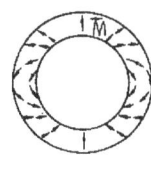

(a)

(b)

FIGURE 13–9.—Possible magnetic events in lunar history as models for magnetization of near-surface regions of the Moon. The Moon is shown before and after each hypothetical event. The vectors **M** indicate the direction of remanent magnetization in the crustal region. (a) Magnetization by a strong solar or terrestrial field. (b) Lunar internal dynamo field.

rents or unipolar currents). A few possible models are discussed briefly, the first two of which are illustrated in figure 13–9.

Solar field. In this model (fig. 13–9(a)), the Sun generates a magnetic field, stronger than presently exists, which magnetizes the lunar material. The model requires a solar field of constant direction with respect to the solar and lunar spin axes (e.g., solar dipole moment vector pointed out of the solar ecliptic plane) rather than the present-day sector-structure geometry.

Terrestrial field. In this case, the Earth possesses a larger field than in the past (not indicated by paleomagnetic studies), or the Moon has an orbit much closer to the Earth (fig. 13–9(a)). For a terrestrial field of present-day magnitude, the Moon would have had to approach to within 2 to 3 Earth radii R_E (close to the Roche limit (ref. 13–22)) to be subjected to a 10^5-gamma field.

Lunar internal dynamo field. An internal lunar dynamo model (fig. 13–9(b)) requires both a hot core and a sufficient spin rate at the time the surface material cools below the Curie temperature. Past mechanisms for cooling the lunar interior and slowing the spin rate to the present value are also necessary to the dynamo model.

Thermoelectric current field. According to this model, large areas of the Moon were covered at one time by hot lava flows; the temperature gradient between surface and subsurface material induced thermoelectric currents across the interface. The magnetic field associated with this current magnetized the surface material as it cooled below the Curie temperature. It should be kept in mind that these currents would be radial; thus the associated fields would be tangent to the surface and, therefore, would not tend to induce a radial component of magnetization.

Magnetosheath induction of eddy currents. As the Moon passes through the magnetosheath of the Earth during each lunation, it experiences fields that have a preferred direction and are stronger in magnitude than those encountered in the free-streaming solar wind. Repeated passage through the sheath causes an average magnetization to be induced in the direction of flow. For this model, the ancient solar field would have had a geometry other than the present-day sector structure, because the field reversals across the sector boundary would have tended to average out the magnetizing field over many lunar revolutions.

Solar-wind v × **B** *induced currents.* The solar wind transports magnetic fields past the Moon at velocities v of approximately 400 km/sec; the corresponding v × **B** electric field causes currents to flow along paths of high electric conductivity (refs. 13–23 to 13–25) such as molten mare regions. The fields associated with these currents magnetize these materials as they cool below the Curie temperature. Because this induction mechanism has the strongest influence while the hot region is sunlit, an average preferred direction is associated with v. The v × **B** induction model also requires some solar-field geometry other than the present-day sector structure.

Preferential selection of any one of the possible models is difficult because available magnetic data represent only a small fraction of the lunar surface. Mapping global geometry and strength of the steady field by future surface and orbital missions should elucidate the "magnetic epoch" of lunar history.

Summary and Conclusions

The Apollo 14 LPM measurements have allowed calculation of vector magnetic fields at two Fra Mauro sites:

(1) Site A: 103±5 gammas, directed 64° down from the horizontal and azimuthally 39° clockwise from due east (fig. 13–7(b))

(2) Site C': 43±6 gammas, directed 22° down from the horizontal and azimuthally 152° clockwise from due east (fig. 13–7(b))

These sites are separated by a distance of 1.12 km, indicating a 54-gamma/km linear surface-field gradient along a line between the two sites. The data show that local magnetization is not confined to the Apollo 12 site but extends over at least a 180-km region of the Moon and possibly over most of the lunar surface.

The steady fields could be due to a variety of types of sources (e.g., a distant dipole or an extended near-surface platelike region that was originally uniformly magnetized but has subsequently been altered by some mechanism such as

meteoroid shock impact). The similarities of magnitudes and directions of the two Apollo 14 measurements and the one Apollo 12 steady-field measurement give preference to the latter model.

References

13–1. DYAL, PALMER; PARKIN, CURTIS, W.; AND SONETT, CHARLES P.: Apollo 12 Magnetometer: Measurement of a Steady Magnetic Field on the Surface of the Moon. Science, vol. 169, no. 3947, Aug. 21, 1970, pp. 762–764.

13–2. STRANGWAY, D. W.; LARSON, E. E.; AND PEARCE, G. W.: Magnetic Studies of Lunar Samples— Breccia and Fines. Proc. Apollo 11 Lunar Sci. Conf., Geochim. Cosmochim. Acta Suppl. 1, vol. 3, 1970, Pergamon Press, Inc., pp. 2435–2451.

13–3. RUNCORN, S. K.; COLLINSON, D. W.; O'REILLY, W.; BATTEY, M. H.; ET AL.: Magnetic Properties of Apollo 11 Lunar Samples. Proc. Apollo 11 Lunar Sci. Conf., Geochim. Cosmochim. Acta Suppl. 1, vol. 3, 1970, Pergamon Press, Inc., pp. 2369–2387.

13–4. DOELL, RICHARD R.; GROMME, C. SHERMAN; THORPE, A. N.; AND SENFTLE, F. E.: Magnetic Studies of Apollo 11 Lunar Samples. Proc. Apollo 11 Lunar Sci. Conf., Geochim. Cosmochim. Acta Suppl. 1, vol. 3, 1970, Pergamon Press, Inc., pp. 2097–2102.

13–5. HELSLEY, C. E.: Magnetic Properties of Lunar 10022, 10069, 10084, and 10085 Samples. Proc. Apollo 11 Lunar Sci. Conf., Geochim. Cosmochim. Acta Suppl. 1, vol. 3, 1970, Pergamon Press, Inc., pp. 2213–2219.

13–6. MIHALOV, J. D.; SONETT, C. P.; BINSACK, J. H.; AND MOUTSOULAS, M. D.: Possible Fossil Lunar Magnetism Inferred From Satellite Data. Science, vol. 171, no. 3974, Mar. 5, 1971, pp. 892–895.

13–7. KEMP, J. C.; SWEDLUND, J. B.; LANDSTREET, J. D.; AND ANGEL, J. R. P.: Discovery of Polarized Light From a White Dwarf. Astrophys. J., vol. 161, no. 2, pt. II, Letters, Aug. 1970, p. 77.

13–8. BRANDT, JOHN C.; AND HODGE, PAUL W.: Solar System Astrophysics. McGraw-Hill Book Co., Inc., 1964.

13–9. NESS, N. F.: Interplanetary Medium. Space Physics, W. N. Hess and G. D. Mead, eds., Gordon & Breach Sci. Pub., 1968, p. 363.

13–10. DYAL, P.; PARKIN, C. W.; SONETT, C. P.; AND COLBURN, D. S.: Electrical Conductivity and Temperature of the Lunar Interior From Magnetic Transient Response Measurements. NASA TM X–62012, 1970.

13–11. SONETT, C. P.; SMITH, B. F.; COLBURN, D. S.; SCHUBERT, G.; AND SCHWARTZ, K.: The

Lunar Electrical Conductivity Profile: Mantle-Core Stratification, Near Surface Thermal Gradient, Heat Flux and Composition. Nature, vol. 230, no. 5293, 1971.

13-12. DOLGINOV, SH. SH.; EROSHENKO, E. G.; ZHUZGOV, L. N.; AND PUSHKOV, N. V.: Investigation of the Magnetic Field of the Moon. Geomagn. Aeron., vol. 1, no. 1, 1961, pp. 18–25.

13-13. DOLGINOV, SH. SH.; PUSHKOV, N. V.; EROSHENKO, E. G.; AND ZHUZGOV, L. N.: Measurements of the Magnetic Field in the Vicinity of the Moon by the Artificial Satellite Luna 10. Dokl. Akad. Nauk SSSR, vol. 170, no. 3, Oct. 31, 1966, pp. 574–577.

13-14. SONETT, C. P.; COLBURN, D. S.; AND CURRIE, R. G.: The Intrinsic Magnetic Field of the Moon. J. Geophys. Res., vol. 72, no. 21, Nov. 1967, pp. 5503–5507.

13-15. NESS, N. F.; BEHANNON, K. W.; SCEARCE, C. S.; AND CANTARANO, S. C.: Early Results From the Magnetic Field Experiment on Lunar Explorer 35. J. Geophys. Res., vol. 72, no. 23, Dec. 1967, pp. 5769–5778.

13-16. BEHANNON, KENNETH W.: Intrinsic Magnetic Properties of the Lunar Body. J. Geophys. Res., vol. 73, no. 23, Dec. 1968, pp. 7257–7268.

13-17. GEYGER, W. A.: Nonlinear-Magnetic Control Devices. McGraw-Hill Book Co., Inc., 1964.

13-18. GORDON, D. I.; LUNDSTEN, R. H.; AND CHIARODO, R. A.: Factors Affecting the Sensitivity of Gamma-level Ring-core Magnetometers. IEEE Trans. Magn., vol. MAG-1, no. 4, Dec. 1965, pp. 330–337.

13-19. BARNES, A.; CASSEN, P.; MIHALOV, J. D.; AND EVIATOR, A.: Permanent Lunar Surface Magnetism and Its Deflection of the Solar Wind. Science, vol. 171, 1971.

13-20. DYAL, P.; PARKIN, C. W.; AND SONETT, C. P.: Lunar Surface Magnetometer Experiment. Sec. 4 of Apollo 12 Preliminary Science Report. NASA SP-235, 1970.

13-21. NAGATA, TAKESI: Rock Magnetism. Maruzen Co., Ltd. (Tokyo) (Plenum Press (New York) exclusive U.S. distributor), 1961.

13-22. INGLIS, F. J.: Planets, Stars, and Galaxies. John Wiley & Sons, Inc., 1968.

13-23. SONETT, C. P.: Principle of Planetary Unipolar Generators. Vol. II of Planetary Electrodynamics, ch. VII-11, Samuel C. Coroniti and James Hughes, eds., Gordon & Breach Sci. Pub., 1969, pp. 373–378.

13-24. SCHUBERT, G.; AND SCHWARTZ, K.: A Theory for the Interpretation of Lunar Surface Magnetometer Data. The Moon, vol. 1, 1969.

13-25. SILL, W. R.; AND BLANK, J. L.: Method for Estimating the Electrical Conductivity of the Lunar Interior. J. Geophys. Res., vol. 75, no. 1, Jan. 1970, pp. 201–210.

ACKNOWLEDGMENTS

The authors wish to express deep appreciation for the effort expended by all the personnel connected with this experiment. In particular, thanks are extended to the subcommittee and steering committee chairmen who modified and added this new experiment to their agenda for consideration. The authors wish to acknowledge, in particular, Carle Privette, Michael Dix, Donald Mulholland, Glen Goodwin, Ernest Iufer, Robert Murphy, and Kathy Stark of the NASA Ames Research Center; and J. B. Thomas of the NASA Manned Spacecraft Center. The diligent efforts of these and many others brought success to this experiment. Special thanks are extended to Dr. David S. Colburn of the NASA Ames Research Center for use of Explorer 35 data.

14. Lunar-Surface Closeup Stereoscopic Photography

T. Gold [a]

Closeup photographs of soil and rock on the lunar surface were obtained with the Apollo lunar-surface closeup camera (ALSCC) (fig. 14–1). The ALSCC stereoscopic camera contains an internal strobe light and an external sundial that can be read in real time for postmission orientation of photographs. The ALSCC is capable of photographing an area 72 by 82.8 mm and of resolving objects as small as 0.085 mm in diameter. Built under contract to the NASA Manned Spacecraft Center, the ALSCC was used during the Apollo 11 and 12 missions (refs. 14–1 to 14–4).

During the Apollo 14 mission, 17.5 stereopairs were obtained of various surfaces (table 14–I). The stereopairs include views of the tracks of the modularized equipment transporter (MET), bootprints, undisturbed lunar soil, the surface of a rock partially obscured by soil, and samples of thermal coatings taken in support of the thermal degradation samples (TDS) experiment.

All the photographs were taken by the commander during the second traverse. The first four stereopairs were taken at a location midway between the lunar module (LM) and station A of the traverse (fig. 3–1, sec. 3), while the crew paused to check their route. The next 11 stereopairs were obtained in the vicinity of station A. One stereopair may have been taken at either station A or B2; and the last 1.5 stereopairs were probably obtained near the LM at the end of the traverse. Although photography at other locations along the traverse had been planned, no additional photographs could be taken because of a shortage of time.

During the Apollo 11 and 12 missions, most of the closeup stereoscopic photographs were ob-

[a] Cornell University.

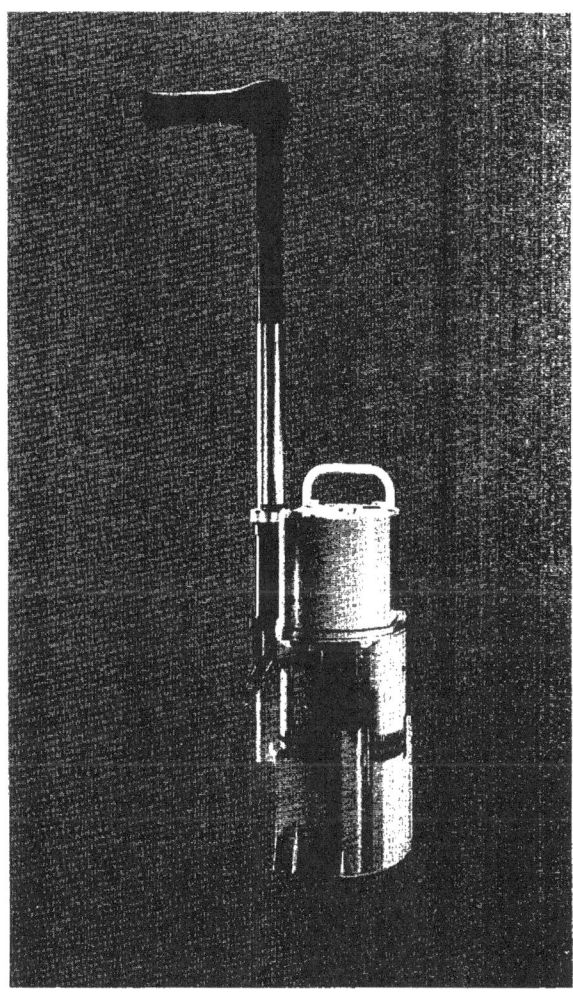

FIGURE 14–1.—ALSCC.

tained near the LM, where the surfaces may have been affected by the engine exhaust of the LM during the landing. All the Apollo 14 mission photographs were obtained at locations distant

TABLE 14–I. *Lunar-Surface Closeup Stereoscopic Photographs*

Description	Figure no.[a]	Photograph no.	Shutter no.
MET track	14–2	AS14–77–10357	301
MET track	14–3	AS14–77–10358	302
Bootprint	14–4	AS14–77–10359	303
Bootprint	14–5	AS14–77–10360	304
Bootprint	14–6	AS14–77–10369	313
Undisturbed soil	14–7	AS14–77–10368	312
Undisturbed soil	14–8	AS14–77–10370	314
Soil-covered rock surface	14–9	AS14–77–10371	315
Soil-covered rock surface	14–10	AS14–77–10372	316
Unexposed			[b] 317
Soil-covered rock surface	14–11	AS14–77–10373	[b] 318
Soil-covered rock surface		AS14–77–10374 (half)	[b] 319
TDS experiment		AS14–77–10361	305
TDS experiment	14–12	AS14–77–10366	310
TDS experiment	14–13	AS14–77–10367	311
TDS experiment	14–14	AS14–77–10362	306
TDS experiment	14–15	AS14–77–10363	307
TDS experiment	14–16	AS14–77–10364	308
TDS experiment	14–17	AS14–77–10365	309

[a] Only one of each stereopair is shown in the figures.
[b] Has not been verified; further investigation is required.

from the LM, where no such effect can be expected.

MET Tracks

The MET has two smooth nitrogen-filled rubber tires, each of which applies a pressure of 0.3 to 0.5 N/cm^2 to the lunar surface when the MET is loaded. The crew reported that the MET did not produce a significant "rooster tail" when pulled across the lunar surface and that the depth of the tracks varied from 1 to 2 cm, which is consistent with the applied loads and the previous estimates of the bearing properties of the lunar soil. An illustration of how the fine-grained material produces a precise imprint from the smooth surface of the wheel and its slightly raised midline is shown in figure 14–2. The compaction of the soil in the track results in a drastic change in the photometric properties of the soil. (This phenomenon is also evident in some Hasselblad photographs.) The compacted soil appeared to crack along lines that are similar to those of brittle fracture.

The MET track shown in figure 14–3 evidently was made in soil firmer than the soil shown in figure 14–2. The track appears narrower, indicating less sinkage, and fine cracks, perpendicular to the direction of motion, are evident. A small white rock, in the upper left corner of the photograph, was probably pushed into the lunar soil by the MET tire. The indentation in the upper right

FIGURE 14–2.—The MET track (AS14–77–10357). The original stereopairs have a higher resolution value than can be reproduced by the method used in the printing of this report.

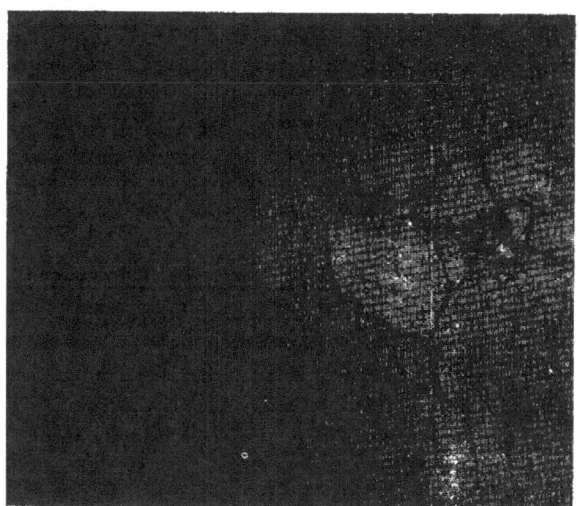

FIGURE 14–3.—The MET track (AS14–77–10358).

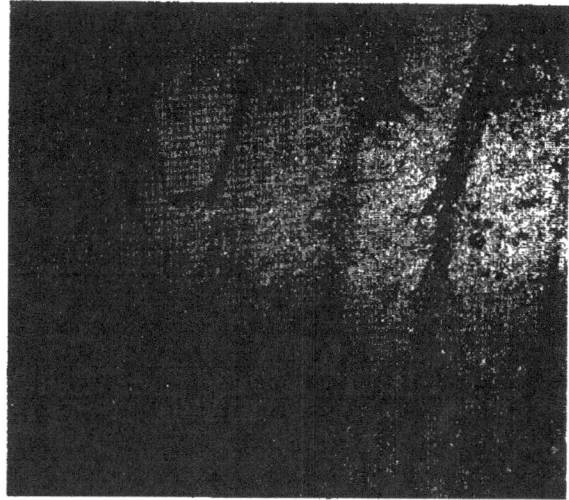

FIGURE 14–5.—Bootprint (AS14–77–10360).

corner of the photograph was probably produced by the edge of the camera while the camera was being positioned. The undisturbed ground to the sides of the track appears to have some fine-scale structure similar to that seen on the pictures of undisturbed ground.

Bootprints

In cross section, the boot treads are trapezoidal in shape and 4.6 mm deep. Standing on one foot, an astronaut applies a pressure of approximately

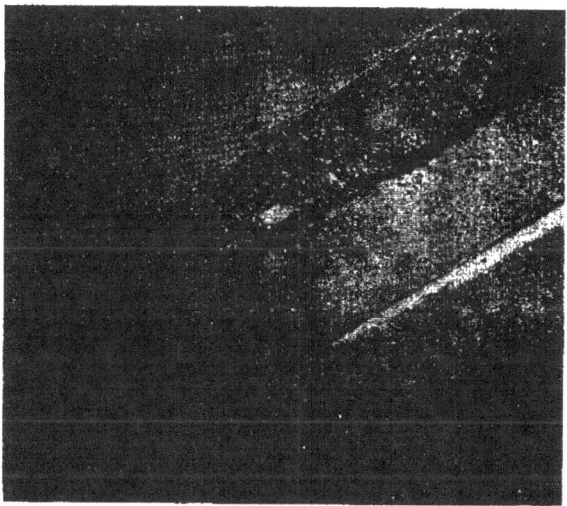

FIGURE 14–6.—Bootprint (AS14–77–10369).

0.7 N/cm^2 on the lunar surface. This pressure generally results in a sinkage of approximately 2 cm—except on the rims of small, fresh craters—where bootprints as deep as 10 cm have been observed.

The lunar soil is able to retain the shape of the bootprint tread (figs. 14–4 to 14–6), including the nearly vertical wall, because the soil is slightly cohesive. The relatively smooth surface over most of the imprint in figure 14–4 indicates that this cohesion is relatively stronger than the adhesion

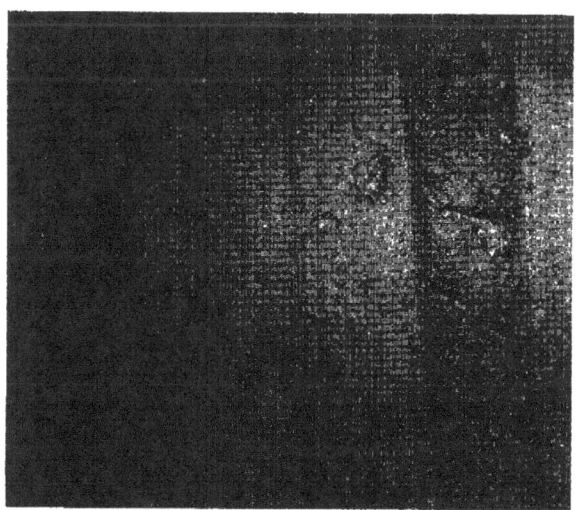

FIGURE 14–4.—Bootprint (AS14–77–10359).

between the soil and the sole of the boot. Although similar soil characteristics were observed in Apollo 11 and 12 mission photographs, the cohesion of the lunar soil shown in the Apollo 14 photography appears to be less than at the other sites. One possible explanation for the difference in the soil cohesion is a coarser grain-size distribution for the soil at the Apollo 14 site. The bulk of the soil particles is smaller than the ALSCC is capable of resolving (0.085 mm); however, more resolvable particles apparently occur in the Apollo 14 bootprint photographs. The nature and cause of the lunar soil cohesion are not fully understood, and more study is required.

Undisturbed Soil

During the Apollo 11, 12, and 14 missions, the astronauts reported that in some places the ground appeared much like a beach after a heavy rain. The observation was interpreted as a characteristic regularity of indentation size and distribution. However, the "raindrop" phenomenon is not evident in the stereopairs obtained by the Apollo 14 crew (figs. 14–7 to 14–10). Even though the definition in the stereopairs is better than that of a crewman standing on the lunar surface, the limited surface area seen may not allow the effect to be observed.

The origin of the raindrop pattern is not known,

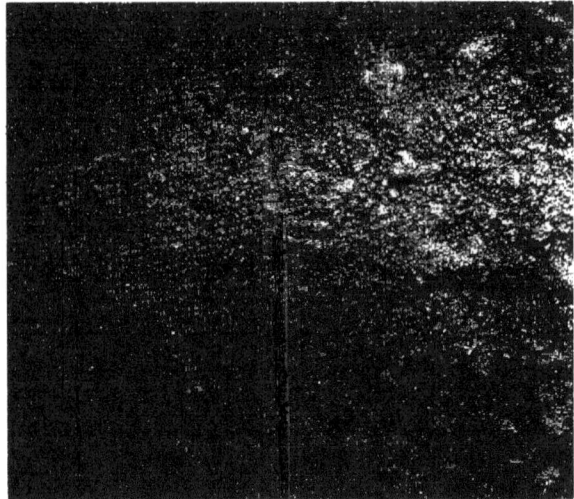

FIGURE 14–8.—Undisturbed soil (AS14–77–10370).

but the pattern is thought to be the result of small craters formed by the impact of meteoritic particles or of craters formed by secondary impacts of fragments from larger events. Attempts to simulate meteoroid impacts, using sand as the receiving medium, have resulted in bowl-shaped craters. However, one of the best-defined craters in figure 14–7 is steep walled rather than bowl shaped. The difference in the shape of the craters may be due to the difference between the mechanical properties of the lunar soil and those of sand, or the crater in the figure may be the result of a low-speed secondary impact.

Another characteristic feature of the lunar soil noted by the Apollo 14 crew was the abundance of lineations. A slight, though clearly visible, pattern of lines rising slightly from left to right is shown in figure 14–7; in figure 14–8, the lineations appear similar to scratch marks and run from the top left center to the center of the photograph. In figure 14–9, the lineations appear nearly parallel and run upward approximately 20° from right to left in the photograph. When viewed stereoscopically, some of the lineations appear as ridges at one point and as depressions at another. These linear features strongly resemble the lineations produced (on approximately the same scale) in insulating powders subjected to electron bombardment.

At the top of figure 14–7, the imprint of the

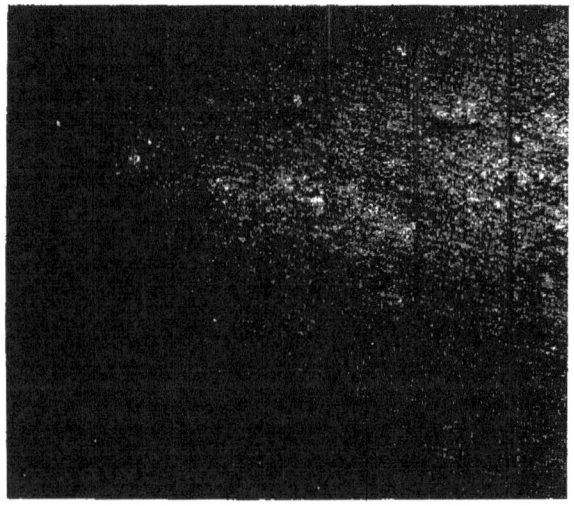

FIGURE 14–7.—Undisturbed soil (AS14–77–10368).

FIGURE 14–9.—Soil-covered rock surface (AS14–77–10371).

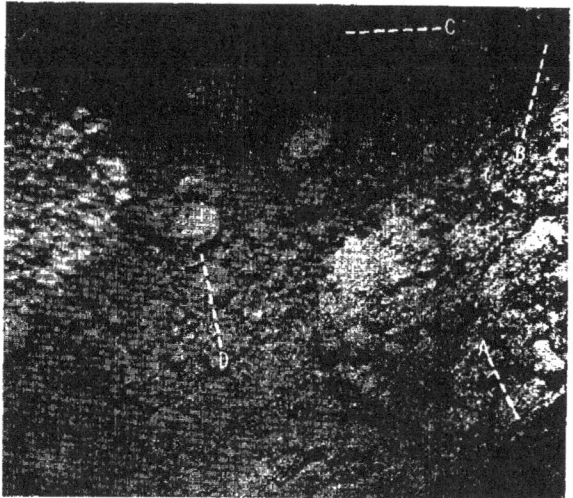

FIGURE 14–11.—Soil-covered rock surface (AS14–77–10373).

edge of the camera indicates that the surface material is a fine powder that can be molded precisely. All the particles visible in figures 14–7 and 14–8 are aggregates of fragmental material ranging in size from 0.1 to 0.6 mm. The only individual fragments resolvable in figure 14–8 are white, equant to elongate, angular fragments that may be crystalline rocks, fragmental rocks, or feldspathic crystals.

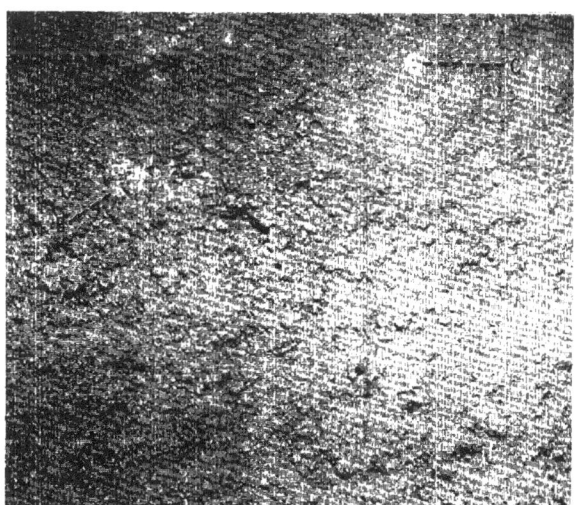

FIGURE 14–10.—Soil-covered rock surface (AS14–77–10372).

As in figures 14–7 and 14–8, the soil in figures 14–9, 14–10, and 14–11 appears to be cohesive with equant aggregates bound into the surface. The aggregates in figure 14–9 range in size from 0.1 to 6 mm.

Rock Surfaces

The rock surface partially visible in figure 14–9 is obscured by soil at the rock base or by soil lapped up onto the rock. The rock is located at station A (frame AS14–9409 or fig. 3–3, sec. 3) and is white and crystalline or clastic. The rock appears to have a relatively smooth surface with a small glass-lined impact pit visible at point A in the figure. Indications of other pits in the rock surface exist, but they are dust covered. The dust cover on the rock ranges up to 3 mm deep, with aggregates as large as 2 mm in diameter adhering to the surface.

The rock surface in figure 14–10 is partially visible in the upper half of the photograph. The pits in this rock may also be due to micrometeoroid impacts. (An example of the pits is given at A in fig. 14–10.) The rock is clastic with clasts visible at points B and C in the figure. The clast at point B appears clastic with elongate white clasts in a gray matrix.

The last frame of the stereoscopic photography

(fig. 14–11) was obtained during the winding up of the film before unloading. The depression in the photograph is several inches deep and may have been made by a crewmember. The light area in the left side of the photograph is the result of sunlight leaking into the area under the hood of the ALSCC.

All the fragments visible in figure 14–11 are 0.2-mm to 2-cm clastic rocks or aggregates. Equant fragments with spinose or very irregular surfaces are aggregates of soil particles (examples at point *A* in fig. 14–11) and angular fragments exhibiting smooth fracture surfaces (examples at points *B* and *C* in fig. 14–11) may be clastic rock fragments. The spiny fragment at *D* may be soil bonded by glass spatter. The needlelike spines around the periphery of this fragment may be projections of glass.

TDS Experiment

The purpose of the TDS experiment was to evaluate the effect of lunar dust on the optical properties (absorptivity and emissivity) of a dozen candidate thermal coatings. The coatings may be used in subsequent lunar missions on such items as the lunar communications relay unit, the lunar roving vehicle, the television camera, and the Apollo lunar-surface experiments package.

Two duplicate arrays (serial nos. 1001 and 1002), each containing samples of the 12 thermal coatings (table 14–II), were taken to the Moon. A series of seven stereopairs of the arrays was obtained for three conditions (table 14–III): when the arrays were pristine (i.e., before exposure to lunar soil), when the arrays were dusted, and after the lunar dust had been brushed off. The arrays were then packaged and returned to Earth for extensive examination and testing.

Array 1001

One TDS array (serial no. 1001), panels 1 to 6 and 9 to 12, is shown, after soil was sprinkled over it, in figures 14–12 and 14–13. Significant differences in the soil coating on the different parts of the array are evident. The edges between the panels and the metal preferentially attracted dust and were densely filled. Some of the indented numbers (0.015 in. deep) in the metalwork were

TABLE 14–II. *Block Descriptions*

Block no.	Thermal coating sample
1	White paint: S–13G
2	White paint: Z–93, zinc oxide/potassium silicate.
3	White paint: Goddard MS–74
4	Vacuum-deposited silver and Inconel on Teflon with Teflon side exposed.
5	Vacuum-deposited silver on quartz with quartz side exposed.
6	White paint: Dow-Corning 92–007, titanium dioxide-silicone.
7	White paint: Cat-a-lac White, titanium dioxide/epoxy.
8	White paint: Minnesota Mining & Manufacturing Co. White Velvet (400 series), titanium dioxide/epoxy polyester.
9	White fabric: Dacron on aluminized Mylar laminate.
10	Oxidized silicon monoxide over vacuum-deposited aluminum on Kapton with oxidized silicon monoxide side exposed.
11	Vacuum-deposited aluminum on Kapton with Kapton side exposed.
12	Anodized 6061 aluminum MIL–A–8625, type II, class I.

filled preferentially (similar to an electrostatic copy process). The screws attracted dust, as did the groove in the metal shown near the top edge of

FIGURE 14–12.—Panels 1 to 6 of TDS 1001 covered with soil (AS14–77–10366).

TABLE 14–III. *Thermal Degradation Samples Experiment*

Figure no.	TDS serial no.	Description	Block no.
....	1002	Pristine blocks	9 to 12
14–12............................	1001	Soil sprinkled on blocks	1 to 6
14–13............................	1001	Soil sprinkled on blocks	9 to 12
14–14............................	1002	Soil sprinkled on blocks	1 to 6
14–15............................	1002	Soil sprinkled on blocks	7 to 12
14–16............................	1002	Soil brushed off blocks	1 to 6
14–17............................	1002	Soil brushed off blocks	7 to 12

figure 14–12. Before the photograph in figure 14–13 was taken, the whole unit was apparently jarred. The resultant impact removed the filling from the indented numbers without, however, completely destroying the shapes. Recognizable numbers formed by the dust were displaced several millimeters from the positions in which they must have originally been formed. This phenomenon indicates that the electric fields around the edges of the indented numbers must have been effective in attracting the dust grains toward them and that the dust must have been sufficiently cohesive to be shaken out later without the formed numbers breaking apart.

Panel 4 in figure 14–12 appears to have an edge of fine powder surrounding each clump. The dust on panel 6 was more evenly scattered and did not bunch together as readily. Several clumps of dust on panel 2 are remarkably high and appear to stand up straight from the surface. The clumps of soil on panel 10 (fig. 14–13) appear to have been surrounded by dust causing a halo effect. The dust on panel 9 was preferentially attracted to the edge of the panel and may have been displaced from the original position. The number denoting panel 8 and the letter *S* are neatly filled with dust that has not been displaced. All these effects noted in figures 14–12 and 14–13 indicate that the manner of settling of the dust is markedly influenced by electrostatic effects, that edges and grooves have significant effects on local electric fields, and that the different panel surfaces have different electrical properties.

Array 1002

The other TDS array (serial no. 1002), panels 1 to 12, is shown covered with soil in figures 14–14 and 14–15. It is evident that the different panels had a different effect on the manner in which the dust settled. Panel 6 has a rather uniform contamination of a thin soil layer, whereas on panels 3 and 5 the dust settled in a more uneven manner. The screw above panel 3 appears to have attracted more dust than the neighboring surface, whereas panels 2 and 4 are remarkably clean compared with the intervening metal surface. In particular, the gap around panel 4 appears to be well filled with dust in some places, while the panel itself is absolutely clean. Some difference exists in the manner of deposition on panel 10 compared with the other panels in figure 14–15. The lower edge of the entire unit appears to have preferentially attracted dust that appears to be

FIGURE 14–13.—Panels 9 to 12 of TDS 1001 covered with soil (AS14–77–10367).

FIGURE 14–14.—Panels 1 to 6 of TDS 1002 covered with soil (AS14–77–10362).

FIGURE 14–15.—Panels 7 to 12 of TDS 1002 covered with soil (AS14–77–10363).

adhering on a vertical metal surface. The indented numbers across the middle of figure 14–15 have attracted dust preferentially, as evidenced by the outline formed by the soil in some parts of the numbers.

The disposition of the dust on the blocks appears to be dependent on electrostatic effects connected with the nature of the surfaces and the details of the shape. This relationship implies that the dust had an electric charge. Whether the charge on the dust was acquired as a result of the photoelectric effect in the sunlight, or whether it was the consequence of having been handled by the astronauts, who, in turn, through friction during walking, would have acquired a substantial charge, is not known. A further investigation and a laboratory simulation of the process with the same surface samples would be desirable.

After array 1002 was sprinkled with soil and photographed, the array was brushed off with a nylon-bristle brush and photographed again (figs. 14–16 and 14–17). Most of the soil was removed as a result of the brushing action. A small amount of soil remained on panels 10 and 12 and on some of the screws. The striations on some panels (figs. 14–16 and 14–17) were evidently caused by the brush. After the photography of the array was completed, the array was packaged and returned to Earth. Further study of the overall effect of the lunar soil on the thermal samples is underway.

Summary

The Apollo 14 mission photography is the first look at definitely undisturbed lunar regolith surfaces. Closeup photography from the Apollo 11 and 12 landing sites was taken in areas that may have been disturbed by LM descent-engine ex-

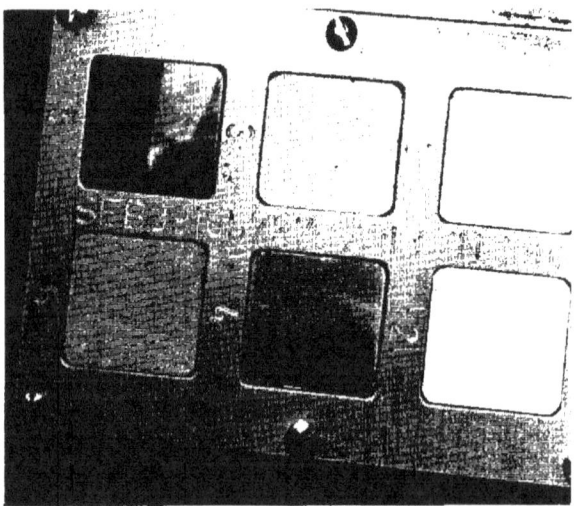

FIGURE 14–16.—Panels 1 to 6 of TDS 1002 after the soil was brushed off (AS14–77–10364).

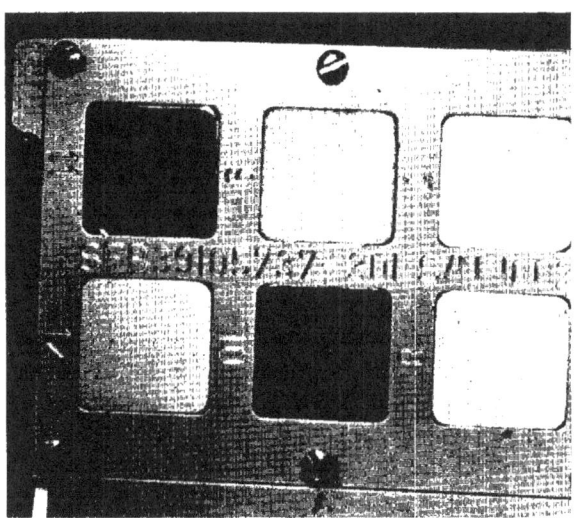

FIGURE 14–17.—Panels 7 to 12 of TDS 1002 after the soil was brushed off (AS14–77–10365).

haust. The crew described the surface at the Apollo 14 site as powdery (visible in 70-mm photography). The surface is covered with loose and partly buried, equant aggregates of soil and fragments of fragmental rocks of less than 0.1 mm to several millimeters in diameter. The equant aggregates give the powdery appearance to the surface. Within the resolution of the photography, no individual grains were visible. The coherence of these aggregates is easily destroyed, as can be seen in the photographs of MET tracks (figs. 14–2 and 14–3) and boot tracks (figs. 14–4 and 14–5).

Although the behavior of the lunar soil at the Apollo 14 site is qualitatively similar to that of the soils at the Apollo 11 and 12 sites, the co-hesion appears to be slightly less. Yet, the cohesion is still relatively stronger than the adhesion to any object placed against it and produces clean, distinct impressions.

The raindrop patterns noticed by the astronauts appear to be pits formed by the impact of micrometeoroids in the lunar soil. The stereopairs provide excellent documentation of the pits and should be studied further by investigators interested in crater morphology.

Photographs of the contact between the soil surface and rocks buried in the soil show the kind of fillet that can form in such cases. Linear patterns are evident in the powdery soil on a scale of millimeters in width and many centimeters in length. The origin of this phenomenon requires further investigation.

References

14–1. ANON.: Lunar Surface Closeup Stereoscopic Photography. Sec. 9 of Apollo 11 Preliminary Science Report. NASA SP–214, 1969.

14–2. GOLD, T.: Apollo 11 Observations of a Remarkable Glazing Phenomenon on the Lunar Surface. Science, vol. 165, no. 3900, Sept. 26, 1969, pp. 1345–1349.

14–3. GOLD, THOMAS: Apollo 11 and 12 Close-up Photography. Icarus, vol. 12, no. 3, May 1970, pp. 360–375.

14–4. GOLD, T.; PEARCE, F.; AND JONES, R.: Lunar Surface Closeup Stereoscopic Photography. Sec. 11 of Apollo 12 Preliminary Science Report. NASA SP–235, 1970.

ACKNOWLEDGMENTS

The assistance of David Carrier, Grant Heiken, Fred Pearce, and Robert Jones of the NASA Manned Spacecraft Center in the preparation of this report is gratefully acknowledged.

15. Gegenschein-Moulton Region Photography
From Lunar Orbit

L. Dunkelman,[a]† C. L. Wolff,[a] R. D. Mercer,[b] and S. A. Roosa [c]

The Apollo 14 investigation was the first phase of this experiment and was considered to be an operational test to determine the feasibility of performing dim-light photography from the Apollo command module (CM) using an available camera equipped with a fast lens. The experiment will be continued on later missions. The 16-mm data acquisition camera equipped with an $f/0.95$ or $T/1$ lens was used. Even though the format of this camera is small for gegenschein photography, it provided an acceptable operational test and produced a preliminary set of exposures. Fifteen exposures were obtained, six more than required.

Vehicle pointing and stability were well within the desired limits of $\pm 1°$ and were much better than had been expected. Star fields were easily identified. The original negatives are being analyzed in detail with a microdensitometer. Analysis of the vehicle pointing and stability and the exposure level has provided the basis for the timeline and exposure sequences for the Apollo 15 mission experiment phase. Basic exposures of 3 and 1 min have been set as a result of the operational test during the Apollo 14 mission.

The Light Called the Gegenschein

The gegenschein (counterglow), which has been known for more than a century, is an extended light source, a very faint patch of light, visible near the antisolar point in the night sky under favorable conditions. This phenomenon can be

[a] NASA Goddard Space Flight Center.
[b] Dudley Observatory.
[c] NASA Manned Spacecraft Center.
† Principal investigator.

mistaken for a high cirrus cloud; however, a simple test can be used if one remembers that this faint patch is always located exactly opposite the Sun (i.e., the antisolar point). Under very favorable conditions, the gegenschein appears to be approximately 10° in diameter but, when the surrounding sky is brighter, it appears smaller or may even be undetectable. The only known aspect of the gegenschein phenomenon is that it is sunlight reflected to the observer by grains of dust in space. The location of the dust has been a continuing mystery. Near the turn of the century, Carl Moulton and H. Gylden theorized that the dust might accumulate, or at least linger, at the libration point L_1 of the Sun-Earth system (fig. 15–1). This theory was widely accepted for the first half of this century. More recently, however, doubts have arisen,

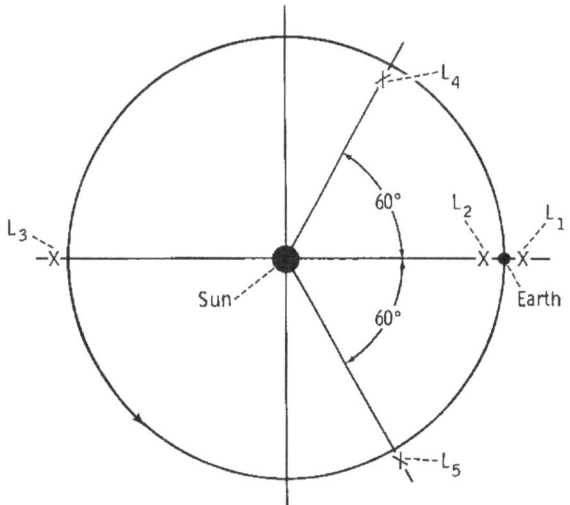

FIGURE 15–1.—The location of the libration points for the Sun-Earth system.

249

and this Apollo experiment is expected to settle the issue of whether any significant portion of the gegenschein is caused by a collection of particles at this libration point.

Libration Points

The libration points were discovered long ago by the French mathematician Lagrange. They are points of equilibrium in a three-body gravitating system, where gravitational forces caused by the primary and secondary bodies are balanced by centrifugal forces resulting from rotation of the system (ref. 15–1). Five such points exist (shown in fig. 15–1 for the Sun-Earth system). They are fixed in a coordinate system rotating about the Sun synchronously with the Earth. The orbit of the Earth is, of course, determined by a balance between centrifugal force and solar gravity. Farther out on the Sun-Earth line, there is an excess of centrifugal force over solar gravity, but there is a point (approximately 1.5 million km from Earth) at which this excess force is just compensated by the additional gravitational attraction of the Earth. This is the libration point L_1. Similar considerations determine the other points.

In theory, the points L_4 and L_5, which form equilateral triangles with the Sun and Earth, are stable equilibrium points and the others are unstable. In practice, the situation is complicated by perturbations caused by other planets and the Moon, and the detailed stability of the points remains controversial. It is known, however, that, in the case of the Sun-Jupiter system, the equilateral libration points contain accumulations of large rocks known as the Trojan asteroids. Furthermore, Earth-based observations of faint clouds at the equilateral points of the Earth-Moon system have been reported; these regions were also photo-graphed during the Apollo 14 mission, but the images have not yet been analyzed (sec. 18). It should be noted that it is possible for clouds of particles to persist even at unstable libration points if the loss of particles is sufficiently slow to be compensated by the influx and capture (through collision processes) of interplanetary dust.

An Alternate Explanation of the Gegenschein

An alternate hypothesis to Moulton's explanation is that the gegenschein is a phase effect (preferential backscatter) from sunlight incident on interplanetary dust that is farther out in the solar system. It is known that the zodiacal light, a luminous band lying in the ecliptic plane near the Sun (and hence only visible just before dawn and after sunset), is due to scattering of sunlight from micrometeoroids. Some scientists believe the gegenschein is merely an intensification of the zodiacal light in the antisolar direction because of the scattering properties of small particles. If this be the case, the light perhaps originates in the asteroid belt between Mars and Jupiter.

Triangulation

Because the orbit of the Moon offsets it from the Sun-Earth line by as much as 384 000 km, observations of the gegenschein from lunar orbit may allow triangulation to be used to determine which of these hypotheses is correct. If the light originates at the Moulton point L_1, it should be displaced approximately 15° from the antisolar direction, as seen from the Moon when it is near quarter phase. If no significant shift in position be observed, the light must come from much farther away (fig. 15–2) and can have no significant connection with the Earth.

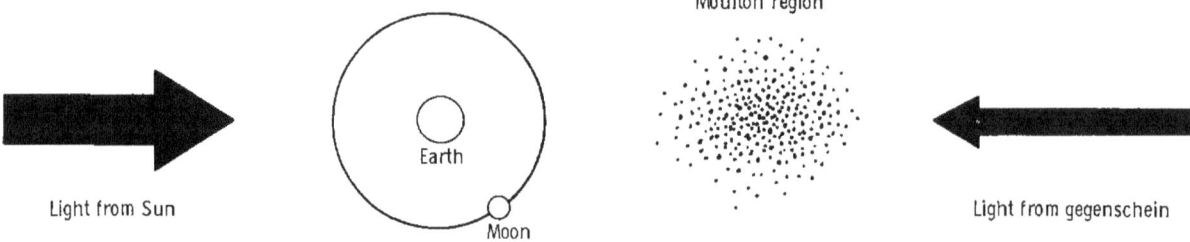

Light from Sun   Earth   Moon   Moulton region   Light from gegenschein

FIGURE 15–2.—The Moulton region.

Background Experiment Information

This experiment is a conclusive observational test of the contribution of dust at the Moulton point to the phenomenon of the gegenschein. In addition to the advantage of the geometry afforded by observations of the antisolar and Moulton region from lunar orbit, the optical environment is improved. The gegenschein has a very low brightness; therefore, it is important that the experiment photography be performed from as dark a region as possible. The Apollo 14 spacecraft in low lunar orbit spent approximately 25 min every 2 hr in the darkest region of the universe that man has reached, where both sunlight and earthshine are blocked by the Moon (a region referred to as the "double umbra"). Making observations from lunar orbit, of course, also removes the problems associated with the airglow of the Earth.

The earliest experiment on zodiacal light and gegenschein from a manned spacecraft occurred on the Project Mercury MA–9 flight (ref. 15–2), on which a specially modified, high-speed camera system was used. Similar investigations were conducted during several of the Gemini program missions (refs. 15–3 and 15–4) using standard flight camera systems containing higher speed films for their better light-collecting characteristics. A Gemini program experiment entitled "Dim-Light Photography" (refs. 15–4 to 15–7) provided partial success in similar investigative attempts. Related studies have been performed from rockets, balloons, and unmanned orbiting vehicles under NASA Goddard Space Flight Center sponsorship.

On previous manned space flights, degradation of data collected from low-intensity astronomical sources could be attributed to several factors. The more obvious of these are spacecraft-window cleanliness and transmission qualities, as well as light-scattering effluents from spacecraft purges, dumps, ventings, and depressurizations, particularly if there is any nearby natural light source such as airglow to illuminate these contaminants. A more subtle problem is the difficulty of avoiding extremely-low-level stray light from the spacecraft cabin and from spacecraft-thrustor firings during attitude changes and stabilization. The flightcrew must literally work in the dark during periods of data collection to avoid the problems that might obscure the phenomenon under study. In some

cases, the contaminating light levels are well below the visual threshold of the crew, and their presence is undetected. The phenomena being photographed are equally as faint, and the crew must use computed attitudes or secondary aiming points. Furthermore, a darkened cabin makes the use of clocks, attitude indicators, and written checklists difficult. This situation complicates the data-collection procedures and hampers the acquisition of necessary supportive data to the accuracy desired.

The Apollo 14 operational techniques were devised to eliminate the use of cabin lights near the camera and the use of thrustors while the camera shutter was open for the most critical data-collection tasks. In addition, a special window shield, designed primarily for this experiment, was installed to isolate the window and camera optics from the faint internal cabin lighting required to set up and time the exposures.

Operational Conduct of the Experiment

Three sets of photographs were required to meet the experiment objectives. Each set was to consist of two 20-sec exposures and one 5-sec exposure taken in quick succession. The first set was pointed near the antisolar direction; the second set, midway between the antisolar position and the computed position of the Moulton point as viewed from the Moon; and the last set, near the position of the Moulton point.

The 16-mm data acquisition camera was used with an 18-mm focal-length lens set at $f/0.95$ and Kodak 2485, black-and-white, high-speed recording film. (This type film had been used successfully to record faint dayglow from a rocket (ref. 15–8).) The camera was bracket mounted in the right-hand rendezvous window with a right-angle mirror assembly attached in front of the lens. A remote-control electrical cable was attached to the camera to allow actuation of the shutter from the lower equipment bay. The flight film had special low-light-level calibration exposures added to it both preflight and postflight. Several days before the time of the Apollo 14 data collection, ground-based photography of the gegenschein was obtained using identical equipment and film. The schedule required that the experiment be performed at 12:26 G.m.t., February 5, 1971, during the 15th orbit of the Moon. Communications between the Mission Control Center and the CM during experiment performance were not possible because of the spacecraft-Moon-Earth geometry.

Results

Preliminary inspection of the flight film has shown that the proper star field was photographed and that the spacecraft maintained a very steady inertial attitude during the time exposures. The analytic procedure consists of studying two sets of images: the flight film and the ground-based pictures taken from the McDonald Observatory nearly simultaneously with the Apollo 14 photography. Each set is measured on a microdensitometer that digitizes and transfers the entire picture to magnetic tape. Then, a computer program processes the tape and automatically removes all the bright stars in the picture. Another portion of the program corrects the picture for the known vignetting effect of the lens. Finally, a Cal-Comp plotter is used to plot equal brightness contours of the true sky brightness in the picture. On these plots, the gegenschein should appear as a small isolated hill on an otherwise-flat plane. Unless the gegenschein appears in the same place relative to the stars on both ground-based and Apollo 14 photography (after correcting for any differences in the times at which the exposures were made), parallax will have been detected. Consequently, the distance to the particle cloud can be computed.

All the computer programs have been written and tested; however, some minor improvements still remain to be added to the plotting program. Approximately half of the ground-based photographs have been processed with the microdensitometer; several have had stars successfully removed, and two have already been plotted. Because no significant difficulties have been experienced in the processing, measurements of the flight film can be started. However, the flight-film images will be mounted semipermanently between glass plates before further handling. It is already evident from the Apollo 14 experiment that the pointing accuracy of the spacecraft is sufficient to permit longer exposures (probably in the range of 1 to 3 min each) on the next flight. Also, a better measure of the exposure time and *f*-stop appropriate to the characteristics of the Kodak 2485 film has been gained. No information exists yet regarding the location of the gegenschein—the prime objective of the experiment—because this determination requires careful microdensitometric measurement.

References

15-1. DANBY, J. M. A.: The Three-Body Problem. Fundamentals of Celestial Mechanics. Macmillan Co., 1962, p. 192.

15-2. ANON.: Mercury Project Summary, Including the Results of the Fourth U.S. Manned Orbital Space Flight. NASA SP-45, 1963.

15-3. NEY, EDWARD P.: Night-Sky Phenomena Photographed from Gemini 9. Sky and Telescope, vol. XXXI, no. 11, Nov. 1966.

15-4. CAMERON, W. S.; DUNKELMAN, L.; GILL, J. R.; AND LOWMAN, P. D., JR.: Man in Space. Ch. 14 of Introduction to Space Science, 2nd ed., W. N. Hess, ed., Gordon & Breach Sci. Pub., 1967.

15-5. DUNKELMAN, L.; AND MERCER, R. D.: Dim Light Photography and Visual Observations of Space Phenomena From Manned Spacecraft. NASA TM X-55752, 1966.

15-6. DUNKELMAN, L.; GILL, J. R.; MCDIVITT, J. A.; ROACH, F. E.; AND WHITE, E. H., II: Geoastronomical Observations. Proc. Manned Space Flight Exper. Symp.: Gemini Missions III and IV (Washington, D.C.), Oct. 18-19, 1965, pp. 1-18.

15-7. ROACH, FRANKLIN E.; DUNKELMAN, LAWRENCE; GILL, JOCELYN R.; AND MERCER, ROBERT D.: Geoastronomical Observations. Sec. 32 of Gemini Midprogram Conference. NASA SP-121, 1966.

15-8. EVANS, D. C.; AND DUNKELMAN, L.: Airglow and Star Photographs in the Daytime From a Rocket. Science, vol. 164, no. 3886, June 20, 1969, pp. 1391-1393.

ACKNOWLEDGMENTS

The authors wish to express their gratitude to R. Scolnik of the Observational Astronomy Branch, NASA Goddard Space Flight Center, for his assistance in the observational program.

16. S-Band Transponder Experiment

W. L. Sjogren,[a]† P. Gottlieb,[a] P. M. Muller,[a] and W. R. Wollenhaupt[b]

The S-band transponder experiment, performed from the orbiting command and service module and lunar module (LM), provides detailed information on the near-side lunar gravity field. The data when completely reduced will encompass a 100-km-wide band along the Apollo 14 lunar-surface track shown in figure 16–1. The experiment uses the same technique of gravity determination employed on the Lunar Orbiter mission data from which the large anomalies called mascons (refs. 16–1 to 16–3) were first observed. No special instruments are required other than the existing real-time radio navigation system. The data consist of variations in the spacecraft speed as measured by the navigation system.

Techniques

The schematic drawing in figure 16–2 shows the basic measuring system. A very stable frequency of 2115 MHz obtained from a cesium reference is transmitted to the orbiting spacecraft. The transponder in the spacecraft multiplies the received frequency by the constant 240/221 (so as to avoid self-lockup) and transmits to the Earth. (The transmitted and received frequencies are within the S-band region.) At the Earth receiver, the initial transmitted frequency, multiplied by the same constant, is subtracted and the resulting cycle-count differences are accumulated in a counter along with the precise time at which differencing occurred. These cycle-count differences are the doppler shift in frequency f_d caused by the radial component V_r of the spacecraft velocity or $(2V_r/c)$ 2300 MHz, where c is the speed of light. At times of high resolution, the counter is read every second, whereas at low resolution, it is read once a minute. Not only is the cycle-count difference recorded but the fractional part of the cycle is measured. This process allows a resolution in the measurements of approximately 0.01 Hz or 0.65 mm/sec. This speed (range rate) observable is often referred to as doppler or line-of-sight velocity.

The raw data represent or contain many components of motion, and these must be removed before gravity analysis can proceed. Factors that must be accounted for include the tracking-station rotation about the Earth spin axis; the spacecraft motion perturbed by accelerations from the Sun, Earth, Moon, and planets; and atmospherics and signal-transit times. All these quantities are known a priori and are removed to accuracies well be-

FIGURE 16–1.—Apollo 14 lunar-surface track.

[a] Jet Propulsion Laboratory.
[b] NASA Manned Spacecraft Center.
† Principal investigator.

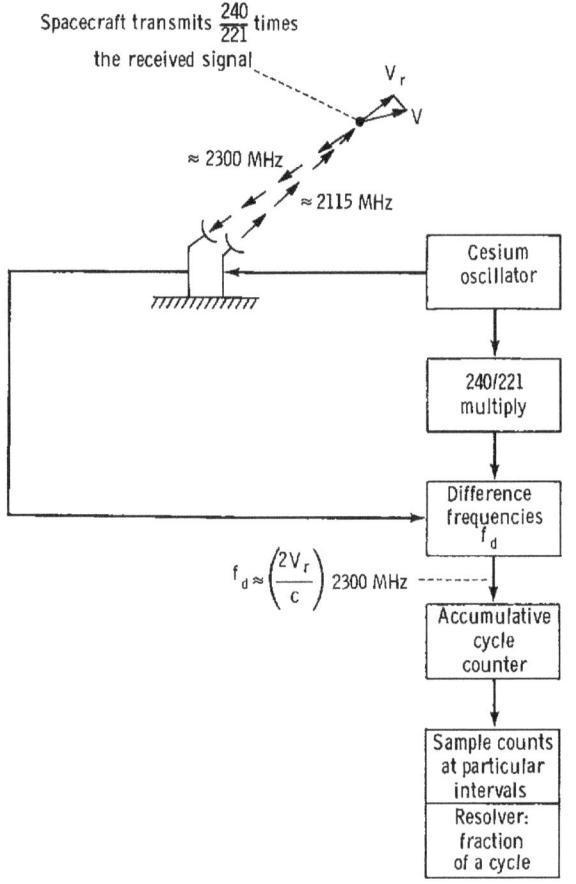

Spacecraft transmits $\frac{240}{221}$ times the received signal

$f_d \approx \left(\frac{2V_r}{c}\right) 2300\ MHz$

FIGURE 16–2.—Simplified schematic drawing of doppler transponder system.

yond those required to evaluate local gravitational effects.

Two approaches in reducing the resulting velocity data are possible. The first approach directly differentiates the velocity observations, and gravitational accelerations are immediately determined. The second approach estimates a surface-mass distribution from a dynamic fit to the observations.

The reduction procedure for the first approach was performed using the Jet Propulsion Laboratory orbit-determination computer program, which contains the theoretical model with all the dynamical constraints and parameters previously mentioned. Each orbit of data ($\approx$65 min) was evaluated independently, with the doppler observations least-squares-fitted using a spherical Moon and solving for (adjusting) only the six state parameters of initial position and velocity. The

resulting systematic residuals (i.e., real observations minus theoretically calculated observations) are then attributed to lunar gravitational effects. The velocity residuals from orbit 7 over Mare Nectaris are shown in figure 16–3. The signature is very definite and far above the noise level of 0.01 Hz. The Nectaris acceleration is predominant. These data were sampled every 10 sec (i.e., cycle-count readings averaged over 10 sec) so the noise level is lower than that on the 1-sec data residuals

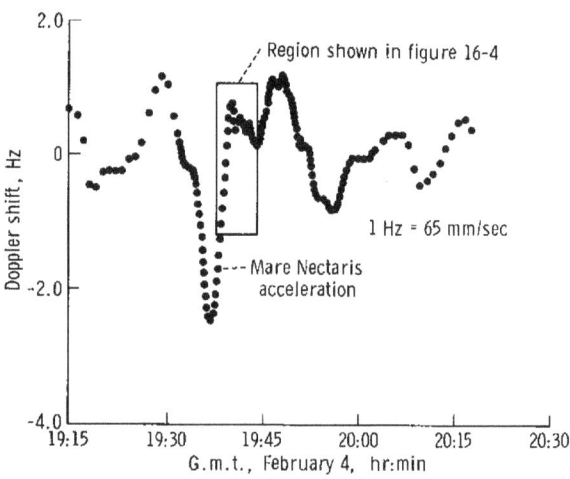

FIGURE 16–3.—Doppler residuals recorded at Madrid tracking station during orbit 7 at 10-sec sample rate.

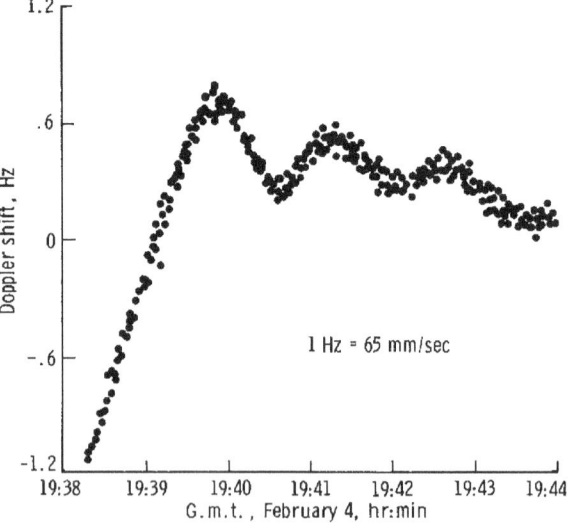

FIGURE 16–4.—Doppler residuals recorded at Madrid tracking station during orbit 7 at 1-sec sample rate.

shown in figure 16–4. It should be noted that figure 16–4 contains just that portion of figure 16–3 between 23 and 29 min past epoch and indeed verifies that the small variations in figure 16–3 are valid and not just noisy points.

Once these residuals are checked for consistency between adjacent orbits and all blunder points are removed, analytic patched cubic splines with second derivative continuous are least-squares-fitted to the residuals. These functions are then differentiated and the line-of-sight accelerations ("gravity") are analyzed and correlated with the subspacecraft lunar track and existing topography. At present, five of the 10 good orbits have been screened; when all are completed, the accelerations as a function of spacecraft lunar latitude and longitude will be plotted on a 1:1 000 000 Mercator projection through $\pm 70°$ of longitude. Because these are line-of-sight, rather than vertical, gravity components, a geometric effect exists that shifts gravity-feature locations toward the limbs and reduces the amplitude slightly. The shift is a little more than 1° and the amplitude reduction is about 30 percent for an object at 50° longitude. Altitudes over the 100-km band vary from 17 km at periapsis ($\approx 5°$ longitude) to 37 km at 50° longitude. No normalization factor has been applied to bring the accelerations to a constant-altitude surface. A plot of the accelerations (≈ 25-km altitude) from five orbits over the Nectaris mascon is shown in figure 16–5. The consistency of the adjacent orbits is clearly evident, and the increase in amplitude is precisely what should occur as the center of the Nectaris mascon rotates closer to the orbit plane.

The second approach is to estimate a dense surface grid of disk masses spaced approximately 1° to 2° apart and lying along the orbit surface track. This estimation will involve determining 200 to 300 masses along with the state parameters from some 20 independent orbits—10 at low altitude (17 km (9.2 n. mi.), periapsis) and 10 at high altitude (110 km (59.4 n. mi), periapsis). These results will be more quantitative than those derived from the first approach because all the geometric and dynamic effects will be accounted for and no spurious effects will result from the least-squares operations (which reduce the absolute amplitude of the residuals and sometimes introduce erroneous negative accelerations (ref. 16–4)). This reduction is in process at the time of writing this report.

The analysis of the impact orbit of the LM should be very interesting because very-low-altitude data were obtained. Resolution will increase and small features may be visible.

Preliminary Results

Preliminary implications from the scant data reduced to date may be deduced from figure 16–5, which shows gravity profiles over the Nectaris mascon. The flat-top characteristic of these curves is highly indicative of a flat surface feature, as was noted by Kane (ref. 16–5). Using models of a deeply buried point mass and a surface plate, he simulated both high-altitude (100 km) and low-altitude (30 km) gravity profiles. The two models appeared much the same at high altitude, but at 30 km the plate had the flat-top appearance. Thus, the first implication is that the mascons are likely to be near-surface features rather than deeply buried inhomogeneities.

Another fact from figure 16–5 is that the longitude of the maximum amplitude does not occur at the physical center (longitude 34° E) of Nectaris as one would expect. A geometric correction of 0.5° exists because of the viewing angle. The gravity center appears to be another degree west.

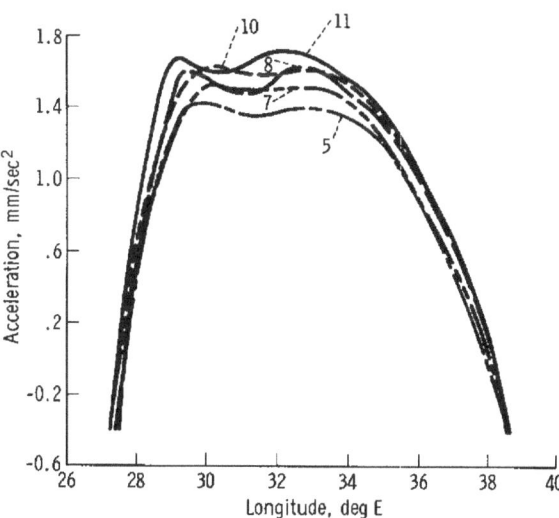

FIGURE 16–5.—Effects of Mare Nectaris mascon, orbits 5, 7, 8, 10, and 11.

This condition may be due to the Theophilus blanket thrown out over the old Nectaris mare. The additional shouldering at longitude 29.5° E may also be caused by the Theophilus blanket. If true, this observation would imply that no significant isostatic adjustment had occurred after the Theophilus event. Although not shown in figure 16-5, Theophilus has a perfectly centered negative-gravity anomaly implying that the Theophilus event caused a loss of mass and no subsequent isostatic adjustment has taken place. Many similar conclusions will no doubt be reached when the complete band of gravity data is finally mapped.

References

16-1. MULLER, P. M.; AND SJOGREN, W. L.: Mascons: Lunar Mass Concentrations. Science, vol. 161, no. 3843, Aug. 16, 1968, pp. 680–684.

16-2. MULLER, P. M.; AND SJOGREN, W. L.: Lunar Gravimetrics. Proceedings of Open Meetings of Working Groups at the 12th Plenary Meeting of COSPAR, Space Research X. North-Holland Pub. Co. (Amsterdam), 1970, pp. 975–983.

16-3. WONG, L.; DUETCHLER, G.; DOWNS, W.; SJOGREN, W. L.; ET AL.: A Surface Layer Representation of the Lunar Gravitational Field. J. Geophys. Res., vol. 76, 1971.

16-4. GOTTLIEB, P.: Estimation of Local Lunar Gravity Features. Radio Science, vol. 5, 1970, pp. 303–312.

16-5. KANE, M. F.: Doppler Gravity; A New Method. J. Geophys. Res., vol. 74, no. 27, Dec. 15, 1969, pp. 6579–6582.

ACKNOWLEDGMENTS

The authors wish to thank B. Hartley, B. Deluca, and R. Miller for expediting the data, and W. Stapper for an outstanding effort of data processing.

17. Bistatic-Radar Investigation

H. T. Howard [a][t] *and G. L. Tyler* [a]

The Apollo 14 bistatic-radar investigation uses radiofrequency electromagnetic scattering from the lunar surface to determine the principal electrical and structural properties of the lunar crust. Transmissions from the orbiting command and service module (CSM) are directed toward the lunar surface, and oblique reflections of these signals are monitored on the Earth.

The received signals are processed in a way that preserves the information (frequency, phase, polarization, and amplitude) contained in them as functions of time. Comparison of the received echoes with the known characteristics of the transmitted signal (through the application of a well-developed scattering theory (ref. 17–1)) yields quantitative information about lunar crustal properties, such as dielectric constant, average slope and slope probability, and small-scale surface roughness. These characteristics are of interest to several disciplines concerned with the problems of lunar history, evolution, and origin. The experiment results will probably prove most useful in defining the processes that tend to modify the lunar surface and in distinguishing between adjacent (and perhaps subjacent) geological units. The experimental observations are also of intrinsic interest to those involved in the study of electromagnetic scattering.

Both the radar technique and the associated theory have evolved rapidly in recent years. The Apollo 14 bistatic-radar experiment is a strong test of the convergence between theory and technique because numerous comparisons exist (ref. 17–2) that can be made with photographic, geological, magnetic, seismic, and sampling experiments conducted within the radar field of view.

[a] Stanford University.
[t] Principal investigator.

An ultimate goal of this work is the remote determination of the vertical crustal structure.

It will soon be possible to extend this technique to investigations of the Earth and other planets— the Earth, for resource investigations; Mars, for similar goals as were attempted for the Moon; and Venus, shrouded in dense clouds, for the only information man may ever gain on the construction of the surface.

For the experiment discussed in this report, it was possible to use continuous-wave transmissions from both the S-band telemetry system (the spacecraft-to-Earth communications link) and the very-high-frequency (vhf) communication system (normally used for voice transmissions and ranging between the CSM and the lunar module in lunar orbit). The techniques employed are similar to those that had been used previously in conjunction with the Lunar Orbiters 1 and 3 and Explorer 35 spacecraft (refs. 17–3 and 17–4). Simultaneous S-band (0.13-m) and vhf (1.2-m) observations were conducted on approximately two-thirds of one near-side pass while the CSM was maneuvered to maintain a predetermined orientation of the spacecraft antennas with respect to the lunar surface. There were four complete passes of vhf data alone. The vhf data were obtained in an inertial attitude hold, which resulted in a constant variation of the illumination with respect to the scattering geometry. Echoes of S-band transmissions were received with the NASA 210-ft antenna located at Goldstone, Calif.; and vhf echoes were received with the 150-ft antenna of the Stanford Research Institute.

Good data were obtained during all the observation periods. These data represent about an order-of-magnitude improvement in the signal-to-noise ratio and surface resolution over the results of previous experiments of this type. In addition, there has also been a significant increase in knowledge of the experiment parameters and controls. The simultaneous S-band and vhf observations form a unique data set.

At the present time, the major portion of the data-reduction process has been completed. Data analysis is awaiting certain ancillary Manned Spacecraft Center (MSC) ephemeris computations required to define the experiment geometry. The theoretical basis for the experiment, the experiment design, and the experiment data are described in subsequent portions of this report. Preliminary results are given in the form of echo power spectra and echo polarization spectra.

Basic Theory

The bistatic-radar echo is composed of the sum of the reflections from the area of the Moon that is mutually visible from the spacecraft and the Earth. Because continuous-wave transmissions are used, echoes from this entire area are received simultaneously. For the purposes of analysis, the echo signal may be considered to be arising from a large number of elemental surfaces, each with an area ds. Then, from the radar equation, the power received from a particular area ds is

$$dP_R = \frac{P_T G_T}{4\pi r_1^2} \sigma_0 \, ds \, \frac{1}{4\pi r_2^2} A \qquad (17\text{-}1)$$

where

dP_R = power received from the elemental area ds

P_T = transmitted power

G_T = gain of the transmitting antenna

r_1 = distance from the transmitter to ds

σ_0 = incremental radar cross section at ds

ds = elemental area on the lunar surface

r_2 = distance from ds to the receiving antenna

A = effective aperture of the receiving antenna

The geometry of the problem is illustrated in figure 17-1. The total received power is obtained by integration over the surface S (the surface that is mutually visible from the spacecraft and the Earth), or

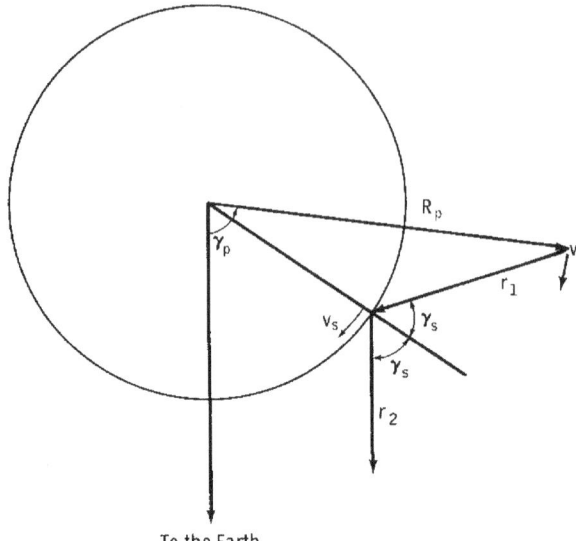

To the Earth

FIGURE 17-1.—Experiment geometry and parameter definition.

$$P_R = \frac{P_T A}{(4\pi)^2 r_2^2} \int_S \frac{G_T \sigma_0}{r_1^2} \, ds \qquad (17\text{-}2)$$

where it is assumed that the variations in r_2 over the area S are negligible and that G_T is constant over the area S. Both G_T and σ_0 are retained within the integral to emphasize the fact that they may vary considerably with the scattering geometry or, in the case of σ_0, with the location of ds. If G_T is constant over S, then the cross section may be separated from the other parameters, or

$$P_R = \frac{P_T G_T A}{(4\pi)^2 R_p^2 r_2^2} \sigma \qquad (17\text{-}3)$$

where

$$\sigma = R_p^2 \int_S \frac{\sigma_0}{r_1^2} \, ds \qquad (17\text{-}4)$$

and R_p is the distance of the spacecraft from the center of the Moon.

In general, the principal contributions to σ arise from a small region about the center of the first Fresnel zone, which is the specular point on the mean lunar surface. In ray-optics terminology, this is the point at which the angles of incidence and reflection are equal. If the Moon were a perfectly smooth sphere, all the echo would originate from a Fresnel-zone-site spot surrounding this point. By roughening the surface through the introduction of large-scale (with respect to a wavelength) topographic undulations, this spot is caused to break up into a number of glints. The location of these glints will correspond to specular reflection from local surface undulations. Echoes that result from this type of surface are designated quasi-specular.

Quasi-specular scattering constitutes the principal scattering mechanism in this experiment. To the first order, the effects of surface material composition and shape are separable; that is, if σ is the radar cross section of a perfectly conducting surface of a particular shape, then the radar cross section of a dielectric surface of precisely the same shape and with reflectivity ρ is

$$\sigma = \rho \hat{\sigma} \qquad (17\text{–}5)$$

or, from the radar equation

$$\rho = \frac{(4\pi)^3 r_2^2 R^2}{P_T G_T A \hat{\sigma}} \qquad (17\text{–}6)$$

Thus, if σ can be measured and $\hat{\sigma}$ is known, then ρ may be determined.

The quantity $\hat{\sigma}$ may be computed on the basis of statistical surface models. The results of one such computation for a gently undulating surface with a gaussian height distribution and autocorrelation function are given for bistatic geometry in figure 17–2. The radar cross section is plotted as a function of the angle formed by the spacecraft, the center of the Moon, and the Earth. The two families of curves (solid and dashed) are parametric in the normalized spacecraft orbital radius measured from the center of the Moon. The dashed curves correspond to a perfectly smooth sphere, and the solid curves correspond to a surface with root-mean-square (rms) slopes of 10°. The effect of the surface slopes on the cross section is second order. The reflectivity of the Moon is inferred from the data by measuring the total echo powers received and by normalizing the results by the use of the theoretical reflectivity values (plotted in fig. 17–2 from the data given in ref. 17–4) for a perfectly conducting sphere. The inferred reflectivity is then compared with the reflectivity of a dielectric surface under oblique geometry. The effective dielectric constant of the surface may be determined directly from the observation of the Brewster angle and indirectly by a quantitative comparison of the reflectivity values.

The bandwidth of the echo depends directly on the surface slope. For the surface model considered previously, the bandwidth is given by

$$\Delta f = 4.9 \left(\frac{v_s}{\lambda} \right) \cos \gamma_s \tan \beta_0 \qquad (17\text{–}7)$$

where

$\Delta f =$ one-half the power bandwidth of the echo spectrum

$v_s =$ velocity of the specular point with respect to the lunar surface

$\lambda =$ wavelength

$\gamma_s =$ angle of incidence on the mean surface at the specular point

$\beta_0 =$ rms surface slope

On the basis of this model, surface slopes may be inferred directly from the orbital parameters and the width of the echo spectrum.

Physically, the spectrum is broadened according to the probability-density function of the surface slopes and the doppler shift. In figure 17–1, reflection from a point ahead of the mean reflecting point will have a greater doppler shift (determined by the angle between $\bar{r}_1$ and $\bar{v}$) than those

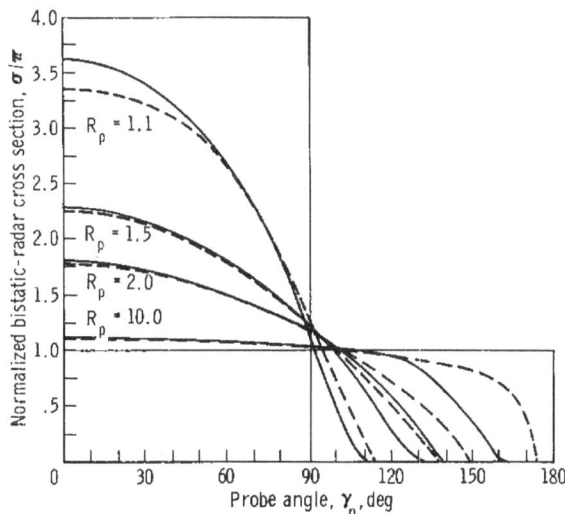

FIGURE 17–2.—Normalized bistatic-radar cross section σ/π as a function of spacecraft-Moon-Earth angle γ_p for a perfectly reflecting sphere with a gently undulating surface.

points behind the mean reflecting point. At every point (on the basis of the quasi-specular model), the probability of obtaining a reflection depends on the probability of finding a local surface undulation with the proper slope. Slopes, then, also determine the surface resolution of the experiment. Reflections are obtained from an area with a radius of 5 to 10 km (which is approximately equal to the rms slope multiplied by the spacecraft altitude).

Inferences based on this model are valid on a set of scales related to the wavelength of the probing wave. Quantitative comparisons of slope distributions inferred from Explorer 35 data and from photogrammetry have been conducted with good results for a limited number of locations on the lunar surface (ref. 17–5). Under fairly broad assumptions, the quasi-specular scattering may be considered to occur at a fictitious surface that is a low-pass-filtered version of the actual surface. Although the bounds on this filter cutoff have been only approximated, it is known that they scale with the length of the probing wave. Typically, slopes on the order of 10 wavelengths (or longer) are expected to be important in the scattering process. Thus, for quasi-specular scattering, bistatic S-band data are sensitive to surface structure on the order of 1.3 m (and larger); and for vhf, 12 m is the lower bound. A more complete discussion of these theoretical concepts and results is available in reference 17–6.

Surface-reflectivity measurements are also sensitive to wavelength. Dry geological materials with approximately the density of the lunar regolith exhibit loss tangents that are independent of the radiofrequency, for frequencies greater than approximately 10 MHz (ref. 17–7). Penetration depths between 10 and 20 wavelengths are typical. Thus, the reflection coefficient inferred from S-band data will be sensitive to vertical structure within the lunar crust to a depth of 1 to 2 m and, from vhf data, to a depth of approximately 10 to 20 m. Such penetration effects have been observed with Explorer 35 data obtained at a wavelength of 2.2 m (ref. 17–8).

Diffuse scattering arises from wavelength-size (and smaller) surface structures and from second-order effects of the gently undulating surface. In the lunar case, the diffuse component of the reflected radiation is normally associated with the presence of large numbers of wavelength-size (or smaller) rock or rock fragments. Very small rocks will be in the Rayleigh regime and will not contribute individually to the echo. Some attempts to provide quantitative descriptions of diffuse scattering in terms of rock distributions from the Moon have been made (ref. 17–9). However, in terms of surface structure, the diffuse scattering is not understood nearly so well as the quasi-specular scattering.

Experimentally, quasi-specular and diffuse scattering can be distinguished by its polarization and coherence properties and by the scattering law (ref. 17–10). Quasi-specular scattering, which by definition originates from those portions of the surface that produce mirrorlike reflections, is deterministically polarized and is the predominant scattering mechanism. Although the echo polarization will change with variations in the polarization of the illuminating wave and the geometry, the polarization is the same as would be produced by a smooth surface of the same material. Diffusely scattered waves are not expected to exhibit this behavior. To the extent that the diffuse component arises from randomly oriented structures on or within the surface, it will be unpolarized. A decomposition of the echo spectrum into the polarized and unpolarized components provides a mechanism for separating the scattering from large-scale (wavelength) surface and small-scale randomly oriented features or roughness.

Equipment Description

Receivers

Schematic block diagrams of the receiving and data-processing systems are shown in figure 17–3. The NASA Deep Space Network 210-ft parabolic antenna at Goldstone, Calif., was used to receive the S-band signals. Both the open- and closed-loop receivers, which were installed for the Mariner spacecraft to be orbited about Mars in 1971, were used for the bistatic-radar experiment. Normal Apollo mission operations are conducted with the regular ground-station receivers. A signal-conditioning unit processes the Mariner receiver 10-MHz intermediate-frequency output. This unit determines the overall system bandwidth for bi-

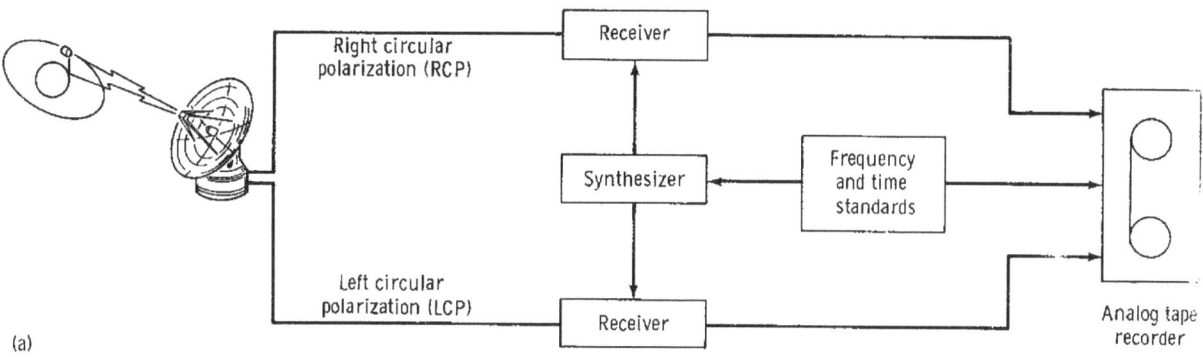

(a)

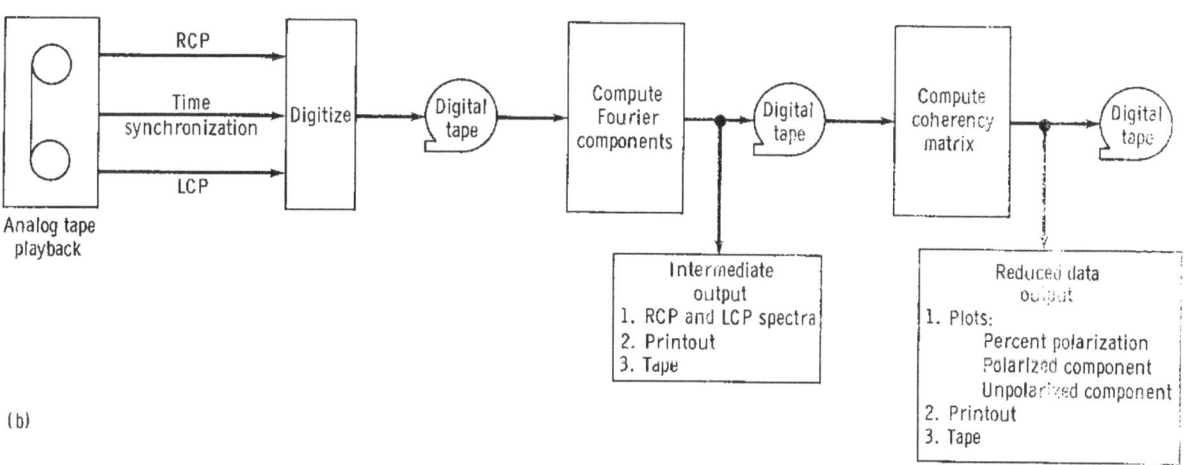

(b)

FIGURE 17–3.—System schematics. (a) Signal-receiving system. (b) Data-processing system.

static echoes, provides signal-level control, and produces an audiofrequency for magnetic-tape recording.

The use of the open- and closed-loop systems provides redundancy and additional operational flexibility. Because the cost of the digital data reduction is directly proportional to the signal bandwidth, it is desirable to keep this bandwidth as small as possible. However, the possibility existed that the CSM-transmitted frequency would change by several times the echo bandwidth during the experiment.

Eighty-kHz filters are used in conjunction with the open-loop receivers. This bandwidth was selected to insure that the echo would be within

the passband at all times during observation with only a single, predetermined frequency setting. The closed-loop (tracking receiver) bandwidth is 20 kHz. This bandwidth is sufficiently wide to insure that the echo is within the passband as long as the receiver is locked on the direct signal from the CSM. The disadvantage of using the closed-loop system alone is that brief periods exist when the direct signal fades below the threshold required for receiver lock. Wideband data can be processed for those periods when lock is lost. On the Apollo 14 mission, only one period of several minutes was experienced during which the direct signal was below the closed-loop threshold.

Both left and right circular polarizations are

received. All receivers are driven from a single frequency source, which enables the relative phase between polarizations to be preserved. System noise temperature in this configuration is near 30° K when the antenna is aimed at the sky alone and 192° K when the Moon fills the beam of the antenna.

The vhf receiving facility was the 150-ft parabolic antenna at the Stanford Center for Radar Astronomy. A complete, open-loop receiver was constructed, which consists of solid-state preamplifiers for 259.7 MHz with frequency conversions to 50 MHz, 10 MHz, and audio. The data bandwidth is determined at 10 MHz by 3.5-kHz-wide multipole crystal filters. This bandwidth is narrower than the bandwidth with S-band, by the ratio of the transmitted frequencies. Only open-loop channels were used. As was the case for the S-band receiving system, left and right circularly polarized signals were received, and the system is coherent. The system noise temperature is approximately 700° K.

Open-loop operation is similar at S-band and vhf. Based on a doppler ephemeris calculated from elements supplied by MSC personnel, the receivers are tuned so that the direct signal will be centered in the passband at the time the CSM crosses the Earth-Moon line. The closed-loop receiver is initially tuned according to the operational-frequency predictions for the CSM. Once lock is achieved, the receiver automatically compensates for doppler effects.

Two magnetic-tape recorders are used simultaneously for data recording. Tapes are started at different times so that overlapping records with no gaps for tape changes are available.

Data Reduction

Data reduction consists of a three-step process that is independent of the data source, S-band (open or closed loop) or vhf. This process is outlined in the signal channel shown in figure 17-3(b). First, the analog tapes are replayed and digitally sampled. The sampled data are converted to weighted Fourier coefficients and spectral estimates. Finally, the weighted Fourier coefficients from the two polarization channels are combined to determine the polarization spectra of the echo. As the analog data are played back, the signals are low-pass filtered to avoid aliasing of the high-frequency tape-recorder noise in the sampling process. Sampling is synchronized with the original recording time through the use of a NASA 36-bit time code and a synchronizing waveform, both of which are multiplexed onto the data tracks of the tape recorder. The two receiver channels, for right and left circular polarization, are sampled simultaneously so that the coherence between channels is preserved. Calibration signals are recorded, sampled, and processed in the same manner as the data.

Weighted Fourier coefficients are computed using fast Fourier transform techniques. Groups of 1024 data samples (from each channel) are multiplicatively weighted with a sine-squared data window, and the Fourier coefficients are computed. Because the three analog-data sources are each sampled at different rates, the corresponding frequency and time resolutions of the spectral estimates are not uniform. When the effects of the data window are considered, spectral resolution of approximately 40 Hz is achieved with the closed-loop S-band data. A spectral resolution of approximately 10 Hz is obtained with the vhf data.

The Fourier coefficients are easily manipulated to provide a variety of data presentations. For example, sums of the squares of successive Fourier coefficient magnitudes yield spectral estimates of the received signals. The time resolution and stability of these estimates may be varied simply by changing the number of terms included in the time average. Spectral estimates for signals in two orthogonal polarizations may be combined with the cross spectra to obtain the polarization properties of the echo.

Results

At the time of this writing, the data obtained simultaneously on S-band and vhf have been sampled, the Fourier coefficients computed, and the preliminary spectra examined. The data have not yet been correlated with position of the reflecting region on the lunar surface except in a most general way. Some typical results are presented in figures 17-4 to 17-7.

An S-band power spectrum is shown in figure 17-4. In this and the following figures, frequency increases to the right and power increases linearly

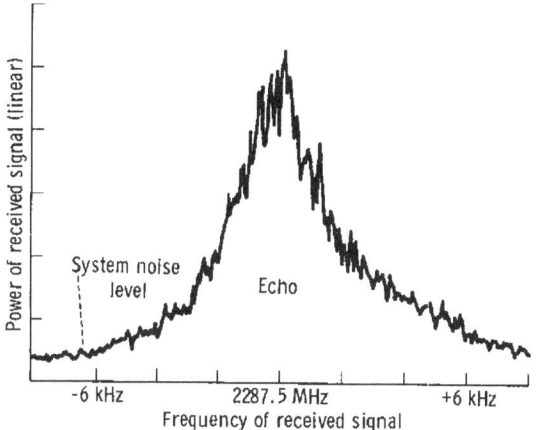

FIGURE 17–4.—S-band bistatic-radar spectrum taken over a smooth mare region.

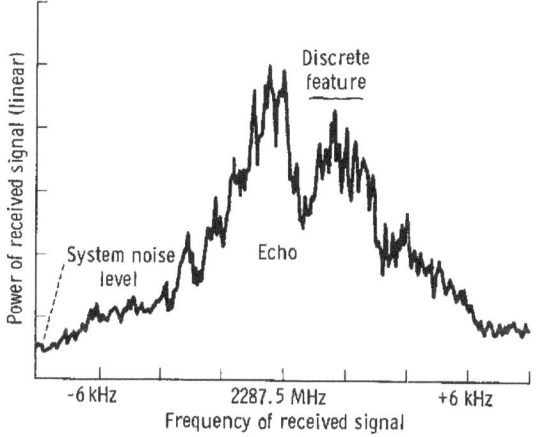

FIGURE 17–5.—S-band bistatic-radar spectrum with a strong discrete feature observable on the high-frequency side of the quasi-specular return.

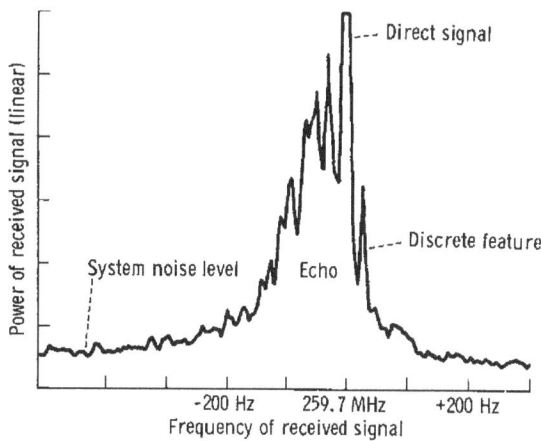

FIGURE 17–6.—A vhf bistatic-radar spectrum with a strong direct signal surrounded by quasi-specular return showing a discrete feature.

with height. The relative smoothness and the slightly asymmetrical shape are predicted by the gaussian model for the lunar mare region from which this signal was reflected. Comparison with figure 17–5 emphasizes this in that the quasi-specular echo from a fairly homogeneous lunar region is still present, but a double-peaked reflection from a large-scale surface inhomogeneity or feature is also present. This secondary reflection persists for approximately 2 min. A vhf spectrum is shown in figure 17–6. The tall, very narrow spike is the direct signal. It is surrounded by the quasi-specular return and several discrete features.

A summary of several minutes of S-band data taken as the CSM passed over the Lansberg Crater is shown in figure 17–7. Frequency increases to the right, and the ordinate is linear in the polarized part of the echo power spectrum. The frequency resolution is approximately 40 Hz. The plot gives total *polarized* power in one of the final data-reduction formats. Initially, the "normal" quasi-specular return dominates the center portion of the spectrum. A small reflection can be seen at the extreme right. As time progresses (2.5 sec between lines from the bottom to the top), this additional feature increases in amplitude, splits into two parts, and shifts downward in frequency. Later, a third, distinct reflection appears to the left of the other two. The two outside reflections are from the crater walls, and their separation is a measure of the crater size. As time progresses, the crater echoes move through the principal reflection zone, at which point the crater clearly dominates the reflection process. As the reflecting region moves past the crater, the normal signal from the homogeneous surface reappears.

Conclusions

The conclusions possible at this stage of data analysis are mainly technical rather than scientific. All facets of the experiment—from the CSM maneuvering through final plotting of polarization parameters—worked well. Experiment resolution

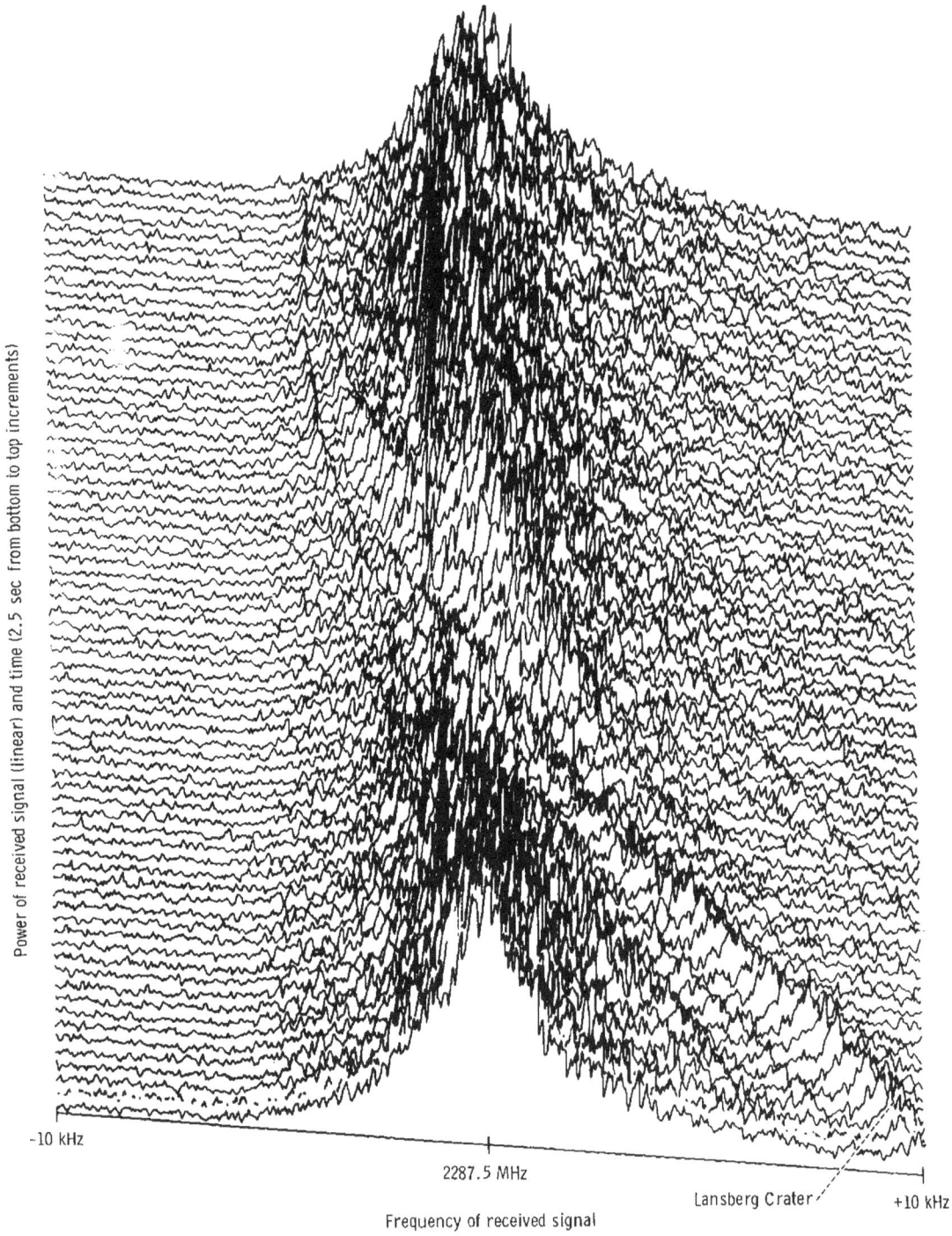

Power of received signal (linear) and time (2.5 sec from bottom to top increments)

-10 kHz

2287.5 MHz

Lansberg Crater   +10 kHz

Frequency of received signal

FIGURE 17–7.—S-band bistatic-radar spectra of Lansberg Crater area.

and signal-to-noise ration sufficient to permit detailed studies of moderate-sized lunar features were achieved. For fairly homogeneous regions of the lunar surface, the data can be used to obtain surface parameters of direct interest to lunar geology. The effects of the bulk surface electrical properties (such as the Brewster angle) are clearly evident at both the vhf and S-band frequencies.

In these cases, the two sets of frequency data will provide information that is either not obtainable in other ways or that can be obtained only through the use of much more laborious methods. Analysis of these data in conjunction with Apollo and Lunar Orbiter photography should provide a more detailed picture of the lunar surface than is possible from either type of observation alone.

All the signal characteristics described in the portion of this report entitled "Basic Theory" have been observed. For example, the general differences between mare and highland areas are clearly distinguishable at both frequencies as are numerous smaller areas within both the maria and highlands. Further, there are many small, localized scattering features, many instances of an observable diffuse component, and large variations in the total received power throughout the data.

Data analysis of the signals is just beginning. The next step is to combine the data with detailed trajectory and echo-locus computations. Normalization according to the predicted bandwidth and radar cross section will permit the two data sets to be compared. In addition, a number of other measures of the echo shape (such as skewness and internal fluctuation statistics) will be used to define lunar-surface characteristics. These steps will be performed for each 10 km of motion of the reflecting area along the lunar surface. The majority of the spectra (such as shown in fig. 17-7) that appear to originate from fairly homogeneous surface regions will be reduced systematically. Other spectra, such as the data obtained from Lansberg Crater, will require a more detailed and specialized interpretation.

Appendix

Data-Processing Formulas

Weighted Fourier transforms. The analog output of the receiving system and the appropriate timing signals are initially recorded on magnetic tape for reduction later. The principal data-reduction steps consist of analog-to-digital conversion, application of a sine-squared data window, and the computation of the corresponding Fourier coefficients with fast Fourier transform techniques. More explicitly, if $\hat{d}_i$ represents one of a long

series of data samples, then these samples may be grouped according to $d_j^n = \hat{d}_i$, where $i = nN + j$ for $j \leq N$ and n, N, and j are all positive integers or zero. The sine-squared weighted complex Fourier coefficients are

$$f_k^n = \sum_{j=0}^{N-1} \sin^2\left(\frac{2\pi}{N}j + \frac{\pi}{N}\right) d_j^n e^{i(2\pi/N)jk} \quad (17A\text{-}1)$$

where $\leq k \leq (N-1)$ and $i = \sqrt{-1}$. In the Apollo data-reduction programs, $N = 1024$ and each set of coefficients corresponds to a time interval $T = 1024$ times the sampling rate. These coefficients may be combined to form spectral estimates

$$F_{k'} = \sum_{m=1}^{M} \sum_{l=1}^{L} |f_{Mk'+m}^l|^2 \quad (17A\text{-}2)$$

where all the indices are integers. The variance of the estimates is approximately $(ML)^{-1}$ times the mean squared value. Summation over l represents integration in time for a period LT, while summation over M is equivalent to broadening the analysis window in frequency. Such summations are equivalent to postdetection averaging in analog spectral-analysis schemes. The resulting values for $F_{k'}$ are proportional to the energy received in a passband with an approximate width MT^{-1} in time TL.

Polarimetry (separation of polarized and unpolarized components). The polarized and unpolarized components of the echo signal may be separated by statistical behavior. The normalized coherency matrix (ref. 17-11) of the signals is

$$\rho_k = \begin{pmatrix} \displaystyle\sum_{n=1}^{N} |_1f_k^n|^2 & \displaystyle\sum_{n=1}^{N} {}_1f_k^n {}_2f_k^{n*} \\ \displaystyle\sum_{n=1}^{N} {}_1f_k^{n*} {}_2f_k^n & \displaystyle\sum_{n=1}^{N} |_2f_k^n|^2 \end{pmatrix}$$

$$\times \frac{1}{\displaystyle\sum_{n=1}^{N} |_1f_k^n|^2 + \sum_{n=1}^{N} |_2f_k^n|^2}$$

$$= \begin{pmatrix} \rho_{11} & \rho_{12} \\ \rho_{21} & \rho_{22} \end{pmatrix} \quad (17A\text{-}3)$$

where $_1f$ and $_2f$ represent signals received on two orthogonal polarizations. It follows that the fractional polarization γ_k is

$$\gamma_k = (1 - 4 \text{ Det } \rho_k)^{1/2} \qquad (17A\text{-}4)$$

When γ_k has been computed, the polarized component of the spectrum can be calculated

$$\gamma_k \sum_{n=1}^{N} (|_1f_k^n|^2 + |_2f_k^n|^2) \qquad (17A\text{-}5)$$

while the unpolarized component is the complement, or

$$(1 - \gamma_k) \sum_{n=1}^{N} (|_1f_k^n|^2 + |_2f_k^n|^2) \qquad (17A\text{-}6)$$

It should be noted that a polarization spectrum is obtained as a function of the frequency index k.

References

17-1. EVANS, J. V.: Radar Studies of Planetary Surfaces. Annual Review of Astronomy and Astrophysics, vol. 7, L. Goldberg, David Layzer, and J. G. Phillips, eds., Annual Reviews, Inc. (Palo Alto, Calif.), 1969, pp. 39-66.

17-2. THOMPSON, T. W.; ET AL.: A Comparison of Infrared, Radar and Geologic Mapping of Lunar Craters. Proc. Geophys. Interpretation Moon, Lunar Science Institute, June 1970.

17-3. TYLER, G. L.; ESHLEMAN, V. R.; FJELDBO, G.; HOWARD, H. T.; AND PETERSON, A. M.: Bistatic-Radar Detection of Lunar Scattering Centers With Lunar Orbiter I. Science, vol. 157, no. 3785, July 14, 1967, pp. 193-195.

17-4. TYLER, G. L.; INGALLS, D. H. H.; AND SIMPSON, R. A.: Stanford Telemetry Monitoring Experiment on Lunar Explorer 35. Stanford Electronics Lab. Final Rept. SU–SEL–69–066, Oct. 1969.

17-5. TYLER, G. L.; SIMPSON, R. A.; AND MOORE, H. J.: Lunar Slope Distributions: A Comparison of Bistatic Radar and Photographic Results. J. Geophys. Res., vol. 76, no. 11, Apr. 1971, p. 2790.

17-6. TYLER, G. L.; AND INGALLS, D. H. H.: Functional Dependences of Bistatic Radar Frequency Spectra on Lunar Scattering Laws. J. Geophys. Res., vol. 76, 1971.

17-7. CAMPBELL, MALCOLM J.; AND ULRICHS, JURIS: Electrical Properties of Rocks and Their Significance for Lunar Radar Observations. J. Geophys. Res., vol. 74, no. 25, Nov. 1969, pp. 5867-5881.

17-8. TYLER, G. L.: Oblique-Scattering Radar Reflectivity of the Lunar Surface: Preliminary Results From Explorer 35. J. Geophys. Res., vol. 73, no. 24, Dec. 1968, pp. 7609-7620.

17-9. THOMPSON, T. W.; POLLACK, J. B.; CAMPBELL, M. J.; AND O'LEARY, B. T.: Radar Maps of the Moon at 70-cm Wavelength and Their Interpretation. Radio Science, vol. 5, no. 2, Feb. 1970, pp. 253-262.

17-10. BECKMAN, PETR; AND SPIZZICHINO, ANDRÉ: The Scattering of Electronic Waves From Rough Surfaces. International Series of Monographs on Electromagnetic Waves, Pergamon Press, 1963.

17-11. BORN, MAX; AND WOLF, EMIL: Principles of Optics. Pergamon Press, 1959.

ACKNOWLEDGMENTS

The observations described herein could not have been carried out without the help of a large number of people from several organizations. The authors gratefully acknowledge the assistance of Patrick Lafferty, Louis Leopold, Stuart Roosa, and Harley Weyer of the NASA Manned Spacecraft Center; Jim Raleigh of Bellcom, Inc.; Allen Chapman and Booth Hartley of the Jet Propulsion Laboratory; and Robert Dow, William Faulkerson, John Williamson, and Barbara Warsavage of the Stanford Center for Radar Astronomy.

18. Orbital-Science Photography

PART A

VOLCANIC FEATURES IN THE FAR-SIDE HIGHLANDS

Farouk El-Baz [a]

The role of volcanism in the formation and modification of lunar-surface features has been a matter of controversy, especially with regard to the lunar highlands. The relatively younger features in the maria are generally well preserved, whereas those in the highlands are somewhat ambiguous. However, in a few cases, geomorphology and spatial relationships identify certain features as being of volcanic origin.

Terrain volcanism has been described in several areas on the near side of the Moon. For example, in reference 18–1, several of the units that form the Kant Plateau materials are attributed to volcanism, mainly viscous flows. Two regional volcanic units, one west of Mare Nectaris and the other west of Mare Humorum, have also been mapped (ref. 18–2). In addition to these units, plains-forming materials and deposits that are associated with some craters and rilles mantle the lunar terrain. These are also attributed to volcanism as summarized in references 18–3 and 18–4.

Landforms of probable volcanic origin are also quite prevalent on the lunar far side. Some were revealed for the first time in Lunar Orbiter photography (ref. 18–5). Others were depicted by Apollo 8 photography; for example, the flow scarps and cones (ref. 18–6). Still others were only discernible in Apollo 10 photography, as in the case of the deposits related to King Crater, as reported in reference 18–7.

To test some of the theories regarding far-side upland volcanism, three photographic strips were planned for the Apollo 14 mission. The photographic target strips, assigned nos. 14, 3, and 15, provided examples of the far-side highlands from 160° E to the eastern limb (fig. 18–1). The photographs were taken on the 14th orbit of the Moon.

The Hasselblad camera with SO–368 color film and a 500-mm lens was used. The camera was hand held and no spacecraft image-motion compensation was attempted. A ringsight on the camera was used for aiming at the targets. All three strips and two additional targets were successfully acquired, and the quality of the photographs is excellent. The three sequences are covered as follows:

(1) Target 14 (west-northwest of Chaplygin Crater), frames AS14–72–9947 to 9959

(2) Target 3 (vicinity of King Crater), frames AS14–72–9961 to 9978

(3) Target 15 (west of Pasteur Crater), frames AS14–72–9979 to 10003, and northward toward the eastern rim of Mare Smythii, frames AS14–72–10004 to 10030

Description of Features

Probable volcanic features displayed in these photographs include two conjugate craters west-northwest of the Chaplygin Crater, a lava lake and flow scarps associated with the King Crater, volcanic flows associated with a 35-km crater called "the bright one" by the Apollo 14 crew, furrowed and grooved terrain west of the Pasteur Crater, and calderalike craters on the east rim of Mare Smythii. Each of these features is discussed in the following paragraphs.

Two Conjugate Craters

Approximately 40 km from the northwestern rim of Chaplygin Crater are two unnamed con-

[a] Bellcomm, Inc.

267

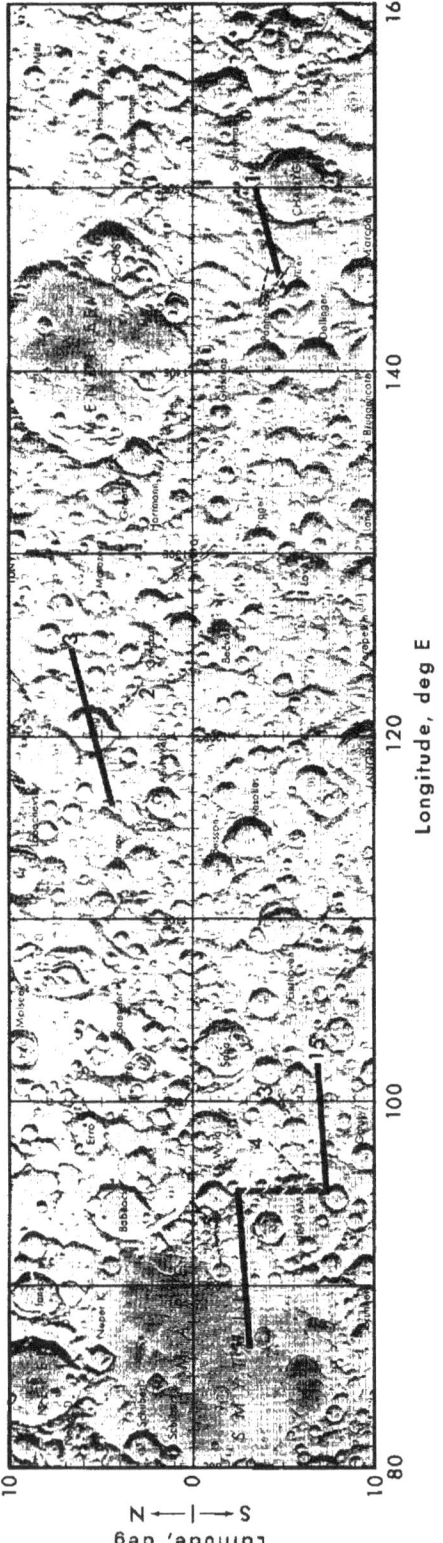

FIGURE 18-1.—Index map of far-side photographic target strips 14, 3, and 15. Two conjugate craters are located at point 1; King Crater, at point 2; a bright-haloed crater, at point 3; and an elongate depression in hilly terrain, at point 4. Multiringed craters on the east rim of Mare Smythii are located at point 5.

jugate craters that display unique characteristics. The crater to the north is approximately 35 km in diameter, and the one to the south is 25 km from rim to rim. Both craters display fairly smooth rims including those portions at the juncture between the two craters. The craters are shallow and the floors are contiguous. The floor material is distinctly different from that in neighboring craters. It is somewhat darker than the surrounding material and displays fractures that appear to be endogenetic in origin. The fractures in the crater to the north are concentric with the rim, whereas

FIGURE 18-2.—Two conjugate craters (point 1 of fig. 18-1) on the lunar far side. The Sun-elevation angle is 18° (Lunar Orbiter 1 photograph H-115).

those in the southern crater form a network that gives a turtleback appearance (fig. 18–2).

The Apollo 14 high-resolution photographs of these two conjugate craters were taken at low Sun-elevation angles (~6°). This situation resulted in the enhancement of contrast and the clear display of the small-scale features. As shown in figure 18–3(a), the floor displays an elongate depression, approximately 5 km in length, and a number of rimless and low-rimmed craters. The floor of the smaller crater that is dissected by a network of fractures or rilles is shown in figure 18–3(b). These display a V-shaped cross section and fairly smooth walls.

The aforementioned characteristics lead one to the conclusion that the two conjugate craters are probably of volcanic origin. The floor material is made of volcanic flows, and the fractures are probably shrinkage cracks produced during the cooling of the lava. This interpretation is also supported by the fact that the craters do not display significant rim deposits and that lobate flow scarps are distinct within the floor material. The morphology of these flow scarps suggests a somewhat viscous flow.

King Crater

King Crater (formerly International Astronomical Union no. 211) is approximately 75 km in diameter and displays a generally round, partly crenulated rim.

The crater is situated in as-yet-undivided highland materials in the general area previously known as the Soviet Mountains (ref. 18–8). The crater exhibits a raised, wavy, and sculptured rim and terraced interior walls, which indicate, although not unequivocally, an impact origin. It is not discernible in the photographs whether the crater is rayed. The presence of an extensive ray system is generally regarded as a strong criterion for the impact origin of the younger lunar craters.

The crater is a few kilometers deep, and the depth of the floor in relation to the rim crest varies with the amount of fill. The crater wall is terraced up to six levels, and the highest terrace is steeper than most, a feature common to craters of similar size. The floor of the crater displays a prominent central peak that is forked. It forms a unique Y-shape (fig. 18–4), with the right arm trending nearly due north.

(a)   (b)

FIGURE 18–3.—Two high-resolution frames of different portions of the conjugate craters shown in figure 18–2. (a) The northwestern quadrant of the northern crater (AS14–72–9959). (b) The eastern half of the southern crater (AS14–72–9954).

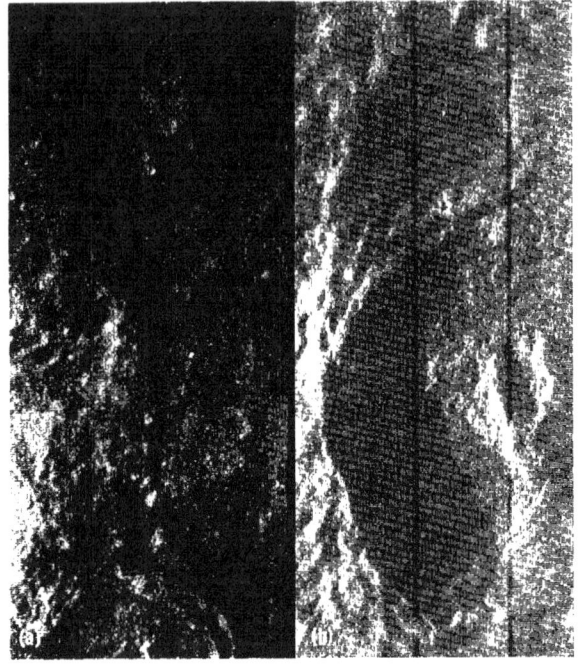

FIGURE 18-4.—Portions of King Crater and the area to the north. (a) The northern rim and area to the north. Dark and light materials are apparent in the rim (AS14-60-8665). (b) The Y-shaped central mountain and dark pool on the north rim (AS14-71-9851).

On the basis of Apollo 10 photography, four different types of materials had already been noted in the walls of the crater (ref. 18-7). These types were distinguished by color, texture, and morphology. Also, some tabular bodies of high albedo materials and wall-like bodies that cross the walls, rim, and floor materials were described as igneous intrusions. The interpretation of these features as igneous intrusions was used as supportive indications of the heterogeneity of lunar materials, as

well as the plausibility of intrusive igneous activity on the Moon (ref. 18-7).

The Apollo 14 high-resolution photographs constitute a useful complement to existing imagery of the crater and environs. The high obliquity of the photographs (approximately 60°) enhances details that were unnoticeable previously. A mosaic of the oblique photographic strip is shown in figure 18-5. This view provides additional evidence to the previously stated interpretation of the dark materials on the north rim of the crater. The flat, relatively smooth appearance of this material and the collapsed sinuous depressions within the unit indicate that the material originated as a lava lake. Some flow scarps are noticeable within and around the borders of this unit.

An additional flow scarp can be seen for the first time to the southwest of the crater. The nature of the lobate flow fronts is made obvious by the obliquity of the view. This particular scarp, however, may represent the terminus of a debris flow or rock glacier.

"The Bright One"

The so-called bright crater is a young (Copernican) crater. Although the crater is only 35 km in diameter, its bright halo dominates a surrounding area of approximately 150 km in diameter (fig. 18-6). Its rays, as well as those of a neighboring, somewhat smaller crater approximately 50 km to the northeast, join the rays from Bruno Crater to the north. This bright area of the lunar far side (at high Sun) was erroneously interpreted as the Soviet Mountains, a mountain range not related to a circular basin. However, Apollo photographs showed that the bright region is due mainly to crater rays (ref. 18-8).

FIGURE 18-5.—Oblique view of King Crater (AS14-71-9967 to 9972).

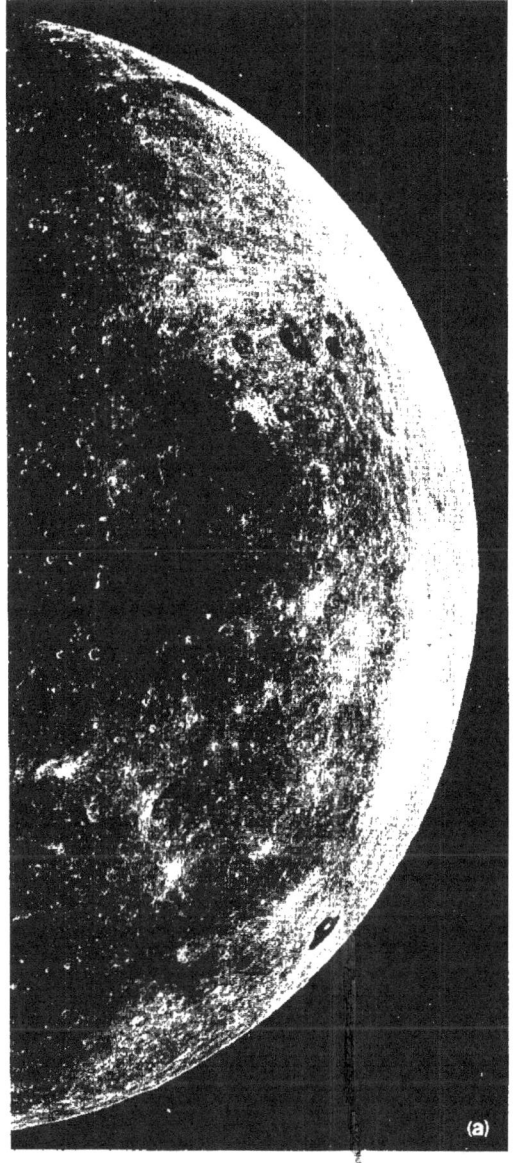

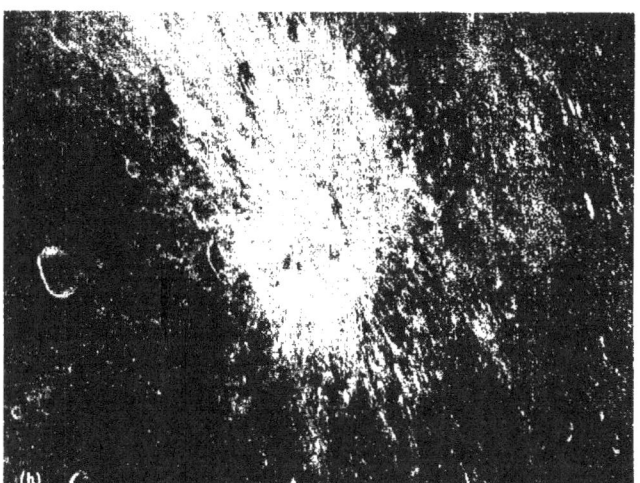

FIGURE 18-6.—Bruno and Tsiolkovsky Craters and a halo around a young crater. (a) The two joining ray systems of Bruno Crater (to the north) and two craters north of the dark-floored Tsiolkovsky Crater (to the south) (AS8-14-2506). (b) A bright halo, approximately 150 km in diameter, that surrounds a young crater located at latitude 5° S, longitude 123° E (AS8-12-2189).

The bright crater displays an irregular shape because of slumping of the wall materials (fig. 18-7). The rim crest is very sharp and the crenulations are wavy. The high-resolution photograph taken during acquisition of target 3 (King Crater) to the north reveals interesting details of the crater interior. As the spacecraft moved over the crater, the command module pilot (CMP) swung the camera down to acquire the photograph and swung it back to complete the King Crater photostrip.

As shown in figure 18-8, the interior of the crater is very hummocky. The lack of a flat floor is striking and the excessive terracing is unique. Prominent flow scarps can be seen, and flow fronts are not always lobate; one major front is unusually straight. Small fractures apparently caused by drag can be seen in the main flow in the upper right portion of the frame. The unusually large quantity of blocks on the terraces and the virtual lack of craters suggest that this is an extremely young crater. These characteristics, when coupled with the gross morphology, provide evidence that King Crater is the youngest lunar crater of its size range ever photographed.

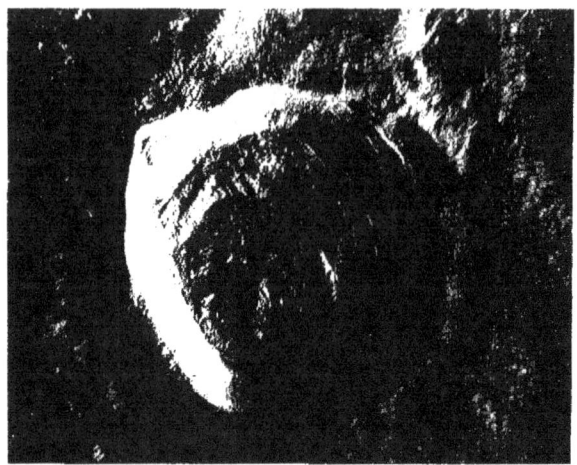

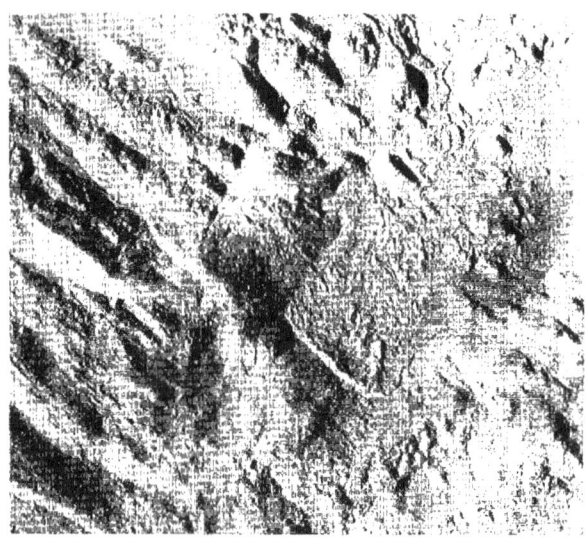

FIGURE 18-7.—View of the bright-haloed crater ("the bright one") shown in figure 18-6 (80-mm lens was used). The low-Sun angle (approximately 17°) is favorable for studying the details of the crater wall and floor and the fine textures on the ejecta blanket (AS14-70-9671).

FIGURE 18-8.—A high-resolution (500-mm lens) photograph of the portion of the hummocky interior of the crater shown in figure 18-7. A raised flow scarp is in the middle of the photograph, and numerous blocks are evident on the terraces (AS14-72-9975).

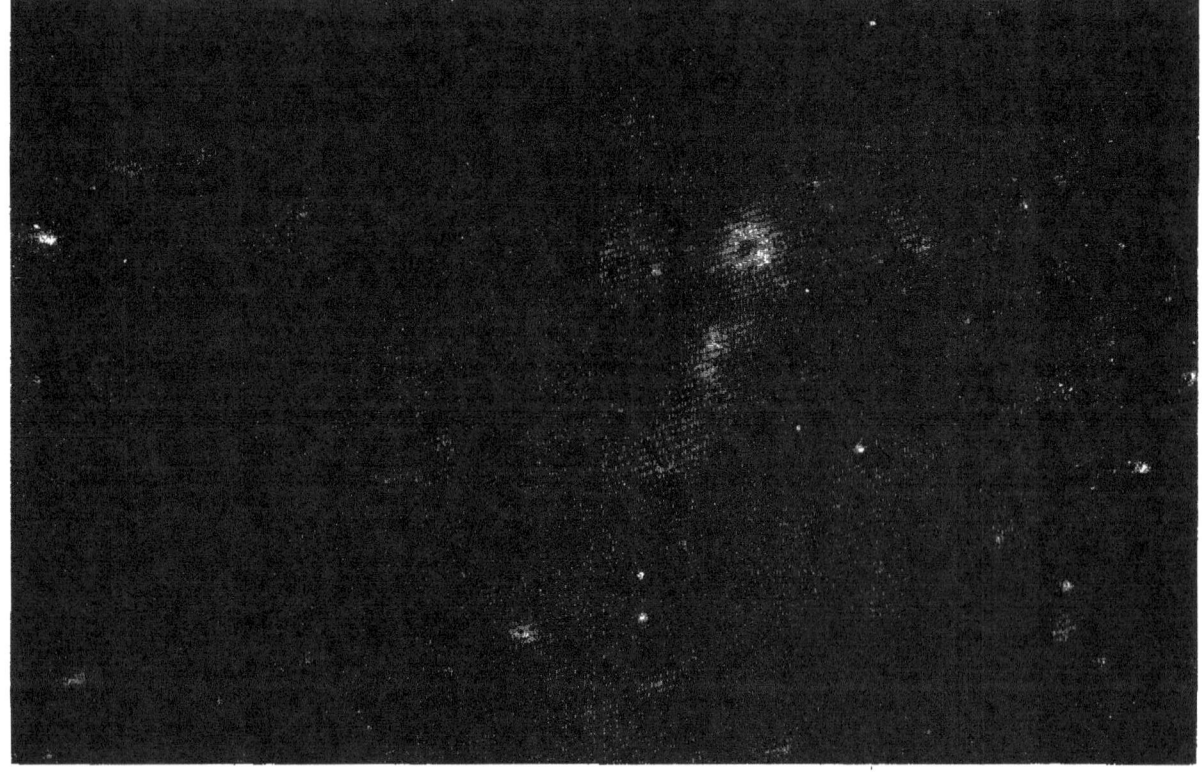

FIGURE 18-9.—Elongate depression, furrows, and grooves in hilly terrain northwest of Pasteur Crater (AS14-72-9993, left: AS14-72-10000, right).

Elongate Grooves

Several small elongate grooves and depressions were noted within the hilly terrain northwest of Pasteur Crater. The grooves and furrows range between 5 and 30 km, as revealed by Lunar Orbiter photography. The exact morphology of these grooves and furrows was, however, not discernible and their origin not known.

The Apollo 14 high-resolution stereophotographs reveal that these elongate features are of probable volcanic origin. The unit in which these grooves are located is somewhat hilly and displays characteristics reminiscent of the Kant Plateau materials (ref. 18-1). Similar units also exist in the far-side highland farther east. Most of the grooves display a raised rim, and many have the appearance of conjugate crater chains (fig. 18-9).

Terrestrial analogs to these features are formed by eruptions of magmas of intermediate composition. Although the nature of the material is not known, the morphology of these lunar features strongly suggests a volcanic origin.

The larger of the furrowlike craters and most smaller grooves are oriented in a north-south direction. Indications are that the craters are structurally controlled and that a tectonic belt with north-south fractures exists in this area. It is further speculated that that tectonic belt may be a physical expression of a dividing line between a thick crust in the far side and a thinner crust in the near side.

East Rim of Mare Smythii

Mare Smythii, a near-circular basin, displays units with varying albedo. Near the eastern rim of the basin are somewhat hummocky units of high albedo. They are somewhat similar to the plains-forming units within large craters and appear to be made of older mare material. One example of the type of terrain is shown in figure

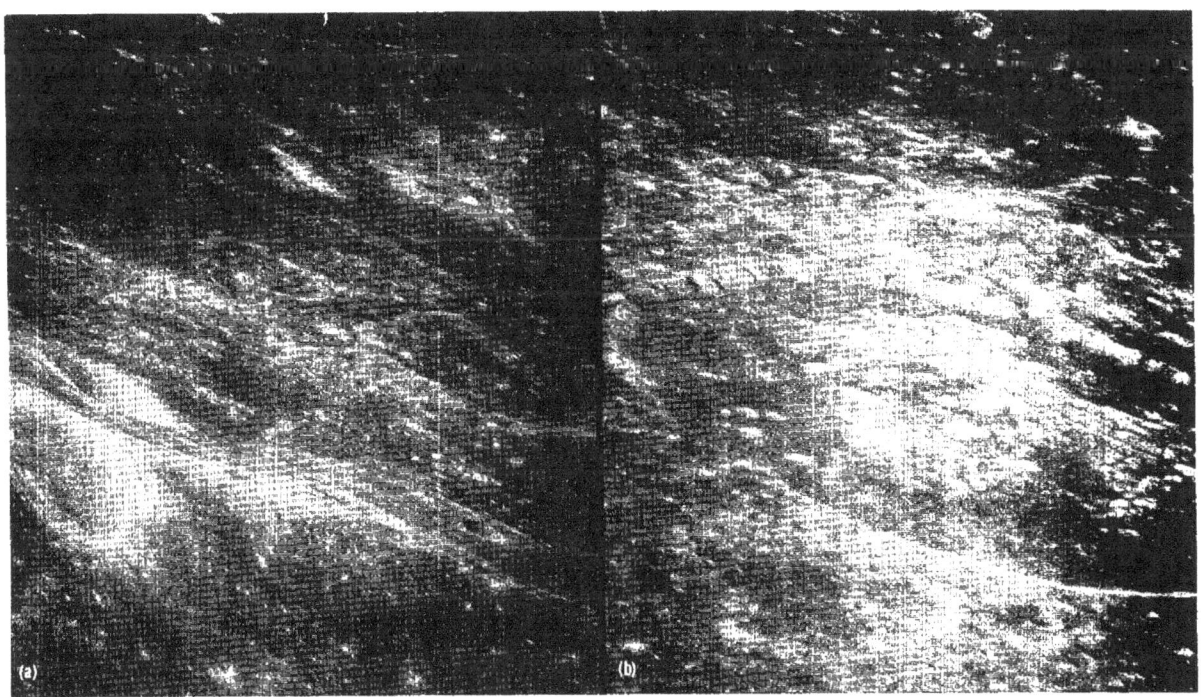

FIGURE 18–10.—Probable volcanic features near the eastern rim of Mare Smythii. (a) A sinuous depression (AS14–72–10008). (b) A multiringed crater pair (AS14–72–10026).

18–10 as a sinuous rille that crosses the middle of the photograph.

Several of the craters in this area appear to have been flooded and almost completely filled by mare-type materials. Although no distinct flows are decipherable, examples of flooded craters exist in the mare itself, as shown in figure 18–10. These multiringed craters appear to have been formed by eruptions and successive collapses much like calderas on Earth. They do not display hummocky ejecta on the rims, and no indications of an impact origin can be seen. Similar, but less distinct, craters on the east rim of Mare Smythii can be seen, and the setting also suggests that these craters are calderas formed by volcanic eruptions.

PART B

PRELIMINARY GEOLOGIC RESULTS FROM ORBITAL PHOTOGRAPHY

D. H. Scott,[a] *M. N. West,*[a] *B. K. Lucchitta,*[a] *and J. F. McCauley*[a]

New geologic features as well as previously unrecognized details on the morphology, structure, and stratigraphy of the Moon were revealed during the preliminary examination of more than 800 Apollo 14 70-mm Hasselblad photographs. The photographs strengthen or, in some cases, refine previous interpretations of the origin and relative ages of many of the geologic units previously mapped on the near side. The new orbital photography has also enhanced knowledge of the processes involved in the emplacement of these units. Certain photographs suggest that some previous interpretations should be reconsidered or modified to accommodate the new relationships shown more clearly by the Apollo 14 orbital photography. Only the most visually dramatic examples of this new and useful collection of photographs have been selected for annotation.

One of the most significant new features, particularly with respect to the site chosen for the Apollo 15 landing (Apennine-Hadley region), is a sinuous rille that has distinct continuous levees extending along the entire length. One sinuous rille with a distinct, but only partly leveed, channel in the Marius Hills region had previously been described (ref. 18–9). The narrow but continuous overbank deposit along the rille, shown in color photography AS14–73–10120 to 10122 (fig. 18–11), has not been seen in previous photographs. This rille is relatively small (approximately 15 km long by 700 m wide) and begins and ends in mare material that embays the southeast rim of Lansberg Crater. The narrow width of the levees (approximately 100 m) contrasts sharply with the relatively broad ridges commonly asso-

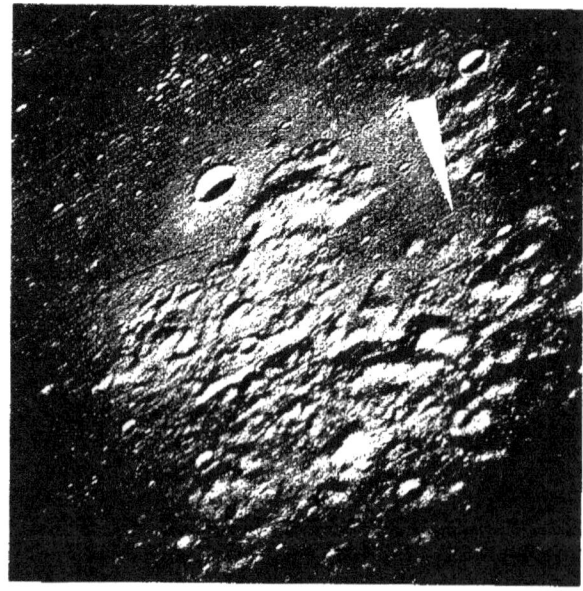

FIGURE 18–11.—Sinuous rille with leveed channels in mare material near Lansberg Crater (north toward upper right; AS14–73–10120).

[a] U.S. Geological Survey.

ciated with terrestrial collapsed lava tubes (ref. 18–10). The levees more closely resemble those associated with stream channels, lava channels, and certain types of highly fluid debris and mud flows. The fluidization-channel hypothesis (ref. 18–11) could be applicable in light of these new photographs. The arguments put forth (ref. 18–12) in support of a collapsed-lava-tube origin for the leveeless Hadley rille do apply to the majority of sinuous rilles previously photographed but are not applicable to this newly noted feature. Three alternative explanations present themselves with regard to sinuous rilles as a result of the three cited photographs.

(1) Two genetically distinct classes of lunar sinuous rilles exist; those without levees are collapsed subsurface lava tubes or flow channels as suggested in reference 18–12.

(2) Rilles with levees or overbank deposits are the results of surface fluid flow in open channels with the fluid medium possibly, but not necessarily, consisting of lavas.

(3) Alternatively, partly leveed rilles may have formed by both open-channel and subsurface lava flow, or some other type of flowage may be involved.

The lunar surface may contain many more leveed rilles that have not been detected in previous photographs because of inadequate resolution or lighting problems. The previously mentioned alternatives are discussed in part G of this section. Orbital photography and the surface studies planned at the Apollo 15 site may help to clarify the origin of these still-enigmatic features.

A distinct color anomaly can be seen in photographs (AS14–73–10142 to 10148) of the eastern wall of Langrenus Crater. A distinct rusty color is evident in one of a group of smooth bulbous domes with single or multiple furrows at or near the crests (fig. 18–12). Except for the dome exhibiting the color anomaly, all have a light-gray, faintly speckled appearance. These domes appear morphologically distinct from the sharper crested, more irregular hills that lie concentrically around the inner wall of Langrenus Crater and that are evidently slump blocks from the crater walls.

The rusty color in the domelike hill is evident in all seven of the cited photographs. A patch of

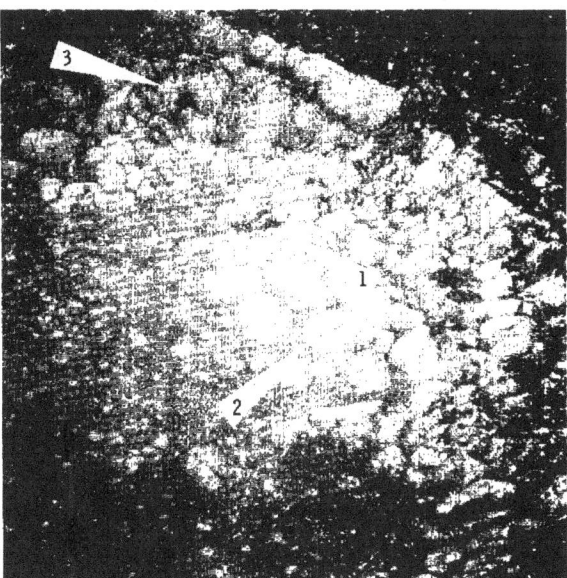

FIGURE 18–12.—East wall of Langrenus Crater. The color anomaly is located at point 1; bulbous domes on crater floor, at point 2; and typical slump blocks from crater walls, at point 3 (north toward upper left; AS14–73–10142).

dark material at the estimated location of this dome can be seen in Apollo photograph AS10–32–4676, thus confirming that the anomaly is not an artifact of the color photographs. The color anomaly may be of little regional geologic significance. However, in reference 18–13, it is pointed out that color differences on the lunar surface are attributable to gross differences in petrologic composition. The location of the dark material among a group of morphologically distinctive bulbous structures suggests that post-Langrenus endogenetic processes may be responsible for both the domes and the color anomaly.

Another, but less distinctive, color anomaly is evident in the young, small, 4-km, dark-halo Beaumont L Crater (fig. 18–13). The crater walls exhibit horizontal, wavy, locally discontinuous, alternating dark and rusty-colored bands, and the crater floor has a brownish hue. The implications of this color anomaly are even less clear than those of the anomaly previously described. The rusty appearance of the crater interior may be attributed to its apparent youth, if the experiments discussed in reference 18–14 are applicable. More recently formed, bright-appearing

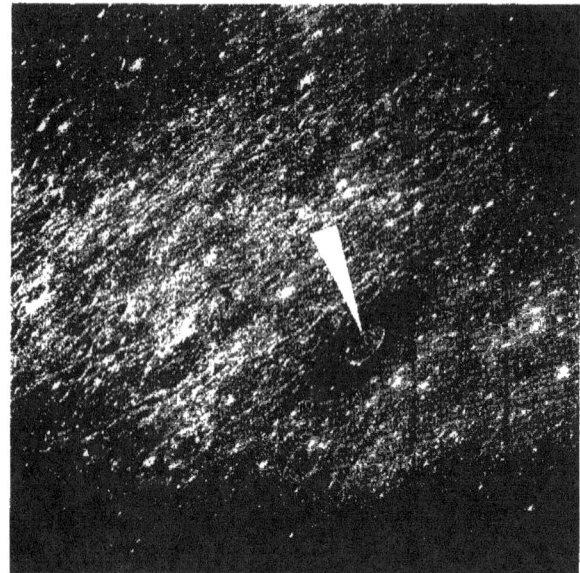

FIGURE 18-13.—Beaumont L Crater (4-km diameter) with dark halo and interior. The arrow points to the color anomaly in walls (north approximately toward right; AS14-73-10040).

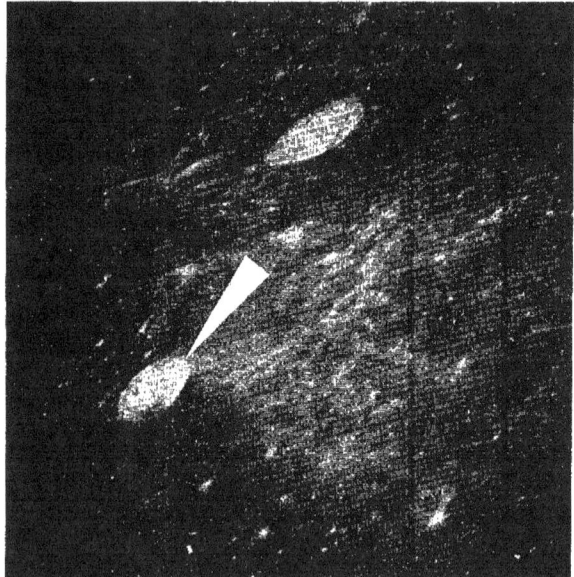

FIGURE 18-14.—Cyrillus G Crater (7.5-km diameter) near Beaumont L Crater (north toward lower right; AS14-73-10047).

craters, as measured from telescope data, are the most colorful of lunar features (ref. 18-13).

Color differences on the Moon in the visible range of the spectrum were previously measured instrumentally (refs. 18-13 to 18-15). The two colored features show that variations of spectral reflectivity are locally intense enough to be recorded on photographic color emulsions. These two areas are thus appropriate targets for further color and multispectral orbital experiments.

The Beaumont L Crater is of additional interest because of the dark halo. In reference 18-1, the nearby 7.5-km bright crater Cyrillus G (fig. 18-14) is considered to be of impact origin; Beaumont L Crater, on the other hand, was considered as being of questionable origin although both craters are obviously of Copernican age because they lie on the ejecta blanket of Theophilus, a Copernican crater. Beaumont L Crater is considered the younger of the two on the basis of fewer superposed impact craters (light speckles) on the dark halo. Cyrillus G Crater appears to have as many small craters on its surrounding ejecta blanket as are present on the surrounding terrain. Similar morphologies of the latter two craters, despite the color and albedo disparities, strongly suggest that both are of impact origin. Cyrillus G appears to penetrate light-hued terrain composed, in this area, mainly of older crater deposits, whereas Beaumont L Crater excavated dark mare material from beneath the Theophilus Crater ejecta blanket.

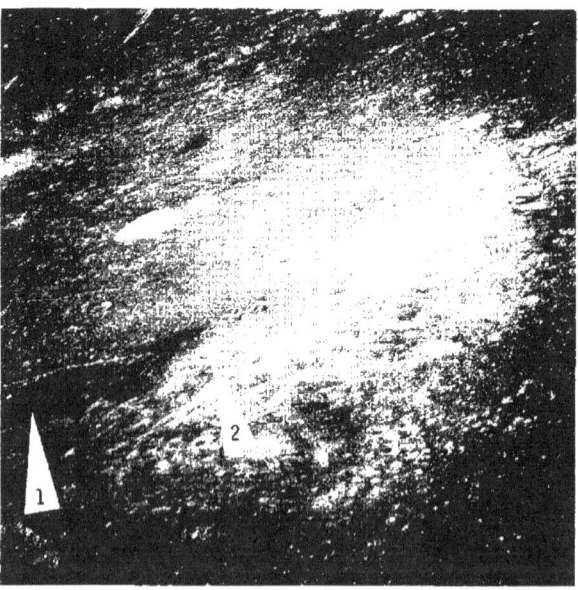

FIGURE 18-15.—Floor of Alphonsus Crater showing a smooth-rimmed dark-halo crater with dark interior walls at point 1 and a typical bright-walled impact crater at point 2 (north toward right; AS14-73-10093).

The origin of dark-halo craters has been a long-standing problem since the early days of telescopic geologic mapping, when all such craters were necessarily considered to be maars (ref. 18–16). The Ranger and Lunar Orbiter mission results subsequently suggested that at least two varieties of dark-halo craters exist on the Moon. The mapping of Alphonsus Crater at the 1:250 000 scale (ref. 18–17) showed that most of the dark-halo craters on its floor were structurally controlled. The mapping of the northeastern part of the floor of Alphonsus Crater at the 1:50 000 scale (ref. 18–18) indicated that the dark-halo craters, particularly the Alphonsus MD Crater, were morphologically distinct from the dark-halo craters present on the ejecta blankets of Theophilus (Beaumont L, for example) and Copernicus craters. The Alphonsus-type craters are gently convex upward from the edge of the recognizable dark blanket inward to the crater lip, whereas the latter are pronouncedly concave upward. In addition, Earth-based full-Moon photography indicated that the Alphonsus-type crater has dark interior walls as opposed to the bright interiors of the Copernicus- and Theophilus-type craters. The relatively poor resolution of the Earth-based photography precluded the development of a strong observational case for these differences in the reflectivity of the crater interiors. An excellent sequence of near-down-Sun oblique pictures of the Alphonsus dark-halo craters (fig. 18–15) are provided in photographs AS14–73–10090 to 10097. In successive photographs, the Sun-facing interior slopes of Alphonsus KC Crater are distinctly seen to be nearly as dark as the surrounding rim deposits. This observation suggests that the same type of material drapes both surfaces, as is the case with well-preserved terrestrial volcanic craters. Apollo 14 photography has thus strengthened the hypothesis that at least two basic varieties of dark craters are indeed present on the Moon. Certain impact events penetrate and excavate intrinsically dark subjacent materials that are deposited mostly within the ejecta blankets; the steeper, more brecciated and unstable interior walls brighten quickly by mass wasting, thus exposing blocks and in some cases the uppermost layers of bedrock. The lower interior slopes are talus covered. Certain volcanic craters bring up

material of uniformly dark albedo and distribute this tephra more or less uniformly over the interior walls of the vent and in the surrounding dark blanket. It appears that, as these volcanic craters age by meteoritic churning of the surface, the surrounding blankets and interior walls become lighter and less distinguishable from the surrounding terrain (on the basis of albedo), but they generally retain the convex upward profiles. These two classes of craters are obviously distinguishable only by use of a variety of both high- and low-Sun-illuminated photographs in addition to the high-resolution obliques.

Color photograph AS14–73–10145 shows inclined layering within the central peak of Langrenus Crater similar to that recognized previously in the famous Lunar Orbiter 2 oblique photograph of Copernicus Crater (ref. 18–5). The Apollo 14 photograph (fig. 18–16) is superior in detail (taken from an opposite viewpoint, the southwest) to the Apollo 10 photographs of Langrenus Crater. The face of a steep-walled ridge in the northern part of the central peak complex shows, from top to bottom, the following discrete units:

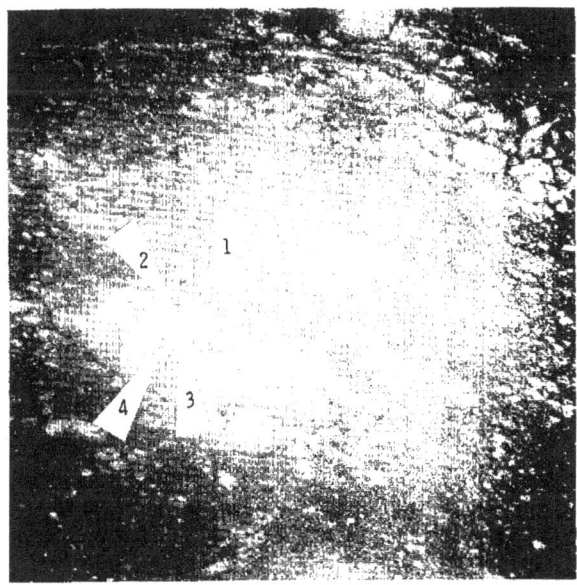

FIGURE 18–16.—Central peak of Langrenus Crater, viewed from the southwest. The curved ridge in the northern part of the peak assemblage measures approximately 13 km in an east-west direction; points 1 to 4 refer to stratification seen on the ridge face as described in text. (North is toward upper left; AS14–73–10145.)

(1) A highly reflective but somewhat streaked uppermost layer with a diffuse lower boundary

(2) A dark, relatively thin, discontinuous unit

(3) Dark and light patches of possibly mixed materials slumped from overlying units. (The dark spots on the hill in the foreground may represent the equivalents of the dark rocks seen on the ridge face.)

(4) Main part of hill (massive, light gray, and vertically streaked in the middle part of the slope (probably talus))

A light mottled bank at the base of the ridge (and at the base of the hill in foreground) may be accumulations of boulders.

Thus, two clearly photographed examples of layered central peaks within large relatively young craters are now available. The existence of this stratification is consistent with an impact origin for the craters and the origin of central peaks by rebound of deep-seated target materials that maintain stratigraphic identity.

Dark material that covers terraces and fresh scarps on the wall of the west rim of Langrenus Crater is shown in figure 18–17. The dark material extends as diffuse inward-facing lobes as far as

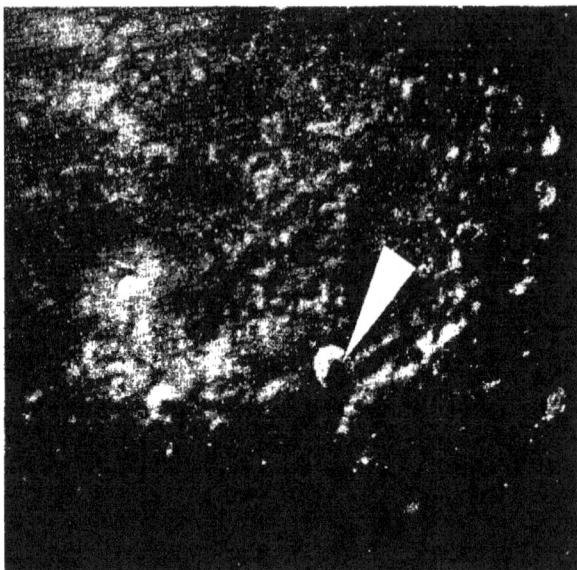

FIGURE 18–18.—Partly buried rim, wall, and floor of Zöllner DC Crater (diameter approximately 5 km) (north left of top center; AS14–70–9783).

40 km and appears to occupy local topographic lows, including the centers of some small craters and linear grooves. The diffuse dark lobes extend down and across the faces of multiple slump scars and apparently terminate on the crater floor. These characteristics indicate postcrater modification by lava flows, pyroclastics, or highly mobile debris flows.

Of interest in terms of surface processes is the young Zöllner DC Crater (fig. 18–18). The Zöllner DC Crater, approximately 5 km in diameter, is superposed on complex terrain materials in the Descartes region. The southeast part of the rim, wall, and floor have been buried by a postcrater deposit that appears to originate near the crestline of a far more ancient and larger crater that lies immediately to the southeast. The material apparently is mass-wasted debris that accumulated shortly after the crater formed. Numerous examples of partial crater destruction on sloping surfaces have been seen in previous orbital photographs. The infilling always takes place from the higher ground, and this phenomenon has been used by a number of authors to explain the relative paucity of craters in the more rugged parts of the terrain. This photograph provides a more dramatic example of crater filling than many previously available photographs.

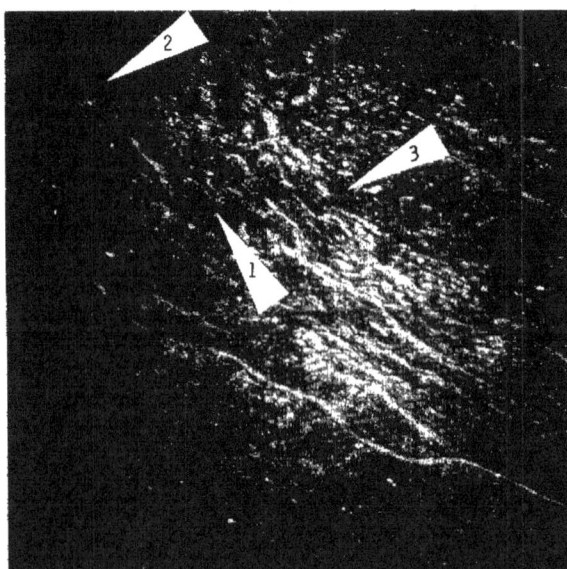

FIGURE 18–17.—Dark materials extending from west rim of Langrenus Crater into the crater floor. The points refer to patches preferentially occupying local topographic lows: (1) center of a small crater; (2) a linear groove; (3) floor of main crater (north toward upper left; AS14–73–10152).

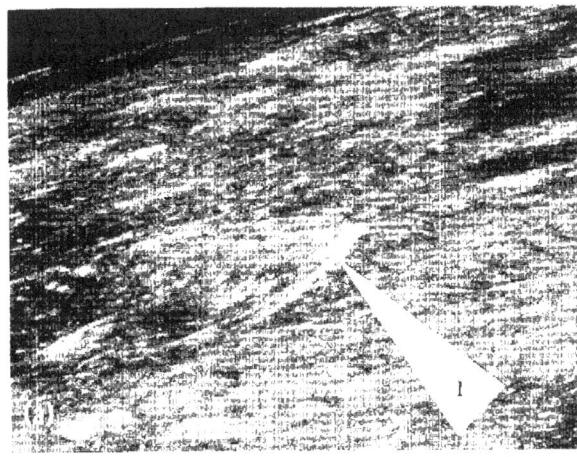 

FIGURE 18-19.—Oblique views of Alpetragius Crater (40-km diameter). (a) Sharp rim crest on north side shown at point 1 (AS14-73-10095). (b) Subdued rim on south side (point 2); unusually large central peak shown at point 3 (AS14-73-10096).

The relative ages of Alpetragius and Arzachel Craters (shown on the nearly vertical Lunar Orbiter 4 photographs) have troubled many photo-investigators. Alpetragius Crater has generally been mapped as the younger of the two, but it belongs to a class of lunar craters with unusually prominent central peaks, of which a dozen or so examples exist on the near side. An oblique view of Alpetragius Crater (fig. 18-19(a)) shows a sharp-appearing rim crest along the northwest wall of the crater. The southeast side of the crater (fig. 18-19(b)) toward Arzachel Crater clearly exhibits a more rounded and subdued rim crest. Thus, despite the apparent youth of, and the apparent superposition on, the rim deposits of Arzachel Crater (as determined from analysis of Lunar Orbiter 4 photographs), Alpetragius Crater might be the older crater with the southeast wall degraded by the later Arzachel Crater event. These observations attest to the utility of repeated photography of the same regions from different angles.

Fine details in the ejecta pattern of rim material around the Copernican crater Lansberg A are shown in figure 18-20. The low-Sun illumination of 8° accentuates the braided and radially ridged texture of this 10-km crater. The pattern bears a striking resemblance to that around the smaller Mösting C Crater as shown in reference 18-5. Craters formed in mare material usually exhibit a remnant rim crest even though they may be deeply embayed by younger mare flows. The small

fresh crater (3 km in diameter) immediately east of Lansberg A Crater shows no evidence of any raised rim deposit, yet it appears to truncate the rim material of Lansberg A Crater. Some of these rimless craters and many rimless bowl-shaped depressions have been noted on previous orbital photographs. An excellent example of a rimless

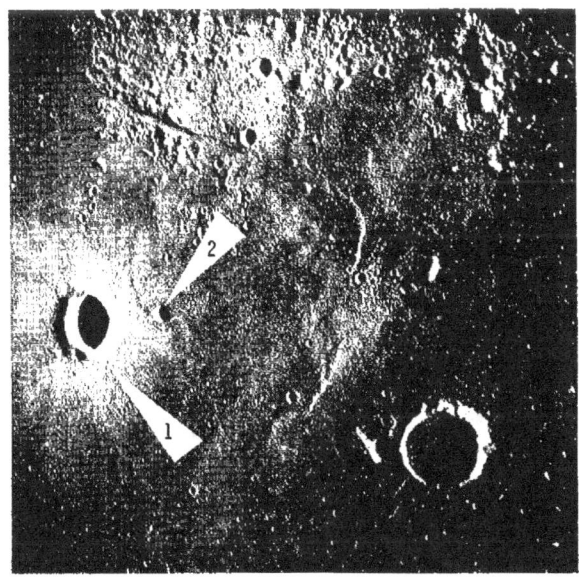

FIGURE 18-20.—Rim deposits of the presumed impact crater Lansberg A, at point 1; small, sharp, rimless crater apparently superposed on rim material at point 2. (North is toward top; AS14-70-9830.)

crater approximately 10 km southeast of Kunowsky Crater is shown in part G of this section. Still another is situated near the Marius Hills, just north of the rille shown in Lunar Orbiter 4 high-resolution photograph 157–H₂ (northeast corner). A slight darkening is sometimes observed around the rims. This type crater is difficult to explain as being of impact origin; such craters may be analogous to certain terrestrial maars in which gas has been the major discharge product, and solid materials, mostly quarried from the vent walls, are subordinate in amount.

A number of elongate, somewhat triangular patches of bright-ray material can be seen in figure 18–21. These elongate patches can be traced eastward and are radial to Lalande Crater outside the photograph. The principle of superposition is well illustrated because the rays from Lalande Crater in this photograph clearly lie across rays and secondary crater clusters identified with Copernicus Crater, located approximately 450 km to the north. Thus, the relative ages of these two craters are established by simple superposition relations.

Davy G Crater is an irregularly shaped crater

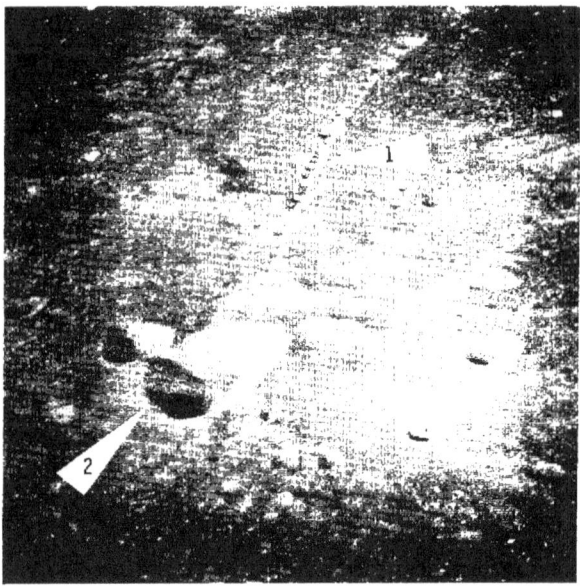

FIGURE 18–22.—Davy Crater chain at point 1 and Davy G Crater at point 2 (north toward right; AS14–73–10102).

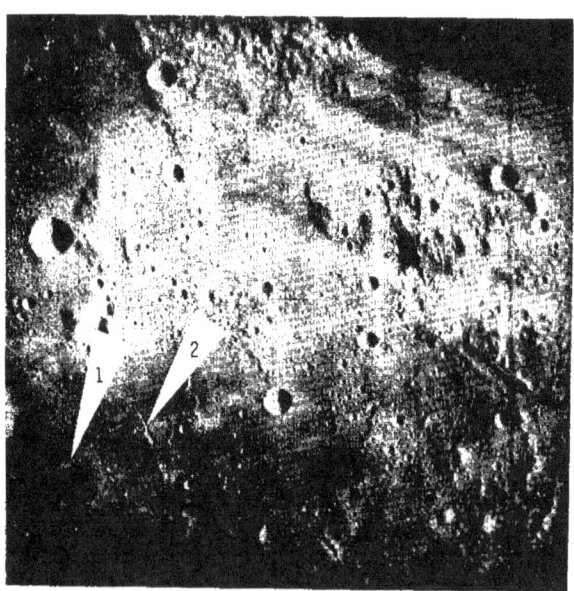

FIGURE 18–21.—Ray material from Lalande Crater superposed on secondary impact craters (points 1 and 2) from Copernicus Crater. Both Lalande and Copernicus Craters are outside of photograph (north toward top; AS14–70–9812).

approximately 15 km long, lying along the eastward extension of a chain of smaller craters referred to as the Davy Crater chain (fig. 18–22). All these craters appear to be relatively young (late- to post-Imbrian) and are considered by some investigators to be of volcanic origin (ref. 18–19). An unusual view of the walls of Davy G Crater under varying illumination angles is afforded in the photograph. At near-grazing Sun incidence, the partly shadowed wall has a rough, lineated appearance wherein the dominant texture consists of grooves extending from rim crest to crater floor. Aside from a slight mottling, the bright sunlit walls appear smooth and resemble those of most young lunar craters viewed under high-Sun angles. This observation is in accord with the high infrared thermal response and radar reflectivity that are generally attributed to the rough and blocky interiors of young craters.

Far-side photographs show exceptional details on the rim, walls, and floor of the bright-rayed crater (latitude 5° S, longitude 123° E) that forms one of the two extensive ray systems in the Montes Sovietici briefly discussed in references 18–8 and 18–20. The crater is best shown under a relatively low illumination angle (17°) in figure

(ref. 18–21), was formed by impact. A different and more highly oblique view of the central ridge of this crater is shown in figure 18–26 in

FIGURE 18–23.—Copernican-age crater (35-km diameter) forming one of the major ray systems on lunar far side. Point 1 indicates uppermost scarp of a series of slump terraces; point 2, a secondary crater loop; and point 3, radial-rim facies (north right of top center; AS14–70–9671).

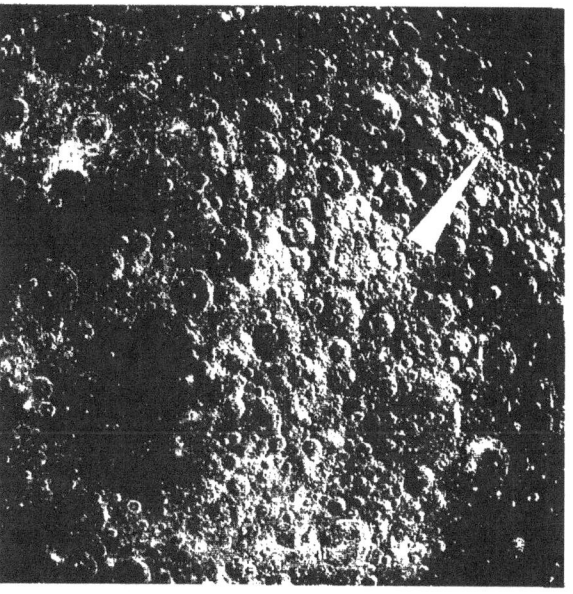

FIGURE 18–24.—Distant posttransearth-injection photograph showing King Crater (75-km diameter) with prominent ridge on floor (north toward upper left; AS14–75–10307).

18–23. The crater has a rim-crest diameter of approximately 35 km, but the general outline is partly distorted by incipient major wall failure and slumping. Looped secondary crater chains, a braided radial facies, finely terraced walls, and major fracture patterns are clearly seen. The interior terraces within this crater generally trend to the northwest, as opposed to the usual concentric terraces seen in craters of similar age in this size range. In this case, it is evident that most of the postcrater gravitational filling occurred by the collapse of the partly shadowed wall.

Distant photographs of the Moon provide regional views in which only the gross morphology of the larger features is accentuated. An unusually prominent ridge (ref. 18–7) extending across the floor of the large 75-km King Crater is shown in figure 18–24. Higher resolution photography (fig. 18–25) shows the ropy structure of the ridge that, in places, seems to be draped over the massive central peak.

A radial facies and secondary craters extend beyond the prominent rim, indicating that King Crater, which appears to be of late Imbrian age

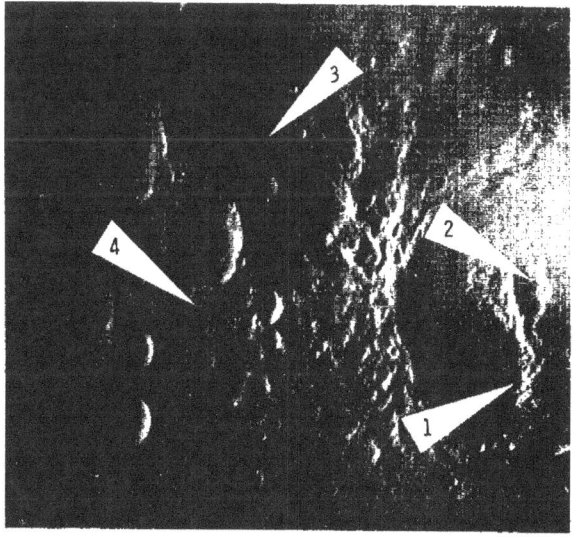

FIGURE 18–25.—High-resolution photograph of King Crater showing central ridge at point 1, central peak at point 2, radial faces at point 3, and secondary craters at point 4 (AS14–71–9851).

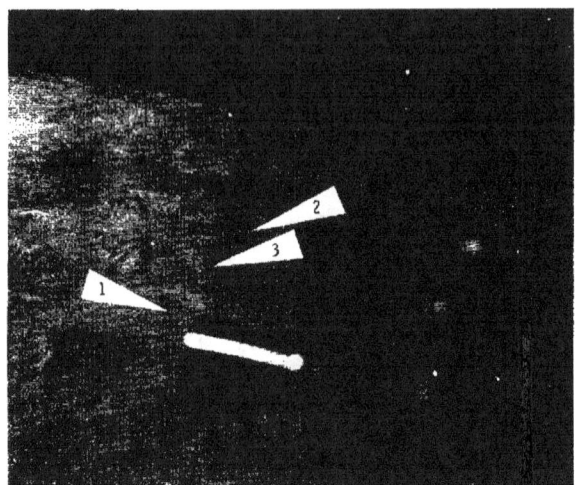

FIGURE 18–26.—Oblique view of King Crater showing slumped near wall at point 1, dark blotches on far wall at point 2, and central ridge and peak at point 3 (AS10–29–4209).

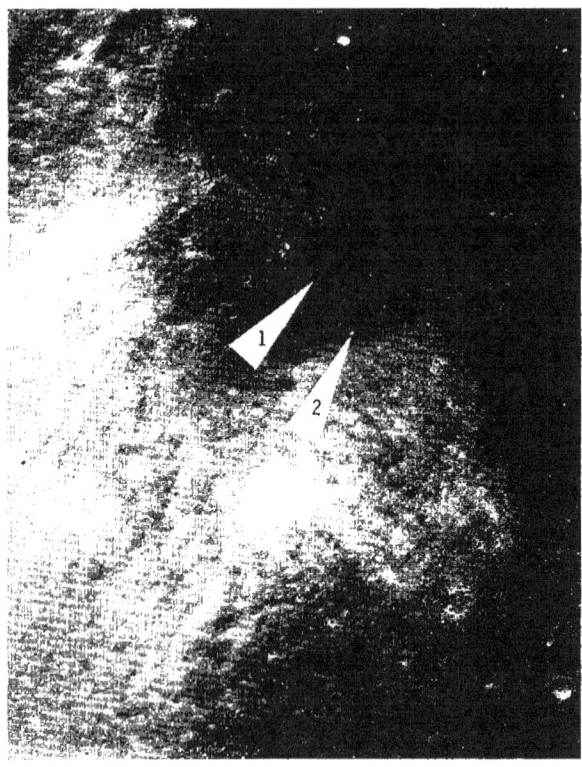

FIGURE 18–27.—Track (point 1) of large boulder (point 2) that rolled down northeast wall of Theophilus Crater (AS14–80–10448).

which the ropy, convoluted character of the ridge material is not as evident; in this photograph, the material more nearly resembles a large slide mass extending from crater wall to central peak. The light-hued inclined band shown at point 3 in figure 18–26 may be an interbedded unit or dike. This band appears to lie across the central peak shown in figure 18–25 and texturally resembles the ropy material. A wide (approximately 10 km) dark band extends across the floor of King Crater and up the far northern wall where it continues for some distance into the hilly terrain in the background. Small, very dark patches occur along the central ridge and peak as well as in the walls and floor of the crater. The patches do not appear stratified and might, therefore, represent local zones of postcrater volcanic modification (ref. 18–7). Additional orbital photography of this crater under a wide variety of lighting conditions would be useful, particularly in the case of the central ridge.

Vertical stereophotography made with the Hycon lunar topographic camera began at the northeast rim of Theophilus Crater and, because of a camera malfunction, ended just east of the candidate Apollo landing site in the Descartes region. The approximate scale of the frames varies from 1:48 000 to 1:37 000 with a recognition resolution

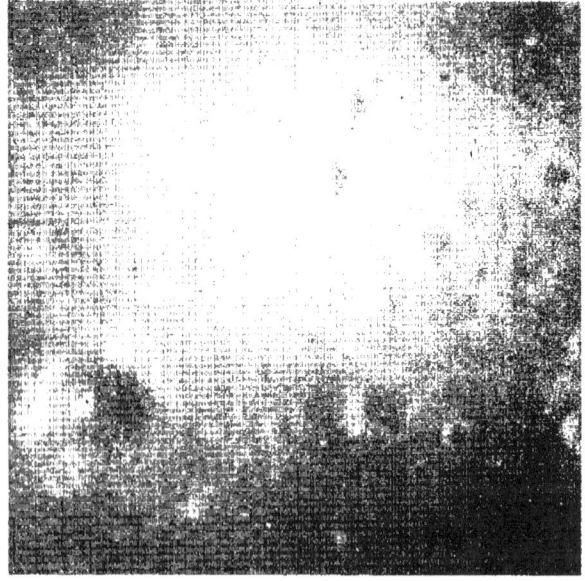

FIGURE 18–28.—Prominent, bright, central peak in small crater on floor of Theophilus Crater (AS14–80–10454).

of 2.5 to 1.9 m (Apollo 14 Preliminary Photographic Index). Interesting details of lunar-surface texture and small-crater morphology are seen in most of the photographs. A track made by a large boulder (approximately 20 m in diameter) as it rolled down the northeast wall of Theophilus Crater is shown in figure 18–27. The boulder track has been studied in detail (ref. 18–22) and was found to be helpful in estimating the soil properties of the lunar surface.

The floor of Theophilus Crater contains a small (800-m) crater that is unusual because of the prominent, smooth-surfaced, bright central peak (fig. 18–28). On a small scale, it resembles the type of crater exemplified by Alpetragius Crater, discussed earlier. Fresh rock may actually be exposed in the upper bright portion of the peak as a result of downslope shedding of fragmental material. Blocks are clearly visible on the crater wall.

PART C

HYCON PHOTOGRAPHY OF THE CENTRAL HIGHLANDS

Farouk El-Baz [a] *and J. W. Head III* [a]

During the Apollo 14 mission, the lunar topographic camera (LTC) recorded the first high-resolution strip photography of the lunar surface. The stereophotographs cover a segment of the central lunar highlands from the eastern rim of Theophilus Crater to a point northwest of Kant Crater. The photography will be of value in detailed studies pertaining to the geologic units of this part of the Moon. The purpose of this paper is to summarize the preliminary geologic analysis by citing examples of terrain characteristics as revealed in the LTC photographs.

The LTC has an 18-in. focal-length ƒ/4 lens. An automatic rocking mount compensates for forward motion of the spacecraft over the lunar surface during exposure by keeping the camera pointed directly at the object being photographed. The photographs discussed in this section were taken during the fourth orbit when the command and service module (CSM) was approximately 18 km above the lunar surface. Black-and-white type-3400 film was used at a 65-frame/min rate and at a shutter speed of 1/200 sec to obtain these photographs. A camera malfunction resulted in overexposure of the last half of the film magazine. The correctly exposed half of the magazine (ap-

proximately 200 frames) is discussed in this section.

The high-resolution stereostrip (approximately 60 percent sidelap) covers an area approximately 4 km wide from latitude 11.3° S, longitude 28.2° E, to latitude 8.3° S, longitude 18.7° E. The contrast is low because the Sun-elevation angle varied between approximately 45° at the beginning of the strip to 35° at the end of the properly exposed frames.

Preliminary Analysis

The highland terrain covered by the photographs (fig. 18–29), at approximately 2-m resolution, includes two major units: Theophilus Crater ejecta and Kant Plateau materials. A ridge was overflown when the camera was malfunctioning, and, consequently, an area equivalent to the coverage of 15 frames was not photographed. This ridge, however, is discernible in Lunar Orbiter 4 and Apollo (Hasselblad) photographs at much lower resolution.

Theophilus Crater materials are represented by the walls, floor, and the western ejecta blanket of the crater. The ejecta blanket may, in turn, be divided into a hummocky unit near the rim crest and a smoother facies farther away. The second major unit is that of the Kant Plateau materials.

[a] Bellcomm, Inc.

FIGURE 18–29.—Area of photographic coverage includes Theophilus Crater ejecta and Kant Plateau materials. Dots show subspacecraft points of photographs AS14–80–10436 to AS14–80–10642.

The plateau is topographically higher than the surrounding terrain and includes plains-forming units and other subdued ridges. Following are descriptions of examples of the two major units.

Theophilus Crater and Ejecta

Theophilus Crater is a relatively young (Copernican-age) 100-km-diameter crater of probable impact origin. Located in the northwest rim of the Nectaris Basin, the crater displays an ejecta blanket, terraced walls, a flat floor, and central peaks, all of which are characteristic of major lunar craters.

The eastern rim of Theophilus Crater is particularly well illustrated in frames AS14–80–10437 and 10438 in which the characteristics of the crater rim, rim crest, and the inner crater wall may be seen. In the right-hand portion of figure 18–30, the crater rim is characterized by relatively smooth, nontextured material that contains a wide spectrum of crater sizes and ages. The crater walls, however, are sloping and highly textured with a treebarklike pattern, evidently indicative of downslope movement of material along the terraces.

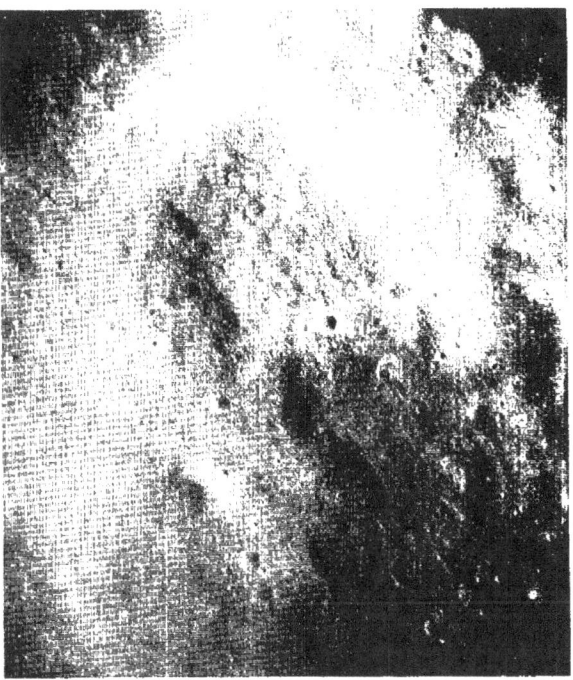

FIGURE 18–31.—Numerous linear arrangements of blocks on the rim of Theophilus Crater (AS14–80–10441).

Numerous large blocks are also seen in the crater walls. The number of craters seen on the crater walls is much lower than that on the crater rim. This distribution is due, in most part, to the differences in slope. Also, virtually all craters on the walls appear to have been modified so that the upslope sides are either smoother and less pronounced or are obliterated altogether. Two 300- to 400-m-diameter craters at the crater-rim crest (fig. 18–30) illustrate this modification well. The crater to the east is slightly modified from the circular form, while the wall of the adjacent crater has been highly modified in an upslope direction.

The crater-distribution variations and the modification of craters on the wall are attributed to downslope movement of debris, which causes filling and ultimate erasure of many craters on the interior walls of Theophilus Crater. Progressive stages of modification are evident in the three large craters (250- to 400-m diameter) located on the Theophilus Crater rim (fig. 18–30). Downslope movement of material is encroaching and filling craters on the terrace edges.

On the crater terraces, numerous linear arrange-

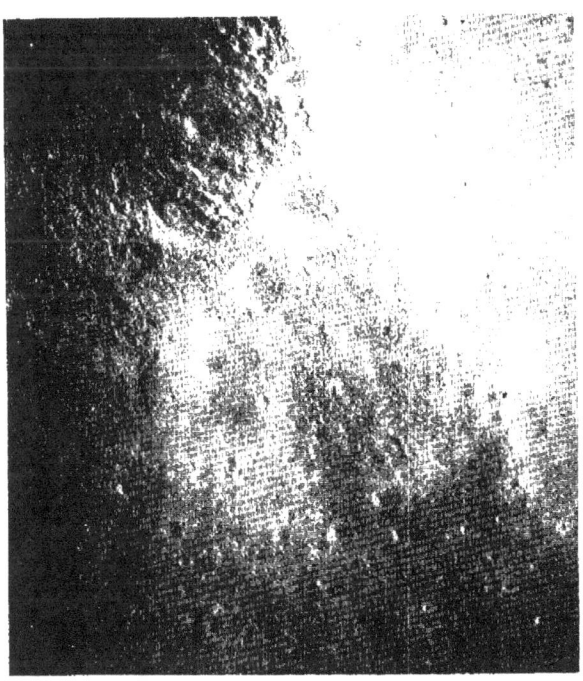

FIGURE 18–30.—Eastern rim of Theophilus Crater (AS14–80–10437).

ments of blocks exist that are interpreted as out-crops of layers or ledges (fig. 18–31). As shown in figure 18–32, blocks up to at least 25 m in diameter are visible in the wall terraces. A spectacular boulder track or trail at least 200 m in length can be seen in the central part of the figure.

The floor of Theophilus Crater is generally smooth east of the central peaks. The craters in this unit display blocky rims (photograph AS14–80–10454), which suggest that the regolith is probably relatively thin. Domed floors are seen in several craters in the 200- to 400-m-diameter range (fig. 18–32). A flat-floored crater approximately 500 m in diameter can be seen in photograph AS14–80–10460.

Several large craters approximately 500 m in diameter occur where the CSM ground track crosses the base of the central peaks. Numerous blocks are visible on these crater rims as well as along the basal slopes of the central peak as shown in figure 18–33.

West of the central peaks, domical hills within the floor materials occur in considerably greater abundance than they do to the east. Typical

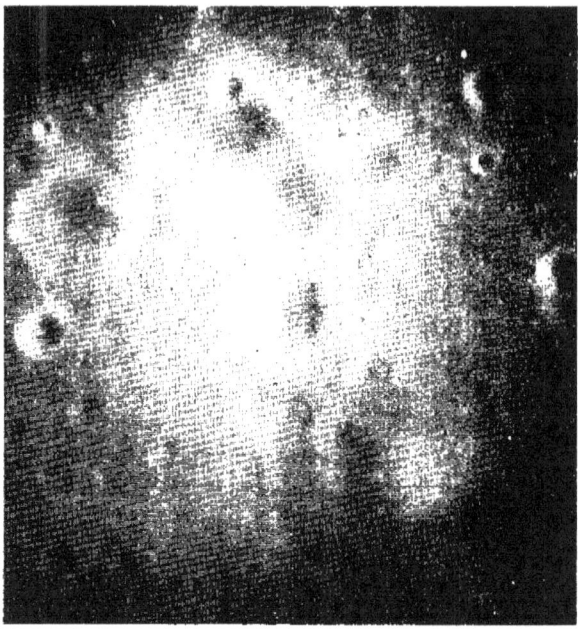

FIGURE 18–33.—Blocks on the crater rims and basal slopes of central peaks of several large craters in floor of Theophilus Crater (AS14–80–10466).

domical hills approximately 500 m in diameter can be seen slightly separated from the central peaks in photograph AS14–80–10470. Other hills abound in the northwest quadrant of the crater floor (photograph AS14–80–10476). Boulders often occur on the slopes of these hills, and the incidence of craters with central peaks and domes

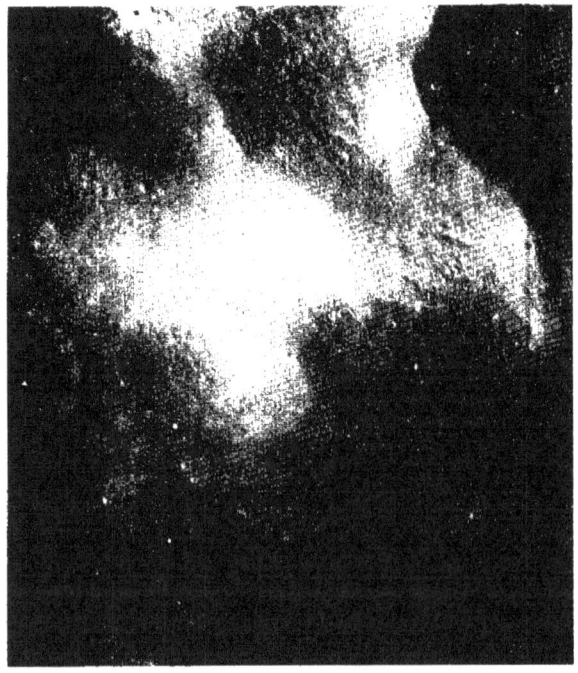

FIGURE 18–32.—Blocks up to at least 25 m in diameter in wall terraces of Theophilus Crater (AS14–80–10448).

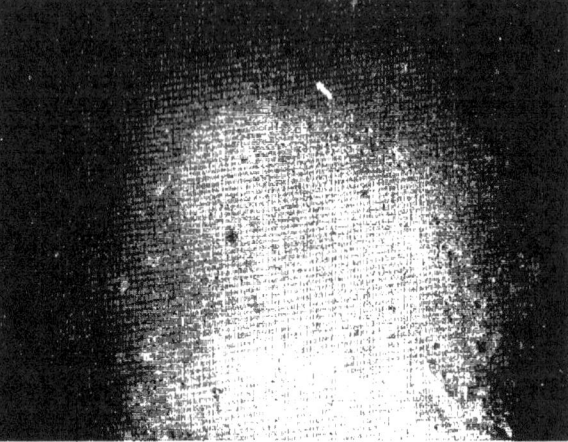

FIGURE 18–34.—The southern rim and a portion of the wall of Theophilus B Crater (AS14–80–10490).

is much greater in this region than to the east, especially in the smaller size ranges.

The western wall of Theophilus Crater is disrupted by the relatively younger Theophilus B Crater and its ejecta. The southern rim and portion of the wall of Theophilus B Crater are shown in figure 18–34. Of particular interest are the large blocks that occur on the interior wall and rim deposits. Apparent layering exists at several areas on the crater interior. Several large boulders, somewhat greater than 10 m in diameter, show well-developed fillets (photograph AS14–80–10493, northeast part).

The western rim crest (photograph AS14–80–10495) of Theophilus Crater displays characteristics similar to those described for the eastern rim, although the difference in illumination of slopes obscures some wall detail in the west.

The characteristic hummocky rim deposits of large craters is less distinctly developed or preserved at Theophilus Crater. However, a generally hummocky rim pattern exists (photograph AS14–80–10497). The smooth, poollike mapped areas (ref. 18–1) on the north rim of Theophilus Crater do not occur on the western rim. A small chain of probable secondary craters of uncertain source is visible on the rim (photograph AS14–80–10503);

other crater clusters and blocky craters can be seen in the hummocky crater-rim deposits (photographs AS14–80–10500 to 10505 and AS14–80–10511).

An old, extremely subdued crater (Cyrillus M) is superposed by the rim deposits of Theophilus Crater. The rim of Cyrillus M Crater is particularly prominent in photograph AS14–80–10519; a crater approximately 2 km in diameter is evident in the Theophilus Crater rim. The area to the east of this crater appears smooth (photographs AS14–80–10511 to 10516) (fig. 18–35). The texture may be the result of ponding of Theophilus Crater ejecta in the floor of Cyrillus M Crater or of post-cratering modifications of low-lying areas.

Kant Plateau Materials

Numerous interesting craters occur on the margins of the hummocky Theophilus Crater ejecta blanket in an area mapped as Kant Plateau (ref. 18–1). These craters appear to be part of a unit described as densely pitted and interpreted as representing craters and associated materials produced by explosive maar-type volcanism. The smaller craters (less than 500 m in diameter) are alined (photograph AS14–80–10525) and are

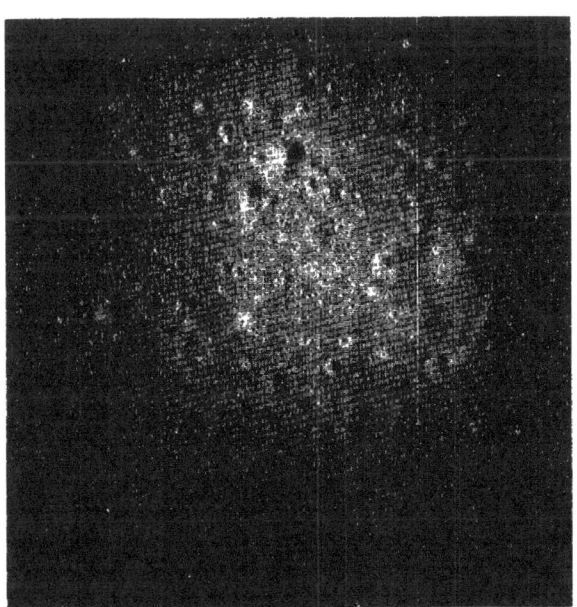

FIGURE 18–35.—Smooth area east of crater in rim of Theophilus Crater (AS14–80–515).

FIGURE 18–36.—The larger craters on the margins of the Theophilus Crater ejecta blanket (AS14–80–10529).

characterized by soft, rounded rims; lack of floors; and by nonterraced walls that are funnel shaped or often Y-shaped. The larger craters, 0.5 to 2 km in diameter (photographs AS14-80-10527 to 10529), have similar characteristics with the non-floored funnel- and Y-shaped aspect more apparent (fig. 18-36). Whatever the original shape of the craters, it is evident that considerable mass wasting has occurred in generally noncohesive material to produce the crater morphologies seen in photographs AS14-80-10527 to 10529. A few kilometers to the east, the largest crater in the area (approximately 2 km in diameter; photograph AS14-80-10535) displays a nearly flat but slightly domed floor, although the crater rim and walls are similar to those of smaller craters.

The terrain toward Kant E Crater to the west is cratered and hummocky. The terrain is generally undistinguished by any major feature and may have been smoothed over by a thin blanket of Theophilus Crater ejecta (photographs AS14-80-15037 to 10557).

A change in regional slope is evident in photographs AS14-80-10561, near the base of Kant C Crater; a corresponding change in surface texture

to the treebark type can also be seen. Of particular interest in this area are the modifications, by mass wasting, of 200- to 400-m craters on the slopes (fig. 18-37). These craters have the Y-shaped interiors characteristic of several craters a few kilometers to the east. They also share the characteristics of craters developed on steep slopes such as those described in the east wall of Theophilus Crater. In this instance, the upslope rims are subdued and the upslope walls are much less steep than the downslope walls. The point of inflection at the crater bottom appears to migrate downslope as the crater ages. Additional movement down the side slopes of the crater may produce additional lines of inflection producing a Y-shaped interior contour. A camera malfunction, for a total of 15 frames, precluded photography of the remainder of this terrain, which includes a ridge of possible pre-Imbrian age.

North of Kant Crater, near Kant M Crater, the terrain is again hummocky and undistinguished with the exception of the rim and interior wall of part of a crater visible in photographs AS14-80-

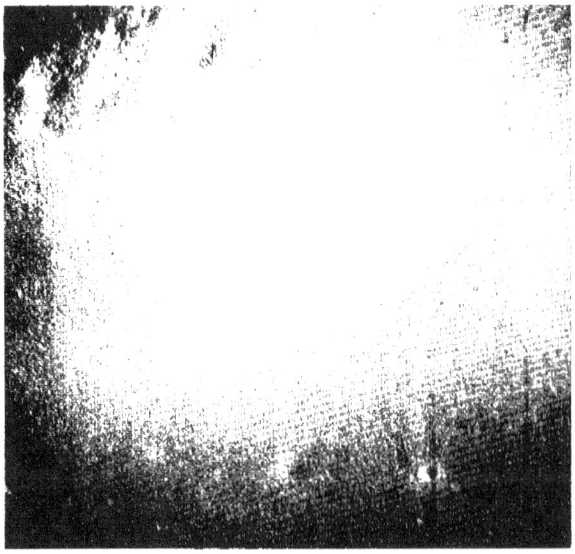

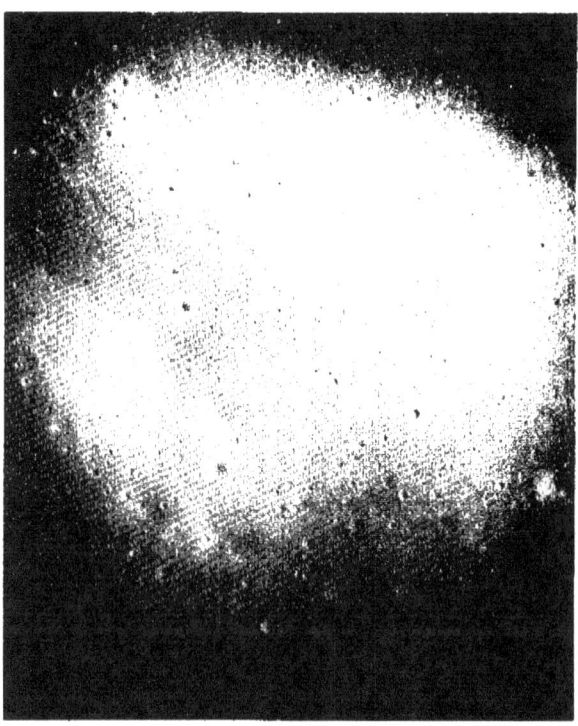

FIGURE 18-37.—Modifications by mass wasting of 200- to 400-m craters on the slopes of Kant C Crater (AS14-80-10563).

FIGURE 18-38.—Broad ridges associated with Kant N and Kant G Craters (AS14-80-10589).

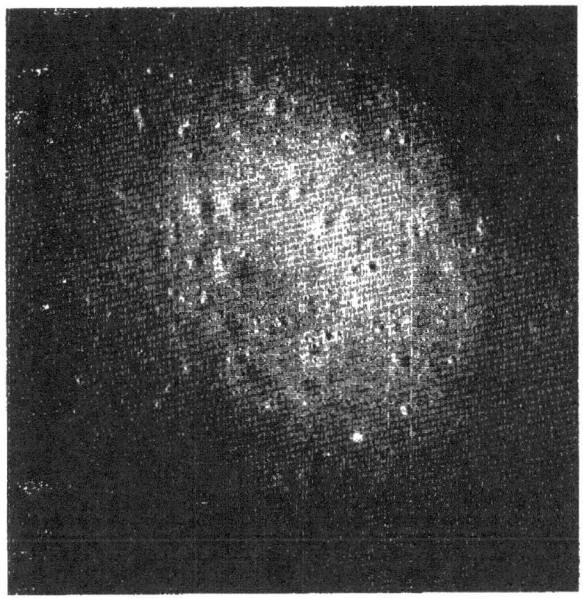

FIGURE 18-39.—Textured wall of, and plains materials near, Kant B Crater (AS14-80-10614).

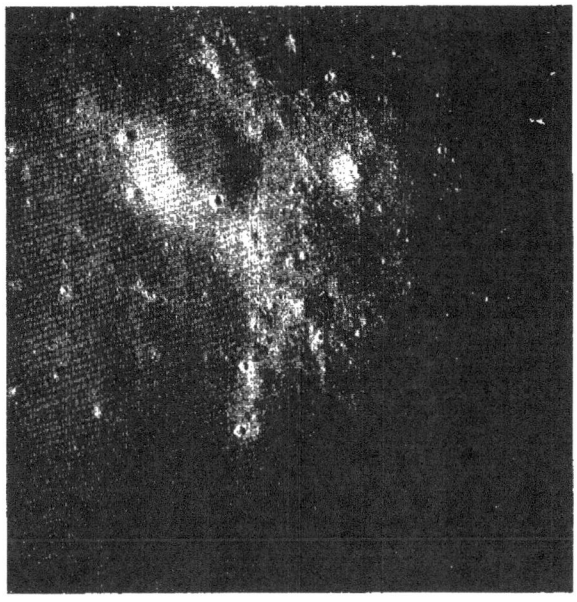

FIGURE 18-41.—Complex elongate depression in plains unit (AS14-80-10637).

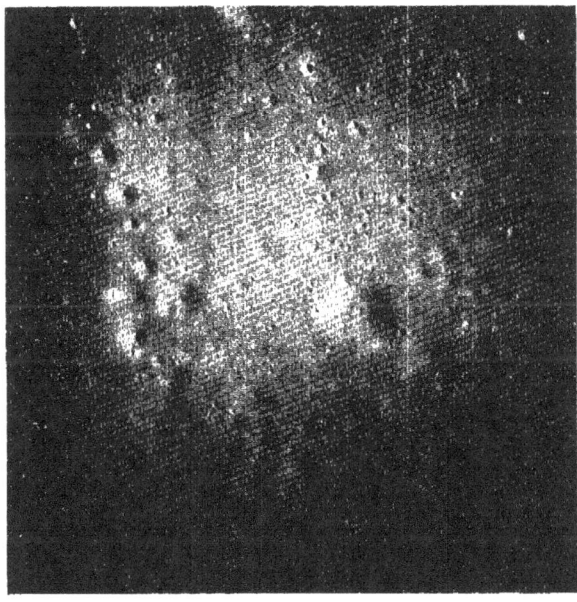

FIGURE 18-40.—Crater chain in the plains unit near Kant B Crater (AS14-80-10620).

10583 and 10584. Several additional broad ridges associated with Kant N and Kant G Craters may be seen in photographs AS14-80-10589 to 10594 (fig. 18-38).

A hummocky plains-type unit (Cayley Formation) has been mapped within Kant B Crater (ref. 18-1). The textured wall of Kant B Crater as well as the smoother plains material is shown in figure 18-39. A crater chain in this plains unit is well illustrated in figure 18-40. An extremely blocky bright-rayed crater within this unit is shown in photographs AS14-80-10628 and 10629. In this region, apparent bedrock can be seen in the floors of both fresh and subdued craters. A complex elongate depression is evident in figure 18-41. The nature and origin of this depression are not yet well understood.

Summary

The Hycon photography of the central highlands is valuable in characterizing certain units that were photographed previously by low-resolution imagery. The value is due to the fact that it is high-resolution (approximately 2 m) stereo-strip photography (with 60 percent sidelap).

The strip covers an area from the eastern rim of Theophilus Crater to the Kant Plateau. Materials of Theophilus Crater are well depicted in the photographs. Details of the rim crest, rock ledges on the wall terraces, blocky craters on the floor, domical hills of the central peaks, and the ejecta blanket of Theophilus B Crater are well displayed. The ejecta blanket of Theophilus Crater may be divided into a hummocky and ridgy unit near the rim crest and a smoother facies farther out. The Kant Plateau materials are represented by plains-forming units with subdued craters and ridges.

Perhaps the most significant contribution of the Apollo 14 strip photography is that it provides necessary information for detailed crater studies. It provides excellent coverage of Theophilus Crater, a 100-km crater. This coverage will allow detailed studies of all units related to that typical, large crater that is probably of impact origin. In addition to this, future analysis of the photographs will provide insight into the nature of smaller craters, mass wasting, and aging. In turn, these studies will provide valuable data pertaining to the nature and thickness of the regolith in this area of the Moon.

PART D

GEOLOGY OF THE REGION AROUND THE CANDIDATE DESCARTES APOLLO LANDING SITE

Daniel J. Milton [a]

A geologic sketch map of approximately 10 000 km² surrounding the candidate Apollo 16 landing site in the Descartes region was prepared from Apollo 14 80-mm Hasselblad (fig. 18–42) and Lunar Orbiter 4 photography. The area of interest (fig. 18–43) is located in the highlands southwest of Mare Tranquillitatis and is approximately 300 km west-northwest of Theophilus Crater. Most of the mapped area is covered by plains-forming material and hilly materials of Imbrian age. Circumbasin materials, such as those surrounding the Orientale and Imbrium Basins, are not distinguishable from the photographs, although thin deposits of ejecta from the Imbrium and Nectaris Basins may be present in the subsurface or perhaps exposed (as shown in the northwest section of the map) in an area associated with Dolland B Crater, the only pre-Imbrian crater (pIc) recognized. Dolland B Crater and the adjacent lineated terrain are cut by Imbrian sculpture and are thus assignable to the earliest period of lunar history. These old surfaces are highly modified—it is uncertain

whether by superposition of a thin blanket of hilly upland material or simply by long, continued, degradation processes. If the age of the old surfaces has been correctly determined, they will be the closest surface exposures of pre-Imbrian materials to the candidate Descartes landing site. Dolland B Crater is superposed incidentally on what may be the vague rim of a still older crater 150 km in diameter.

The hilly units mapped as Ih (formerly referred to as materials of the Kant Plateau) and plains unit (Ip) of the Cayley Formation form clearly distinct terrains, although some transitional zones are present. Division of these two major units, both of which are of regional extent and significance (ref. 18–2), is more difficult and subjective. Imbrian sculpture is absent or expressed only as vaguely defined lineations; consequently, these units are interpreted to be no older than the Imbrian age. The moderate crater densities further suggest these units are of Imbrian age but older than the late Imbrian mare of nearby Mare Tranquillitatis (of which the Apollo 11 rocks are presumably a representative sample). Embayment

[a] U.S. Geological Survey.

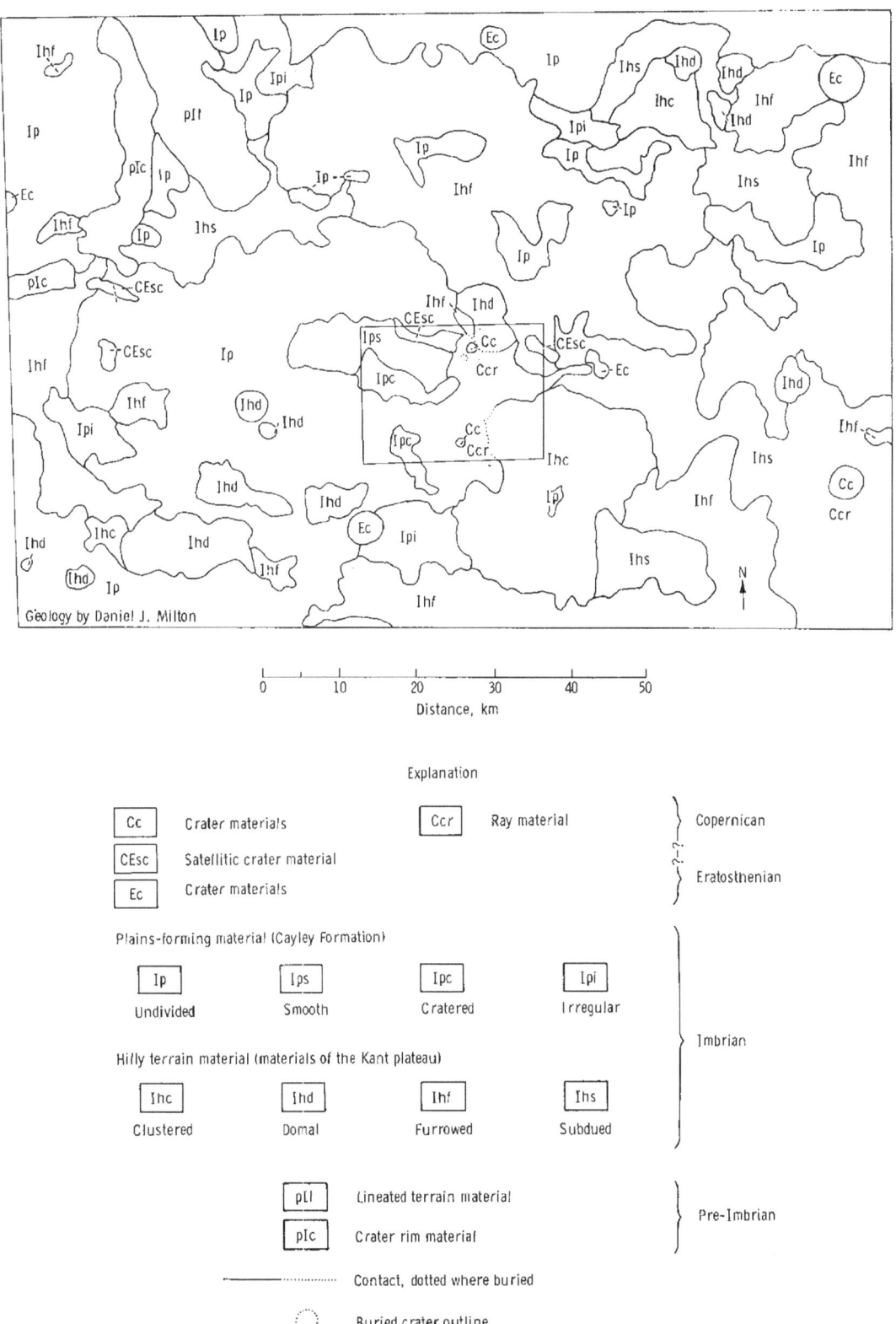

0   10   20   30   40   50

Distance, km

Explanation

Cc	Crater materials	Ccr	Ray material	}	Copernican
CEsc	Satellitic crater material			?-?	
Ec	Crater materials			}	Eratosthenian

Plains-forming material (Cayley Formation)

Ip	Ips	Ipc	Ipi
Undivided	Smooth	Cratered	Irregular

Hilly terrain material (materials of the Kant plateau)

Ihc	Ihd	Ihf	Ihs
Clustered	Domal	Furrowed	Subdued

} Imbrian

pll   Lineated terrain material

plc   Crater rim material

} Pre-Imbrian

——————········ Contact, dotted where buried

Buried crater outline

FIGURE 18–42.—Geology of the region around the candidate Descartes landing site.

FIGURE 18–43.—Uncontrolled photomosaic of the Descartes region made from Hasselblad (80-mm lens) photographs (AS14–70–9784, 9785).

relationships suggest that unit Ip is younger than Ih, but this observation may be illusory. Apollo 14 photography of Zöllner DC Crater, east of the mapped area, shows a crater in Ip at the base of a steep slope half buried by Ih material, but it cannot be determined with certainty whether the crater was buried during original emplacement of unit Ih or by later downslope migration of debris. (Zöllner DC Crater is described in pt. B of this section.)

The clustered hilly unit (Ihc) is characterized by a patternless array of steep-sided equidimensional hills 3 or 4 km across. Broader individual convex shieldlike hills are distinguished as the domal hilly unit (Ihd). The areas of furrowed material (Ihf) are characterized by positive relief forms of similar areal extent to the clustered hills and domes, but with lower relief, flatter tops, and gentler slopes, so that they grade texturally into broad rolling surfaces. Negative features, particularly irregular furrows along the crests of many of the elongated hills, are more apparent in this unit. The subdued unit (Ihs) is similar but has less local relief and is transitional with the light plains units.

The positive features of the hilly units are Ihc and Ihd and are regarded as volcanic landforms. Craters visible near the summits of a few mounds and domes (the northeasternmost dome is a good example) are probably volcanic craters or cal-

deras. Materials of the other hilly subunits, which typically occur at greater distances from the presumed eruptive centers, probably are composed of lavas originally having lower viscosity than those that accumulated closer to the vents.

Generally, the plains-forming units are characterized by moderately bright level surfaces marred only by the presence of small craters. Patches with anomalously few craters are distinguished as unit Ips, as are patches with anomalously abundant craters (Ipc) where these craters occur in large numbers. It is not known whether these craters are intrinsic to the area or satellitic to distant craters. Areas of plains-forming materials that have discernible positive relief features are recognized and mapped as unit Ipi. The latter unit may be a thin deposit of plains-forming material incompletely masking hilly material. The plains-forming materials may be lava flows (determined by comparison of the general characteristics to the maria), or they could be fragmental debris that has been transported into topographic lows by mass wasting and sustained churning of the surface by impact. Alternatively, they could have been extruded from multiple vents within the present plains or could be a very mobile lunar magmatic fractionation product that originally spread from eruptive centers in the hilly areas.

Most craters younger than the plains and the hilly units are arbitrarily considered Eratosthenian in age (Ec) if rays are lacking or Copernican in age (Cc) if rays are present. The more irregular and grouped craters are probably secondary craters, perhaps from the Cyrillus and Theophilus Craters outside the mapped area. Their ages cannot presently be ascertained; accordingly, they are indicated by the symbol CEsc.

Structural grain is not strongly developed in the area. Imbrian sculpture is exhibited in the pre-Imbrian terrain in the northwest and elsewhere may locally have had some influence on the pattern of volcanism in the hilly unit. A poorly expressed north-south lineation is also evident, one example of which is the gentle slope trending northward between the two bright-rayed craters near the candidate landing site.

PART E

SKETCH MAP OF THE CANDIDATE DESCARTES APOLLO LANDING SITE

Carroll Ann Hodges [a]

A preliminary geologic map (fig. 18–44) of the candidate Apollo 16 landing site near Descartes was prepared from photography (fig. 18–45) taken with the Hasselblad camera, equipped with a 500-mm focal-length lens, during the Apollo 14 flight. The area of interest encompassing the landing site includes two prominent bright-rayed Copernican craters approximately 11 km apart. These two craters are informally called North and South Craters. Plains and terrain materials of Imbrian age and crater materials of Imbrian and younger ages are present within the mapped area. Pre-Imbrian units occur outside the mapped area (outlined on the regional geologic maps in fig. 18–42, pt. D).

Terrain materials are mapped according to topographic expression as furrowed (Ihf), domal (Ihd), and clustered (Ihc) hills units. At the scale of the photographs (approximately 1:40 000) used for compilation of the original map, the domical unit is indistinguishable from the hilly and furrowed units. This contact, therefore, was taken from the regional geologic map (fig. 18–42).

The plains materials represent part of the Cayley Formation (ref. 18–23) and are subdivided according to topographic characteristics more readily observed on the smaller scale photographs. The contacts of the cratered plains unit (Ipc) were also taken from the regional map. With the exception of the smooth plains unit (Ips), the plains are rather evenly marked by numerous shallow depressions resulting in an undulating surface. These depressions, many of which appear to be rimless and are shown by short dashed outlines on the map, are probably subdued Imbrian-age craters but are older than other more distinctly defined Imbrian-age craters.

Craters have been assigned to lunar-time stratigraphic systems on the basis of morphology and albedo. Craters with rays are in the Copernican (Cc) system; further subdivision by relative age within this system is made mainly according to degree of ray prominence. Rayless craters are classified as Eratosthenian (Ec) or Imbrian (Ic) depending upon the degree of subdual; all crater-age determinations are preliminary. The clusters of shallow craters mapped as CEsc are probably secondary craters formed by ejecta from Theophilus Crater approximately 300 km east-southeast of the mapped area.

Apollo 14 orbital photographs used during preparation of the Descartes area map revealed a number of observations that may be applicable to surface exploration.

Craters having convex floors seem to be unusually numerous. Most of them are of Eratosthenian age and approximately 1 km in diameter. The two largest Copernican-age craters apparently do not have convex floors. Floor convexity may be the result of local isostatic adjustment (rebound over an extended period of time) or of a strength discontinuity at depth and, thus, locally reflect depth to bedrock. Several craters have ledges in the walls; a near-surface ledge seems to be present along the east rim of a shallow trough trending south from North Crater. The ledge appears to extend into the crater; large boulders on the crater floor further suggest that outcrops may occur in the vicinity of North Crater. A linear ridge extends northward from the ejecta blanket of South Crater. Several small, young craters have ledges near the bottom of the walls and may have penetrated bedrock.

Irregular dark patches are visible, but not mapped, in several photographs of plains materials. Dark streaks in the ray materials of North and South Craters may have been caused by a local absence of bright ray material or by the ejection of dark materials from the crater. An accumulation of dark materials that form an an-

[a] U.S. Geological Survey.

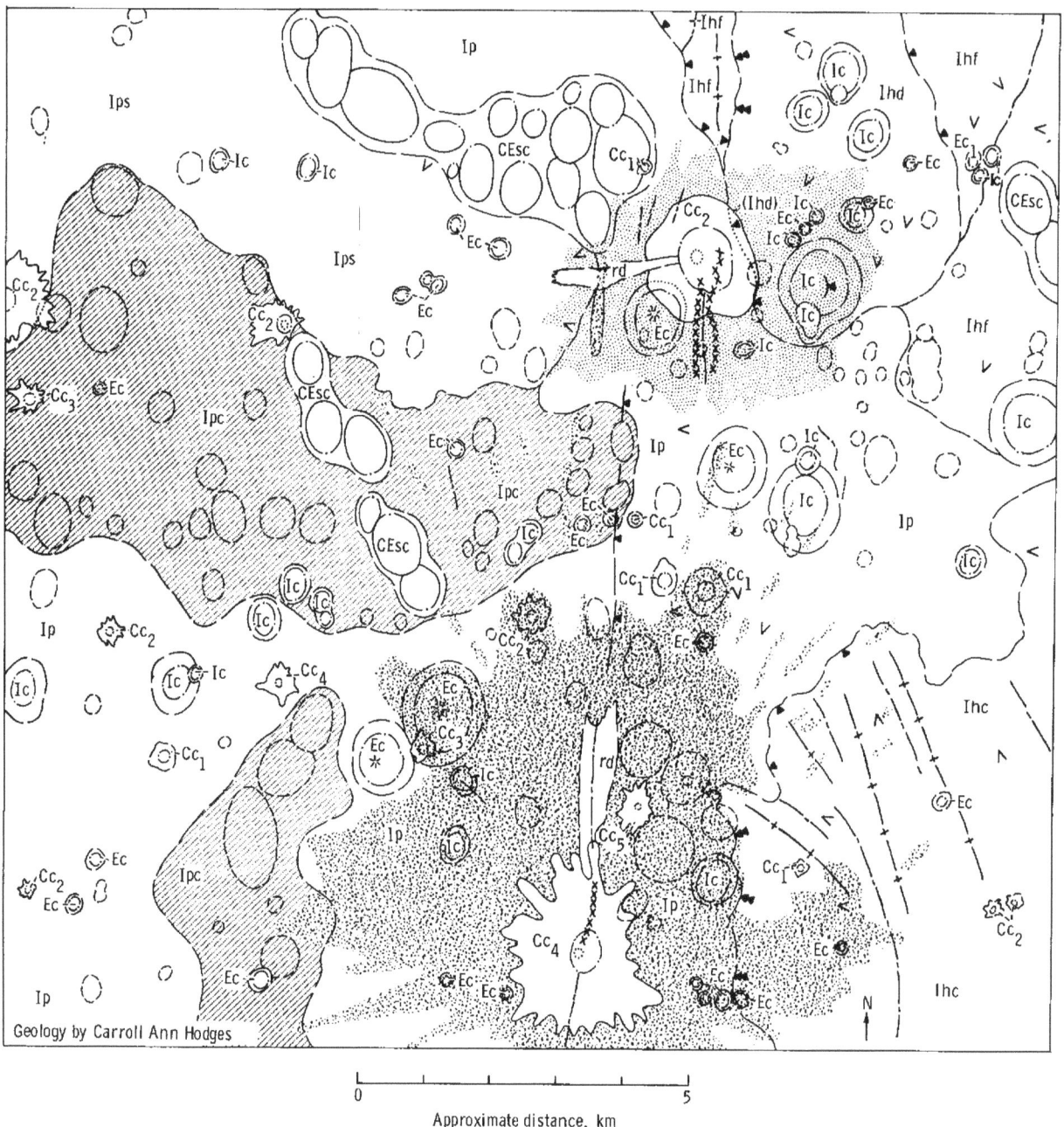

Geology by Carroll Ann Hodges

0   5
Approximate distance, km

FIGURE 18-44.—Candidate Descartes landing site.

nular ring around the floor of South Crater may be debris mass wasted from the crater walls. Differences in petrologic composition may be responsible for most of the albedo variations discussed in part B of this section.

Obvious differences exist in the density of craters superposed on the various units mapped in this area. Most notable is the lack of craters on the clustered hills unit (Ihc) as compared with the plains and other terrain units. Craters are

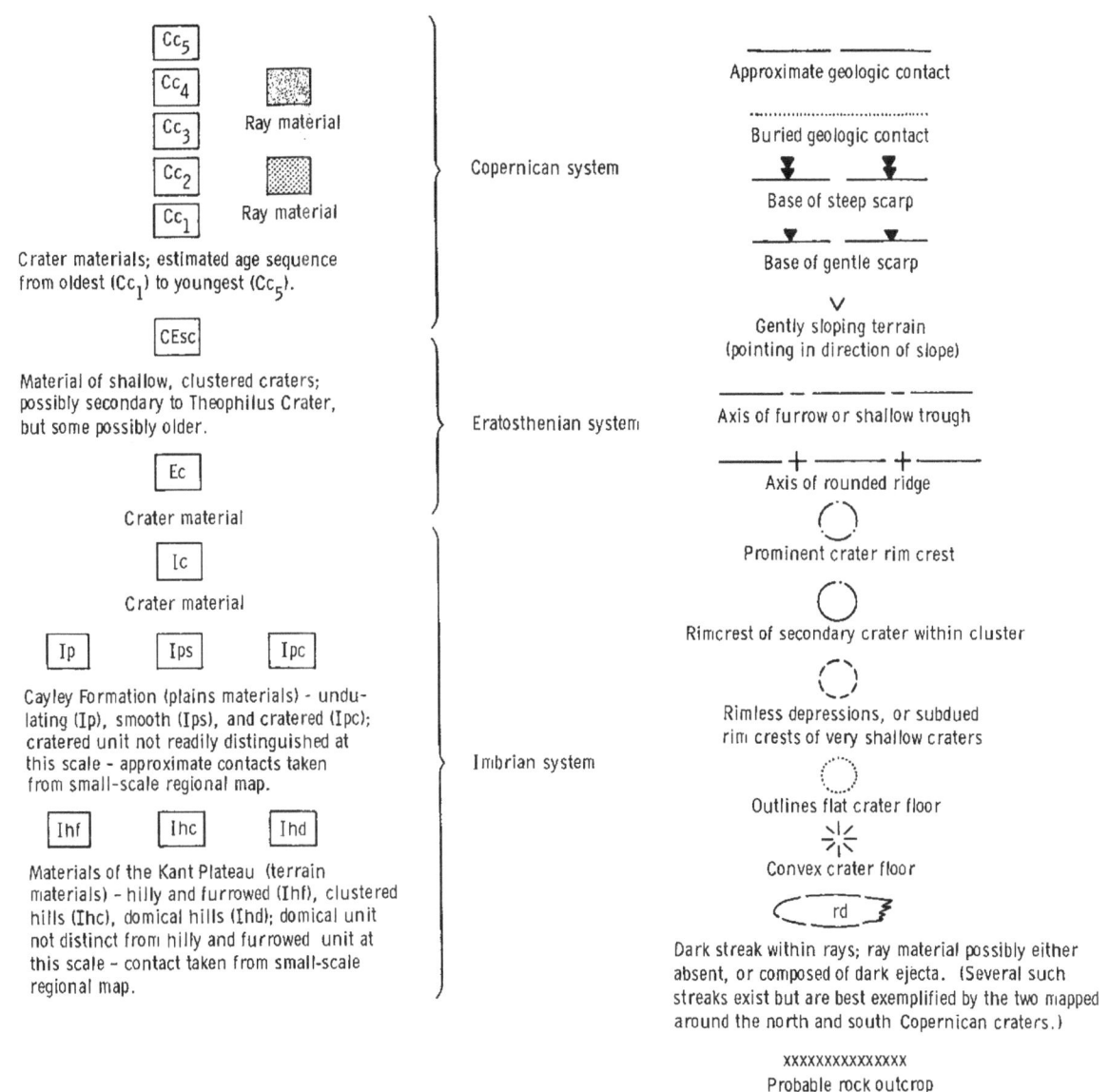

Key to figure 18-44.

more rapidly obliterated by mass-wasting processes on steep slopes than on flat surfaces, and the prevalence of steep slopes in the clustered hills unit may account for the scarcity of craters. Some differences in crater densities may be explained by the age of the units; older units, expectedly, will be more densely cratered than younger ones. In the more densely cratered plains units, some craters may be of endogenetic origin.

In terms of surface operations, the highest

priority might be assigned to investigating and sampling the Cayley Formation because of its abundance in the terrain; the next highest priority might be assigned to investigating the probable volcanic materials of the Kant Plateau. Also of interest is ejecta from North and South Craters because both may quarry pre-Imbrian rock from beneath the inferred local volcanic cover. The ridge that crosses the wall and rim of North Crater is also of interest because it may be only thinly mantled by regolith.

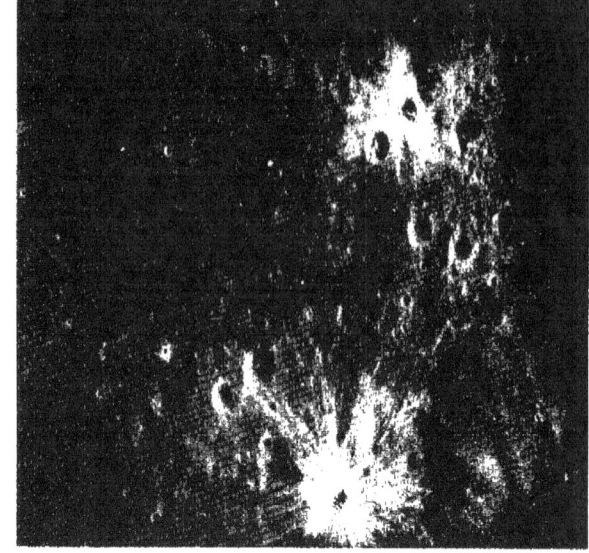

FIGURE 18-45.—Photograph of the candidate Descartes landing site taken from Hasselblad (500-mm lens) (AS14-69-9531).

PART F

CORRELATION OF THE ZERO-PHASE BRIGHTNESS SURGE (HEILIGENSCHEIN) WITH LUNAR-SURFACE ROUGHNESS

H. A. Pohn,[a] R. L. Wildey,[a] and T. W. Offield [a]

Microdensitometry of selected Apollo 8 and 10 photography revealed for the first time a brightness increase in the phase-angle region from 1.5° to 0° of 19 percent on plains materials (ref. 18–24) and an increase of 7.2 percent on mare materials (ref. 18–25). These observations were not coupled with any explanation at the time of these reports. Further reflection on this problem, visual inspection of selected Apollo 14 photography, and application of current regional geologic knowledge now indicate an apparent direct correlation between the brightness surge (heiligenschein effect) and the subresolution roughness of the lunar surface as suggested by empirical experiments (ref. 18–26).

Variations of the heiligenschein pulse (strength of the observable retroflection at or near zero phase) as a function of terrestrial terrain and

vegetation types have been noted by many individuals and, since 1963, have been discussed intermittently by U.S. Geological Survey Center of Astrogeology personnel. The visual observations and discussions were the consequence of numerous light plane trips, including astronaut training flights, in the vicinity of Flagstaff, Ariz., and over varied desert terrain on flights to Pasadena, Calif., particularly along airway radial V–12 from Prescott to Palmdale, Calif. Flight altitudes varied from approximately 500 m to a maximum of 1500 m. Although of considerable interest at the time to the Ranger and Surveyor flights, no follow-on work resulted. More recent visual observations during 1970 of heiligenschein variations on terrestrial terrain types from commercial jet aircraft between Phoenix, Ariz., and Denver, Colo., at altitudes in excess of 7500 m indicate an ordering of the brightness surge. The surge appears to de-

[a] U.S. Geological Survey.

crease in the following order: coniferous forests, deciduous forests, barren rugged crystalline rocks, grass-covered soil, and, finally, bare soil. These observations for the terrestrial case support a probable strong dependence on the heiligenschein on the fine-scale roughness of the surface observed.

Apollo 14 photographs were the first to show heiligenschein on a variety of lunar-terrain types. The greatest observed heiligenschein surge is in photographs AS14–75–10229 to 10234, in which the zero-phase point is coincident with ray material of the Copernican crater near Love and Prager Craters on the far side (fig. 18–23). The surge progressively decreases from the rugged upland materials near Ptolemaeus Crater (photograph AS14–75–10259), to the light plains nearby (photograph AS14–75–10257), to a low in the dark mare materials east of Fra Mauro (photograph AS14–75–10263).

Because the heiligenschein effect is considered the general result of shadowing in diffuse reflection (ref. 18–27), it is reasonable to expect more intergranular shadowing and relatively more retroflection near zero phase as the fine structure of any surface becomes rougher.

The potential geologic applications of this phenomenon are quite direct. When sufficient zero-phase orbital photographs have been accumulated and measured, it may be possible to use the magnitude of the heiligenschein pulse as an additional descriptor of geologic units. Investigation has already shown that a measurable correlation may exist between measurable photometric properties as seen on high-resolution oblique photographs and the relative age of small craters (ref. 18–28). The near-zero-phase reflectivity of large craters has long been used as a relative age indicator (ref. 18–29). Similarly, the brightness surge of terrestrial materials can be determined from aircraft by relatively simple instruments and might prove useful as a method of determining fine-scale surface roughness that may be relatable to geologic conditions, soil porosity, and moisture content as well as type and condition of local vegetation.

PART G

NEAR-TERMINATOR PHOTOGRAPHY

J. W. Head [a] *and D. D. Lloyd* [a]

For many years, it has been widely accepted that an examination of the lunar surface under near-terminator lighting conditions (low-Sun elevation) is extremely valuable to geologists and other scientists. Before the era when photography could be obtained from spacecraft in lunar orbit, a large percentage of the telescopic observation of the Moon (both directly and photographically) was conducted when the lunar point of interest was under near-terminator lighting conditions. When unmanned spacecraft were flown with the prime mission objective of lunar photography (e.g., Lunar Orbiter), the mission parameters were selected to produce photography near the termina-

tor. One reason is that, under near-terminator conditions, small changes in slope produce greater contrast changes than at high-Sun-elevation angles. A related desirable phenomenon is that, at low-Sun elevation, the shadow is longer than the object is high, thus increasing certain information about the object. For example, at a 20° Sun angle when the exaggeration is 2.75 (cot 20°), detectability and morphologic identification are enhanced.

Historically, it has been difficult to obtain photography any nearer the terminator than approximately 8° without severe underexposure. The Lunar Orbiter and the panoramic camera (planned for the Apollo 15, 16, and 17 missions) were optimized for photography at 20° or above in maria. (In the lunar highlands, the cameras can

[a] Bellcomm, Inc.

operate at lower Sun elevations because of the higher albedo). Neither type camera can produce the desired midrange exposures when operating nearer the terminator than 8°. However, some Lunar Orbiter 4 far-side photography was quite useful in certain regions nearer the terminator.

Faster film could be used to obtain photography nearer the terminator than 8°, but such fast film costs a resolution penalty that, for most unattended camera systems, would be paid for in all the photography obtained during the mission; that is, the film selected for photography approximately 0.5° from the terminator would also have to be used for all other photography.

The ability of an astronaut to change film in the Hasselblad cameras provided an opportunity to use very-high-speed film. Although no image-motion compensation is normally available for these cameras, it seemed possible to select film capable of photography within 0.5° of the terminator at a resolution that might provide photography of special geological interest. (Image-motion compensation can be obtained by rotation of the spacecraft.) The shadow length of 0.5° from the terminator would be greater than the height of any object by a factor of 114.6 (cot 0.5°). Slight variations (less than 0.5°) in slopes near the horizontal would produce significant variations in scene contrast.

Technical Discussion

Operation

A sequence of photographs was taken on orbit 19 of the Apollo 14 mission a few minutes before crossing the sunrise terminator and continued past the terminator. These photographs were taken at the request of the Photographic Team. The location of the area photographed is shown in the chart in figure 18–46.

A Hasselblad data camera with the 80-mm lens was used. The timing sequence was set to provide approximately 60 percent forward overlap by exposing at 20-sec intervals. The camera was set at f/2.8 and the exposure was 1/60 sec. A very-high-speed black-and-white film (Kodak 2485) was used. The spacecraft was oriented for vertical photography.

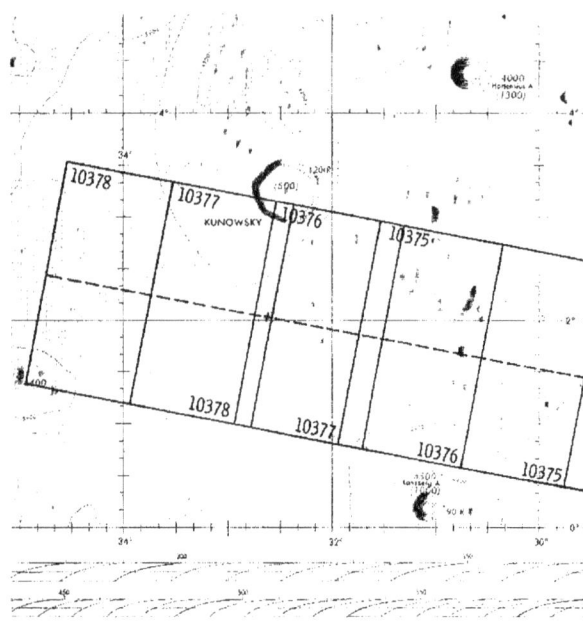

FIGURE 18–46.—Photographic footprint of area discussed, showing coverage of Hasselblad photographs AS14–78–10375 to 10378.

Results

Four photographs (AS14–78–10375 to 10378) are of special significance. The field of view of each covers an area approximately 2.5° square (80 km east to west by 76 km north to south). Figures 18–47 and 18–48 are particularly impressive. The terminator is at the left side of figure 18–47, which depicts an area to 2.5° east of the terminator. The center of figure 18–48 is at approximately latitude 2° N, longitude 32°30″ W. The terminator passes approximately midway through Kunowsky Crater, the southern half of which appears at the top center. Near-terminator photography (approximately 0.5° Sun angle) occupies the middle third of the figure. The area covered by these two photographs can be viewed stereoscopically.

If the photographs were to be examined without prior knowledge of the area south of Kunowsky Crater, the lunar surface would probably be described as undulating and rough. But human perceptual filters must be recognized as controlled by recognition models formed by experience that has never included useful photography this close

FIGURE 18–47.—Near-terminator photograph of area southeast of Kunowsky Crater (AS14–78–10376).

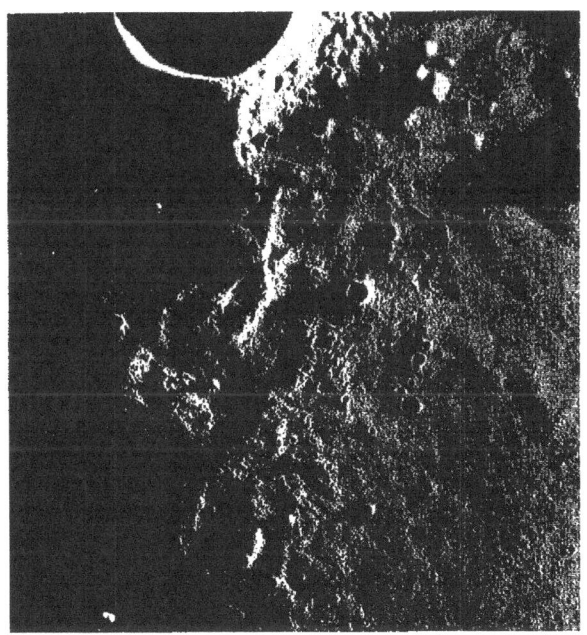

FIGURE 18–48.—Photograph of area south of Kunowsky Crater, bisected by terminator (AS14–78–10377).

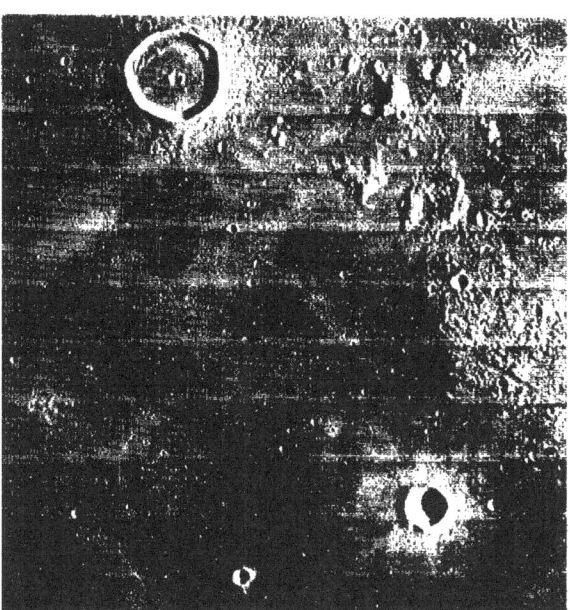

FIGURE 18–49.—Lunar Orbiter 4 high-resolution photograph of area.

to the terminator. The areal coverage of the photographs is shown in figure 18–49, a Lunar Orbiter 4 high-resolution photograph of the same general region. The area south of Kunowsky Crater is seen to be maria, and only a minor ridge pattern

(or higher albedo suggestion of a ray) is seen in much of the area.

The area covered by the near-terminator photography is located in the south-central portion of Oceanus Procellarum in the vicinity of Kunowsky Crater and approximately 210 km southeast of Kepler Crater. Several types of terrain are observable, including Kunowsky Crater, a hexagonal flat-floored crater approximately 20 km in diameter with a characteristic hummocky rim and central peaks; the hills and domes to the east of Kunowsky Crater; the flat mare surface that surrounds Kunowsky Crater and the hills and domes; and the craters, crater clusters, and rays superposed on the whole area.

The geologic relationships of these terrain types (ref. 18–30) suggest the following historical sequence of events. The earliest history recorded in the surface features was the origin of the multiringed Imbrium Basin and the deposition of a corresponding ejecta blanket known as the Fra Mauro Formation. This ejecta blanket, probably deposited over the entire area, was characterized by a hummocky and hilly texture. Subsequent lava flooding during the filling of Oceanus Procellarum covered the low-lying parts of the Fra Mauro Formation and isolated the higher portions into islands and groups of hills. Either before or during the lava flooding of Oceanus Procellarum, Kunowsky Crater was formed. Its age relative to the mare is indicated by the incursion of dark mare material onto the crater ejecta blanket. After the mare flooding, rays and secondary craters from the two postmare Kepler and Copernicus Craters were formed in the area and produced the northwest-southeast and northeast-southwest albedo changes in the mare.

Of the terrain just described, the near-terminator photography provides the best results in the near-level mare regions, because optimum enhancement occurs there. Features of interest fall in three broad categories; detailed studies are being made of each group of features.

(1) *Mare features.*—In the near-terminator photography, extensive flow fronts are visible that are barely (if at all) perceptible on Lunar Orbiter photography. In particular, a single flow front has been tentatively mapped for more than 50 km in the area southwest of Kunowsky Crater. Mare domes and features emanating from them are also evident in these newer photographs. Other mare flow fronts, ridges, and structures are also enhanced in this photography.

(2) *Craters.*—Crater rims are enhanced in the near-terminator photography, and rimless craters are particularly evident. A rimless mare crater exceeding 1-km diameter approximately 20 km southeast of Kunowsky Crater is a particularly good example. Secondary crater clusters and associated ray material are also enhanced.

(3) *Surface lineaments.*—Subtle surface lineaments are visible on these photographs and are being mapped and compared to regional structures.

Other features of interest are evident, including the upland terrain and its associated structures as well as the deposits of Kunowsky Crater.

Conclusions

Four unusual photographs were obtained that showed lunar-surface areas within 0.5° of the terminator and were certainly of significant interest. Many geological features stand out in a manner not usually observed in conventional lunar photography. For a distance of approximately 0.5° from the terminator, variations in surface slopes that are within 0.5° of the horizontal produce significant changes in contrast that enable the brain to perceive these variations. Throughout the area 2.5° east of the terminator shown in figure 18–47, significant geological details are evident. Craters and other features produce shadows and contrast changes that readily enable recognition, although on first impression, these shadows seem to exaggerate the height of the features. The flow fronts shown in figures 18–47 and 18–48 have shadows that are approximately 100 times their height, thus enhancing detection and recognition. The absence of a shadow west of the 1-km rimless crater delimits its rim. Significant contrasts, and even shadows, are produced by minor surface lineaments.

This special Apollo 14 near-terminator photography was considered experimental. The extent to which such photography will be planned for future missions and how it will be optimized will depend on further scientific analysis of these first efforts.

PART H

ASTRONOMICAL PHOTOGRAPHY

R. D. Mercer,[a]† L. Dunkelman,[a]† and S. A. Roosa [b]

Photographic observations of astronomical interest (other than lunar photography) included

[a] Cornell University.
[b] NASA Manned Spacecraft Center.
† Investigator.

those of the zodiacal light and the lunar libration clouds. These investigations, while not formal experiments, are being conducted under the guidance of the Apollo Orbital Science Photographic Team to facilitate analysis of the Apollo 14 photographic data. As in the case of the gegenschein experiment on this mission (sec. 15), many of the photographic observations of faint astronomical phenomena were considered to be of an operational-test nature to determine the feasibility of performing these difficult tasks from the command module (CM) using the currently available 16-mm data acquisition camera. (Additional information on the background of dim-light photography is contained in sec. 15.)

The investigations were an unqualified success with regard to flight- and ground-support operations. Scientific results cannot be obtained until original flight film is made available for analysis. The CMP reported that all equipment functioned properly and resulted in more exposed photographs than requested prior to flight (table 18–I).

TABLE 18–I. *Comparison of Number of Photographs Requested With Number Obtained*

Target	Exposures	
	Requested	Obtained
Zodiacal light...............	25	30
Galactic light................	6	13
Libration region light L₄......	4	4
Earth dark side..............	6	9
SIVBᵃ	6	7

ᵃ Real-time request for photography through the sextant.

Although the data photographs, as well as the preflight and postflight calibration photographs, will require microdensitometric measurements and isodensitracing for proper analysis of scientific results, very preliminary results can be summarized. All star fields have been readily identified, and camera pointing appears to be within 1° of the desired aiming points with less than 0.33° of image motion for fixed positions. This accuracy is well within the limits requested, and conclusively indicates that longer exposures will be possible for these studies on future Apollo missions. The zodiacal light is evident to the unaided eye on at least half the appropriate frames. Therefore, further analysis is expected to produce quantitative results. The galactic-light-survey and lunar-libration photographs are faint and will require careful

work. The Earth dark-side photographs of lightning patterns, as well as the SIVB photographs, required that the camera be attached to the automatically pointed CM sextant. Scattered light in the sextant optics, from the sunlit portion of the Earth, and, perhaps, from portions of the docked lunar module during translunar coast, obscured the scenes being photographed. Consequently, the data were unusable for further analysis. Light scattering in the optics during transearth coast also rendered the Earth dark-side photography unusable.

References

18–1. MILTON, D. J.: Geologic Map of the Theophilus Quadrangle of the Moon. U.S. Geol. Survey Misc. Geol. Inv. Map I–546, 1968.

18–2. WILHELMS, D. E.; AND MCCAULEY, J. F.: Geologic Map of the Near Side of the Moon. U.S. Geol. Survey Misc. Geol. Inv. Map I–703, 1971.

18–3. MUTCH, T. A.: Geology of the Moon: A Stratigraphic View. Princeton Univ. Press, 1970.

18–4. WILHELMS, D. E.: Summary of Lunar Stratigraphy; Telescopic Observations. Contributions to Astrogeology. U.S. Geol. Survey Proj. Paper 599–F, 1970.

18–5. KOSOFSKY, L. J.; AND EL-BAZ, FAROUK: The Moon as Viewed by Lunar Orbiter. NASA SP–200, 1969.

18–6. EL-BAZ, F.; AND WILSHIRE, H. G.: Landforms. Initial Photographic Analysis. In ch. 2 of Analysis of Apollo 8 Photography and Visual Observations. NASA SP–201, 1969, pp. 32–33.

18–7. EL-BAZ, FAROUK: Lunar Igneous Intrusions. Science, vol. 167, no. 3914, Jan. 2, 1970, pp. 49–50.

18–8. WHITAKER, E. A.: Comparison with Luna III Photographs. Initial Photographic Analysis. In ch. 2 of Analysis of Apollo 8 Photography and Visual Observations. NASA SP–201, 1969, pp. 9–10.

18–9. MCCAULEY, J. F.: Preliminary Small-Scale Geologic Map of the Marius Hills Region, 1968. Preliminary Exploration Plan of the Marius Hills Region of the Moon. U.S. Geol. Survey Interagency Rept., Astrogeol. 5, Feb. 1968.

18–10. OBERBECK, V. R.; QUAIDE, W. L.; AND GREELEY, RONALD: On the Origin of Lunar Sinuous Rilles. Modern Geol., vol. 1, 1969, pp. 75–80.

18–11. SCHUMM, S. A.: Experimental Studies on the Formation of Lunar Surface Features of Fluidization. Geol. Soc. Amer. Bull., vol. 81, Sept. 1970, pp. 2539–2552.

18-12. HOWARD, K. A.: Geologic Map of Part of the Apennine-Hadley Region of the Moon, Apollo 15 Premission Map. U.S. Geol. Survey Map I-723, 1971.

18-13. McCORD, T. B.: Color Differences on the Lunar Surfaces. J. Geophys. Res., vol. 74, no. 12, June 1969, pp. 3131-3142.

18-14. HAPKE, B. W.: Optical Properties of the Moon's Surface. The Nature of the Lunar Surface, W. N. Hess, D. H. Menzel, and J. A. O'Keefe, eds. Proc. 1965 IAU-NASA Symp. (Baltimore, Md.), Johns Hopkins Press, 1966, pp. 141-154.

18-15. WHITAKER, E. A.: The Surface of the Moon. The Nature of the Lunar Surface, W. N. Hess, D. H. Menzel, and J. A. O'Keefe, eds. Proc. 1965 IAU-NASA Symp. (Baltimore, Md.), Johns Hopkins Press, 1966, pp. 79-98.

18-16. SHOEMAKER, EUGENE M.: Interpretation of Lunar Craters. Physics and Astronomy of the Moon, Zdenek Kopal, ed., Academic Press, Inc., 1962, pp. 283-359.

18-17. CARR, M. H.: Geologic Map of the Alphonsus Region of the Moon. U.S. Geol. Survey Map I-559, 1969.

18-18. McCAULEY, J. F.: Geologic Map of the Alphonsus GA Region of the Moon. U.S. Geol. Survey Map I-586, 1969.

18-19. HOWARD, K. A.; AND MASURSKY, HAROLD: Geologic Map of the Ptolemaeus Quadrangle of the Moon. U.S. Geol. Survey Map I-566, 1968.

18-20. WILHELMS, D. E.; STUART-ALEXANDER, D. E.; AND HOWARD, K. A.: Preliminary Interpretations of Lunar Geology. Initial Photographic Analysis. In ch. 2 of Analysis of Apollo 8 Photography and Visual Observations. NASA SP-201, 1969, pp. 16-21.

18-21. POHN, H. A.; AND OFFIELD, T. W.: Lunar Crater Morphology and Relative-Age Determination of Lunar Geologic Units. Pt. 1, Classification. U.S. Geol. Survey Proj. Paper 700-C, 1970, pp. C153-C162.

18-22. MOORE, H. J.: Estimates of the Mechanical Properties of Lunar Surface Using Tracks and Secondary Impact Craters Produced by Blocks and Boulders. U.S. Geol. Survey Interagency Rept., Astrogeol. 22, July 1970.

18-23. MORRIS, E. C.; AND WILHELMS, D. E.: Geologic Map of the Julius Caesar Quadrangle of the Moon. U.S. Geol. Survey Misc. Geol. Inv. Map I-510, 1967.

18-24. POHN, H. A.; RADIN, H. W.; AND WILDEY, R. L.: The Moon's Photometric Function Near Zero Phase Angle From Apollo 8 Photography. Astrophys. J., vol. 157, Sept. 1969, pp. L193-L195.

18-25. WILDEY, R. L.; AND POHN, H. A.: The Normal Albedo of the Apollo 11 Landing Site and Intrinsic Dispersion in the Lunar Heiligenschein. Astrophys. J., vol. 158, Nov. 1969, pp. L129-L130.

18-26. HALAJIAN, J. D.: Photometric Measurements of Simulated Lunar Surfaces. NASA CR-65169, 1965.

18-27. IRVINE, W. M.: The Shadowing Effect in Diffuse Reflection. J. Geophys. Res., vol. 71, June 15, 1966, pp. 2931-2937.

18-28. WILDEY, R. L.: Limited Interval Definitions of the Photometric Functions of Lunar Crater Walls by Photography From Orbiting Apollo. Icarus, vol. 14, no. 1, 1971.

18-29. SHOEMAKER, EUGENE M.; AND HACKMAN, ROBERT J.: Stratigraphic Basis for a Lunar Time Scale. The Moon, Zdenek Kopal and Zdenka Kadla Mikhailov, eds. (Proc. Symp. 14, IAU), Academic Press, Inc., 1962, pp. 289-300.

18-30. HACKMAN, R. J.: Geologic Map and Sections of the Kepler Region of the Moon. U.S. Geol. Survey Misc. Inv. Map I-355, 1962.

ACKNOWLEDGMENTS

The authors extend their gratitude to the following persons who contributed their efforts to this report: George Esenwein and Floyd Roberson of NASA Headquarters; Nat Hardee, Helmut Kuehnel, Dale Denais, Andrew Patterson, and L. C. Wade of NASA Manned Spacecraft Center; G. W. Colton of the U.S. Geological Survey; and F. J. Doyle, Ewen Whitaker, Harold Masursky, Don Light, Larry Schimmerman, and Leon Kosofsky, members of the Apollo Orbital Science Photographic Team.

Work by the U.S. Geological Survey was supported in part by NASA contract T-65253G and is partially the result of studies conducted under NASA contracts R-66 and W-13, 130.

APPENDIX A

Glossary

achondrite—a stony meteorite devoid of rounded granules

agglutinate—a deposit of originally molten ejecta

anhedral—pertaining to mineral grains that lack external crystals

anorthite—a calcium-rich variety of plagioclase feldspar

anorthosite—a granular, plutonic, igneous rock composed almost exclusively of a soda-lime feldspar

apatite—any of a group of calcium-phosphate minerals that occur variously as hexagonal crystals, as granular masses, or in fine-grained mass as the chief constituent of phosphate rock

aphanite—a dark rock of such close texture that the individual grains are invisible to the unaided eye

aphyric—not having distinct crystals

augite—one of a variety of pyroxene minerals that contain calcium, magnesium, and aluminum; usually black or dark green in color

bleb—a small bit of particle of distinctive material

breccia—a rock consisting of sharp fragments embedded in a fine-grained matrix

bytownite—a calcium-rich variety of plagioclase feldspar

calcic—derived from or containing calcium

chondrite—a meteoritic stone characterized by the presence of rounded granules

clast—a discrete particle or fragment of rock or mineral; commonly included in a larger rock

clinopyroxene—a mineral that occurs in monoclinic, short, thick, prismatic crystals and that varies in color from white to dark green or black (rarely blue)

coherent—a term used to describe two or more parts of the same series that are in contact more or less adhesively but are not fused

conchoidal—a term used to describe a shell-like surface shape that has been produced by the fracturing of a brittle material

cristobalite—an isometric variety of quartz that forms at high temperatures (SiO_2)

dendrite—a crystallized arborescent form

devitrification—the change of a glassy rock from the glassy state to a crystalline state after solidification

dunite—a peridotite that consists almost entirely of olivine and that contains accessory chromite and pyroxene

eucrite—a meteorite composed essentially of feldspar and augite

euhedral—pertaining to minerals the crystals of which have had no interference in growth

exsolution—unmixing; the separation of some mineral-pair solutions during slow cooling

fayalite—an iron-rich variety of olivine (Fe_2SiO_4)

feldspar—a group of abundant rock-forming minerals

feldspathic—a term used to describe a material that contains feldspar as a principal ingredient

gabbro—a granular igneous rock of basaltic composition with a coarse-grained texture

holocrystalline—consisting wholly of crystals

ilmenite—a mineral rich in titanium and iron; usually black with a submetallic luster

indurated—a term used to describe masses that have been hardened by heat; baked

intersertal—a term used to describe the texture of igneous rocks in which a base or mesostasis of glass and small crystals fills the interstices between unoriented feldspar laths

lamella—a layer of a cell wall

lath—a long, thin mineral crystal

leucocratic—a term used to describe light-colored rock, especially igneous rocks that contain between 0 and 30 percent dark minerals

lithic—of, relating to, or made of stone

lithification—consolidation and hardening of fines into rock

lithology—the physical character of a rock, as determined with the unaided eye or with a low-power magnifier

magcon—a magnetized concentration

magma—molten rock material that is liquid or pasty

maskelynite—a feldspar found in meteorites

melanocratic—a term used to describe dark-colored rocks, especially igneous rocks that contain between 60 and 100 percent dark minerals

metamorphic—a term used to describe rocks that have

formed in a solid state as a result of drastic changes in temperature, pressure, and chemical environment

microlite—small lath-shaped minerals, commonly plagioclase feldspar, occurring as minute phenocrysts in basalt

mosaic—a term used to describe the texture sometimes seen in dynamo-metamorphosed rocks that have angular and granular crystal fragments and that appear like a mosaic in polarized light

norite—a type of gabbro in which orthopyroxene is dominant over clinopyroxene

olivine—an igneous mineral that consists of a silicate of magnesium and iron

ophitic—a rock texture characterized by lath-shaped plagioclase crystals enclosed in augite

pigeonite—a variety of pyroxene

plagioclase—a feldspar mineral composed of varying amounts of sodium and calcium with aluminum silicate

plutonic—pertaining to igneous rock that crystallizes at depth

poikilitic—a term used to describe the condition in which small granular crystals are irregularly scattered without common orientation in a larger crystal of another mineral

pyroxene—a mineral occurring in short, thick, prismatic crystals or in square cross section; often laminated; and varying in color from white to dark green or black (rarely blue)

regolith—the layer of fragmental debris that overlies consolidated bedrock

schlieren—tabular bodies that occur in pluton; generally several centimeters to several meters long

slickensides—a polished and striated surface that results from friction along a fault plane

spall—a relatively thin, sharp-edged piece of rock that has been produced by exfoliation

spinel—a mineral that is noted for its great hardness ($MgAl_2O_4$)

subhedral—pertaining to minerals that are intermediate between anhedral and euhedral

tephra—a collective term for all clastic volcanic materials that are ejected from the volcano and transported through the air

troilite—a mineral that is native ferrous sulfide

vermicular—a term used to describe a group of platy minerals that are closely related to the chlorites and montmorillonites

vesicle—a small cavity in a mineral or rock, ordinarily produced by expansion of vapor in a molten mass

vug—a small cavity in a rock

zircon—a mineral, $ZrSiO_4$; the main ore of zirconium

APPENDIX B

Acronyms

ALSCC—Apollo lunar-surface closeup camera
ALSEP—Apollo lunar-surface experiments package
ALSRC—Apollo lunar sample return container
amu—atomic mass unit
ASE—active seismic experiment
ASP—Apollo simple penetrometer
BRN—Brown & Root-Northrop
CCGE—cold-cathode-gage experiment
CDR—commander
CM—command module
CMP—command module pilot
CPLEE—charged-particle lunar environment experiment
CSM—command and service module
DAC—data acquisition camera
dc—direct current
dpm—disintegrations per minute
DU—digital units
ESRO—European Space Research Organization
e.s.t.—eastern standard time
EVA—extravehicular activity
G.m.t.—Greenwich mean time
KREEP—potassium, rare-Earth elements, and phosphorus
LCP—left circular polarization
LM—lunar module
LMP—lunar module pilot
LP—long period
LPM—lunar portable magnetometer
LPX, LPY—long-period horizontal component (seismometer)
LPZ—long-period vertical component (seismometer)

LRL—Lunar Receiving Laboratory, NASA Manned Spacecraft Center
LRRR—laser ranging retroreflector
LRV—lunar roving vehicle
LSS—lunar soil simulant
LTC—lunar topographic camera
MET—modularized equipment transporter
MIT—Massachusetts Institute of Technology
MSC—Manned Spacecraft Center
NASA—National Aeronautics and Space Administration
ppm—parts per million
PSE—passive seismic experiment
RCL—Radiation Counting Laboratory
RCP—right circular polarization
rms—root mean square
RTP—reverse tie point
SESC—special environment sample container
SIDE—suprathermal ion detector experiment
SIVB—third stage (IVB) of Saturn launch vehicle
SM—solar magnetospheric
SP—short period
SPECS—switched proton electron Channeltron spectrometer
SPS—service propulsion system
SPZ—short-period vertical component (seismometer)
SWC—solar-wind composition
TDS—thermal degradation sample
USGS—U.S. Geological Survey
vhf—very high frequency
WES—Waterways Experiment Station (U.S. Army)

APPENDIX C

Units and Unit-Conversion Factors

In this appendix are the names, abbreviations, and definitions of International Systems (SI) units used in this report and the numerical factors for converting from SI units to more familiar units.

Names of International Units Used in This Report

Physical quantity	Name of unit	Abbreviation	Definition of abbreviation
Basic Units			
Length	meter	m	
Mass	kilogram	kg	
Time	second	sec	
Electric current	ampere	A	
Temperature	kelvin	K	
Luminous intensity	candela	cd	
Derived Units			
Area	square meter	m^2	
Volume	cubic meter	m^3	
Frequency	hertz	Hz	sec^{-1}
Density	kilogram per cubic meter	kg/m^3	
Velocity	meter per second	m/sec	
Angular velocity	radian per second	rad/sec	
Acceleration	meter per second squared	m/sec^2	
Angular acceleration	radian per second squared	rad/sec^2	
Force	newton	N	$kg \cdot m/sec^2$
Pressure	newton per square meter	N/m^2	
Work, energy, quantity of heat	joule	J	$N \cdot m$
Power	watt	W	J/sec
Voltage, potential difference, electromotive force	volt	V	W/A
Electric field strength	volt per meter	V/m	
Electric resistance	ohm	Ω	V/A
Electric capacitance	farad	F	$A \cdot sec/V$
Magnetic flux	weber	Wb	$V \cdot sec$
Inductance	henry	H	$V \cdot sec/A$
Magnetic flux density	tesla	T	Wb/m^2
Magnetic field strength	ampere per meter	A/m	
Luminous flux	lumen	lm	$cd \cdot sr$

Derived Units—Continued

Physical quantity	Name of unit	Abbreviation	Definition of abbreviation
Luminance	candela per square meter	cd/m^2	
Illumination	lux	lx	lm/m^2
Specific heat	joule per kilogram kelvin	J/kg $\cdot$ K	
Thermal conductivity	watt per meter kelvin	W/m $\cdot$ K	

Supplementary Units

Plane angle	radian	rad	
Solid angle	steradian	sr	

Unit Prefixes

Prefix	Abbreviation	Factor by which unit is multiplied
giga	G	10^9
mega	M	10^6
kilo	k	10^3
centi	c	10^{-2}
milli	m	10^{-3}
micro	μ	10^{-6}
nano	n	10^{-9}

Unit-Conversion Factors

To convert from—	To—	Multiply by—
ampere/meter	oersted	1.257×10^{-2}
candela/meter^2	foot-lambert	2.919×10^{-1}
candela/meter^2	lambert	3.142×10^{-4}
joule	British thermal unit (International Steam Table)	9.479×10^{-4}
joule	Calorie (International Steam Table)	2.388×10^{-1}
joule	electron volt	6.242×10^{18}
joule	erg	1.000×10^{7} a
joule	foot-pound force	7.376×10^{-1}
joule	kilowatt-hour	2.778×10^{-7}
joule	watt-hour	2.778×10^{-4}
kelvin	degrees Celsius (temperature)	$t_C = t_K - 273.15$
kelvin	degrees Fahrenheit (temperature)	$t_F = \%\, t_K - 459.67$
kilogram	gram	1.000×10^{3} a
kilogram	kilogram mass	1.000×10^{3} a
kilogram	pound mass (pound mass avoirdupois)	2.205×10^{0}
kilogram	slug	6.852×10^{-2}
kilogram	ton (short, 2000 pound)	1.102×10^{-3}
lumen/meter^2	foot-candle	9.290×10^{-2}
lumen/meter^2	lux	1.000×10^{0} a
meter	angstrom	1.000×10^{10} a
meter	foot	3.281×10^{0}
meter	inch	3.937×10^{1}
meter	micron	1.000×10^{6} a
meter	mile (U.S. statute)	6.214×10^{-4}

Unit-Conversion Factors—Continued

To convert from—	To—	Multiply by—
meter	nautical mile (international)	5.400×10^{-4}
meter	nautical mile (U.S.)	5.400×10^{-4}
meter	yard	1.094×10^{0}
meter/second²	foot/second²	3.281×10^{0}
meter/second²	inch/second²	3.937×10^{1}
newton	dyne	1.000×10^{5} ª
newton	kilogram force (kgf)	1.020×10^{-1}
newton	pound force (avoirdupois)	2.248×10^{-1}
newton/meter²	atmosphere	9.870×10^{-6}
newton/meter²	centimeter of mercury (0° C)	7.501×10^{-4}
newton/meter²	inch of mercury (32° F)	2.953×10^{-4}
newton/meter²	inch of mercury (60° F)	2.961×10^{-4}
newton/meter²	millimeter of mercury (0° C)	7.501×10^{-3}
newton/meter²	torr (0° C)	7.501×10^{-3}
radian	degree (angle)	5.730×10^{1}
radian	minute (angle)	3.438×10^{3}
radian	second (angle)	2.063×10^{5}
tesla	gamma	1.000×10^{9} ª
tesla	gauss	1.000×10^{4} ª
watt	British thermal unit (thermochemical)/ second	9.484×10^{-4}
watt	calorie (thermochemical)/second	2.390×10^{-1}
watt	foot-pound force/second	7.376×10^{-1}
watt	horsepower (550 foot-pound force/second)	1.341×10^{-3}
weber	maxwell	1.000×10^{8} ª

ª An exact definition.